NEW TRENDS AND DEVELOPMENTS IN AUTOMOTIVE INDUSTRY

Edited by **Marcello Chiaberge**

New Trends and Developments in Automotive Industry
Edited by Marcello Chiaberge

Published by InTech

Janeza Trdine 9, 51000 Rijeka, Croatia

Publishing Process Manager Iva Lipovic

Technical Editor Teodora Smiljanic

Cover Designer InTech Design Team

Additional hard copies can be obtained from orders@intechopen.com

New Trends and Developments in Automotive Industry, Edited by Marcello Chiaberge
p. cm.
ISBN 978-953-307-999-8

For EU product safety concerns:
IN TECH d.o.o., Prolaz Marije Krucifikse Kozulić 3, 51000 Rijeka, Croatia,
info@intechopen.com or visit our website at intechopen.com.

Meet the editor

Marcello Chiaberge is assistant professor at Electronic Department of Politecnico di Torino and vice-director of Centro per i Servizi di Prototipazione of Politecnico di Torino, where he is also responsible and coordinator of the Electronic Group at Mechatronics Laboratory (LIM) of Politecnico di Torino. His research interests are: hardware implementation of neural networks and fuzzy systems, design and implementation of reconfigurable computing architectures for hard-real-time control systems, design and implementation of hybrid control systems based on programmable state-of-the-art devices (DSP, FPGA and FPAA), implementation of fault-tolerance communication networks based on plastic optical fibers (POF). Another research field is the design and implementation of non-conventional power driver systems for special actuaction systems in the automotive, industrial and space fields of application. He is also involved in the design and control of small rovers for space exploration, archeological prospecting and humanitarian/anti-terrorism application in collaboration with Piedmont Region, Thales Alenia Space and Italian Space Agency (ASI). Chiaberge is an IEEE member, the author of more than 60 articles accepted in international conferences and reviews and a co-author of 7 international patents.

Contents

Preface

The automotive industry is experiencing a considerable "stress period", which can lead to important changes in the whole industry. Many aspects contribute to this situation, starting from the global recession (unemployment rates, slowing growth, etc.) to credit meltdown (dependency of car sales on credit, OEM refinancing, etc.) and finishing with globalization aspects (global sourcing, foreign investments, etc.) and "green challenges" (both for the industry and the final products).

Moreover, the global market developments are influencing the whole automotive industry in different fields (volumes, technologies, regional aspects), while the environmental compatibility of car power-trains will lead to huge investments needs for the innovation of many different technologies.

In this complex scenario, regional environmental regulations (both on production process and on final product) will have great influence on further technological developments. For example, Japan and Europe are world leaders with the most severe standards in terms of gas, fuel and oil economy, while other emerging economies are facing right now this kind of problem. This obviously means that the approaching techniques to new vehicle standards are quite different, and emission targets and measures vary heavily depending also on industry/consumers incentives that will play a big role in the future E-mobility scenario.

In this new scenario the automotive industry will not only be characterized by original and historical OEMs, but entirely new players will enter this industry area. Non-OEM players will introduce skills related to information systems and connectivity, new components, new systems or innovation related to energy supply in order to provide solutions directly connected with the end customer (just think about electric vehicles).

Customers are becoming the main factor of this small "revolution" that will lead OEMs to defeat their original market position and will be placed at the same level of standard suppliers in order to provide new solutions for the final customer. A clear example of this new perspective is the E-mobility scenario, where common projects between OEMs and power suppliers are driving innovation and new products.

In this new perspective, different solutions will help OEMs to innovate this important industry and to face the challenges that new markets, regulations, standards and technologies are introducing. The possible areas of interest will span from engine cooperation, platform and process sharing, development of new business fields, to joint ventures with electric companies and important strategic partnerships with suppliers.

This book is divided in five main parts (production technology, system production, machinery, design and materials) and tries to show emerging solutions in automotive industry fields related to OEMs and no-OEMs sectors in order to show the vitality of this leading industry for worldwide economies and related important impacts on other industrial sectors and their environmental sub-products.

Thanks to KPMG for important data and industrial analysis.

Marcello Chiaberge
Mechatronics Laboratory – Politecnico di Torino
Italy

Part 1

Industrial Production Technology

1

Data Mining and Intelligent Agents for Supporting Mass Customization in the Automotive Industry

Efthimia Mavridou[1,3], Dimitrios Tzovaras[1], Evangelos Bekiaris[2], Pavlos Spanidis[2], Maria Gemou[2] and George Hassapis[3]

[1]*Informatics and Telematics Institute, Centre for Research and Technology Hellas, 6th km Charilaou – Thermi Rd., P.O. Box: 60361, P.C.: 57001, Thermi, Thessaloniki,*

[2]*Hellenic Institute of Transport, Centre for Research and Technology Hellas, 6th km Charilaou – Thermi Rd., P.O. Box: 60361, P.C.: 57001, Thermi, Thessaloniki,*

[3]*Department of Electrical and Computer Engineering, Faculty of Engineering, Aristotle University of Thessaloniki, P.C.:54124 , Thessaloniki,*

Greece

1. Introduction

Mass customisation has been said to be the new frontier in business competition (Pine, 1992). The objective of mass customisation is to deliver goods and services that meet individual customers' needs with near mass production efficiency (Tseng & Jiao, 2001). Currently, only few automotive industries have deployed mass customisation systems in their product design and manufacturing processes. In the current paper, we present such a mass customization system, designed as an agent-oriented architecture which proposes to the vehicle customers (of car and truck segments) personalised vehicle configurations according to their personal affective needs.

Design for performance (i.e. functional design) and design for usability (i.e. ergonomic design) no longer empower a competitive edge because product technologies turn to be mature, or competitors can quickly catch up (Khalid & Helander, 2004). Affective design has become very important in prescribing that designed objects have a meaning that goes beyond their functional needs (Khalid et al., 2006). Customers actively seek design features that are important for their emotional satisfaction, and vehicle design must therefore address customer affective needs. Affective needs are defined as user requirements for a specific product, driven by emotions, sentiments and attitudes (Khalid et al., 2006). Understanding customer affective needs is important to ensure a good fit of affective and functional requirements to design parameters.

Several pieces of research have been presented for supporting affective design such as Kansei engineering which has been well recognized as a technique of translating consumers' subjective impressions about a product into design elements (Nagamashi, 1989). (Ishihara et al., 1995) apply neural network techniques to enhance the inference between Kansei words and design elements in Kansei design systems. (Matsubara & Nagamachi. 1997) propose to develop hybrid expert systems for Kansei design support. (Jiao, 2007) proposes an affective design framework based on ambient intelligence techniques to facilitate decision-making in designing customized product ecosystems. In the current paper, a new research focus and perspective that integrates cognition/thinking and emotion/affect in uncovering customer needs is deployed, the Citarasa Engineering (CE) (Khalid et al., 2006). It is developed for the purpose of supporting affective design as an alternative to existing methods such as Kansei Engineering (Nagamashi, 1989). Citarasa refers to a Malay word which means emotional intent or a strong desire for a product. For the purpose of discovering the mapping relationship between customers' affective needs, defined by their citarasa, and the design parameters that characterize the design elements of vehicles, data mining techniques were deployed.

Data mining (DM) enables efficient knowledge extraction from large datasets, in order to discover hidden or non-obvious patterns in data (Witten et al., 2005). Our motivation for using DM was based on the hypothesis that the application of the appropriate DM technique on customer surveys could form a suitable mechanism for the knowledge extraction representing the correlation between customer affective needs and design parameters related to the various design elements of vehicles. The extracted knowledge was then used for the provision of personalised recommendations to customers in collaboration with the agent-based framework developed and via the web and VR based interfaces developed in the context of the CATER - STREP project (Annex I-"Description of Work", 2006). The latter constitutes the second part of the work held. The agent - based system developed interacts with different modules of the overall integrated system developed in CATER, in order to support the mass customisation supply chain including suppliers, factories, subcontractors, warehouses, distribution centres and retailers.

2. Mining of customer survey data

2.1 Data mining process

The aim of the data mining process was to identify the mapping relationship between customer affective needs and vehicle configurations, with final goal to propose to new customers' vehicle configurations according to their personal affective needs. Affective needs are described by the use of citarasa descriptors (*Cd*), which are keywords extracted through probe elicitation surveys and semantic based methods conducted in the scope of CATER (Annex I-"Description of Work", 2006).

We consider a vehicle configuration V as a set of design elements: $V = [de_1, de_2, ..., de_n]$ .The term design element (de_i) refers to the customizable vehicle parts such as steering-wheel, wheel-rim, mirrors etc. Each design element de_i is characterized by a set of design parameters (dp_{ij}) such as color, shape etc. Thus, a design element de_i is represented as a set of design parameters, $de_i = [dp_{i1}, dp_{i2}, ..., dp_{in}]$. Each dp_{ij} has a set of possible values. For example the $dp_{11} = material$ of the $de_1 = steering - wheel$ has the set of values: $[vinyl, alu\min ium, wood]$. Different values of the design parameters result in different versions of the design elements, and consequently in different vehicle configurations. We construct a classification mechanism for predicting the values of each of the design parameters that satisfy customer affective needs. Specifically, we construct a classification mechanism for each of the design parameters (dp_{ij}). Then, by the assistance of the agent-based framework (section 3) we can propose to the customer vehicle configurations that correspond to the predicted design parameters, and therefore to the customer affective needs. We deploy a classification approach based on association rules. Association rule discovery refers to the discovery of the relationships among a large set of data items (Agrawal et al., 1994), while classification focuses on building a classification model for categorizing new data. Let $I = [i_1, i_2, \ldots, i_n]$ be a set of items and let D be a set of records, where each record R is a set of items such that $R \subseteq I$. An association rule is an implication of the form $X \rightarrow Y$, where $X \subset I$, $Y \subset I$ and $X \cap Y = \varnothing$. X is the head of the rule and Y is the body. The confidence c of a rule is defined as the number of records that contain X and also Y ($count(X \cap Y)$) divided by the number of records in D that contain X ($count(X)$):

$$c = \frac{count(X \cap Y)}{count(X)} \tag{1}$$

Confidence can be interpreted as an estimation of the probability of $P(X \mid Y)$. The support s of a rule is defined as the number of records that contain X and also Y ($count(X \cap Y)$) divided by the total number of records in D ($count(R)$).

$$s = \frac{count(X \cap Y)}{count(R)} \tag{2}$$

Classification based on association rules (also known as associative classification, AC), is a relatively new classification approach integrating association mining and classification. Several studies (Li et al., 2001; Yin & Han, 2003 & Sun et al., 2006) have provided evidence that AC algorithms are able to extract classifiers competitive with traditional classification approaches such as C4.5. The main steps of an AC classifier are the following (Thabtah, 2007):

Step 1. Discovery of all frequent rules.

Step 2. The production of all class association rules (CARs) that have confidences above the minimum confidence threshold from frequent rules extracted in Step 1.

Step 3. The selection of one subset of CARs to form the classifier from those generated at Step 2.

Step 4. Measuring the quality of the derived classifier on test data objects.

In our framework we deploy a variation of the CBA (Liu et al., 1998) algorithm, which is a typical associative classifier. CBA first generates as candidate rules all the class association rules exceeding the given support and confidence thresholds using the A-priori algorithm (Agrawal & Srikant, 1994). After the rule generation, CBA prunes the set of rules using the pessimistic error rate method (Quinlan, 1987). More specifically if rule's pessimistic error rate is higher than the pessimistic error rate of rule then the rule is pruned. In the testing phase, the best rule whose body is satisfied by the test object is chosen for prediction. We use a variation of the CBA presented in (Coenen, 2004, b) which replaces the Apriori algorithm with the Apriori-TFP (Coenen et al., 2004, a) which utilizes a tree structure for more effective mining of the association rules.

In the following section, we present a case study on the application of the presented data mining process on data of car customer surveys.

2.2 Case study on car customers

The customer surveys which were conducted in the context of CATER project provided the data for our study. Those included interview surveys of 140 truck drivers and 261 car drivers from Europe and Asia (China, Finland, France, Germany, Greece, India, Italy,

Malaysia, Netherlands, Singapore, Sweden, Switzerland, the UK). We present a case study on the car customer surveys data.
Each individual car customer was asked to select among different versions of various design elements. The case study focused on the 4 design elements that the car customers were more interested to customize. Table 1 includes the design elements (de_i (1st column) and their related design parameters (dp_{ij}) (2nd column) that were included in this case study.

Design elements	**Design parameters**
$de_1 = wheels$	$dp_{11} = material,$ $dp_{12} = number\ of\ spokes$
$de_2 = seats$	$dp_{21} = material,\ dp_{22} = shape$
$de_3 = steering - wheel$	$dp_{31} = material,$ $dp_{32} = number\ of\ spokes$
$de_4 = side\ mirror$	$dp_{41} = shape$

Table 1. Design elements and their related design parameters

For each customer we were provided the citarasa descriptor (*Cd*) that described his/her affective needs, information regarding his/her selections on specific versions of design elements (and thus in specific values of design parameters) and demographic information such as the gender, the age, and the geographic region which according to the citarasa method should also be taken into account. Table 2 includes the respective variables. A snapshot of our complete dataset is presented in Table 3. Each row corresponds to an individual car customer response. For example, row 1 corresponds to a male car customer who comes from Asia, his age is above 55 and his affective needs are described by the

Name	Values
Region	Europe, Asia
Gender	Male, Female
Age	18-24,25-54,55-above

Table 2. Demographic information variables for car customers

Region	*Gender*	*Age*	*Cd*	dp_{11}	dp_{12}	dp_{21}	dp_{22}	dp_{31}	dp_{32}	dp_{41}
Asia	Male	55-	Classic	Alumi-nium	Five	Poly-ester	Flat	Wood	Three	Angu-lar
Asia	Male	55-	Classic	Alumi-nium	Five	Poly-ester	Flat	Wood	Four	Angu-lar
Asia	Female	25-54	Modern	Alumi-nium	Multi-ple	Canvas	Wide	Alumi-nium	Three	Cur-vy
Asia	Female	25-54	Cool	Alumi-nium	Multi-ple	Canvas	Cur-ved	Vinyl	Three	Recta-ngular
Europe	Female	25-54	Cool	Alloy	Six	Canvas	Cur-ved	Vinyl	Multi-ple	Recta-ngular

Table 3. Snapshot the car customers data set

citarasa descriptor (Cd) *Classic*. The rest of the columns correspond to his selection on specific design elements and design parameters. For example, in column 5 and row 1 the customer's selection on the material of the wheels (*Aluminium*) is included.

For each design parameter dp_{ij} a classification based on association rules was constructed. As a result, 7 classification mechanisms were constructed to provide a mapping between customers' affective needs and the specific design parameter of a design element. Towards this direction, the customer survey data set was divided to 7 subsets, each one related to a design parameter, which were provided as training data to the CBA algorithm. The support s and confidence c thresholds were set to 10% and 50% respectively. Table 4 includes the number of rules generated by the CBA algorithm for each design parameter dp_{ij}. The 2nd row refers to the whole set of the generated rules while the 3rd row to number of rules that above the support and confidence thresholds.

Design parameter	dp_{11}	dp_{12}	dp_{21}	dp_{22}	dp_{31}	dp_{32}	dp_{41}
Numbers of rules generated	141	97	144	84	123	44	110
Number of rules above thresholds s=10% and c=50%	27	31	24	27	27	19	24

Table 4. Number of rules generated for each design parameter

Besides the classification purposes, the rules generated provided also a meaningful overview of the associations among data. Table 5 includes the rules generated for the dp_{32} (which refers to the number of spokes of the steering-wheel) that were above the support

No. rule	**Rule**	**Confidence**
1	Region=*Europe* and Gender =*Female* and Cd =*Classic* -> *Three*	100.0%
2	Region=*Europe* and Gender=*Female* and Cd=*Sporty* -> *Three*	100.0%
3	Region=*Asia* and Gender=*Female* and Cd=*Classic* -> *Four*	100.0%
4	Region =*Europe* and Age =*18-24* and Cd =*Cute* -> *Three*	100.0%
5	Region =*Asia* and Gender = *Male* and Cd =*Cool* -> *Three*	100.0%
6	Region=*Asia* and Age =*18-24* Cd =*Classic* -> *Four*	100.0%
7	Region =*Asia* and Gender=*Male* and Cd= *Modern* -> *Multiple*	100.0%
8	Region=*Asia* and Age =*18-24* and Cd =Sporty-> *Multiple*	100.0%
9	Age=*55-above* and Cd=*Classic* -> *Three*	100.0%
10	Age =*55-above* and Cd=*Sporty* -> *Three*	100.0%
11	Age =*55-above* and Cd =*Cute* -> *Multiple*	100.0%
12	Gender =*Male* and Age=*18-24* and Cd =*Cute}* -> *Three*	100.0%
13	Region=*Europe* and Cd =*Classic* -> *Three*	91.66%
14	Region =*Europe* and Cd= *Sporty* -> *Three*	83.33%
15	Region =*Female* and Age=*25-54* and Cd =*Cool* -> *Four*	83.33%
16	Region =*Europe* and Age =*18-24}* -> *Three*	80.0%
17	Region=*Europe* and Gender =*Female* and Cd =*Cute* -> Three	80.0%
18	Region=*Asia* and Gender=Female and Age=*18-24* -> Four	80.0%
19	Default -> *Three*	0.0%

Table 5. Rules generated for the design parameter dp_{32}

and confidence thresholds. For example, rule 1 implies that a female customer who comes from Europe, and her affective needs are described by the citarasa descriptor *Classic* she would be satisfied with a steering-wheel with three spokes.

2.3 Evaluation

The accuracy of the classifiers was assessed by a k-fold cross validation (Kohavi *et al.* 1995) process. According to this method, the dataset is divided into k subsets. Each time one of the k subsets is used as the test set and the other $k-1$ form the training set. The advantage of this method is that it does not depend on how the data gets divided as each one of the data instances takes part in the test set once and in the training set $k-1$ times. The most commonly used value for k ,which is used in our study, is 10. The accuracy (AC) of the classifiers is measured by the proportion of the total number of items that were correctly classified. It is determined using the equation (3):

$$Accuracy = \frac{TP + TN}{TP + TN + FP + FN} \tag{3}$$

The *TP* (*True Positive)* is the number of positive cases that were correctly classified. And the *FP (False positive*) is the number of negatives cases that were incorrectly classified as positive. In proportion, the *TN (True negative*) is defined as the number of negatives cases that were classified correctly and the *FN (False negative*) is the number of positives cases that were incorrectly classified as negative. Figure 1 includes the calculated predictive accuracy of the classifiers generated for each design parameter dp_{ij} .

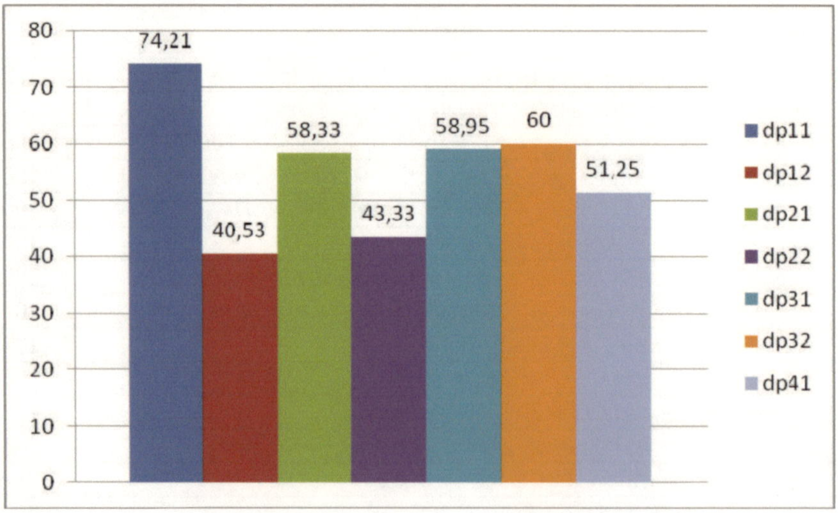

Fig. 1. Predictive accuracy of classifiers

As it is depicted in Figure 1, most of the classifiers have achieved a level of predictive accuracy above 50%. The average accuracy of all classifiers is 55,23%. The generated classifiers form the prediction mechanism which generates for each design parameter a specific prediction based on the generated rules. Table 6 shows the predicted values for an individual customer. The example refers to a female car driver from Europe, who belongs to the age range of 25-54 and would like to have a "Cool" car.

Design parameter	Predicted Values	Design parameter	Predicted Values
dp_{11}	Aluminium	dp_{31}	Wide
dp_{12}	Multiple	dp_{32}	Four
dp_{21}	Aluminium	dp_{41}	Angular
dp_{22}	Canvas		

Table 6. Predicted design parameters for a customer

The predicted values are provided as input to the agent-based framework developed (see following Chapter) and are "interpreted" to configuration elements by the use of the configuration ontology. Finally, the complete vehicle recommendation is then presented visually to the user via web and VR based user interfaces.

3. Agent-based framework

3.1 Agent technology

The agent - based system has been developed with a new technology of JADE which is called Web Service Integration Gateway (WSIG). The objective of WSIG is to expose services provided by agents and published in the JADE framework as web services, though giving developers enough flexibility to meet specific requirements. The process involves the generation of a suitable WSDL for each service-description registered with the Data Framework and also the publication of the exposed services in a UDDI registry. The Web Services are becoming one of the most important topics of software development and a standard for interconnection of different applications.

The WSIG add-on of JADE supports the standard Web Services stack, consisting of WSDL for service descriptions, SOAP message transport and a UDDI repository for publishing Web Services using Models (Jade WSIG Guide 2008). As shown in Figure 2, WSIG is a web application composed of two main elements:

- the WSIG Servlet, and,
- the WSIG Agent.

The WSIG Servlet is the front-end towards the internet world (Jade WSIG Guide 2008) and is responsible for:

- Serving incoming HTTP/SOAP requests;
- Extracting the SOAP message;
- Preparing the corresponding agent action and passing it to the WSIG Agent Moreover once the action has been served;
- Converting the action result into a SOAP message;
- Preparing the HTTP/SOAP response to be sent back to the client.

The WSIG Agent is the gateway between the Web and the Agent worlds (Jade WSIG Guide 2008) and is responsible for:

- Forwarding agent actions received from the WSIG Servlet to the agents actually able to serve them and getting back responses.
- Subscribing to the JADE DF to receive notifications about agent registrations / de-registrations.
- Creating the WSDL corresponding to each agent service registered with the DF and publishes the service in a UDDI registry if needed.

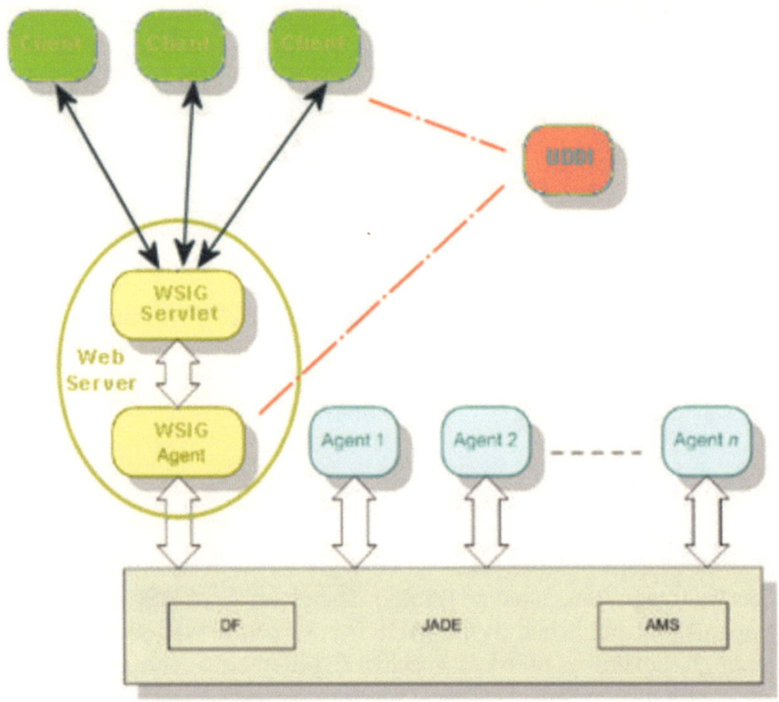

Fig. 2. WSIG Architecture (Jade 2008)

Two main processes are continuously active in the WSIG web application:

i. The process responsible for intercepting DF registrations/de-registrations and converting them into suitable WSDLs. As mentioned, this process is completely carried out by the WSIG Agent.

ii. The process responsible for serving incoming web service requests and triggering the corresponding agent actions. This process is carried out jointly by the WSIG Servlet (performing the necessary translations) and the WSIG Agent (forwarding requests to agents able to serve them).

The FIPA (Foundation for Intelligent Physical Agents) compliant JADE/LEAP platform (Jade 2008) adopted allows for an architecture that is:

- Distributed (different platforms);
- Standards based (FIPA, HTTP, XML, RDF);
- Process centric (agents);
- Widely used in ICT (Information and Communication Technologies);
- Open source (possibility of features addition);
- Cross-platform (Operating System, e.g. Linux);
- Variety of message transport protocols.

JADE (Java Agent DEvelopment Framework) is a software framework fully implemented in Java language (Jade 2008). It aims at the development of multi-agent systems and applications confirming to FIPA standards for intelligent agents. It includes:

- A runtime environment where JADE agents can "live" and that must be active on a given host before one or more agents can be executed on the host.
- A library of classes that programmers can use to develop their agents.

- A suite of graphical tools that allows administrating and monitoring the activity of running agents.

Each running instance of the JADE runtime environment is called a 'Container' as it can contain several agents. A set of active containers is called a 'Platform'. A single special Main Container should always be active in a platform and all other containers register with it when they start. The Main Container differs from normal containers in the ability of accepting registrations from other containers. This registration can be done by the two special agents that start when the main container is launched. These are:

- The Agent Management System (AMS) that provides the naming service and represents the authority in the platform. The Agent Communication Channel (ACC) is the agent that provides the path for basic contact between agents inside and outside the platform.
- Standards The Directory Facilitator (DF) that provides a Yellow Pages service by means of which an agent can find other agents providing the required services. The standard specifies also the Agent Communication Language (ACL). Agent communication is based on message passing, where agents communicate by formulating and transmitting individual messages to each other.

3.2 Agents in the overall CATER architecture

As it has already been mentioned, the CATER architecture is based on agents. Figure 3 shows the connectivity of the agents with the rest modules of the system. More analytically, the agent is interconnected with three main modules. These are namely: a) the Web interface, b) the Citarasa engine and c) the DIYD engine of the system. On the Web interface, the agent allows the user to register and/or login him/herself to the system. Specific entries are requested by the user, such as the name and surname of the user, desired username and password and also the occupation, the region, the age and the gender.

The occupation, the region, the age and the gender in specific, constitute the input that is required by the *Citarasa* engine in order to predict a vehicle configuration, customized to the specific user, classified also per *Citarasa Descriptor* (i.e. "Cute", "Cool", "Classic", etc.). The prediction of the most suitable vehicle configuration is performed through the classification

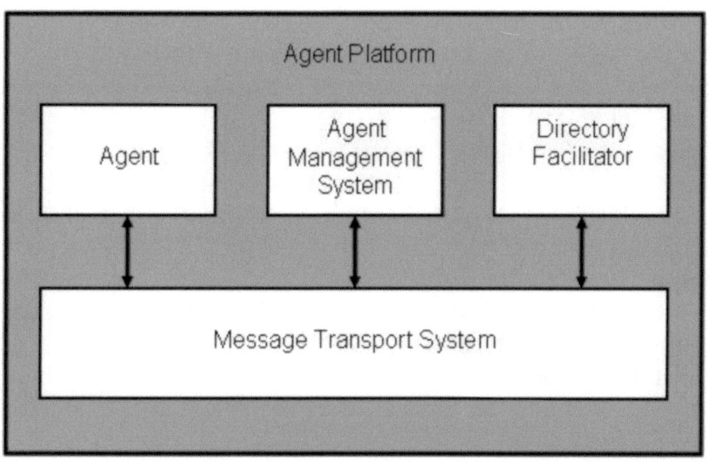

Fig. 3. Agent Platform internal diagram

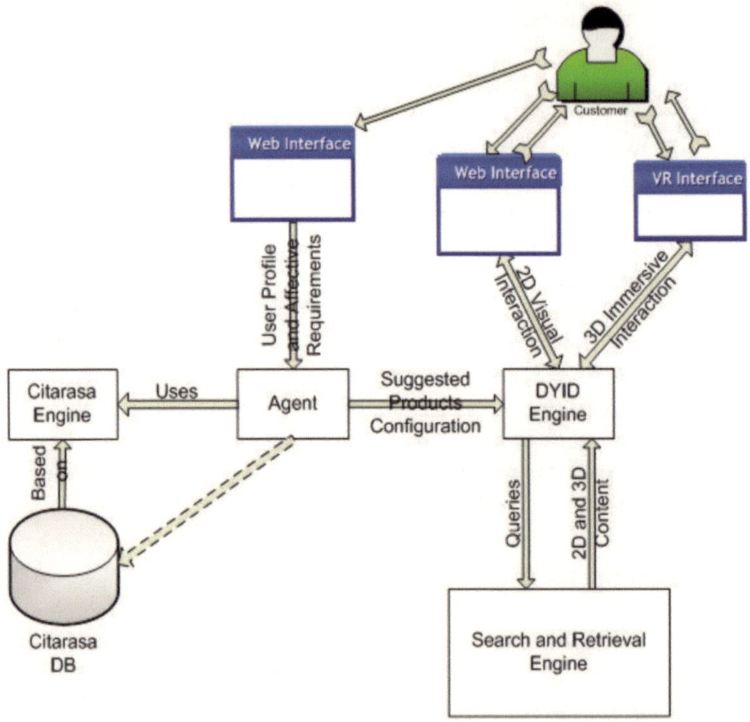

Fig. 4. Agent's conceptual architecture

models generated by the data mining process (as described in section 2) which constitute the knowledge base of the Data Mining module of the *Citarasa* engine. All requested entries by each registered user are collected in the *Citarasa* engineering database.

The agent is interconnected with the DIYD engine that constitutes the interface of the user with the Citarasa engine. The DIYD engine requests from the user (via the web interface) to choose among a list of *Citarasa Descriptors* (i.e. "Cool", "Cute", "Classic", etc.) (Figure 5) that according to his/her opinion characterise in the most felicitous way his/her overall preference regarding the vehicle s/he wishes to view and further configure. The DIYD engine, utilising the output of the *Citarasa* engine, finally provides to the user a suggested vehicle configuration, customised to his/her profile and declared preference, by the means of web or VR based interfaces (Figure 6).

The agent - based system consists of several functions. Each function is responsible for a particular activity and these activities are accessible through a special XML file, the WSDL file. In this file, the client is able to find all the available activities that the agent can perform. Table 7 below contains the list of the major actions that the CATER agent performs.

These functions are available through the World Wide Web. Any module that needs to interact with the CATER system has to follow the rules of the above functions in order to retrieve the required/requested results.

It should be noted that the agent has been designed in such a way so as to support also the self training of the system. Every time a user completes his/her vehicle configuration process, the CATER agent stores this information. A specific number of new entries on the database trigger the update of the knowledge base of the DM module that is responsible for the vehicle configuration prediction recommended to the user.

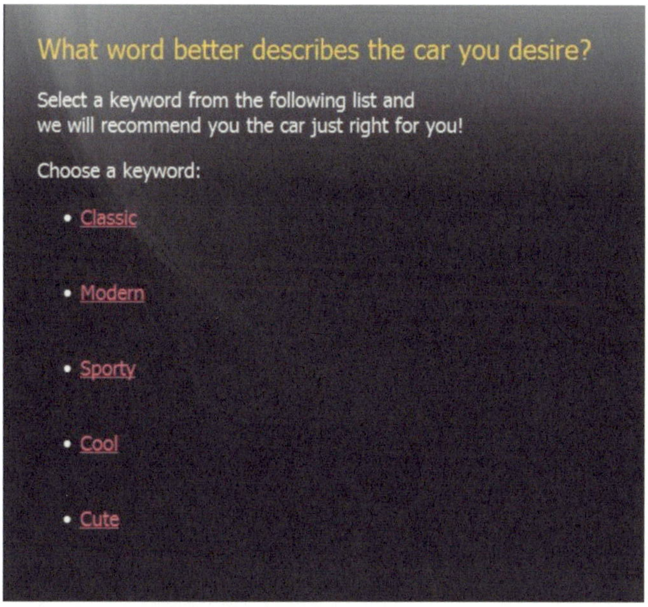

Fig. 5. Selection among a list of Citarasa Descriptors via the web interface

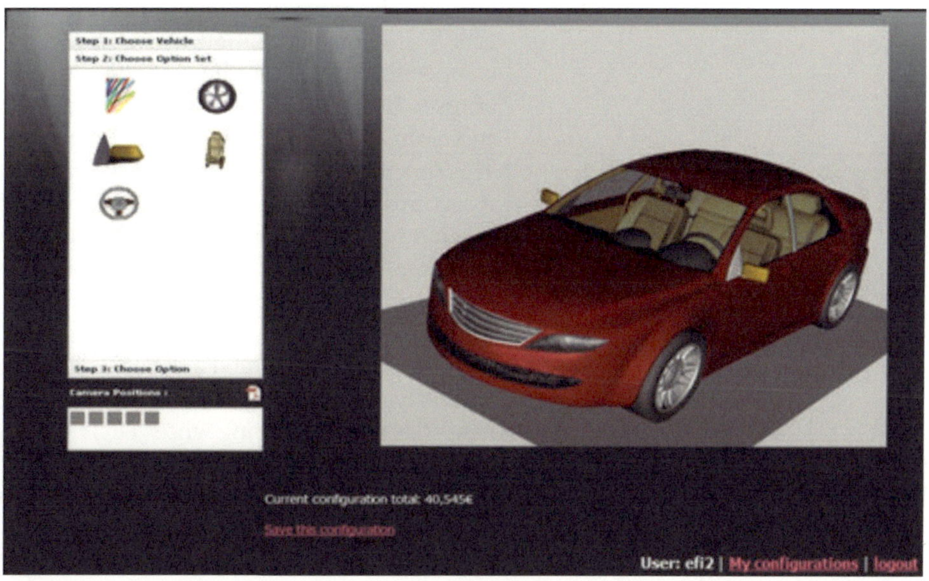

Fig. 6. Proposed configuration via the web interface

Activity	**Function**	**Description**
Registration	setUser( )	This function has a list of attributes as input (name, surname, username, password, region, occupation, age, gender) and outputs "0" or "1" (false or true) which indicates the successful addition of the data in the database.

Activity	**Function**	**Description**
Registration	getUsernameExistance( )	This function examines if a specific username already exists in the database. It has one attribute as an input, the username and it outputs "1" or "0" for true or false respectively.
Log In	getUserId( )	This function provides the ID number of the user. It has two attributes as input (username and password) and outputs the ID number of the user when the log in is successful or "1" if there is an error with the username and/or password.
Log In	updateUser( )	This function is responsible for the user profile update. It gets one value as input (the username) and it returns "0" or "1" (false or true) which indicates the successful addition of the data in the database.
Accessing CATER list	setComponentValue( )	This is a function that stores the user's history. It keeps the CATER components database updated depending on the choices-selections of the user. The function stores the user's updates on a specific component with input attributes user ID, region, occupation, descriptor, component, component ID, attribute and outputs "1" or "0" (false or true) indicating the successful addition of the data in the table.
Accessing CATER list	getUserAttributes( )	This function collects all the attributes of a user. It requires the ID number of the user as input and it outputs a vector (an array of data) with the name, surname, username, password, region, occupation, age, gender.
Accessing CATER list	getComponentId( )	A function that returns the ID number of the vehicle-component combination according to the vehicle type (i.e. car or truck) and the component name (i.e. mirror, steering wheel, etc.).
Accessing CATER list	getActualAttribute( )	This function uses semantics (guided by ontology) for the mapping of predicted design parameters to actual elements which are available in the DIYD engine. It has two input attributes (the component ID and the extracted output) and returns the corresponding element of the DIYD Engine.
Accessing CATER list	getStatistics( )	This is a function that provides the percentage of the available choices of a descriptor according to the entered input (region, gender, age range and descriptor).

Table 7. List of functions of the agent - based system of CATER

4. Conclusion

This paper presented a data mining and agent-based framework based on *citarasa* principles. The methodology followed provided a mapping mechanism of customer affective needs described by their *citarasa* to design parameters related to vehicle design elements. Results derived on the application of the methodology on customer survey data showed that the framework is capable of providing recommendations to the customers based on the generated mechanism. However, the need for more customer data and larger training datasets will be always a desirable option because it results in improvement of the data mining outcome and hence accuracy of user recommendations. Future experiments will be conducted in order to evaluate the generated mechanism and measure the improvement introduced, compared to the initially evaluated rules.

In addition, the design and the development of a specific module of the CATER integrated system, namely the agent - based system of CATER which is responsible for the interconnection and interface of different modules of the system, aiming, finally, at proposing a personalised vehicle configuration to the customer is being presented in the current paper. It should be noted that the customer is able to further elaborate the proposed by the system vehicle configuration through the CATER configurators, if wishes so.

The agents' technology deployed is a FIPA compliant JADE/LEAP platform technology which has been indicated as the best solution for Client - Server communications (Jade 2008). The functionality of the agent required the use of another add-on application of JADE which is called WSIG (Jade WSIG Guide 2008). This add-on transformed the agent's functionality to web service in order to be available to anyone through the World Wide Web and made feasible the interface of CATER engines output through a web interface.

The personalisation enabled by the CATER agent lies in the output of the data mining process described in this paper, and associates in practice the user profile, in terms of age, region, gender and the user needs (reflected through the *Citarasa Descriptors*) in order to predict the most suitable vehicle configuration per se (user).

The communication of this result to the user, via web and VR interfaces, is again a responsibility of the CATER agent system. Finally, a valuable advantage of the CATER agent system is the ability to store the history of each user vehicle configuration tried. The history records are then utilised for the update of the prediction rules, and as such, of the progressing improvement of their accuracy.

5. References

Agrawal, R. & Srikant, R. (1994). Fast algorithms for mining association rules. *Proceedings of the 20th International Conference of Very Large Data Bases*, pp. 487-499, Santiago de Chile, Sept. 1994, Morgan, Santiago de Chile

Annex I-"Description of Work", CATER project , CN. 035030, Sixth Framework Programme, Priority 2-IST, Information Society Technologies (2006). [Online].Available at: http://www.cater-ist.org

Coenen, F.; Leng P. & Ahmed S. (2004). Data structures for association rule mining: T-trees and P-trees, *IEEE Trans. Knowl.,Data Eng.*, Vol. 26, No 6, June 2004, pp. 774-778, 1041-4347

Coenen, F. (2004). The LUCS-KDD Group, Department of Computer Science, The University of Liverpool, UK [Online].Available at: http://www.cSc.liv.ac.uk/~frans/KDD

Cross, N. (2000). Engineering design methods: Strategies for product design (3rd ed.).Wiley, 978-0471872504, Chichester, UK

Jade Agent Development Framework - an Open Source platform for peer-to-peer agent based applications (2008). [Online] Available at: http://jade.tilab.com

Jade Web Services Integration Gateway (WSIG) Guide (2008). [Online] Available at: http://jade.tilab.com (add-ons area)

Jiao, R.J.; Xu Q.; Du J., Zhang Y.; Helander, M.; Khalid, H.; Helo, P. & Cheng, N. (2008). Analytical affective design with ambient intelligence for mass customization and personalization. *International Journal of Flexible Manufacturing Systems,* Vol. 19, No 4, February 2008, pp. 570-595, 0920-6299

Ishihara, S.; Ishihara, K.; Nagamashi, M., & Matsubara, Y. (1995). An automatic builder for a Kansei engineering expert system using selforganizing neural networks. *International Journal of Industrial Ergonomics,* Vol. 15, No. 1, January 1995, pp. 13–24, 1936-6590

Khalid, H. M. & Helander, M. G. (2004). A framework for affective customer needs in product design. *Theoretical Issues in Ergonomics Science,* Vol.5, No. 1, January 2004, pp. 27–42, 1463-992X

Khalid, H.M. and Helander, M.G. (2006). Customer emotional needs in product design. *International Journal on Concurrent Engineering: Research and Applications,* Vol. 14, No. 3, pp. 197-206, 1531-2003

Kohavi, R. (1995). A study of cross-validation and bootstrap for accuracy estimation and model selection, *Proceedings of fourteenth International Joint Conference on Artificial Intelligence,* pp. 1137-143, Montréal, Aug. 1995, Morgan Kaufmann, Québec, Canada.

Li, W.; Han, J. & Pei, J. (2001) CMAR: Accurate and efficient classification based on multiple-class association rule, *Proceedings of the International Conference on Data Mining (ICDM'01),* pp. 369–376, San Jose, November 2001, IEEE Computer Society Press,San Jose California

Liu, B.; Hsu, W. & Ma, Y. (1998). Integrating Classification and Association Rule Mining, *Proceedings of the 4rd International Conference Knowledge Discovery and Data Mining (KDD-98),* pp. 80-96, New York, Aug. 1998, AAAI Press, New York USA

Nagamashi, M. (1989). *Kansei Engineering,* Kaibundo Publisher Co Ltd, Tokyo

Matsubara, Y. & Nagamachi, M. (1997). Hybrid Kansei engineering system and design support. *International Journal of Industrial Ergonomics,* Vol. 19, No. 2, pp. 81–92, 0169-8141

Quinlan, J. (1987). Simplifying decision trees. *International Journal of Man–Machine Studies, Vol. 27,* pp. 221–248.

Sun, Y.; Wong, A. & Wang, Y. (2006). An overview of associative classifier, P*roceedings of the 2006 International conference on Data mining,* pp. 138-143, Las Vegas, June 2006, Nevada

Thabtah, F. (2007). A review of associative classification mining, *The Knowledge Engineering Review,* Vol 22, No.1, pp. 37-65, 0269-8889

Witten, I.H. & Frank, E. (2005). *Data Mining: Practical Machine Learning Tools and Techniques (Second Edition),* Morgan Kaufmann, 0-12-088407-0, San Francisco, CA, USA

Yin, X. & Han, J. (2003). CPAR: Classification based on predictive association rules, *Proceedings of the Third SIAM International Conference on Data Mining,* pp. 331–335, San Francisco, May 2003, SIAM, CA USA

Proceedings of the Third SIAM International Conference on Data Mining, pp. 331–335, San Francisco, May 2003, SIAM, CA USA

2

Automotive Testing in the German-Dutch Wind Tunnels

Eddy Willemsen, Kurt Pengel, Herman Holthusen, Albert Küpper, et al
German-Dutch Wind Tunnels DNW
The Netherlands

1. Introduction

The Foundation German-Dutch Wind Tunnels (DNW) was jointly established in 1976 by the Dutch National Aerospace Laboratory (NLR) and the German Aerospace Centre (DLR), as a non-profit organisation under Dutch law. The main objective of the organisation is to provide a wide spectrum of wind tunnel tests and simulation techniques to customers from industry, government and research. DNW owns the largest low-speed wind tunnel with open and closed test section options in Europe. Also the major aeronautical wind tunnels of the DLR and NLR are fully integrated and managed by the DNW organisation. The wind tunnels are grouped into two Business Units "Noordoostpolder/Amsterdam" (NOP/ASD) and "Göttingen und Köln" (GUK).

DNW provides solutions for the experimental simulation requirements of aerodynamic research and development projects. These projects can originate in the research community (universities, research establishments or research consortia) or in the course of industrial development of new products. Most of the industrial development projects come from the aeronautical industry, but the automotive, civil engineering, shipbuilding and sports industries have also benefited from DNW's capabilities.

For efficient and flexible operations, DNW operates in a decentralised structure under a unified management and supervision. The seat of its Management is in Marknesse, at the location of its largest wind tunnel, the DNW-LLF. DNW's Board, the supervisory body of the Foundation, consists of representatives of the parent institutes NLR and DLR, and is complemented by representatives of the relevant ministries from Germany and the Netherlands.

In order to assure the compatibility of DNW's development strategy with the long-term needs of the research and development market, an Advisory Committee consisting of high-level representatives of participants in the market provides strategic advice and information to DNW.

2. The wind tunnels of DNW

The eleven wind tunnels of DNW include low speed, high speed, transonic and supersonic facilities. They are distinguished with three-lettered names. The following wind tunnels are also used for testing ground and rail vehicles:

KKK Cryogenic wind tunnel at Cologne, Germany. A closed circuit, continuous, low-speed wind tunnel with a closed wall test section

LLF Large low-speed facility at Marknesse, the Netherlands. A closed circuit, atmospheric, continuous low-speed wind tunnel with three closed-wall exchangeable test sections and an open jet

LST Low-speed wind tunnel at Marknesse, the Netherlands. A continuous, atmospheric, low-speed wind tunnel with exchangeable test sections

NWB Low-speed wind tunnel at Braunschweig, Germany. A continuous, atmospheric, low-speed wind tunnel with optionally a closed or a slotted test section with an open jet

2.1 The DNW-LLF

The LLF (Fig. 1) is situated near Marknesse, approximately 100 km north-east of Amsterdam. It is an atmospheric, single return wind tunnel with two exchangeable test section arrangements. It can also be operated in the open-jet mode, whereby the complete test section is removed and acoustic measurements in an anechoic environment can be executed. Each test section configuration exists of three elements: a nozzle which forms the connection between the contraction at the end of the settling chamber and the test section, the actual test section itself and the transition which forms the connection between the test section and the diffuser part of the wind tunnel. These three elements have an overall length of about 44 m.

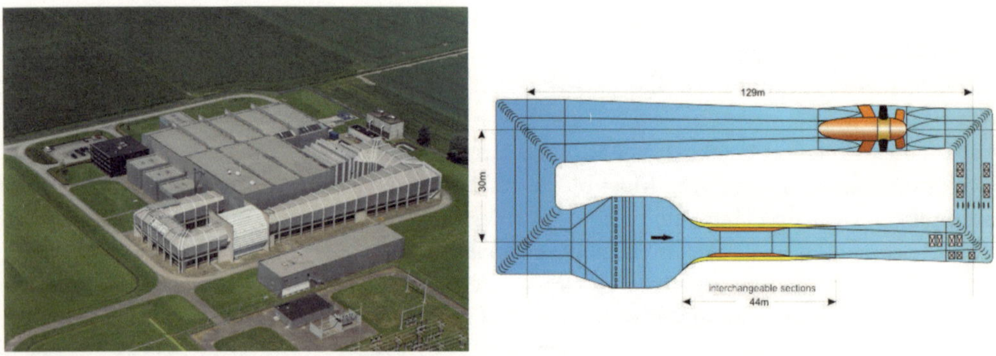

Fig. 1. Aerial view and layout of the LLF circuit

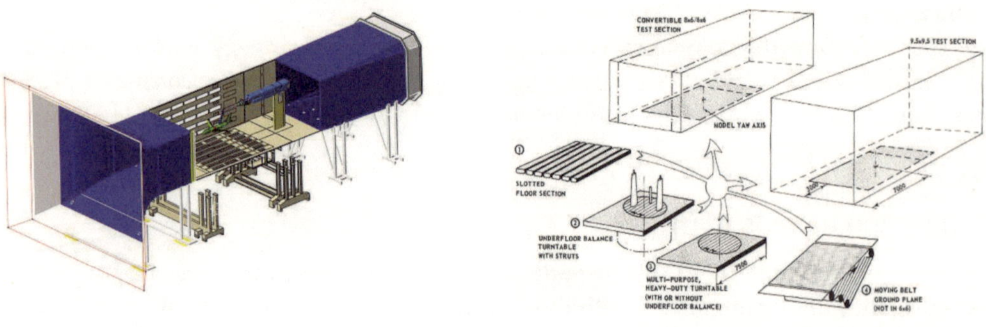

Fig. 2. Test section arrangement (left) and overview of interchangeable floor sections (right)

There are two test sections with different dimensions. The largest one has a cross section of 9.5 m x 9.5 m and the smallest one is convertible between a cross section of either 8 m x 6 m or 6 m x 6 m. These three test sections are further on referred to as the 9.5x9.5, the 8x6 and the 6x6 test section, respectively. Of the convertible test section all four walls have adjustable slots which when opened to their maximum results in 12% porosity. For the open jet arrangement the 8x6 nozzle is used together with the 9.5x9.5 transition. Converting the 8x6 to 6x6 is achieved by mounting inserts in the 8x6 nozzle and moving in the two sidewalls. The 9.5x9.5 and the 8x6 test section have a length of 20 m; the 6x6 is 15 m long. Both test sections rest on air cushions for easy exchangeability.

The air flow in the tunnel is generated by a single stage fan of 12.35 m diameter with eight fixed blades. The fan is driven by a variable speed synchronous motor located in the nacelle with a nominal maximum power of 12.65 MW at 225 rpm.

The maximum wind speed in the 9.5x9.5 test section is 62 m/s, in the closed 8x6 test section 116 m/s, in the open 8x6 test section 78 m/s and in the 6x6 test section 152 m/s.

The test sections have removable and exchangeable floors (Fig. 2). One of the floor sections has slots (floor #1), one has a 5. 5m diameter turntable for use with the external balance (floor #2) and one has a 5 m diameter multi-purpose turntable (floor #3).

The flow quality of the LLF is very high. The turbulence levels vary with wind speed and are different for the different test sections. At 150 km/h (42 m/s) the 8x6 test section has a turbulence level of less than 0.1 percent and the 9.5x9.5 test section of 0.25 percent. This is illustrated in more detail in Figure 3.

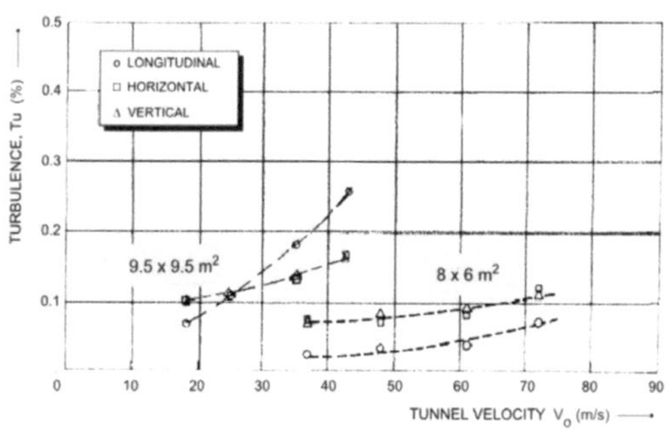

Fig. 3. Turbulence levels

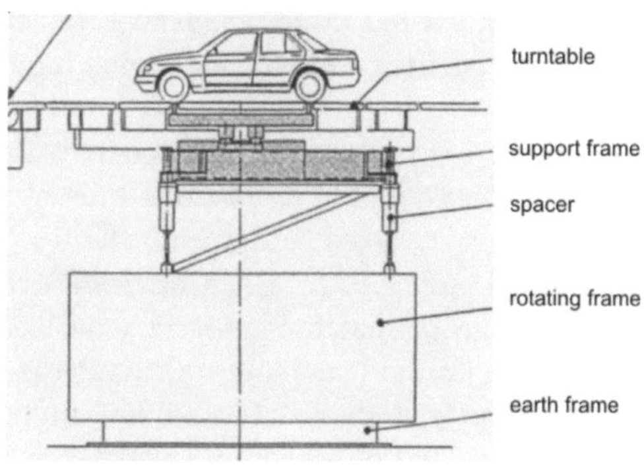

Fig. 4. Setup with the external balance

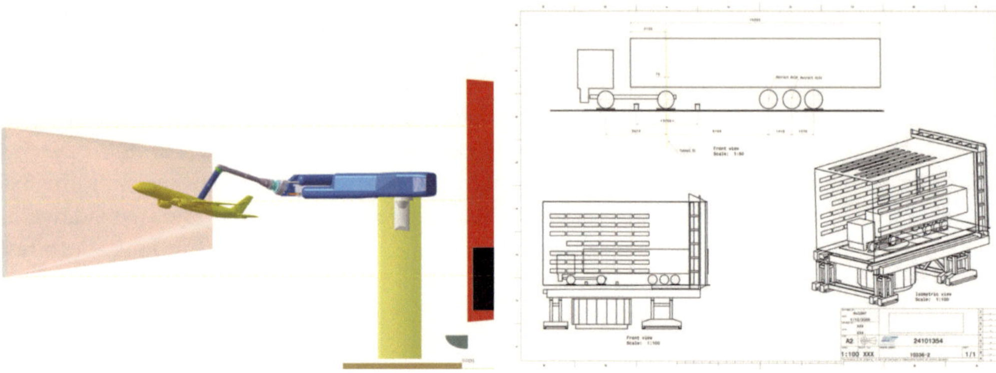

Fig. 5. Sting support system (left) and truck setup in the 9.5x9.5 test section (right)

LLF has two standard support systems (Fig. 4 and 5):

- A sting support system, to which models are mounted via an internal balance. This is mostly used for aircraft and helicopter testing, but is also necessary for cars in combination with the moving belt (moving ground plane)
- An external under-floor platform type balance with a model support by pads or special adapters. This is mostly the primary choice for cars and trucks

3. Force and moment measurements

Two test sections of 20 m length each are suitable for automotive testing at DNW.
Full-scale cars and vans are often tested in the 8x6 closed test section, where up to 400 km/h wind speed can be reached. A car platform with adjustable pads is linked to an external underfloor balance for the measurement of six stationary load components, i.e., drag, side force, lift, and the moments in roll, pitch and yaw. The platform can be used in combination with a tunnel floor boundary layer control system which injects pressurized air into the sublayer upstream of the test vehicle to improve the road simulation conditions and can be used in both test sections.
A sting system for internal balance (six-component) supported cars can be used in combination with a moving belt system for perfect road simulation and rolling wheels.
Full-scale trucks and buses use the 9.5x9.5 closed test section, where the maximum wind speed of 200 km/h is more than enough. A truck platform is available in combination with loose air cushion elements to support the large and heavy vehicles. This setup provides three stationary load components, i.e. drag, side force and yawing moment by means of the external underfloor balance.
The external balance (EXB) is a six-component platform balance, equipped with three horizontal load cells with a resolution of 0.15 N and three vertical load cells with a resolution of 0.30 N. This balance is installed underneath the test section. The test vehicle rests with its wheels on four small pads which are flush with the floor of the turntable and can be adjusted over a wide range to match track and wheelbase. The wheel pads are incorporated in a rigid supporting frame (car platform), which is connected to the metric part of the balance. When use is made of the 8x6 test section, the vertical distance between supporting frame and balance is bridged by a spacer. The balance assembly can rotate over ± 180° in increments of ± 0.1°.

Fig. 6. Model on sting above moving belt

Fig. 7. Truck in 9.5x9.5 test section

4. Road simulation

During wind tunnel tests the relative motion between vehicle and road and the rotation of the wheels is often disregarded. The road is then represented by the rigid floor of the test section and the vehicle rests with stationary wheels on the pads of the balance platform.
The flow pattern around such a configuration is principally different from that on the road due to the grown boundary layer along the wind tunnel floor. The boundary layer thickness near the model may reach a thickness of half the ground clearance of a standard passenger car. This will affect the aerodynamic phenomena around the car.
The effects from various ground simulation techniques at automotive testing in a wind tunnel have been discussed in years around 1990 in various papers of the Society of Automotive Engineers (SAE). Mercker and Knape (1989) discussed the ground simulation with a moving belt or with tangential blowing. Mercker and Wiedemann (1990) compared the results obtained at different ground simulation techniques.

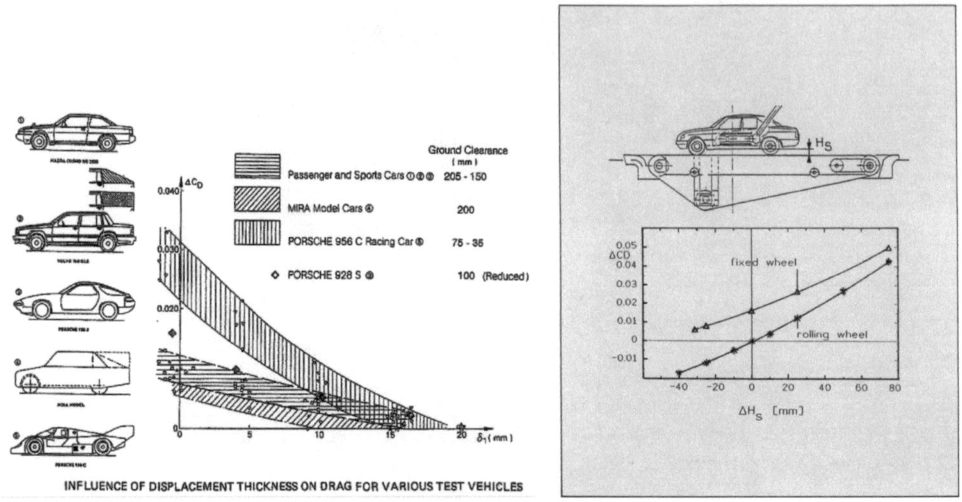

Fig. 8. Effects from ground clearance (left) and rotating wheels (right)

Another deficiency arises from the stationary wheels. Rolling wheels not only affect the flow over and around the wheels but also the overall flow pattern around the car.

As long as the drag of the tested vehicle is relatively high, the error on the drag from the fixed ground floor and the stationary wheels is generally negligible. For vehicles with lower drag this error is not negligible, especially when the low drag is achieved by measures at the underbody of the vehicle. These effects are illustrated in figure 8.

The results from wind tunnel tests can become more realistic when the effect from the boundary layer development along the wind tunnel floor is reduced by boundary layer suction or tangential blowing. Further improvement may be obtained with rotating wheels.

4.1 Tangential blowing

With tangential blowing, air is blown in the direction of the wind through a narrow slot in the wind tunnel test section floor. The blowing device consists of a slot adjustment mechanism and a tubular settling chamber. The principle is that so much air is added to the boundary layer that the momentum deficit in the boundary layer of the wind tunnel floor is reduced to zero. This can be reached exactly at only one downstream distance from the blowing slot.

The system of the LLF is located 4.5 m upstream off the balance centre. The slot spans a length of 6 m and has a variable width between 0 and 5 mm. In order to arrive at a homogeneous spanwise velocity distribution at the exit of the slot, the settling chamber is divided into six individually controlled lateral sections, each provided with a porous smoothing plate. Pressure and temperature are monitored at the centre of the chamber.

The momentum of the thin layer of blown air must be balanced with the momentum of the airflow in the tunnel. From calibration of the blowing device the required ratio V_j/V_0 of the jet velocity V_j at the slot exit to the free stream velocity V_0 is determined as function of the V_0.

Figure 9 shows the setup and calibration results of the tangential blowing system. With tangential blowing good results can be obtained, even for small ground clearances. However, some deformation in the velocity profile remains and the effects from rotating wheels are still ignored.

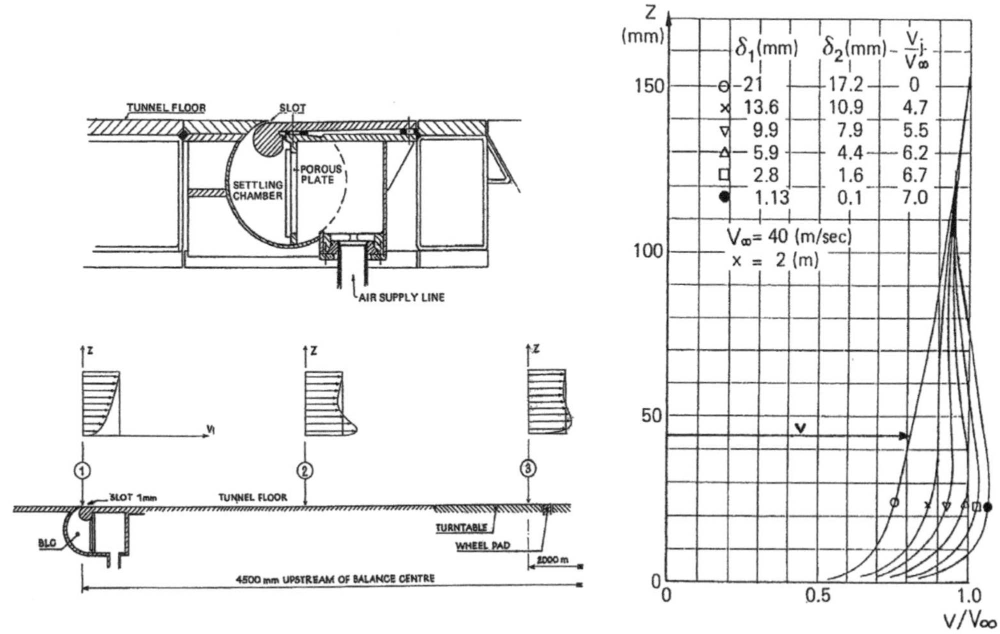

Fig. 9. Tangential blowing system with setup (left) and calibration results (right)

4.2 Moving belt ground plane

The most appropriate way to simulate the full-scale conditions is with a moving belt ground plane. The momentum loss in the boundary layer is almost completely absent over the complete length of the rolling floor and the rotation of the wheels can be enabled as well. However, the disadvantage is that the suspension of the vehicle in the test section is no longer simple.

The moving belt at the LLF has a maximum width of 6.3 m and a flow exposed length between the two main rollers of 7.6 m. Two variable-speed drives of 85 kW each give the belt a maximum speed of 40 m/s. The belt is tensioned and tracked by means of a third roller. Its flatness is monitored during testing by a video camera and with the aid of a laser beam on the wind tunnel side wall. Even under most severe conditions when the vehicle exerts lift, the belt must remain very flat. To reduce friction, pressurized air is fed between belt and support plate. The upstream ground floor boundary layer is scooped off by raising the whole assembly 200 mm above the tunnel floor. The extracted air re-injects automatically at the rear of the belt assembly and through the test section breathers.

The influence of the scoop and belt motion on the boundary layer properties has been calibrated by measuring the velocity profiles at the front and the rear of the belt. Beside the removal of the boundary layer it is essential that the static pressure in longitudinal direction remains constant over the length of a full size passenger car to avoid buoyancy effects in the drag data. This is effectively controlled by adjusting the rearward flaps of the belt. Figure 10 shows the setup and some flow characteristics of the moving belt ground plane at the LLF.

4.3 Rotating wheels

The effects of rotating wheels on the measured aerodynamic forces can be investigated in combination with a moving belt ground plane that drives the wheels by friction. The car is mounted to the available sting and forces are measured with an internal balance between sting and car.

Shock absorbers and springs of each wheel are replaced by a dual-action pneumatic cylinder to counterbalance the wheel's weight but still be able to create an effective downward force.

Rolling resistance of the wheels is determined from the internal balance measurement without wind.

In general, the drag is reduced by wheel rotation, but the magnitude of the drag reduction depends strongly on the type of car and on the ground clearance of the vehicle. The closer to the ground, the more pronounced the effect of the rolling wheels will be; see figure 8.

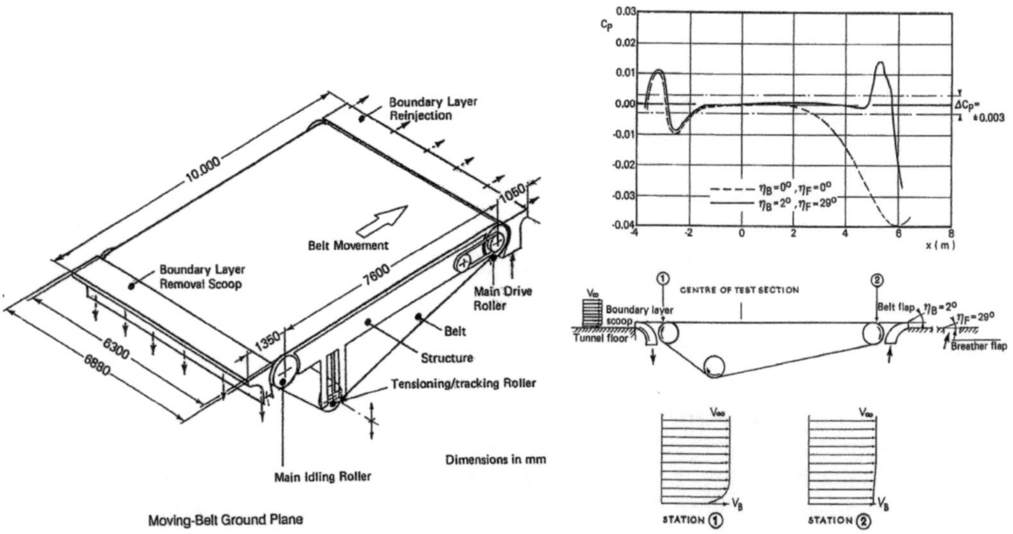

Fig. 10. Setup (left) and velocity profiles (right) of the moving belt ground plane

5. Acoustic measurements

The classical type of acoustic measurements with trucks in wind tunnels is based on the measurement of the noise inside the cabin, as induced by the airflow around the truck and measured with a small number of microphones or with a so-called acoustic head. These cabin noise measurements can be used for the assessment of the acoustic comfort for the truck driver.

The transfer mechanism of the noise from outside the cabin towards cabin interior is often very difficult or impossible to determine. Typical exterior structures are mirrors, the sunscreen above the front window, wind shields, antennas and various spoilers.

Instead of measuring interior cabin noise levels at various exterior configurations, it is more straightforward to measure the exterior sound production. This can be realized with an acoustic mirror or with an array of a large number of microphones.

In an acoustic mirror system a single microphone is mounted in the focal point of a parabolic or elliptic acoustic mirror. Single-microphone measurements give overall noise levels and do not distinct between different noise sources. In a phased microphone system the location and strength of different noise sources can be measured by a phased array technique, whereby on software level the time series of the microphones are analyzed. Similar techniques are applied in radar technology and ultrasonic imaging. A description of array signal processing is given by Johnson and Dudgeon (1993). A description of applications in a wind tunnel environment is given by Underbrink and Dougherty (1996), Piet and Elias (1997), Sijtsma (1997), Dougherty (1997) and Sijtsma and Holthusen (1999). Figure 11 illustrates the principles of the acoustic mirror and the phased microphone array.

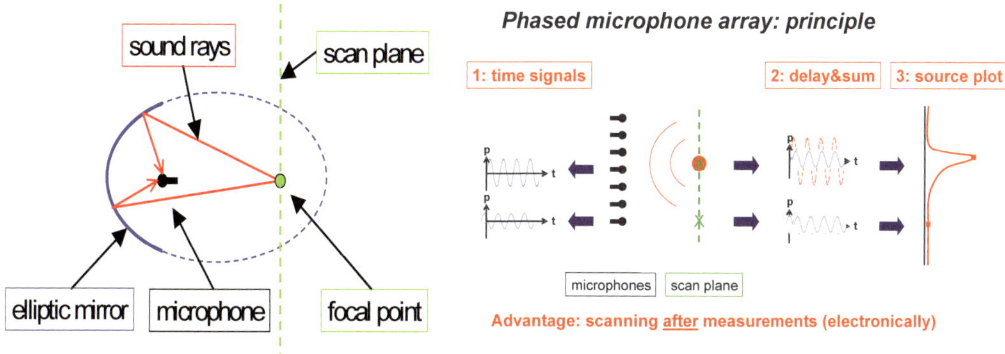

Fig. 11. Acoustic source localization measurement techniques, mirror (left) and acoustic array (right)

Microphone arrays or acoustic mirrors have become popular in wind tunnel measurements as a tool to locate sound sources. Microphone arrays have the advantage over acoustic mirrors of a higher measurement speed. Mirrors have to scan the whole test object point by point, while microphone arrays only need a short time to record the signals from which the aero-acoustic characteristics in a measuring plane can be determined. The process of scanning through possible source locations is performed afterwards by appropriate software running on powerful computer hardware.
An additional advantage of a microphone array is the application inside the flow or in the wall of a closed test section. These in-flow measurements with microphone arrays are possible, when the self-noise of the array microphones, caused by the turbulent boundary layer above the array, is sufficiently suppressed. With a mirror, in-flow measurements are practically speaking impossible.

Fig. 12. Test setup examples: full-scale wing in the LLF (left) and scaled truck model in the LST (right)

In a typical test an array of 1 m diameter, containing about 140 sparsely distributed microphones, may be mounted in or on the wall of the test section.
The microphone array technique can be successfully applied in full-scale tests as well as model tests; see figure 12 for some examples.
The array processing delivers as its main result so-called noise maps. Figure 13 presents some results for a scaled truck and for a full-scale truck. The two dimensional contour maps show the distribution of noise sources in a scanned area near the truck. The noise levels are represented by different colors. The noise maps deliver the location, frequency characteristics and relative strength of the noise source. Additionally the array processing delivers power spectra and overall power levels by integration over the scan area. Several of such scan areas can be defined and processed. One scan area could cover the whole model and other areas could only cover small details, like an outside car mirror, to allow detailed comparisons between different model configurations.

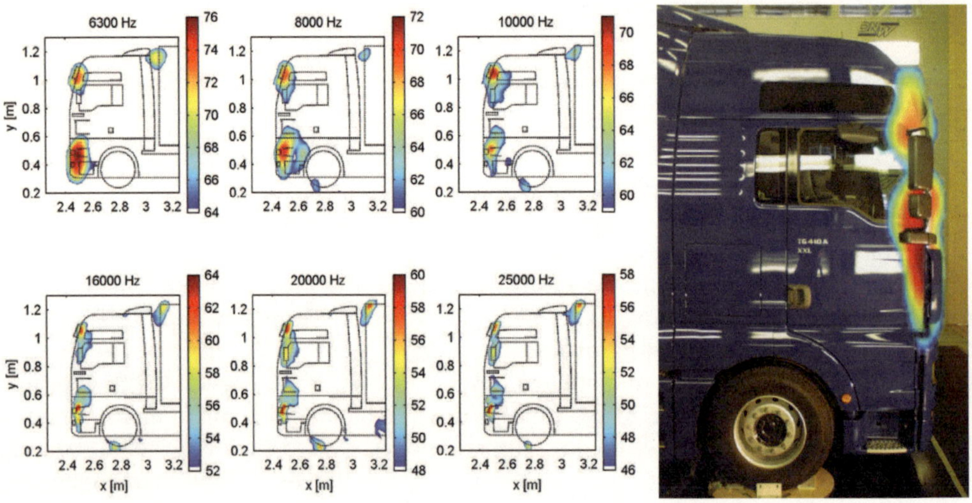

Fig. 13. Microphone array tests results for a scaled truck model (left) and full-scale truck with projected noise map (right)

6. Flow field measurements with a traversing rake of five-hole probes

Quantitative flow field measurements can be executed by means of a traversing rake of multiple five-hole probes. Each five-hole probe can measure the local 3-D wind velocity vector. At each position of the rake the wind speed vector is measured at all probe positions. At DNW there are 18 probes at 15 mm stitch; so each time the data are read out information on a line of 255 mm length are gathered. The rake is normally mounted vertically and connected to a traversing mechanism which is moving at such a low speed that the local flow field is not affected. By repeating the readings during the scan after say every 7.5 mm displacement of the rake and repeating the scan at a vertical displacement of the rake of also 7.5 mm, a block of measuring points is filled with a horizontal and a vertical stitch of 7.5 mm. This is enough to observe flow phenomena on a rather small scale. Software tools may provide additional information, like the strength of the vorticity in the flow. A single scan of about 1 meter at a low traversing speed requires a measuring time of about 10 minutes.

The test technique is providing very nice results in as well a quantitative as a qualitative way. Only close to the surface of the test object the flow may become disturbed by the presence of the rake body close to the object.
An example of the setup in the LST wind tunnel is shown in figure 14, together with some test results behind the wing tip of an aircraft model.

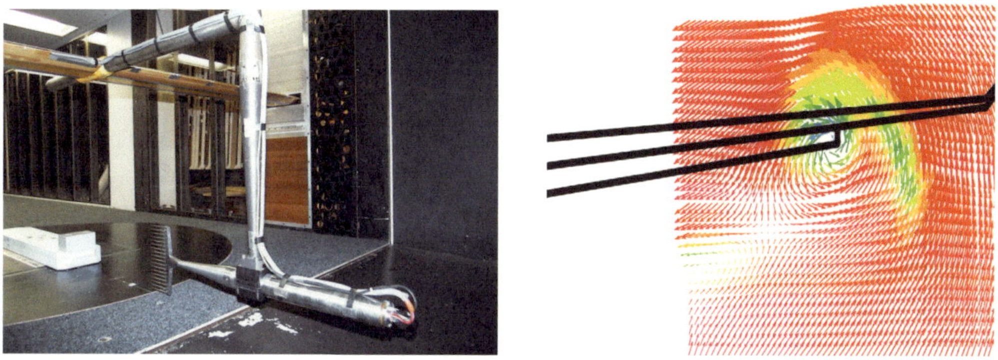

Fig. 14. Test setup in the LST (left) and test results behind the wing tip of an aircraft model (right)

7. Flow field measurements with PIV

The flow field in the vicinity of a test object can be measured by means of Particle Image Velocimetry (PIV); see figure 15.

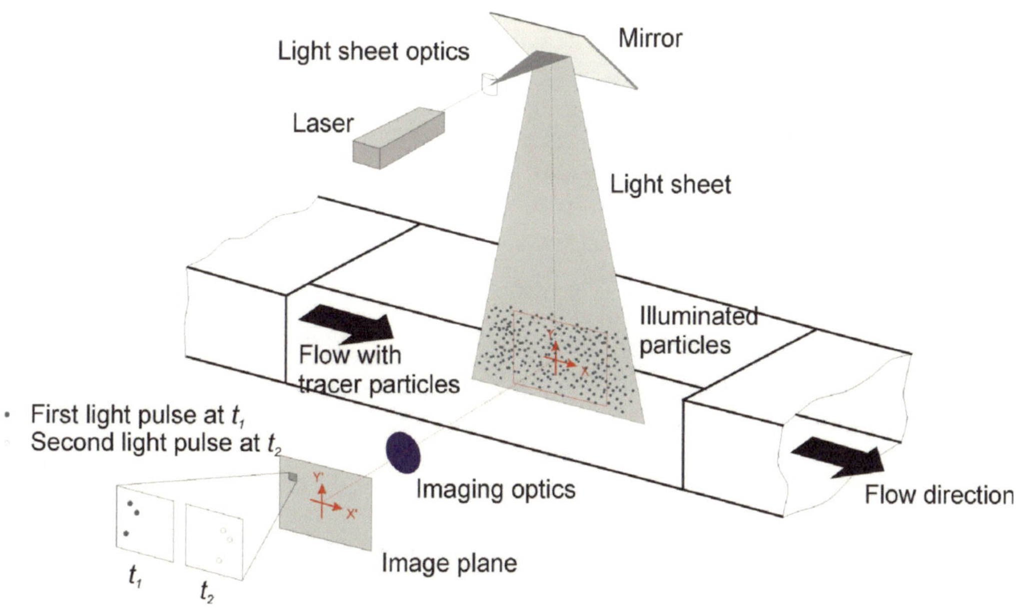

Fig. 15. Experimental setup for PIV

During PIV measurements the flow is seeded with small particles with a diameter in the order of 10 to 100 micrometers (a kind of a light smoke). With a laser two flashing light

planes are created shortly after each other, whereby the light is reflected by the particles. The images are analyzed with a software algorithm, identifying the location of the separate particles during the two images. Once this displacement is established, the corresponding flow speeds in the laser light plane can be calculated. From these wind vector data other characteristic parameters can be calculated, like the vorticity.
The technique and application for a wind tunnel environment became to growth in the last decade of the 20th century. Various authors gave a general description of the principles and possible application at aerodynamic research in wind tunnels, like Willert and Gharib (1991), Adrian (1991), Hinsch (1993), Willert et al. (1996), Willert (1997), Kähler et al. (1998), Raffel et al. (1998), Ronneberger et al. (1998) and Kompenhans et al. (1999).
Figure 16 shows a setup as applied for a wind turbine. A stereoscopic set-up of the cameras enables the determination of the three dimensional flow field characteristics. One camera was directed from above to the horizontal light sheet, the second camera was looking from underneath. This configuration was fixed and could be moved as a whole from one location to another. This fixed set-up of cameras and light sheet allows a system calibration in advance outside the wind tunnel and avoids time consuming re-calibration.

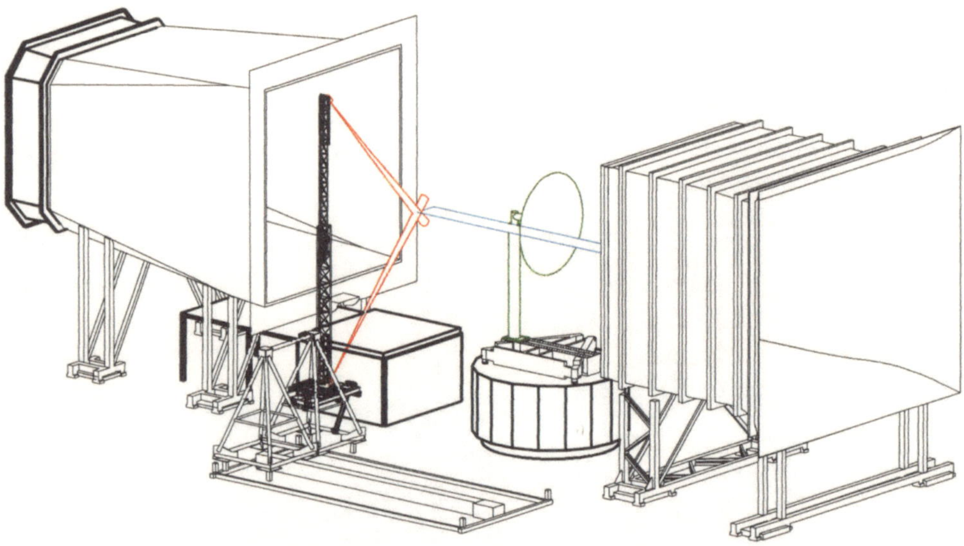

Fig. 16. PIV set-up on a wind turbine

PIV measurements result in vector maps of the velocities in the area where the cameras are focused to. Figure 17 shows some test results at two different setups: underneath a military aircraft and behind the tip of a wind turbine rotor.
The application of PIV in large wind tunnels gives some specific challenges:

- large observation areas requested,
- large observation distances exist between camera and light sheet,
- much time needed for the setup of the PIV system,
- strict safety measures required for laser and seeding,
- high operational costs of the wind tunnel.

In spite of these stringent requirements, the PIV technique is very attractive in modern aerodynamic research. It helps in understanding unsteady flow phenomena such as shear and boundary layers, wake vortices and separated flows. PIV enables spatially resolved measurements of the instantaneous velocity field within a very short time and allows the detection of large and small scale spatial structures in the flow. The PIV method can further provide the experimental data necessary to the validation of an increasing number of high quality numerical flow simulations.

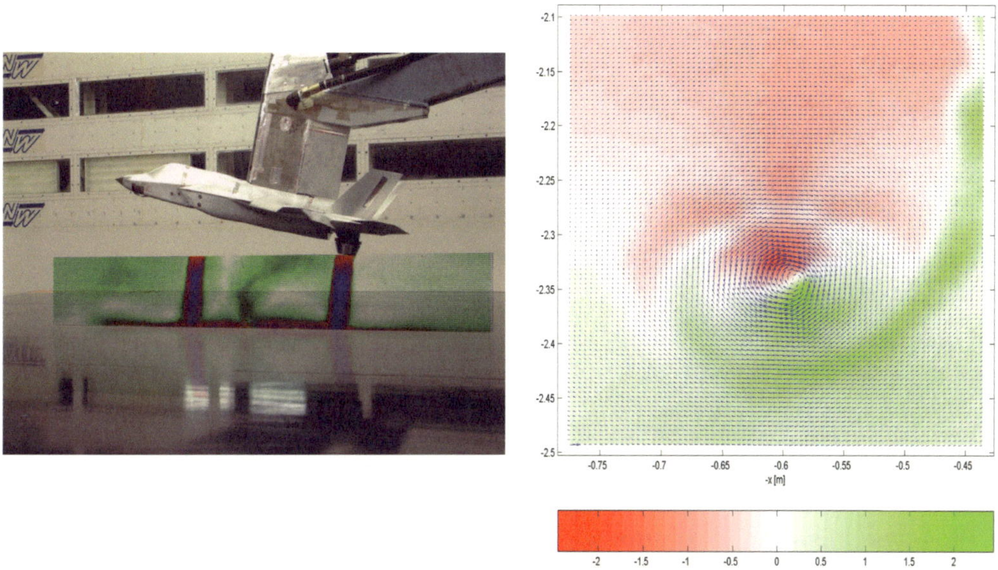

Fig. 17. PIV measurements: exhaust flow of a fighter engine (left) and tip vortex behind a wind turbine rotor (right)

8. Deformation measurements

Within wind tunnel investigations model deformation measurements are possible with techniques like Projection Moiré Interferometry (PMI), Projected Grid Method (PGM) and Stereo Pattern Recognition (SPR) system.
SPR is an optical, non-intrusive method and requires a stereo setup of cameras; it is based on a three-dimensional reconstruction of visible marker locations by using stereo images. Stereo imaging and 3D-reconstruction can be used to determine object locations and their motion with time. There are two possible approaches. If the positions of two (or more) cameras and their optical characteristics are known exactly, a three-dimensional reconstruction is very straightforward. From two images of a certain marker on the object in three-dimensional space by two cameras, the location of this marker can be determined by regarding the images as a result of certain translations and rotations and a final projection on the camera image plane. After calculating transformation matrices for both cameras, derived from the exact set up of the camera positions, the transformation equations for each marker image can be constructed, resulting in an equation system that can be solved by a "least square"-method. A disadvantage of this direct method is that normally the camera positions are not known very exactly. Especially the direction of the optical axis of the cameras and the rotation about this axis can only be measured approximately. In this case a different approach can be used; see figure 18.

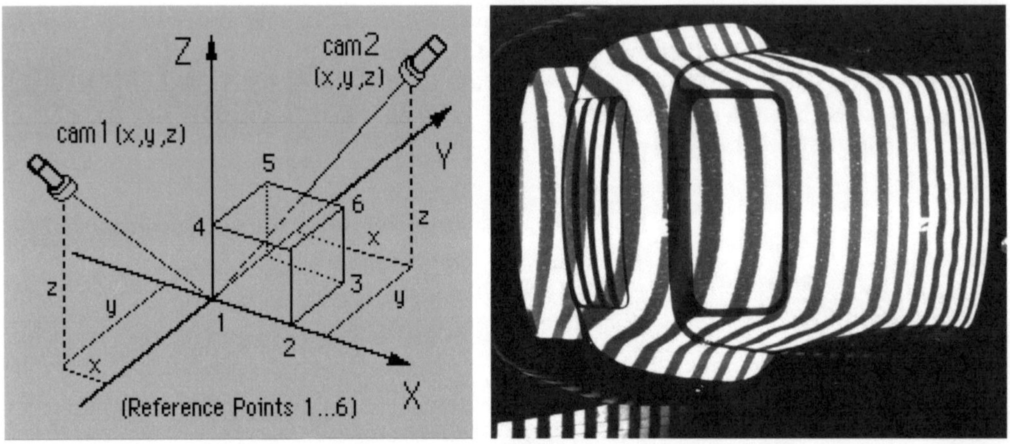

Fig. 18. Setup for SPR with six markers (left) and car roof deformation measurement results from PGM (right)

The transformation matrices can be calculated if the locations of at least 6 markers are known exactly in 3-dimensional space and their images can be detected in both camera views.

Tests have shown that an accuracy of 0.01% of the complete object space can easily be obtained. Measurement accuracy is better than 0.4 millimetres.

9. Flow visualisation techniques

A commonly used flow visualization instrument is a hand-held smoke rod. Oil is ejected through a heated, small tube, whereby the oil is evaporated. Using non-coloured oil results in white smoke, that follows the major flow streamlines and fills wakes and separation zones with smoke.

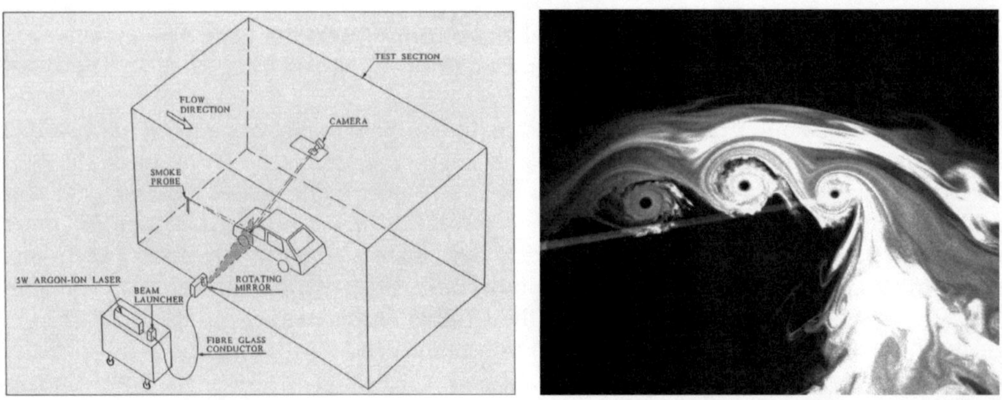

Fig. 19. Laser stroboscope technique; setup with rotating mirror (left) and frozen flow field showing vortices (right)

In combination with a laser light sheet the flow structure is made visible within that plane. A laser light sheet can be created when a laser beam is diverged through a circular cylindrical lens. It is also possible to reflect the laser beam on a rotating mirror. The continuously rotating laser beam gives the same effect as a laser light sheet, provided that the rotational speed is high enough. By varying the rpm value of the mirror, stroboscopic effects are achieved and periodic flow phenomena can be analyzed. Figure 19 shows a sketch of the setup of the laser light sheet by means of a rotating mirror and an example of vortices visualized with this technique.

Other techniques to visualize the flow are using tufts or oil on the surface of the test object. Tufts are small filaments of cotton or plastic, which are mounted on the surface with magic tape or alike. Tufts follow the local streamlines along the body or behave like small waving flags in separated flow regions. Depending on the material, the tufts may reflect ultraviolet light. Tufts are easy to mount and provide useful basic information. They can be used in a continuous way when the flow direction is changed.

Another technique to visualize the flow is by using a kind of oil on the surface. Depending on the applied oil, transition zones from laminar to turbulent flow may become visible or the separation zones of the flow. Certain oils also reflect ultraviolet light, enhancing the pictures.

Disadvantages of using oil are among others the contamination of the wind tunnel and the test time needed to establish a well-developed oil pattern.

10. Wind tunnel blockage corrections

Testing vehicles in a wind tunnel introduces disturbing effects from the finite dimensions of the airflow. In case of a ¾ open test section the flow from the exit nozzle may have some divergence, leading to a streamline divergence near the vehicle which is somewhat larger than in the unconfined real condition. This results in too low wind loads. In case of a closed test section the streamline divergence near the vehicle is reduced because of confinement by the wind tunnel walls. This results in an increase of the kinetic pressure at the tested object and thus an increase of the measured wind loads.

Corrections are needed, especially for closed test sections and relative large vehicles compared to the wind tunnel cross section dimensions.

In case of a closed test section it is possible to correct by measuring wall pressures in the vicinity of the vehicle. This is however rather elaborate and not common practice. More usual is to correct the reference kinetic pressure analytically, e.g. by a formula that incorporates the measured drag. This latter correction method is basically a base-pressure correction method that started with the work of Maskell (1963), Gould (1969) and Awbi (1978). An empirical blockage correction for trucks in the LLF wind tunnel is described by Willemsen and Mercker (1983). A description of a blockage correction method for automotive testing in a wind tunnel with closed test section is described by among others Mercker (1986).

11. References

Adrian, R. J. (1991), Particle-imaging techniques for experimental fluid mechanics, *Annual Reviews Fluid Mechanics*, Vol. 23, pp. 261-304.

Awbi, H.B. (1978), Wind tunnel wall constraint on two-dimensional rectangular section prisms.

Dougherty, R.P. (1997), Source location with sparse acoustic arrays; interference cancellation, presented at the *First CEAS-ASC Workshop: Wind Tunnel Testing in Aeroacoustics*, Marknesse.

Gould, R.W.F. (1969), With blockage corrections in a closed wind tunnel for one or two wall-mounted models subject to separated flow, *Aeronautical Research Council Reports and Memoranda*, no. 3649.

Hinsch, K.D. (1993), Particle image velocimetry, *Speckle Metrology*, Ed. R.S. Sirohi, pp. 235-323, Marcel Dekker, New York.

Johnson, D.H., Dudgeon, D.E. (1993), Array Signal Processing, Prentice Hall.

Kähler, C.J., Adrian, R.J., Willert, C.E. (1998), Turbulent boundary layer investigations with conventional- and stereoscopic particle image velocimetry, *Proceedings 9th International Symposium on Application of Laser Techniques to Fluid Mechanics*, Lisbon, paper 11.1.

Kompenhans, J., Raffel, M., Dieterle, L., Dewhirst, T., Vollmers, H., Ehrenfried, K., Willert, C., Pengel, K., Kähler, C., Schröder, A. and Ronneberger, O. (1999), Particle image velocimetry in aerodynamics: technology and applications in wind tunnels, *Journal of Visualization*, Vol. 2.

Maskell, E.C. (1963), A theory of the blockage effects on bluff bodies and stalled wings in a closed wind tunnel, *Aeronautical Research Council Reports and Memoranda*, no. 3400.

Mercker, E., Knape, H.W. (1989), Ground simulation with moving belt and tangential blowing for full-scale automotive testing in a wind tunnel, *SAE Paper* 890367, Detroit.

Mercker, E., Wiedemann, J. (1990), Comparison of different ground simulation techniques for use in automotive wind tunnels, *SAE Paper* 900321, Detroit.

Mercker, E. (1986): A blockage correction for automotive testing in a wind tunnel with closed test section, *Journal of Wind Engineering and Industrial Aerodynamics*, 22.

Piet, J.F., Elias, G. (1997), Airframe noise source localization using a microphone array, *AIAA Paper* 97-1643.

Raffel, M., Willert, C., Kompenhans, J. (1998), Particle image velocimetry - a practical guide, Springer Verlag, Berlin.

Ronneberger, O., Raffel, M., Kompenhans, J. (1998), Advanced evaluation algorithms for standard and dual plane particle image velocimetry, *Proceedings 9th International Symposium on Application of Laser Techniques to Fluid Mechanics*, paper 10.1, Lisbon.

Sijtsma, P., Holthusen, H. (1999), Source location by phased array measurements in closed wind tunnel test sections, NLR-TP-99108.

Sijtsma, P. (1997), Optimum arrangements in a planar microphone array, presented at the *First CEAS- ASC Workshop: Wind Tunnel Testing in Aeroacoustics*, Marknesse.

Underbrink, J.R.; Dougherty, R.P. (1996), Array design of non-intrusive measurement of noise sources, Noise-Conference 96, Seattle, Washington.

Willemsen, E., Mercker, E. (1983), Empirical blockage corrections for full-scale automotive testing on straight trucks in a wind tunnel, NLR TR 83065 L.

Willert, C., Raffel, M., Kompenhans, J., Stasicki, B., Kähler, C. (1996), Recent applications of particle image velocimetry in aerodynamic research, *Flow Measurement and Instrumentation*, Vol. 7, pp. 247 -256.

Willert, C. (1997), Stereoscopic digital particle image velocimetry for application in wind tunnel flows, *Measurement, Science and Technique*, Vol. 8, No. 12., pp. 1465 – 1479.

Willert, C.E., Gharib, M. (1991), Digital particle image velocimetry, *Experiments in Fluids*, Vol. 10, pp. 181-183.

3

Monitoring and Fault Diagnosis in Manufacturing Processes in the Automotive Industry

Roberto Arnanz Gómez, María A. Gallego de Santiago, Aníbal Reñones Domínguez, Javier Rodríguez Nieto and Sergio Saludes Rodil
CARTIF Technology Centre
Spain

1. Introduction

At present production systems in car manufacturing processes are under high demand requirements and maintenance plans are of great importance in order to achieve the production objectives. The main goal of the maintenance is to increase the operativity of the plant and the machines involved in the manufacturing process, avoiding all unexpected stops. Preventive maintenance has been the solution adopted by most factories for years. Based on past experience or on machines suppliers specifications, the maintenance manager decides when to check or replace the machines or some of their components to guarantee their operation without faults until the next maintenance stop. This implies two kinds of costs for the factory: checking a lot of equipment (time and staff costs) and replacing components that may be in good conditions.

That is why knowing the actual state of the different parts and machines of the factory is so important for a good management of the plant. The increasing automation of the plants allows to acquire, store and visualize lots of variables of the process. Most factories have nowadays SCADA systems that allow supervision of processes and equipment giving a valuable information about them. However it is not easy to manage this great amount of information for different reasons. First of all the sample rate of these variables usually hides their dynamic behaviour. Also the complexity of the processes makes it difficult to identify all the relations and dependencies between variables, so it is not possible to determine a wrong operation looking only the variation of a few variables without taking into account how the rest are changing. The number of variables and data acquired in the whole factory makes it impossible for a human supervisor to process all that information, relate it to past data and try to find out if something is going wrong. Although his experience will allow him to detect some problems it is evident that he needs some help to succeed in his work.

Predictive maintenance is a methodology that improves systems availability and contributes to cost reduction and increase of useful life of production assets. It comprises different techniques to process acquired data from the factory to determine machines state and predict how they will work in the future. The variety of problems that must be solved makes the design of a predictive maintenance system be a very complex task where different

knowledge areas must be integrated. It is very important to know the state of the art in all of them and sometimes introduce innovations for applying the solutions to particular cases.
Next sections explain the main components of a predictive maintenance system and how it was implemented in real industrial problems of the automotive industry. An effort has been made in order to choose case studies that offer a wide range of the possible techniques to use, combining classical solutions with newer ones.

2. Structure of a fault detection system for the automotive industry

The core of any predictive maintenance system is a fault diagnosis system able to detect failures not only when they are happening, but also a pre-failure behaviour. It is an advanced solution for the supervision level of the factory where in most cases only SCADAs and alarms based on variables values are considered. One of the main advantages of predictive maintenance is its ability to provide useful information to the human supervisor showing what the real state of a plant or machine is and helping him in the planification of the factory operation. It is also capable of substituting the human operator in some systems taking decisions such as stopping the operation in case of a critical fault or scheduling maintenance operations.
The three main components for any fault diagnosis system are data acquisition, signal processing and decision making. These three components must be designed jointly because the requirements or outputs of one of them will affect the others. Their complexity level will depend on the application and how the symptoms of the faults can be found.
Data acquisition is the first stage of every diagnosis system. This component consists of all the sensors, signal transmission systems, acquisition devices and storage equipment. Sensors are a key component of the fault detection system because they provide all the information the system will have to deal with, although in some cases information coming from production management systems can be useful. In some cases those sensors can be shared with other tasks such as control or supervision and they are included in the machine or plant during its design. But in most cases predictive maintenance is not taken into account during the design of the machines and new sensors are usually required. This occurs specially when predictive maintenance must to be applied to old machines because they start to be a bottle neck in the plant due to their unexpected faults. Electric current, voltage, accelerometers and temperature sensors are of common use for diagnosis systems. Some applications require more specific sensors, like photodiodes and spectrometers. The selection of the appropriate sensor and acquisition system can be determinant for the success of the application because they must guarantee that the collected data have the information of the state of the machine. Capture and synchronize data from sensors of different nature and variables with different dynamics can be an interesting problem to solve and sometimes requires specific programming or storaging methods designed ad-hoc.
In the signal processing stage, signals acquired and/or stored by the data acquisition component are processed. This includes common signal treatment like filtering that is used to eliminate noise. However, the most important part in signal processing is feature extraction. Feature extraction consists in looking for a particular behaviour in the signals that allows to identify the faulty or pre-faulty states. There are a wide variety of feature extraction techniques and the one used depends on the problem at hand. For example, one of the most common feature extraction techniques is the Fourier Analysis, which gives information on the distribution of energy power associated to different frequency ranges in

the signals acquired by the sensors. This content changes when a fault occurs or is close to occur. Besides this, feature extraction techniques in the time domain are also useful. Some problems require the use of very specific feature extraction techniques, like the estimation of electron temperature. The final stage is the decision making where the features that have been extracted from the signals generated by the sensors, have to be classified in order to determine the state of the system. The classification is the base of the decision making process, so it has an important role in the fault diagnosis scheme. In some cases, classification can be done by merely checking the features values against a threshold, although selecting the threshold value could be a hard problem to solve. In other cases, more sophisticated non-linear classifiers, like neural networks, neuro-fuzzy systems or support vector machines have to be used. Besides this, features time evolution is also of great importance because it allows to perform trend analysis, which is one of the basis of the fault predictive capabilities of the fault detection systems. The lack of historical data is the main problem that must be solved when designing the decision making component. It can be sometimes a problem to decide what is the optimal classification method to use, and it is always an added difficulty to fix the parameters of the system. Usually conservative strategies are used. This leads to a great number of false alarms during the initial phases of the predictive maintenance system implementation. Human experts supervision and knowledge is one of the main supports for a good design of the decision making system and its configuration.

3. Case studies

3.1 Case study 1: Multitooth machine tool

Machine tools represent one of the main examples of highly automated components (Altintas, 2000). In spite of this automation, the cutting process has an inherent degradation (Astakhov, 2004), which is one of the main problems to be overcome. Other aspects to consider are workpiece tolerance deviations, ensuring a correct evacuation of the chips, changing of worn tools and, if necessary, stopping the machine if abnormal working conditions appear (for example chatter). So, to achieve the desired level of autonomy for this kind of machines, it is necessary to develop the monitoring and diagnosis of the cutting process. Many different kinds of machine tools are used in the automotive industry. Among them, the so called Multitooth Machine Tools represent the most challenging ones, from the diagnosis point of view, due to high number of inserts susceptible to break, and the different machining operations integrated within the same tool. The tools analyzed in this chapter are used in the car industry for mass production of different mechanical parts, such as the crankshaft or the camshaft of car engines. These tools are complex *ad hoc* devices built with many cutting inserts (up to 250, depending on the machine) of different kinds (roughing and finishing) presented in Fig.1(b) and for different operations (turning, milling or broaching) within the same tool, as shown in Fig.1(a). The configuration of the tool is based on multiple tool holders specially designed for the particular operation of the mass production line. Such complexity is necessary to achieve the required high metal removal rate.

3.1.1 Data acquisition in machine tool environment

Regarding the main three components of a fault detection system (data acquisition, signal processing and decision making) an optimal selection of sensors is of paramount importance to obtain valuable information from the environment of the machine tool that should be correlated with the abnormalities to be detected. Different signals susceptible of

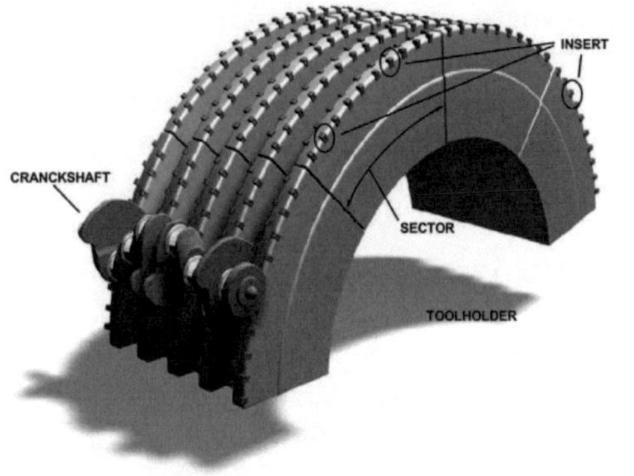

(a) Layout of multitooth tools used in the car industry

(b) Different inserts in the multitooth tool

Fig. 1. Multitooth tools used in the car industry

having correlation with tool wear and the breakage of inserts in the multitooth tool, are shown in Fig.2. Among others the following are the most common in the literature:

Noise: can be measured in the environment of the tool using microphones (Fig. 2(a)). Although noise gathers information coming from the whole machine tool environment, this measure can be very valuable for the first analysis of the machining cycle through the analysis of the time-frequency representation like the spectrogram.

Vibration: measured with accelerometers in one of the main shafts of the machine tool (Fig. 2(b)). As the wear increases in the tool an abnormal increase in the vibration also occurs and can often lead to bad surface quality.

Temperature: the increase of the tool wear causes an increase in the temperature due to an excessive friction. Using sensors like pyrometers, the temperature of the machined surface can be easily measured after the machining has been completed (Fig. 2(c)).

Electrical power consumption: can be measured from the output signals of the frequency converters (for the usual case of AC drives) for every motor that moves the multitooth tool and moves the workpiece (usually rotation movement). Fig. 2(d) depicts the example of rms electrical power consumption of the two drives of an example tool: feed and rotation of the tool holder. These kind of signals show clearly the different parts of the cycle and the grouped attack of the inserts in the tool.

In order to analyze the sensitivity of every recorded signal, the measurements have to be done over the useful life of several consecutive tools. After that, every set of signals is statistically analyzed to extract global information for comparison and to decide whether there is a correlation with the degradation of the tool, or other abnormalities that could have been recorded. In (Reñones, Rodríguez & Miguel, 2009) are presented the results of such analysis that lead to choose the electrical power consumption as the most appropriate signal for use in the diagnosis of the multitooth tool. This signal showed the best signal-to-noise ratio for the evolution of the wear and was the most cost-effective measure: non-invasive, moderate sensor cost (inexpensive if appropriate signals are available at the drive converter) and high reliability of the measure in comparison with other measures like noise and vibration, because of the high influence of the sensor location. Fig. 3 shows the evolution in electrical power consumption in a particular zone of the analyzed tool. It is clear the increase of the power due to the wear and the abrupt decrease after the tool reaches its useful life and it is changed by a new one.

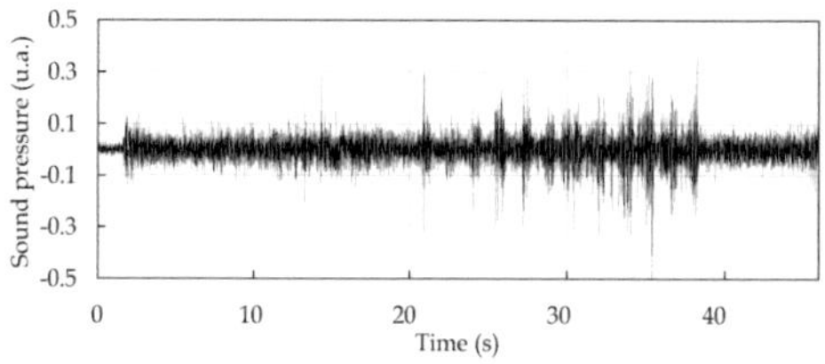

(a) Sound pressure recorded during the machining of a car crankshaft supports

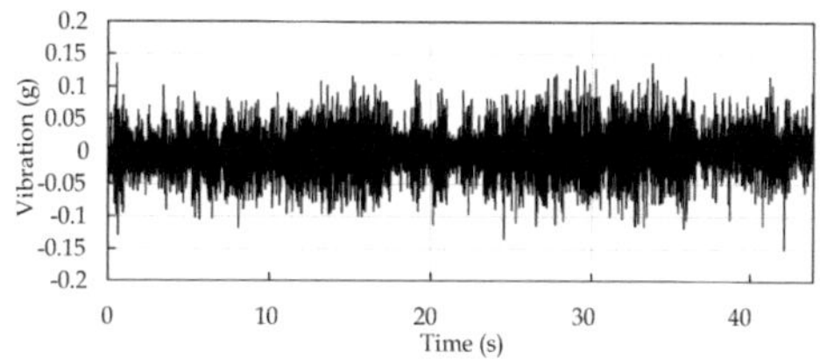

(b) Vibration amplitude of the main rotation tool axis recorded during the machining of a car crankshaft supports

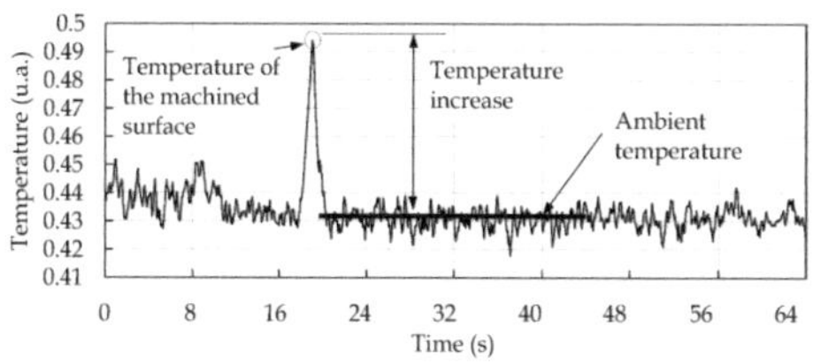

(c) Temperature support scanned after the machining cycle

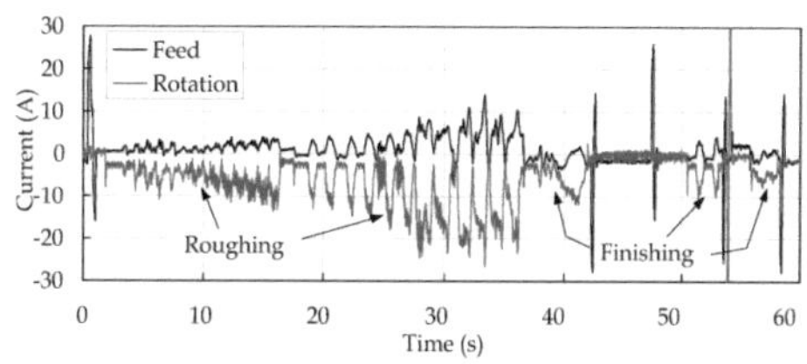

(d) Electrical power consumption of the feed and rotation electrical drives

Fig. 2. Signal more common in the literature

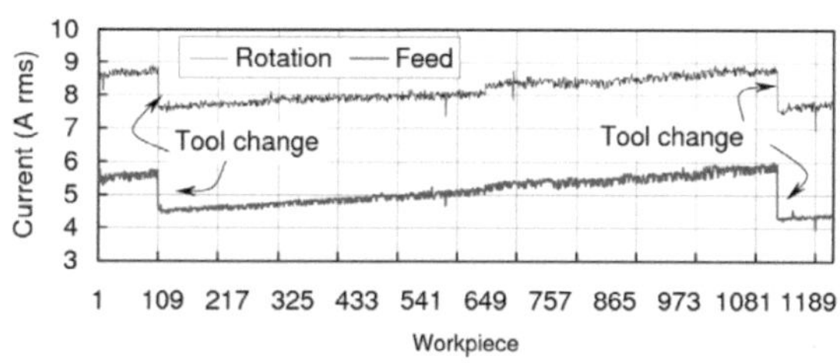

Fig. 3. Evolution of the tool wear using the electrical power consumption.

3.1.2 Signal processing of the electrical power consumption

Once the electrical power consumption has been chosen as the desired signal for diagnosis of the tool, it is time to extract the part of the electrical power consumption that belongs to each insert or every group of inserts that attacks the workpiece simultaneously. This process is known as signal segmentation and can be formulated as the automatic decomposition of a signal into stationary or transient pieces with a length adapted to the local properties of the signal (Basseville & Nikiforov, 1993).

Firstly, the number of segments that must be extracted has to be defined, taking into account different aspects of the machining process, such as the different kinds of cutting inserts, the workpiece material, changes in the cutting conditions, changes in the PLC programming, different mechanized zones of the workpiece and the layout of the tool.

Among the different alternatives for making the segmentation of the electrical power consumption signals, the use of auxiliary signals not directly affected by the machining, such as, for example, the sampled speed reference of the machining cycle, or its acceleration, ensures a reliable segmentation avoiding false alarms in the detection of a fault in the tool.

Once the different signal segments are extracted, the next step is to obtain the model for every segment that should be sensitive to electrical power consumption changes caused by a fault in the tool. Another goal of this step is to reduce the amount of information used in the following steps of the diagnosis scheme. There are different methods available to make such reduction (Reñones, Rodríguez & Miguel, 2009). Among them, the calculation of statistical parameters is a straightforward reduction of information. Only those which presented a greater sensitivity to the variations produced by the failures in the tool must be chosen in order to reduce the amount of data for detection of failures in each group of the tool. With the appropriate statistical parameters chosen, the change detection problem can be stated as the detection of a change in a set of random variables. The change detection is usually carried out using a so called *stopping rule*, as presented in (1); that is, a function of the random variables y_k that exceed a preset threshold λ in case of abrupt change. The parameter t_a represents the estimated time of change at which the stopping rule is true for the first time (Basseville & Nikiforov, 1993).

$$t_a = \inf\{n : g_n(y_1,\ldots,y_n) \geq \lambda\} \quad (1)$$

This problem is frequently solved from a statistical point of view. In Fig.3 an example of abrupt change that must be detected can be seen. The following requirements must be taken into account to solve this change detection problem:

- The segmentations or electrical power consumption trends are non-stationary, so an adaptive detection scheme is needed.
- The changes must be reliably detected, and the false alarms due to occasional electrical power consumption changes must be avoided.
- A mean time between false alarms (MTFA) must be fixed.
- The change detection must be fast enough to avoid serious damage to the whole tool and machine.
- The changes can be abrupt decreases (in case of breakage) but also abrupt increases due to the loss of an insert or an abnormal wear rate caused by the breakage of previous inserts.

Among the different alternatives that can be used to detect abrupt changes (Reñones, Miguel & Perán, 2009), the algorithm based on an adaptive local linear model of electrical power consumption showed the best performance in terms of reliability, and an extremely low computational cost. The algorithm is based on the detection of linear regression outliers. In the present case, the outliers are recorded points with an electrical power consumption out of normal variation due to a breakage (abrupt decrease) or abnormal wear rate (abrupt increase).

Due to the fact that the evolution of the electrical power consumption trends are not linear as the wear increases, this detection scheme must be implemented using a moving data window, let's say of size L.

The outlier detection algorithm is done through the calculation of statistical parameter t_i defined in (2). This statistical parameter follows a Student's t-distribution. Under no fault in the tool and hence no change in the electrical power consumption, the residuals t_i should remain in the interval $\pm t(1 - \alpha/(2L), L - 3)$ of confidence α. These bounds of the interval are also known as the critical level or threshold.

$$t_i = \frac{e_i}{\hat{S}_{R(i)}\sqrt{1 - v_{ii}}} \quad (2)$$

In order to adjust the algorithm in an optimal way, some performance measures must be done and it must be taken into account the variation range of the different parameters for the algorithm (window size L and the critical level or threshold). To make the detection robust, an additional parameter can be added, such as the amount of consecutive detected outliers.

In quality control this is called a *run test*. In fact, this robust mechanism is not particular to this detection scheme and can be applied to other detection algorithms.

In order to optimally adjust the parameters of the detection change algorithm, performance measures must be done, such as (Gusstafson, 2000): MTFA (Mean Time between False Alarms), MTD (Mean Time to Detection) and, MDR (Missed Detection Rate).

The optimal algorithm adjustment is performed by fixing either the performance measure MTFA or MTD, and the parameters of the algorithm are chosen to minimize the other performance measures. The presented algorithm have been evaluated with data coming from the machining of more than 30000 workpieces. As the exploration of the whole range of parameters for the change detection algorithm is unapproachable, some restrictions and assumptions were added to cope with the problem. For the window size L, it seems reasonable to choose a value lower than the mean time between faults. For the test set used, it is approximately 300 workpieces, then the interval for this parameter was set as [40,100] workpieces.

The *run test*, represented as R, influences the speed of detection. After studying historical data and taking into account the protection of the tool, an interval of [2,6] workpieces seems reasonable. The threshold interval was [2,7] and for the residuals was fixed as an interval with a confidence level from 0.1 to 0.001. Two tests have been done to study the relationship between the different parameters, where the threshold is varied in the preset interval and the other two parameters are fixed at the midpoint of their own interval.

In Fig. 4 is presented an example of such performance measures. Detailed analysis of these graphics can be found in (Reñones, Miguel & Perán, 2009). It is straightforward to see that an increase in the threshold (horizontal axis of the graphics) leads to a more reliable detection (higher MTFA) but fewer faults are detected as shown in the third row of graphics. This exploration of parameters influence let to finally make an optimal adjustment in the parameters for the change detection in the different electrical power consumption trends for the different zones of the multitooth tool.

The result of this step is a list of thresholds for every zone of the tool. Positive thresholds can be adjusted to detect abrupt increases of the electrical power consumption due to an abnormal wear rate (called as overload), and also abrupt decreases due to a breakage of one or more inserts in the tool.

3.1.3 Decision making process for the machine tool diagnosis

The last step of the methodology used to detect faults in the multitooth tool is the so-called *Decision-making* process as presented in section 2. In this step, using the information coming from the change detection algorithm and other information of the state of the system, an effective declaration of the fault in some zone in the tool is done. That means, for example, that the machine tool will be stopped at the end of the current cutting cycle, and the operator will fix the problem based on the information of the diagnosis system: the faulty zone of the tool and the type of fault (overload or breakage).

The electrical power consumption signal gives the best signal-to-noise ratio to detect faults, as was presented in section 3.1.1. On the other hand, this signal exhibits sensitivity (abrupt changes in the signal) to other phenomena that may cause false alarms which must

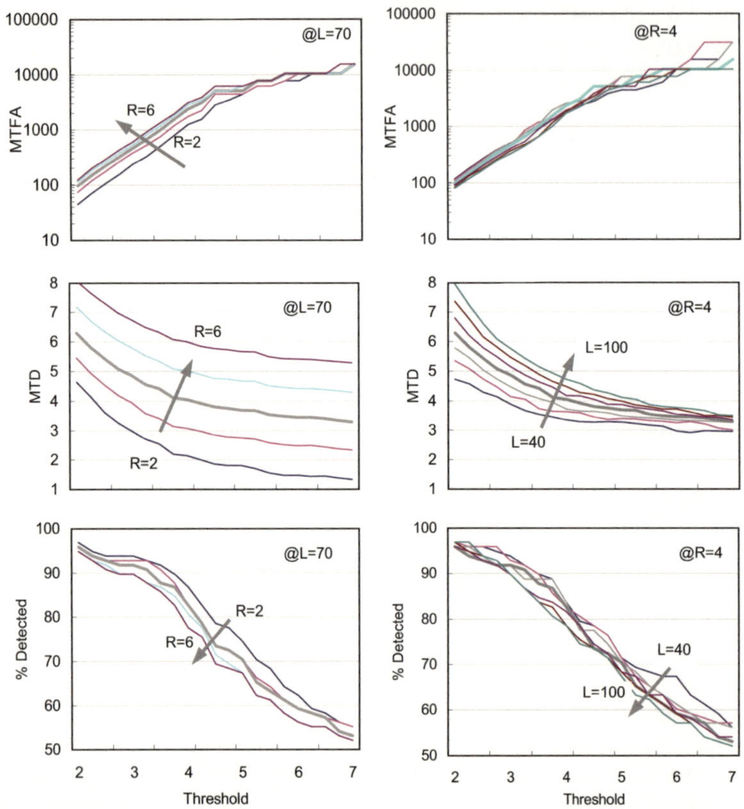

Fig. 4. Performance measures for the linear regression change detection algorithm with parameters: $L \in [40,100]$, $R \in [2,6]$ and $\lambda \in [2,7]$.

be taken into account, such as a tool changed by a new one, changes in the material of the workpieces (foundry or steel), compensation adjustments in the inserts made by the operators to achieve the desired tolerances, the warm up process after a long stop, etc. To prevent false alarms caused by any of these events, it is necessary to protect or disable the change detection algorithm. Protective measures that can be taken to avoid false alarms are to use output signals from the PLC governing the machine tool (new tool, material change, etc), or to inhibit change detection when changes affect the whole recorded signal or there are sample points separated too much time.

3.2 Case study 2: Car painting cabinet

This case study shows a predictive maintenance system currently operating in an assembly car factory, specifically in painting cabinets section. It has been working for thirteen years now and serves as a valuable tool for anticipating to breakdowns all along the plant, optimizing equipment performance and reducing unplanned shutdowns and incidents. This predictive maintenance system is based on mechanical vibrations analysis techniques applied on the motor-fan sets operating in painting cabinets.

The predictive maintenance for this kind of installations can be performed in two ways. With online analysis systems or with hand-held, walk-around vibration analyzers. For extremely large operations and/or very expensive equipment, the first approach is the most cost effective and has repeatedly shown to saving money.

The main advantage of an on-line dynamic vibration monitoring system is that the data acquisition is made continuously. This allows to check past values and to know the evolution of the state of the machine, providing a more reliable diagnosis that off-line data acquisition systems cannot offer. Most of the on-line systems use some kind of acquisition system architecture that involves input channels multiplexing many vibration sensors. This results in a scan rate that varies according to the system scheduler. Another advantage to an on-line dynamic vibration monitoring system is that there is no labour cost to acquiring the data and minimal labour cost for identifying machine faults.

The disadvantages of these systems are that they are the more costly systems to implement and maintain as they include maintaining a full time vibration analyst, and installing a wired network to get the signals from the sensor to the analysis system. Furthermore, the software and hardware that make up the system typically require an extensive maintenance contract as well.

Hand-held, walk around vibration analyzers only provide trending information to identify that a potential problem exists, and do not provide the detailed information necessary to determine the cause of the problem. The supervision is done only at specific moments and it does not provide a trend of vibration levels. Moreover, they require skilled vibration analysts to interpret the data and, without continuous monitoring, problems in between rounds could be costly.

3.2.1 Problem description

The plant under study consists of a series of motor-fans that keep painting cabinets under very strict temperature and humid conditions. In some cases air must be put into these cabinets and in some others air is taken out of them. The target is to keep working atmosphere under control in such a way that safety and sanitary conditions are guaranteed for the staff. Moreover, in order to achieve a good production quality, it is required that air inside the cabinets is at the right temperature, filtered and keeping an adequate relative humidity that prevents varnish thinners from evaporation. It is also necessary to extract the air from the cabinet, in order to eliminate polluting elements.

For each motor-fan the fan is driven by an electric drive whose rotation movement is transmitted to the fan through a couple of pulleys, one attached to the fan and the other one to the drive, together with a belt. Both the electric drive and fan are mounted on an elastic structure that keeps the set isolated from the high frequency excitations of the structure and at the same time, this base structure is not affected by the mechanical vibrations coming from the electric drive and fan.

This assembly plant is able to produce around 1.200 cars every day along three shifts, depending on demand needs. To achieve this, it is mandatory to ensure that every machine is working under optimal conditions avoiding unexpected breakdowns which could lead to stops and subsequent lost of production. Therefore, a predictive maintenance system is needed. A thorough analysis of the related machines has led us to consider the following sources of mechanical vibration that could be the cause of potential failures:

1. Defect related mechanical vibrations: Unbalance, misalignment, looseness, defects in bearings, blade breakage and defects in belts.
2. Mechanical vibrations related to natural frequencies: Natural frequencies of the base structure, natural frequencies from any part of the machine structure and natural frequencies from other elements outside the machine.

3.2.2 Predictive maintenance system

The system consists of an industrial computer in charge of data acquisition, communication protocols and the calculation of spectra and alarms (DCS station in Fig. 5) to which up to four nodes are connected through a LAN. They are multiplexors and receive signals from accelerometers placed on the machines. Analysis and diagnosis tasks are carried out by means of a PC (MD station). This PC has a communication module that allows remote access to the data, so that it is possible to perform the same tasks from a remote computer, outside the factory. Fig. 5 shows the layout just described.

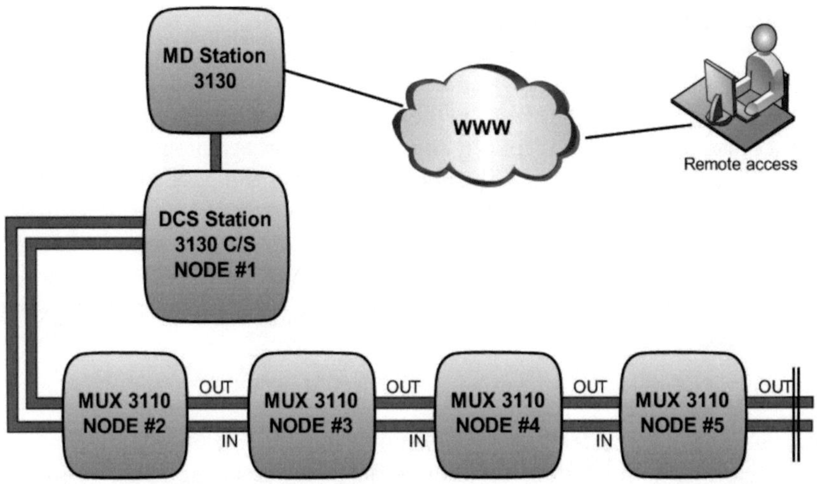

Fig. 5. Predictive maintenance system schema

In each motor-fan two accelerometers have been placed to register mechanical vibrations from the electric drive and the fan, which is the most sensitive part to be monitored in this case. The related bandwidth is 20 kHz, which is enough for the application under study. They have been placed in radial position, as close as possible to the bearings near the pulleys.

The signals from the accelerometers reach one of the four multiplexors (nodes from Fig. 5) inside which they are displayed along 32 channels, and finally get to the industrial computer where they are registered and sent to the PC for further analysis. As soon as an abnormal value is detected, an alarm shows up on the screen so that subsequent actions can be taken in order to solve the problem arising. This scheme is the same for every motor-fan being monitored.

The system is automatically registering data on a daily basis. At the same time, mechanical vibration levels, process variables and alarm levels are being checked for the plant.

It is possible to register three kinds of data: gross scan, spectrum and time signal:

Gross Scan: These data constitute a unique signal taken from a DC stationary signal o calculated from an AC dynamic signal, as for example, a RMS one.

The gross scan measurement from each sensor is compared to a reference value that serves as an alert. After this, the measurement is used to update the related maximum and minimum values that will be finally registered in the database. Whenever any gross scan measurement exceeds the alert value, it is first registered in the database, then it is updated for the DCS, and finally, the related spectrum and time signal are recorded as alarm related data for the specific sensor.

Spectrum: They are calculated from a related time signal and further processed in order to get specific information at certain frequencies associated with the potential defects of the machine being monitored. In this case, this is done through what we call Analysis Parameters Set (APS). Gross scan data are registered for every signal once per each data registered cycle. Spectrum data are recorded for one or two signals once per each data acquisition cycle and time signals are registered simultaneously to the spectra.

Time signal: This software allows the user to visualise data in the time domain, which can be very useful is specific situations, though no further analysis is being performed in the case under study.

The system allows to define up to 12 parameters directly related to the frequencies or selected ranges of them that are of interest in order to characterize (detect) certain types of machine failures. There are many possibilities to choose different types of Analysis Parameter Sets and the most widely used are briefly described next:

Total Energy: This value represents all of the energy of a signal. Because of the nature of the FFT, the first two points of the spectrum are excluded from this summation.

Energy within a Frequency Range: The energy between the two specified points of a spectrum will be summed.

Non-Synchronous Energy within a Frequency Range: The energy between the two specified points, which is not an integer multiple of turning speed, is summed.

Synchronous Energy within a Frequency Range: The energy between the two specified points, which is an integer multiple of turning speed, is summed.

Synchronous Peak: The signal is synchronously sampled to determine the energy at harmonic of running speed. In order to use this parameter, the sensor must have a tach pulse defined for it.

HFD (5k-20kHz) High Frequency Detection: An additional collection of vibration data is made from which the energy from 5.000 Hz to 20.000 Hz is computed. HFD sometimes useful in detecting bearing faults at an early stage.

RPM: This field displays the RPM for this sensor. A good way to ensure that the tachometer definition is set up correctly and the RPM ratio is correct is to compare the reported RPM with the expected value.

3.2.3 Practical example

Next, an example of a fault is showed. In this case, a progressive defect in the fan bearings has been detected. As soon as the pre-alarm level is reached all the related parameters are supervised, and once the system indicates the alarm level has been exceeded, the faulty bearings are replaced. This kind of fault is best detected using the energy within a frequency range parameter. For this kind of defect several frequency ranges have been selected in order to assess the degree of severity of the fault. When a bearing defect is first detected (just within a unique frequency range), the machine will still be able to work under acceptable conditions long before it is advisable to replace the damaged bearings. Therefore, when some ranges are affected simultaneously the fault is considered severe enough so as to recommend the replacement of the faulty pieces. Fig. 6 shows the trend followed by mechanical vibrations for five related consecutive frequency ranges. They all have the same performance, giving precise information on the very moment when the failure first appeared. Then, it became more and more important until the alarm level was reached, and finally it can be seen the level of vibration once the faulty bearings were replaced by means of a planned intervention, not affecting production by any means.

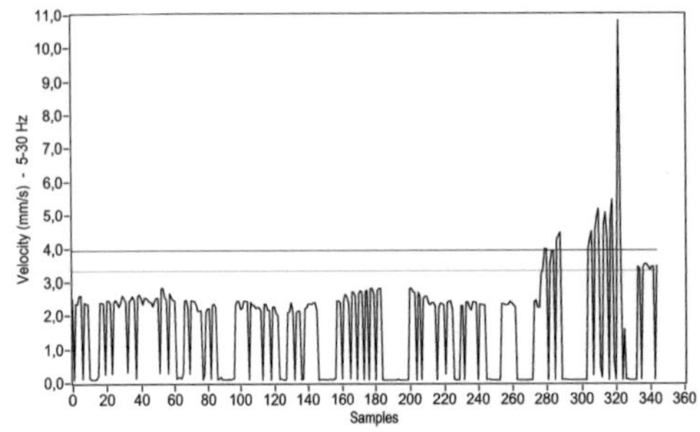

(a) Frequency band 5-30 Hz

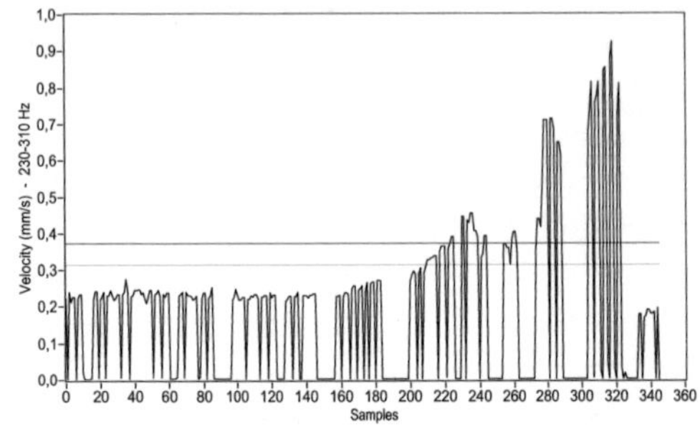

(b) Frequency band 230-310 Hz

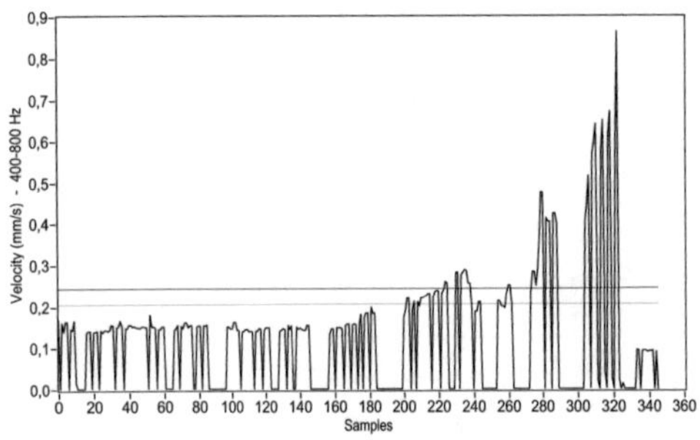

(c) Frequency band 400-800 Hz

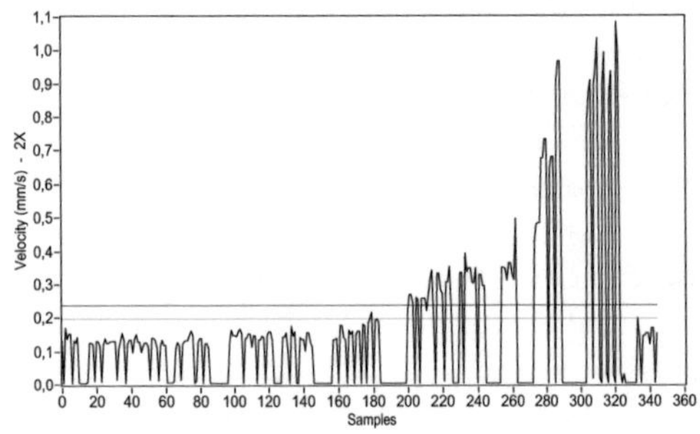

(d) Second Harmonic (2X)

Fig. 6. Fault detected in the bearings of the fan

3.3 Case study 3: Electric motors diagnosis in non-stationary processes

3.3.1 Predictive maintenance of electrical motors

Electrical motors are one of the most crucial components of production, and many of them are of vital importance for factories to be operational. For this reason a great number of diagnosis methods have been developed during years in order to detect motor faults. Some of this methods can only be applied off-line because the motor needs to be disconnected and isolated. This is the case of hipot analysis, partial discharges, isolation test or surge comparison testing. These are well-known techniques in the field of maintenance of electrical motors and are widely used in industry, especially for high power machines. There are another group of techniques that can be used on-line such as thermography or vibration and spectral current analysis. All of them can be considered as predictive methods because allow to detect incipient faults and predict the time until a critical fault is declared. The problem with the off-line methods is that a fault can produce damages in the system before it is detected. This happens when its evolution is faster than the period between analysis. On the other hand, spectral analysis methods (current and vibration) allow on-line detection of mechanical faults besides electrical ones. Bearings faults, mechanical unbalance, eccentricity, windings or coils short-circuits and electrical unbalance are the faults than can be diagnosed using vibration or current spectrum. To obtain good results with these methods it is important to have the adequate precision in the analized spectrum, what is related mainly with the data acquisition rate, acquisition time and speed variation. Though there exist processing techniques to use spectral analysis in case of speed variation, they require the use of an encoder and have a limit in speed variation.

In this section, two industrial applications will be presented:

- Diagnosis of DC motors of stamping presses
- Diagnosis of master-slave synchronized AC motors in metal cutting machine

In both cases it will be explained why it is not possible to use any of the previous detailed methods and how other fault detection techniques can be used instead. It is intended to show the application in the industry of methods validated in laboratory and widely present in scientific literature.

3.3.2 Diagnosis of DC motors of stamping presses

Stamping presses are machines used for metal processing with an important role in the automotive industry. They usually work forming a line of stamping presses in which the piece of metal is sequentially processed along it to acquire its final shape. The movement of the press punch is generated with an electrical motor and transmitted trough several gears that transform the rotation of the motor in a lineal displacement of the punch with the appropriate speed and force to process the metal. The high power of the motor makes that in many cases, specially in old machines, it be a DC motor. In these cases it is not possible to apply current spectral analysis because fault frequencies appear as side bands of the fundamental frequency of the AC motor. Vibration analysis could be used to diagnose faults in bearings or other mechanical faults but electrical faults need another diagnosis method.

In this case a model-based diagnosis system were used to detect faults in motor windings. Model-based diagnosis uses the differences between the real system and a model of it to detect possible faults and locate their origin. Since it was first proposed by (Chow & Willsky, 1984), model-based diagnosis has been object of a great number of publications. Many theoretical and practical studies have been carried out along these years, but it is not easy to find it in the industry. The main reason for this is the complexity of most systems and machines and the difficulty to obtain a model that represents them in all the operating conditions. Multiple techniques and solutions have been proposed to solve non-linearity problems or model uncertainties. The advantage of applying model-based diagnosis to a DC motor is that it has a well-known linear model. In this case the difficulty is the identification of the model, because in an industrial environment it is not easy to develop all the required experiments and only production data were available.

The motor model is defined using two electrical equations, one for field winding and another for armature winding:

$$U_f = R_f \cdot i_f + L_f \frac{di_f}{dt} \tag{3}$$

$$U_a = E + R_a \cdot i_a + L_a \frac{di_a}{dt} \tag{4}$$

being U the source voltage, i the current through the winding, R the winding resistance and L its inductance. Subscripts f and a refers to field and armature windings respectively. Finally, E is the electromotive force and it is proportional to the field current and motor speed ω:

$$E = K \cdot \omega \cdot i_f \tag{5}$$

Identifying a closed-loop system is difficult due to correlation between inputs and outputs what makes impossible to use some of the usual identification methods of linear systems. In this case, the armature and field source voltage are generated with a controlled rectifier so the feedback between output (speed) and input (voltage) is made controlling the firing angle. This means that during the period between commutations of the power electronic switches an RL circuit is established and it can be seen as an open-loop system between voltage and current. Fig. 7 shows measured voltage and current for field and armature windings. Induced voltage E can be easily calculated because it is the value of armature voltage when current armature is zero. From Equation 5, K can be obtained using measured field current and speed. The identification of R and L in each of the windings is made considering intervals of operation when a RL circuit between voltage and current can be assumed. In these intervals the relation between output (current) and input (voltage) is a first order system that can be easily identified calculating the attenuation and lag between signals. A mean of all the values of R and L is obtained as DC motor parameters. For parameter armature identification only data with $i_a > 0$ is used. In the case of field winding the continuity in i_f allows to use all the acquired data for identification. Parameter values are those showed in Table 1.

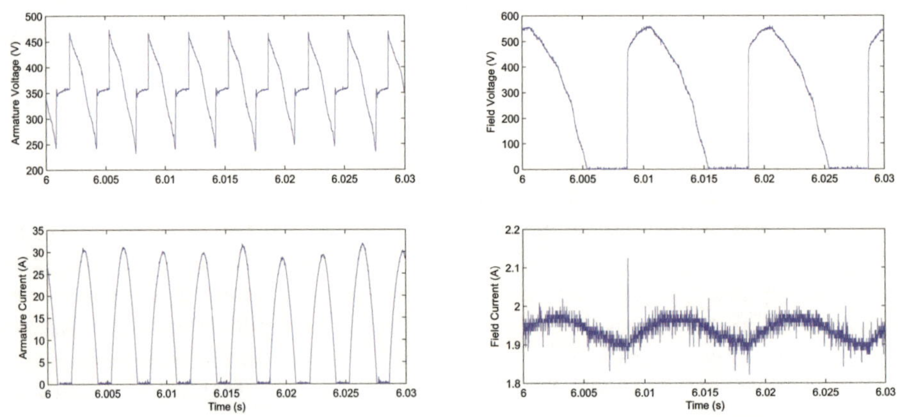

(a) Armature variables (b) Field variables

Fig. 7. Identification data for DC motor

R_a	0.425Ω
L_a	0.00233 H
R_f	138.67Ω
L_f	27.57H
K	1.358

Table 1. DC motor parameters

Using the identified model it is possible to define two equations, called residuals, that take a value different from zero when a variation in the model happens. This two equations are:

$$r_1 = U_f - R_f \cdot i_f - L_f \frac{di_f}{dt} \tag{6}$$

$$r_2 = U_a - K \cdot \omega \cdot i_f - R_a \cdot i_a - L_a \frac{di_a}{dt} \quad (7)$$

The system could have been completed with the mechanical equation of the motor including ω as a variable. As only electrical faults are going to be studied, it is assumed that there will be no faults in the encoder. The considered faults are:

- Brushes faults: can be modelled as a decrease in the armature voltage source respect the measured voltage
- Armature winding short-circuit: this can be turn-to-turn or commutator bar-to-bar faults. In both cases RL circuit change its parameters
- Field winding shot-circuit: also a change in RL circuit is the result of the fault
- Fault in armature voltage rectifier: one of the power switches fails and remains opened
- Fault in field voltage rectifier: one of the power switches fails and remains opened

These five faults have been simulated using the identified model of the motor fed with a controlled rectifier in each of the circuits. The simulation allows to observe how the residuals change with each of the faults. Six and seven intervals have been defined for the values of r_1 and r_2 respectively. The limits of intervals have been fixed using simulation results allowing the use of this two residuals as directional residuals to isolate four type of faults. This is shown in Fig. 8.

Fault	r_1	r_2
No fault	0	0
R_a increase	0	$-1, -2$
R_a decrease	0	$+2$
L_a increase	0	-1
L_a decrease	0	$+1$
Brushes fault	0	-2
R_f increase	-2	$+1$
R_f decrease	$+2$	-1
L_f increase	$+1$	0
L_f decrease	-1	0
Armature thiristor up	0	$+3$
Armature thiristor down	0	$+3$
Field thiristor up	$+1$	$+1$
Field thiristor down	$+1$	$+1$

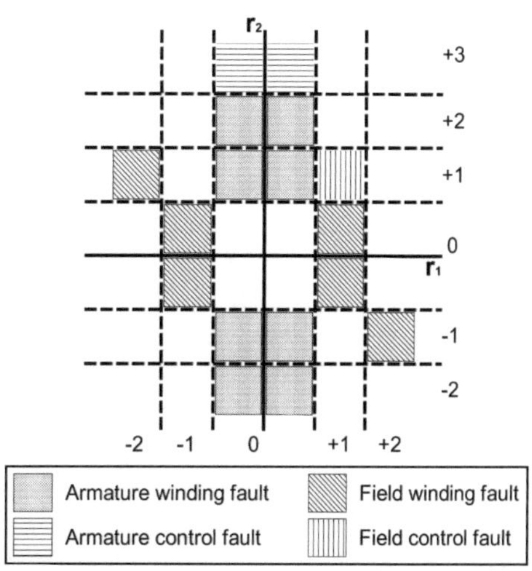

Fig. 8. Structural residuals for DC motor diagnosis

A DC motor diagnosis system was also presented in (Isermann, 2006) using different approaches. Four structured residuals were defined to identify and isolate sensor and motor faults. The limitations in the system identification are the main difference between both case studies. So a different identification method has been proposed in this case and only two residuals have been included in the diagnosis system. This imply that sensor faults cannot be considered.

3.3.3 Diagnosis of AC motors using space current vector

In the case of AC motors, the use of a model-based diagnosis method is more difficult due to non-linearities. But other signal analysis techniques can substitute current spectral analysis

when this cannot be used. Next example studies an AC motor in a cutting machine where speed variation is so high in such a short time that Fast Fourier Transform (FFT) cannot differentiate spectrum lines for fault detection. The cutting machine has two tilting knifes (one at the bottom and one at the top of the machine) that allow cutting trapezoidal pieces alternating between two angles of the knife. The reference position must be reached in one or two seconds. In this time both knifes must change their speed form zero to maximum speed and to zero again. Each knife is moved with an AC motor that are known as master and slave. The knifes are mechanically joined so the motors must be synchronized and generate always the same torque to avoid problems in the mechanical joint. Master motor receive the speed reference that makes possible to achieve the required angle in the specified time. This speed reference is prefixed as a function of the rotating angle and line speed (time to achieve the required angle), but there is no feedback of the knife angle during the movement. The controller of the master motor generates a torque reference -equal to the torque it is producing- that is used in the control of the slave motor. If the torque of both motors is not the same, it will originate medium-term mechanical faults. But the most obvious problem will be the oscillation in the knife control and the uncertainties in the cutting angle that this imply.
To detect problems in the motor windings or in the inverter that controls the motors, current space vector analysis is used. Space vector is constructed from the three phase currents using the next equation:

$$\overline{i}_S = \frac{2}{3}(i_R + a \cdot i_S + a^2 \cdot i_T) \tag{8}$$

being $a = e^{j \cdot \frac{2\pi}{3}}$. The result is a rotating vector that for a balanced system has a constant modulus equal to the amplitude of the current of each phase and whose rotating frequency is the frequency of the currents. When an electrical fault occurs in any of the windings it will produce an electrical unbalance whose effect is that current space vector will not be centered in origin or will loose constant modulus. The fault can be detected using the spectral analysis of the space vector modulus (Cardoso et al., 1999; Acosta et al., 2006) or pattern recognition of the space vector representation during one or several cycles (Nejjari & Benbouzid, 2000; Diallo et al., 2005).
Fig. 9(a) shows a capture of the master and slave motor angle during 50 seconds of cutting process. In Fig. 9(b) a detail of the negative angles can be seen. This difference between angles is a repetitive pattern during the production of this type of piece. To find the origin of this problem current space vector is analized during the movement of the knife at t = 240s and t = 280s. Fig. 10(a) and 10(b) presents the three currents of master motor in each of the cases, Fig. 10(c) and 10(d) the current space vectors and Fig. 10(e) its modulus. As the desired movement of the knife is always the same (constant time and angle references) it is expected that the control actions were identical for every piece. This means that current consumption pattern during the movement of the knife should be repeated continuously. Two points have been selected along this movement to compare current space vector. These are noted as points C and D in Fig. 10(c), 10(d) and 10(e). Points A and B are the start and end of the movement in both cases. It can be seen that when the reference point is at the left side of the plane, the modulus of the current space vector is higher that when it is at the right side. This imply that for the same reference, the generated current and then the generated torque are different. The problem is that the expected torque in both cases is the

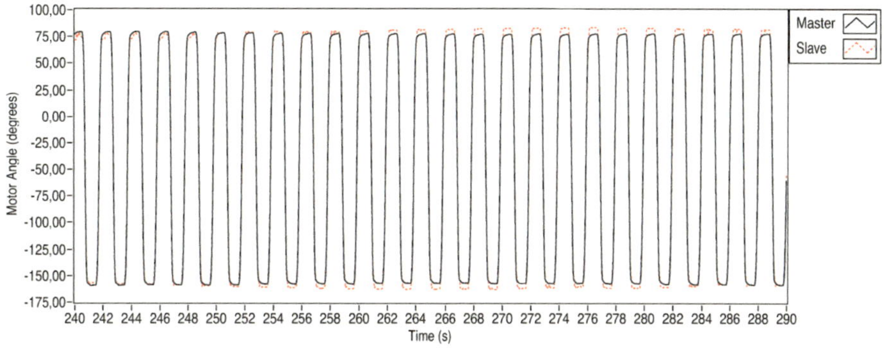

(a) Rotor angle of motors

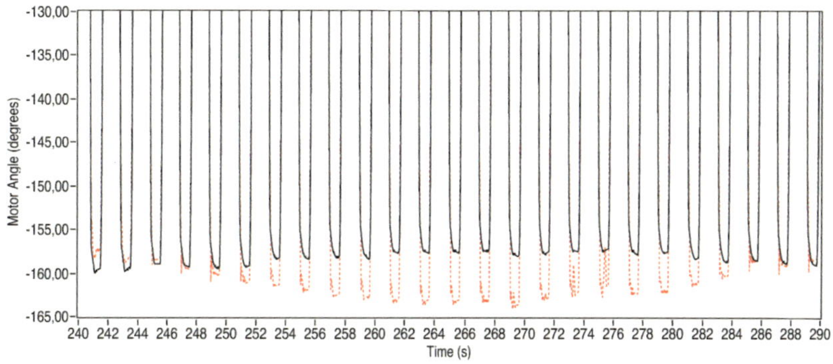

(b) Detail of negative angles

Fig. 9. Master and slave angle for 25 pieces production

same but not the actual torque. That is why the difference between master and slave angles varies along time depending on the origin and final point of the current space vector.

The detected fault is caused by the unbalance of the electric circuit that can be caused by a fault in the motor windings or in the voltage source inverter. The other possible cause that is a fault in the current sensor is rejected because the sum of the three currents is zero. To identify the origin of the fault it would be necessary to find a constant speed and constant torque operation of the motor and then compare the pattern of the current space vector with known fault patterns.

An example of operation with constant torque can be seen in Fig. 11 where it is constant during almost a cycle of the current signal. The current space vector (Fig. 11(d)) is again displaced to the left side of the plane.

Taking into account all the data shown, it can be concluded that the origin of the fault is located in the voltage source inverter. Probably the actual duty cycle of one of the switching devices is slightly different to the desired, what makes that the voltage generated is not balanced.

Looking at this example it can be understood that the main challenge to use current space vector for diagnosis is the automation of the method that could allow using it without expert supervision. If the process under research would be stationary, the task will be only a pattern recognition problem. In a case like the showed tilting knife, the pattern recognition should also have into account the torque variation.

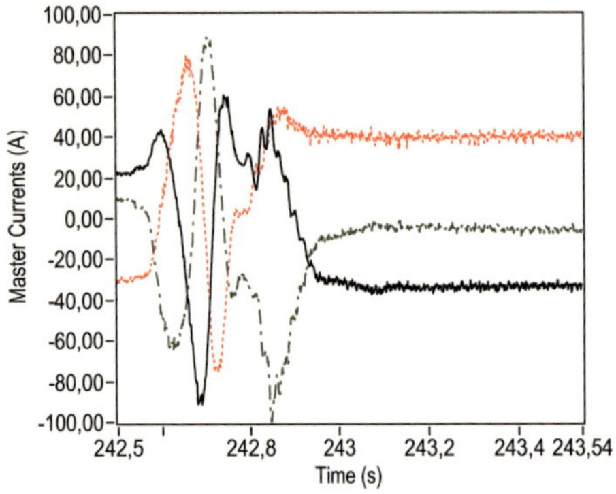

(a) t = 240 s

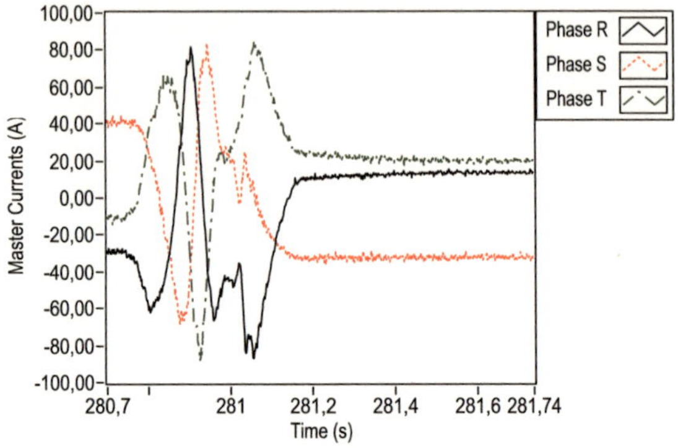

(b) t = 280 s

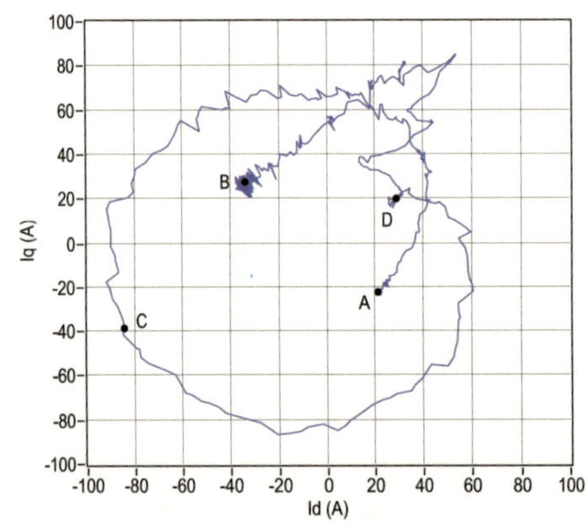

(c) t = 240 s

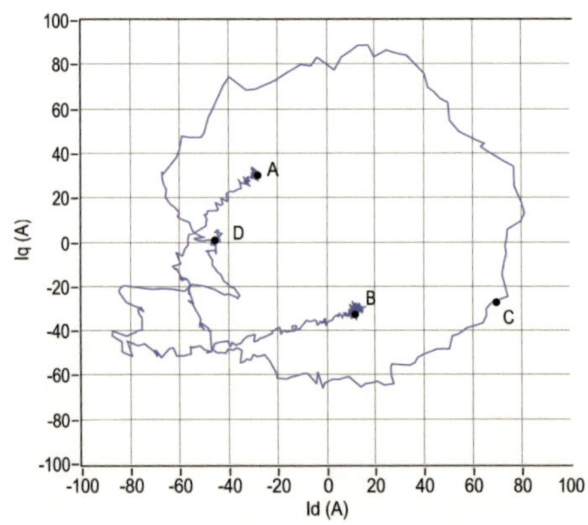

(d) t = 280 s

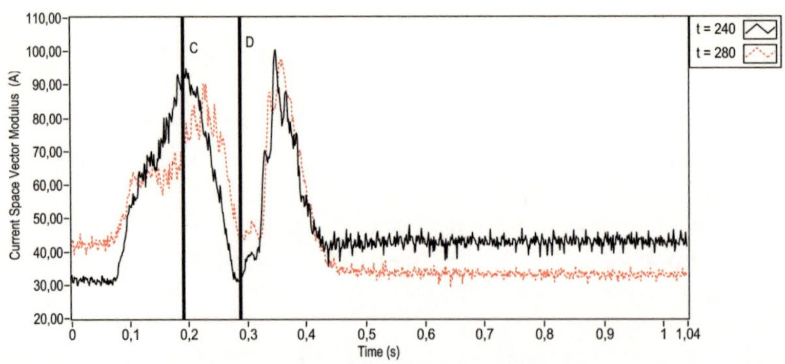

(e) Current Space Vector Modulus

Fig. 10. Current Space Vector Analysis of master motor at different times

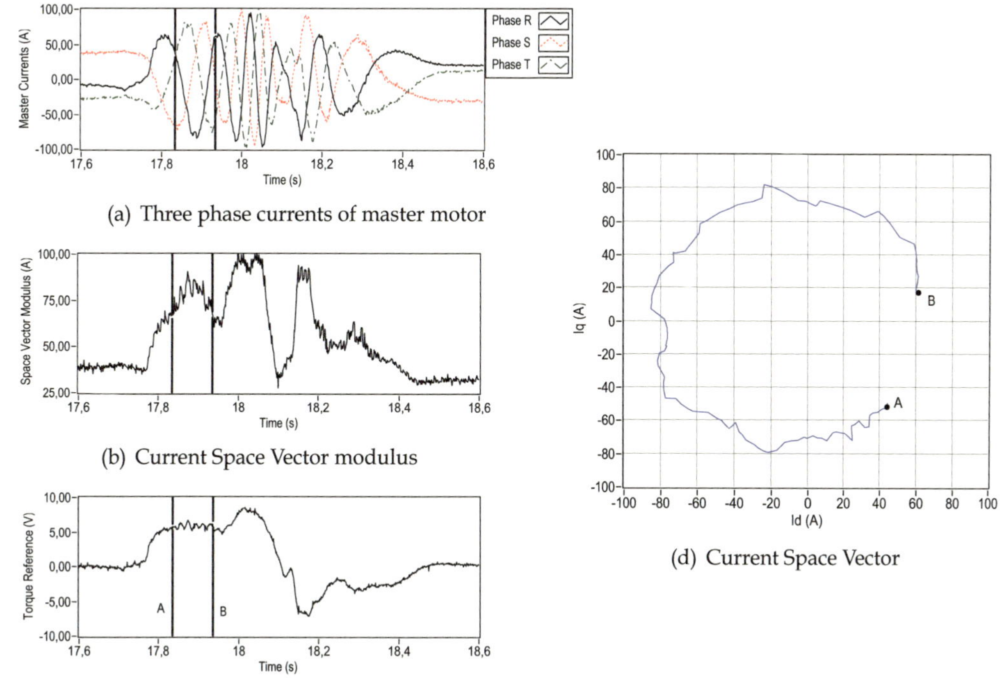

(a) Three phase currents of master motor

(b) Current Space Vector modulus

(c) Torque reference generated in master control

(d) Current Space Vector

Fig. 11. Example of constant torque reference

3.4 Case Study 4: Laser welding defect detection

In this section, two approaches to the problem of defect detection in laser welding are presented. The first is based on analyzing the signal generated by a photodiode in both the time and frequency domain. The second consists of relating variations in the plasma electron temperature with weld quality.

The methods presented have been tested in an industrial facility under real production conditions, exposing them to conditions more requiring than those found in laboratory experimentation. Detailed description can be found in (Saludes et al., 2010).

3.4.1 Problem description

Laser welding is used to weld the tailored welded blanks due to its advantages: a high processing speed, flexibility, low heat input and ease of automation. However, it is possible that some defects could appear in a laser welded seam that can also appear in seams welded using other techniques.

The defects that have to be detected are lack of penetration, pores, inner pores, holes and drop-outs.

The methods described here have been tested on an industrial facility equipped with a Trumpf Turbo 8000 CO_2 laser with output power of up to 8000*W* and operated in a continuous-wave regime. The installation is completely automated and capable of welding up to 20,000 seams a day.

The specimens welded in this installation were galvanized steel sheets whose thicknesses were different and, in both cases, less than 1 *mm*. Taking into account the sheets thickness and according to (ISO, 1997), the minimum size of the defects is 200 *μm*. Beam–on–plate welding was carried out at a power ranging from 6 to 8 *kW*. The welding head displacement speed was between 6 and 10 *m/min*. The shielding gas used was Helium at a flow rate of 40 *l/min*.

3.4.2 Radiation based methods

Two 1.5 *mm* diameter optical fiber EH 4001 type were used to collect and transmit the plasma–emitted and melted–emitted radiation to two different photodiodes. The first was a Siemens SFH203FA IR sensor, sensitive to the 800–1100 nm range, intended to detect variations in the shape of the pool of molten material. The second was a Centronic OSD5,8-7 Q UV and visible light detector, sensitive to the range 200–1100 nm. The signals generated were amplified by means of two Femto LCA-400K-10M amplifiers. A National Instruments PCI 6034E data acquisition board was used to measure and collect data using a PC with a sampling frequency of 10000 *Hz*. The detectors' visual line was 25° above horizontal.

3.4.2.1 Time domain method

As the measured radiation is related to the melting of the welded metals, it is expected that defects in the welding process will produce changes in the signal to be analyzed. If the width and depth of the keyhole is constant, and the laser power is also constant, the quantity of melted metal at each point will be the same and the radiation produced will be constant throughout the process. In the case of a lack of penetration or porosity occurs at any point of the seam, the radiation will instantaneously decrease.

Defect detection will be based on the idea that the changes in the signals generated by the photodiodes are related to the defects. Thus, the location of changes in the signals can lead to defect detection. This issue can be included in what is called detection of abrupt changes (Basseville & Nikiforov, 1993b).

The algorithm used in this case is a CUSUM RLS adaptive filter that combines an adaptive least squares (LS) filter with a CUSUM test for change detection (Gusstafson, 2000).

The time domain fault detection method is intended for finding small defects that can be present in the seam. These faults are typically holes, both trespassing and not trespassing, with sizes ranging from 0.5 *mm* to 2 *mm*.

In order to simulate such kinds of defects, small scrapes have been removed from the edge of the thinnest of the workpieces to be welded. These scrapes have been done in such a way that they are not visible when the workpiece is looked at from above, i.e., from the side the laser hits the workpiece. Then, the workpieces have been welded under normal conditions. Afterwards, visual inspection has been carried out. Finally, the visual inspection findings have been compared to the ones obtained through the time–domain algorithm. The ratio of detected holes versus induced holes is 55.1% and the ratio of false alarms is 2.04%. The detected holes ratio seems to be very low but this can be explained by considering how the detection algorithm works. As it is based on a polynomial fit of the signal, to decide if a signal change is a fault or not, the number of valleys in the signal corresponding to holes will affect the threshold used. So the presence of various defects with great changes in the same signal can move the polynomial to a limit for which small holes with low changes do not overpass. If the number of seams with some hole detected is counted instead of every detected hole, the ratio of faulty seams detected is 100% and the false alarm ratio is 0%.

3.4.2.2 Frequency domain method

The authors found in previous work that, in the frequency domain, the signal energy decreases significantly in the case of a partial penetration fault (Rodríguez et al., 2003). Based

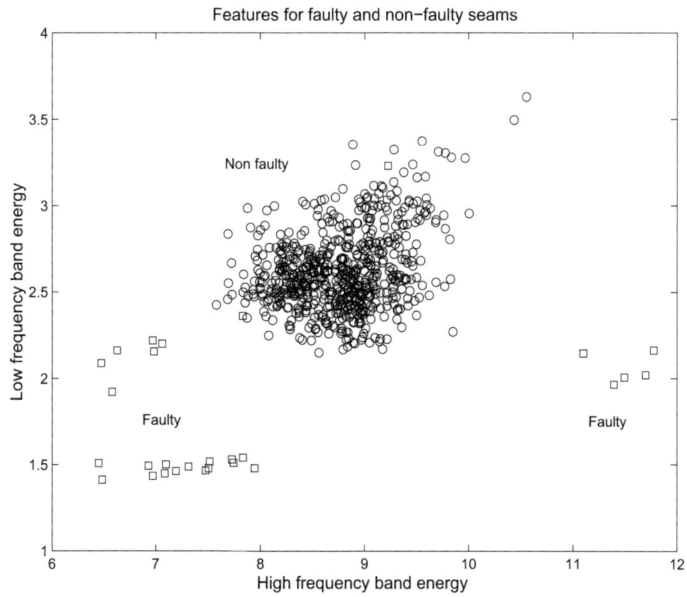

Fig. 12. Features associated to faulty and non-faulty seams

on this result, a method for detecting lack of penetration has been developed. The method comprises two parts. In the first, some features are extracted from the signals generated by both photodiodes. In the second, these features are classified by means of a multilayer perceptron neural network. The two steps are summarized below.

1. Feature extraction. The signal coming from both sensors is divided into N equal-size segments and the Fast Fourier Transform (FFT) is used to perform a frequency domain transformation for each segment. Then, the RMS value for four frequency bands is obtained. Also, the RMS for the whole frequency range is computed. The bands range from 500 *Hz* to 1500 *Hz* and from 4000 *Hz* to 5000 *Hz*. The features can be seen in Fig. 12. Finally, a normalization for each segment is done obtaining the relative harmonic distribution for each frequency band. After all this calculation, four parameters for each sensor and for each segment are obtained: normalized and noise-free data of RMS values for the two frequency bands, global weld RMS and global noise RMS.
2. Decision making. The extracted features are classified using a multilayer perceptron neural network (Haykin, 1999).

The results obtained show that 93.9% of the normal seams were classified as normal and 97.1% of the faulty seams were classified as faulty.

3.4.3 Plasma electron temperature based method

During laser welding, a plasma is formed inside the keyhole. The electron temperature is related to the energy of the electrons that are in the plasma. In the following sections, the estimation of the electron temperature and how to correlate it with weld quality is explained.

3.4.3.1 Electron temperature estimation

Plasma electron temperature T_e can be determined by using the Boltzmann equation (Griem, 1997), which allows the population of an excited level to be calculated by means of the equation (9):

$$N_m = \frac{N}{Z} g_m \exp\left(\frac{-E_m}{kT_e}\right) \tag{9}$$

where N_m is the population density of the excited estate m, N is the total density of the state, Z is the partition function, g_m the statistical weight, E_m the excitation energy, k the Boltzmann constant and T_e the plasma electron temperature. Equation (9) can be used when the plasma is in local thermal equilibrium (LTE), a condition that is assumed to be valid when (Griem, 1997)

$$N_e \geq 1.6 \times 10^{12} T_e^{1/2} (\Delta E)^3 \tag{10}$$

where N_e is the electronic density and ΔE is the largest energy gap in the atomic level system. Equation (10) can be determined by considering that a necessary condition for LTE is that the collision rate has to exceed the spontaneous emission by a factor of ten. The assumption of LTE implies that the different particles within the plasma have Maxwellian energy distributions.

In optically thin plasmas, the intensity of a given emission line I_{mn} induced by a transition from level m to level n, can be related to the population density of the upper level N_m through

$$I_{mn} = N_m A_{mn} h\gamma_{mn} \tag{11}$$

where A_{mn} is the transition probability, and $h\gamma_m$ is the energy of such a transition.

Combining equations (9) and (11), T_e can be obtained from the following expression:

$$\ln\left(\frac{I_{mn}\lambda_{mn}}{A_{mn}g_m}\right) = \ln\left(\frac{hcN}{Z}\right) - \frac{E_m}{kT_e} \tag{12}$$

The plot resulting from using various lines from the same atomic species in the same ionization state and representing the left-hand side of equation (12) versus E_m has a slope inversely proportional to T_e. This technique is usually referred to as a Boltzmann-plot.

3.4.3.2 Spectroscopic lines identification

There are several conditions spectral lines must fulfil in order to be valid candidates for electronic temperature estimation. Selected lines must verify that $\Delta E > kT$ on the upper energy levels to ensure they don't belong to the same multiplet. Moreover, the line must be free of self-absorption; one can prove that this condition has been fulfilled by verifying that the optical depth (Griem, 1997) τ of the plasma for the selected spectral lines is $\tau < 0.1$.

Measurements were performed during normal welding. Radiation emitted by plasma plume was gathered by means of a 3 mm diameter optic fiber. This optic fiber fed light to a high resolution Oriel MS257 spectrometer fitted with an Andor ICCD-520 camera. The spectral lines suitable for electronic temperature estimation found in this way are shown in table 2.

All the spectral lines shown in table 2 come from iron electronic transitions. The wavelength, transition probability, low level energy and its degeneration are all shown in this table. Wave- length is a measured feature, while the remainder come from the NIST (National Institute for Standards and Technology) atomic spectra database.

The spectrometer used during on-line monitoring was an Ocean Optics HR4000, fitted with a 2400 *lines/mm* diffraction grating and a 5 *μm* aperture slit. The spectrometer features a 3600 pixels CCD, a 0.05 *nm* spectral resolution and an 80 *nm* spectral range. Due to that the

λ (*nm*)	A_{mn} (s^{-1})	E_k (cm^{-1})	g_k
411.85	$5.80 \cdot 10^7$	53093.52	13
413.21	$1.20 \cdot 10^7$	37162.74	7
414.39	$1.50 \cdot 10^7$	36686.16	9
425.01	$2.08 \cdot 10^7$	43434.63	7
426.05	$3.20 \cdot 10^7$	42815.86	11
427.18	$2.28 \cdot 10^7$	35379.21	11
430.79	$3.40 \cdot 10^7$	35767.56	9
432.58	$5.00 \cdot 10^7$	36079.37	7
438.35	$5.00 \cdot 10^7$	34782.42	11
440.48	$2.75 \cdot 10^7$	35257.32	9
441.51	$1.19 \cdot 10^7$	35611.62	7
452.86	$5.44 \cdot 10^7$	39625.8	9

Table 2. Spectral lines associated to Fe I

device is able to take data at a rate of 200 *Hz* and the welding speed ranges from 6 *m/min* to 10 *m/min*, the distance travelled between spectra is two or three times the size of the smallest defect that must be detected. Since at least one spectrum must be gathered during a defect occurrence, this will be a drawback of the proposed method unless a strategy based on the synchronization of several spectrometers is adopted.

3.4.3.3 Results

The defect detection method based on electronic temperature has been tested in the industrial facility described in section 3.4.1. The conditions under which experiments were carried out are the same as those found during normal industrial production: electrical noise, mechanical vibrations and steel sheets to be welded covered by an oil film. During experiments, the laser power was set to 8000 *W* and welding speed was 4.5 *m/s*.

Experiments can be classified into two classes: Those that have been performed during normal operation and those in which defects have been forced.

Experiments carried out during normal operation are those in which the manufacturing cadence was the usual in the car factory where the experiments were done. The purpose of these experiments were twofold: to estimate the electronic temperature during normal operation and to observe its variation between seams.

The electronic temperature variation between seams can be seen in Fig. 13(a), in which the electronic temperature of 70 consecutively welded seams is shown. The electronic temperature represented is the mean value of the temperatures estimated in 180 points along each seam. Moreover, the standard deviation is also represented by means of error bars. All the welds were made with the same process parameters. Worth to be noted is the sudden increment in the mean value of the electronic temperature in seam number 21, which decreases in seam number 40. The standard deviation remains constant along all the seams, although it can be seen that it is greater between seams numbers 39 and 40, just during a drop in the electronic temperature. The seams numbers 1 and 28 presents a huge standard deviation, but no differences were found in the seams, with respect to the other seams, that can explain this behaviour. A decreasing trend can be observed, specially from seam number 40. Again, no differences in quality terms, penetration depth in this case, were found. Since no changes in the process parameters were introduced, these fluctuations can only be related to some internal state of the laser welding machine.

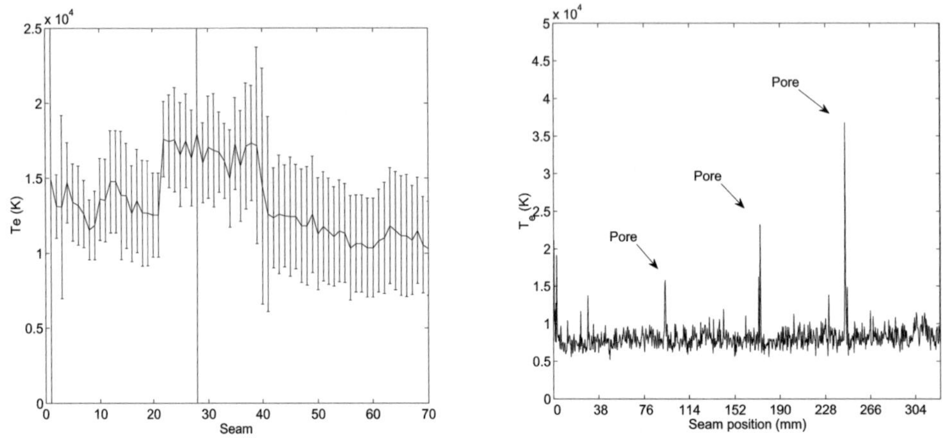

(a) Electronic temperature variation between seams

(b) Electronic temperature along a seam in which three holes have been produced

Fig. 13. Electron temperature results

Besides the estimation of electronic temperature during normal welding, some experiments intended to generate defect have been carried out. The class of detect used in such experiments was holes and pores. The difference between a pore and a hole is that the former does not go through the seam while the latter does.

In Fig. 13(b), the electronic temperature associated to a seam in which three holes were forced is shown. Worth of be noted are the three peaks that appear at the same positions the holes were induced. They can be seen at positions between 76 *mm* and 114 *mm*, 152 *mm* and 190 *mm* and 228 *mm* and 266 *mm* in figure 13(b).

4. Conclusions

Fault detection methods in the automotive industry have a great complexity due to the differences between the different machines and processes involved. This complexity makes difficult or even impossible the human supervision of all the processes, although the available technology are of great help in this task. The difficulties found in process supervision came from the huge amount of variables that have to be taken into account and the overwhelming information available.

Nowadays, the correct operation of any plant is more than keeping all the devices in good shape. It also means to know the state of all the devices and machines in order to avoid disruptions in manufacturing production originated by faults or unexpected stops.

In this chapter, it has been shown that predictive maintenance can be applied to very different equipment. This maintenance approach provides the operator with valuable information about equipment status and its future behaviour. The implementation of any predictive maintenance strategy is subject to the importance of the process to be supervised. This also will determine the diagnosis to be performed. Moreover, the economical analysis of the design and implementation of the diagnosis system will determine the adoption of any predictive maintenance strategy.

Any diagnosis system can be broken down into three main modules: data acquisition, signal processing and decision making. Through the case studies presented in this paper, several implementation ways of each component have been presented.

In this way, data acquisition has been illustrated by the case of a machine tool in which the data needed to perform diagnosis is the same data the controller commanding it uses. In this case no more sensors are required. The opposite situation is found in the case of laser welding. In this case, very specialised sensors, like spectrometers, are required to gather data. In the other two study cases, conventional sensors have been installed. Current transducers and accelerometers are common in industrial applications. Their costs depends on precision, range and other requirements. Acquisition hardware to which sensors will be connected is not usually a critical element. This is due to the variety of devices commercially available. However, it could be necessary to develop tailored solutions for specific applications, although it will never be the most critical step in the implementation of a diagnosis system.

Through the case studies, several approaches to the signal processing module are shown. They range from classical frequency analysis to plasma physics. Also, complex techniques have been used to process signal in the time domain or to detect abrupt changes. The most suitable technique is always determined by the pursued target. In same cases it would be possible to chose between several techniques that pursues the same objective. This is the case of defect detection in bearings, where vibration analysis and current analysis are both suitable. Nevertheless, usually only one technique provides the information required to detect the defects. For this reason, the designer has to have a deep knowledge of the processing techniques in order to find the most suitable for the problem at hand. In some cases this will not be enough, and the designer has to develop the processing techniques. This is the situation in the study case related to the machine tool, where segmentation techniques had to be developed in order to find the exact defect location.

Decision making usually is the most difficult step, due to the lack of information about system behaviour when it is in faulty state. This information can be gathered along time once the data acquisition and signal processing modules are installed. The most simple case presented is the motor-fans in a car painting cabinet. In this case, the decision making is carried out by means of a threshold set whose values are set through observation. This is a process that has to be repeated every time a major maintenance task is done. A very different situation is found in the case of laser welding, where decision making is performed by a machine learning method, like neural networks, whose training is done only when significant information has been collected. In this case there is no need for an operator performing supervision tasks.

It is important to note that process expert knowledge is basic in the design of any diagnosis system. A deep understanding of the physical principles involved in the process is the main clue to choose the best strategy to extract features indicating the presence of a fault. The expert is who will be able to know or to deduce which signals are the most affected by the presence of a fault and how they can change in this situation. For example, part of the failures will have an effect on the signal harmonic content, while others will affect the evolution in the time domain. Moreover, they will play a key role when assessing any other kind of dependencies among the data. Frequently it is advisable to analyse correlations among variables or along the evolution of any variable in the time domain. This can be done by means of mathematical methods that can also offer information on the changes associated with failures. The expert will be able to confirm if that information is relevant or is just a mathematical result coming from particular cases.

To sum up, automotive industry can improve their processes through predictive maintenance and the automatic defect detection methods that can be integrate into it. The vast majority of these techniques have reached a mature state and have been successfully implemented. There are also new promising techniques that can improve new processes in the automotive industry, like laser material processing. The implementation of any of these techniques needs of qualified technicians whose knowledge and expertise will make possible success in their implementation.

5. References

Acosta, G. G., c. J. Verucchi & Gelso, E. R. (2006). A current monitoring system for diagnosing electrical failures in induction motors, *Mechanical Systems and Signal Processing* 20(4): 953–965.

Altintas, Y. (2000). *Manufacturing Automation: Metal cutting mechanics, machine tool vibrations and CNC design*, Cambridge University Press.

Astakhov, V. (2004). The assessment of cutting tool wear, *The International Journal of Machine Tools and Manufacture* 4: 637–647.

Basseville, M. & Nikiforov, I. (1993b). *Detection of abrupt changes: theory and application*, Information and system science series, Prentice Hall.

Cardoso, A. J. M., Mendes, A. M. S. & Cruz, S. M. A. (1999). The Park's vector approach: New developments in on-line fault diagnosis of electrical machines, power electronics and adjustable speed drives, *The 1999 IEEE International Symposium on Diagnostics for Electrical Machines, Power Electronics and Drives Record*, Gijon, pp. 89–97.

Chow, E. & Willsky, A. (1984). Analytical redundancy and the design of robust failure detection systems, *IEEE Trans. on Automatic Control* 29(7): 603–604.

Diallo, D., Benbouzid, M. E. H., Hamad, D. & Pierre, X. (2005). Fault detection and diagnosis in an induction machine drive: A pattern recognition approach based on Concordia stator mean current vector, *IEEE Transactions on Energy Conversion* 20(3): 512–519.

Griem, H. R. (1997). *Principles of Plasma Spectroscopy*, Cambridge Monographs on Plasma Physics, Cambridge University Press.

Gusstafson, F. (2000). *Adaptive Filtering and Change Detection*, John Willey & Sons.

Haykin, S. (1999). *Neural Networks. A Comprehensive Foundation*, 2nd edn, Prentice Hall.

Isermann, R. (2006). *Fault-Diagnosis Systems: An Introduction from Fault Detection to Fault Tolerance*, Springer-Verlag.

ISO (1997). Welding. Electrons and laser beam welded joints. Guidance on quality levels for imperfections. Part 1: Steel. (ISO 13919-1:1996), *Technical report*, International Organization for Standardization.

Nejjari, H. & Benbouzid, M. E. H. (2000). Monitoring and diagnosis of induction motors electrical faults using a current Park's vector pattern learning approach, *IEEE Transactions on Industry Applications* 36(3): 730–735.

Reñones, A., Miguel, L. J. & Perán, J. R. (2009). Experimental analysis of change detection algorithms for multitooth machine tool fault detection, *Mechanical Systems and Signal Processing* 23(7): 2320–2335.

Reñones, A., Rodríguez, J. & Miguel, L. J. (2009). Industrial applications of a multitooth tool breakage system using motor electrical power consumption, *International Journal Advanced Manufacturing Technology* 46(5–8): 517–528.

Rodríguez, F., Saludes, S., Miguel, L. J., Aparicio, J. A., Mar, S. & Perán, J. R. (2003). Fault detection in laser welding, *Proc. of the SAFEPROCESS Symposium*,Washington.

Saludes, S., Arnanz, R., Bernárdez, J. M., Rodríguez, F., Miguel, L. J. & Perán, J. R. (2010). Laser welding defects detection in automotive industry based on radiation and spectroscopical measurements, *The International Journal of Advanced Manufacturing Technology* 49(1–4): 133–145.

Part 2

Industrial System Production

4

The Concurrent Role of Professional Training and Operations Management: Evidences from the After-Sales Services Information Systems Architecture in the Automotive Sector

Nouha Taifi[1,2] and Giuseppina Passiante[3]
[1]*Center for Business Innovation, University of Salento, Lecce,*
[2]*Ecole Mohammadia d'Ingénieurs, University Mohammed V, Agdal, Rabat,*
[3]*Department of Innovation Engineering and Center for Business Innovation, University of Salento, Lecce,*
[1,3]*Italy*
[2]*Morocco*

1. Introduction

The automotive industry is one of the most successful and dynamic environments playing a key role in the economic market. The actors involved in it assume different functions by which they collaborate and interact for the operations and processes of production and innovation. As a part of the product development process, the manufacturing and the after-sales services are indeed connected and inter-dependencies occur among them for product development, customer satisfaction and economic performance. The management of the operations among the manufacturing and after-sales services is one of the most important bases for the development and life cycle management of the products.

The operations and their management are established according to the purposes of collaboration and are updated continuously by creating and adding elements to improve them and to achieve continuous positive outcomes. Besides, the human capital involved in the operations also needs competences and skills improvement in order to be aligned with the operations and its management. Thus, in this chapter, we question the role of professional training that is leading to the development of the competences of the people involved in the operations and their management, and investigate on the concurrent role of both through the presentation of the information systems architecture connecting a large firm in the automotive industry with its after-sales services partners.

First, we present the literature concerning the professional training to define it and to show the major role it plays as competences enhancer and developer, and the operations management to show that it is the focus of engineering and development. Then, we present the information systems architecture subject of the study, but before that we present the conceptual framework and research methodology. After that, we explain the need for strategic changes in the information systems architecture of the after-sales services, some issues for the changes, and initial proofs of the concurrent role of professional training and operations management. Then, we derive a matrix model showing the concurrent role of professional training and operations management, and finally as a conclusion we further model their concurrent role for after-sales optimization, and we propose some future directions for research.

2. Professional training as competences developer and enhancer

The professional training is related to the development of the learners' capacities, competences and skills in the industrial and business environments. Those are managers, technicians, experts in specific fields and need to enrich their capacities for the ongoing high economic performance of their organizations. The professional training is subject to strategic regulations, learning strategies and actions that make it an important developer of competences and capacities (Paton et al., 2005); the professional training is a function of its regulations, structures, learning processes and experts (Figure 1). The training regulations are focused on the contribution to the economic performance of the organizations, thus, there are regulations leading to the intellectual capital protection and to the investigation on the needs of the trainees for the provision of the right professional training and according to their functions and roles. Besides, the organizations providing professional training must have the adequate resources in terms of human capital -experts- and technological infrastructures for a fast and high developed provision of learning (Allan & Chisholm, 2008), and the learners must have good absorptive capacities to assimilate new knowledge and information to develop their competences (Cohen & Levinthal, 1990).

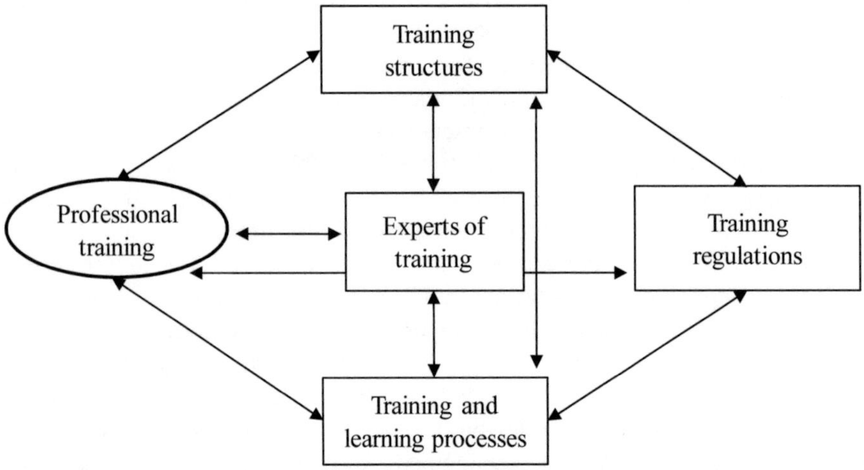

Fig. 1. The professional training success factors.

The professional training can be either internal to the firms through the creation of corporate universities that are entities in charge of fostering individual and organizational knowledge (Allen, 2002) and developing the individual competences and skills (Baets & van der Linden, 2003), and the creation of competences development projects and systems (Scalvenzi et al., 2008; Corallo et al., 2010c) leading to the optimization of the engineering skills and to the competences' development and enhancement of the managers at different levels in the business and industrial environments, or external to the firms through collaboration with higher education institutions, universities and competence centers for the participation of

the managers and technicians, as learners, in executive, technical and training programs dedicated to the development of new competences (Maglione & Passiante, 2009).
Besides, the factors leading to the need for a continuous professional training is the dynamism and complexity of the industrial and business environments. Firms must have dynamic capabilities to improve (Teece et al., 1997) the managerial, engineering and technological competences according to the changes in the economic markets in order to pace with the complexity of information and knowledge for high competitive advantage and economic performance. The changes follow specific procedures and processes in which the human capital is also subject to competences improvements and development accordingly (Lepak & Snell, 1999). There are organizational learning processes (Argyris & Schon, 1996; Nonaka & Takeuchi, 1995) for the effective human capital management and development (Bontis & Serenko, 2009) and the professional training is one of them consisting of specific types of systems and tools. Technology-enhanced learning is one of the most important methods of learning based on information and communication technologies by which people can learn and acquire knowledge in an optimal manner (Ambjörn et al., 2008).

3. Operations management as the focus of engineering

The processes of interactions and communications taking place among different actors for the work activities are directly related to the operations management which consists in ensuring the ongoing processes of production of products and services and using different resources as technological supports for the purpose of customer satisfaction and high economic performance (Slack et al., 2007). The operations management differs from one type of work activity to the other since there are various functions involved in the product development process and lifecycle. This makes it multi-faceted and heterogeneous, and increases the interest of scholars on the investigation on its systems' differences and similarities; There are for instance the ones dedicated to the design phase (Cisternino et al.,

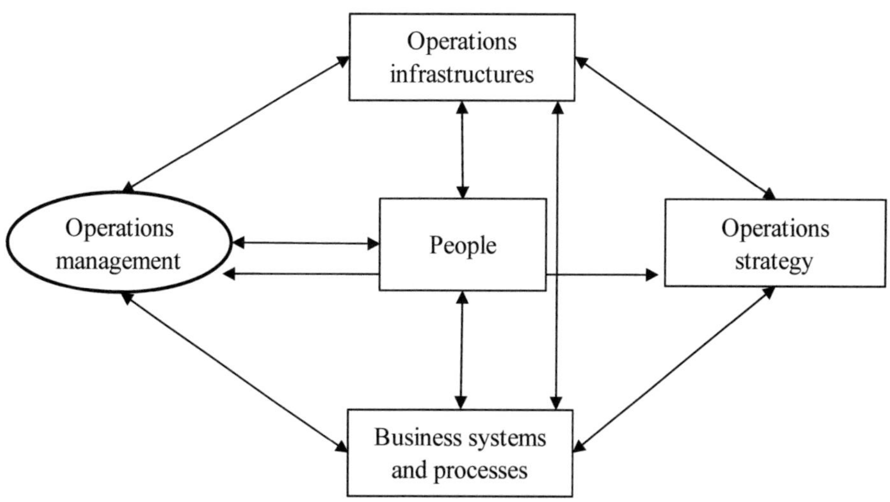

Fig. 2. The operations management success factors

2008, Corallo et al., 2009, Corallo et al., 2010a), to the manufacturing (Kundra et al., 1993), and the ones integrating different phases of the product development process (Corallo et al., 2010b; Taifi et al., 2012).

Various research organizations and centers study operations management and continuously provide standards of business operations, processes, and system engineering (Blanchard, 2004). Operations management is a function of its strategy, infrastructure, business processes and systems, and the people involved in it (Angell & Klassen, 1999) (Figure 2). Roadmaps and strategic guidelines are created (Adam & Swamidass, 1989) and used among professional, business and industrial networks developing or applying and collaborating through or for the operations management. For instance, as specific strategies, action research (Coughlan & Coghlan, 2002) and case research (Voss et al., 2002) provide diverse contributions to the systems and industrial engineering and by this participate to the development of the operations and their management. Also, the operations management uses the right technological infrastructures to achieve fast and optimal productions (Salvendy, 2001) as computer-aided tools for manufacturing or electronic data interchange systems or web and knowledge portals and applications, and the right organizational structures (Lucertini et al., 1995) connecting people involved in the operations and their management. For instance, there can be cross- functional groups or communities of practice (Taifi et al., 2011) dedicated to the operations and as mentioned above networks of research centers and universities for the development of the operations management.

4. Research into the concurrent role of professional training and operations management in the automotive industry

The purpose of this chapter is to investigate on the concurrent role of the professional training and operations management in dynamic environments –more precisely the automotive industry and to provide evidences about that. We focus on the interactions taking place among a large automotive company and its after-sales services partners –the dealers' network- to achieve our research goals. The idea behind the choice of this type of interactions is the strategic role the after-sales services plays for their proximity with the customers, their market positioning (Alexander et al., 2002), and the rich environment surrounding them in terms of systems, tools and technologies. Besides, the after sales services are seen as a relevant resource of revenue and economic performance (Saccani et al., 2007). In general, designing service systems requires a great knowledge about the specific details of each type of services to engineering them (Sakao & Shimomura, 2006), a services engineering strategy (Aurich et al., 2004) and a system strategy as well (Ramaswamy, 1996; Morelli, 2002).

Also, in the research on service operations management, there is a need for further focus on it through the creation of a research agenda and framework (Roth & Menor, 2009) and linking it to other areas as human resources management (Johnston, 1999). Thus, the investigation on the professional training and the services operations management in this context of after-sales services and their concurrent role is more than a strategic topic, leading to awareness about the importance of connecting these two subjects in the industrial and business environment, and in the fields of research on competences development and services operations management for the after-sales services optimization (Figure 3).

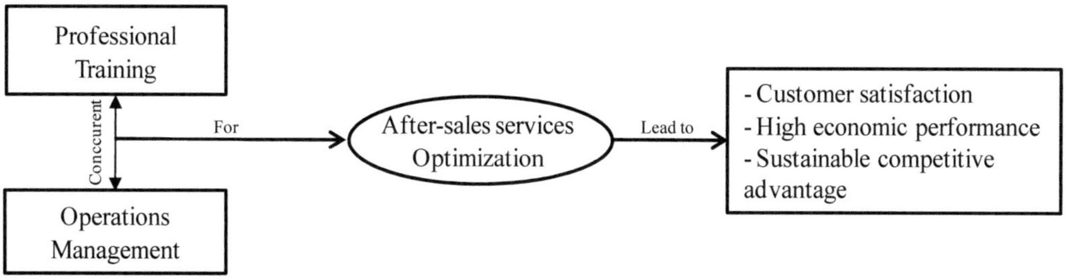

Fig. 3. Research into the concurrent role of the professional training and operations management in the after-sales services.

As the focus of services operations management in this research, the after-sales services are analyzed, explained in terms of operations and their management that is the operations management for the after-sales services and the professional training dedicated to the after-sales services is also detailed, presented and explained. It is possible to investigate on the professional training dedicated to other functions and roles in the product development process but the focus of the chapter is the after-sales services to present the link among them; There will be some links presented in the operations management of the after-sales services with the manufacturing since for the provision of some after-sales services there is the need for the manufactured products' components but we precise that we are investigating on the direct connections and not indirect connections of the operations management and professional training for the after-sales services optimization and thus customer satisfaction, high economic performance and sustainable competitive advantage.

Thus, in the automotive industry, in year 2007, we first interacted with the managers of a large automotive industry for the acquisition of data through interviews concerning the information systems architecture in which the professional training and operations management is represented - the managers are key informants (Yin, 2003) and play a critical role in the success of of the research. They are the head of the Professional Training Department, the head of Quality-Technical Services Department, the manager responsible of the Product Support Unit and the manager responsible of the Service Engineering Unit. Second in order not to become dependent on a key informant, through a questionnaire, dedicated to a significant sample from the population of the after-sales services -the dealers, we investigated on their point of view concerning the professional training, the operations and their management and all dedicated to the after-sales services. We adopted these types of research methods and data collection methods so to have multiple sources of evidences and allow their convergence (Yin, 2003; Creswell, 2003).

5. The information systems architecture among the large automotive firm and the after-sales services

The basis of communication among the large automotive company and its after-sales services partners is a complex information systems architecture on which they share data and information about the after-sales services and the automotive products for a wide range of purposes that are knowledge codification, knowledge sharing, knowledge acquisition and knowledge creation. The information systems architecture consists in systems, tools, applications and web portals. These are mainly supported by information and communication technologies (ICT) and also face to face means of communication so the objective here is to see what these IT systems and tools are and how they are organized to achieve the purposes of collaboration for the after-sales service and professional training, and to have a clear idea about the architecture. Thus, we present the types of ICTs used and the face to face mechanisms for the operations of after-sales services and for their management, and for the professional training, we present the areas, types of trainings provided, and the systems used.

Following are the types of systems and tools used for the after-sales services operations management (Table 1):

- The problem-diagnosis IT-tool: dedicated to the problem-diagnosis in the products' repair activities. This tool supports the technicians in their work; it provides a diagnosis about a problem occurring in the product and the technicians can consult the adequate problem-solving instructions in the e-manuals or the insights in the e-services news –IT-applications.
- The problem-solving IT system: dedicated to the products' problem-solving in the products' repair activities. An efficient procedure is followed in which the dealer communicates, through an IT-application- on the integrative IT-system knowledge portal of the firm-, the product's problem to the problem-solving IT system and this latter gives feedbacks to the dealer including repair packages consisting of the necessary equipment and special tools to be used.
- The e-manuals: are digital manuals comprising the instructions for the repair solutions. The dealers can consult and download them on the IT-systems through the homepage of the service portal of the firm in order to repair the products of the firm and by this provide the after-sales services. The e-manuals contain technical data, procedures descriptions and diagnosis tests instructions.
- The e-services news: as a digital regular newspaper, it provides news about new after-sales services mechanisms or new products' repair solutions, thus, spreading technical knowledge in the dealers' network.

ASS Operations Management	Definition
Problem-diagnosis tool	An IT-based tool supporting the technicians in the products' problem-diagnosis for the after-sales services.
Problem-solving system	An IT-based systems supporting the technicians in the products' problem-solving for the after-sales services.
E-Manuals	Digitals manuals that can be downloaded on the service portal of the firm to support the technicians in the after-sales services.
E-services news	Up-dated electronic informations that can be consulted on the service portal of the firm for the after-sales services.
Spare parts system	An IT-based system supporting the technicians in the ordering of the necessary spare parts for the after-sales services in the products.
Direct support	Technical support provided by the large automotive company to the technicians through personnel.

Table 1. The types of systems and tools for the ASS operations management

- The spare parts system: consists in the ordering of the spare parts for the after-sales services on the service portal of the large automotive company. Then, this latter delivers it to the after-sales services firms.
- Face-to-face support: consists in technical support provided to the dealers by the automotive firm for the investigation on repair solutions and in managerial support provided for the development of the after-sales services activities in general.

Concerning the professional training, which is the other critical element of collaboration among the large automotive company and the after-sales services partners, the technicians are the main focus. The professional training gathers the learners in a face to face manner through courses with instructors and trainers or IT-based manner through web-based courses and tutorials. Thus, the technicians -the learners- acquire new capacities through the acquisition of knowledge about new concepts and make practical experiments in laboratories before starting the after-sales services activities. The main IT-support used for professional training is the e-Learn system which is located on the service portal of the 'Training Academy' - The Professional Training Department of the large automotive firm and on which the technicians are subject to courses and online-learning. Besides, from the e-Learn system, the after-sales services partners can also download documents they can can read as either a support for the face-to face/online theoretical courses or for the practical experiments.

Professional training	Definition
Technical competences	Seminars, courses and practical experiments related to the development of the technical competences for the after-sales services.
Products' components	Seminars, courses and practical experiments related to the development of the technical competences regarding the products' components of the products.
Characteristics of the products' components of each brand	Seminars and courses related to the development of the technical competences regarding the characteristics of products components of each brand.

Table 2. The types of professional training for the after-sales services

The professional training is taking place when new or complex products and after-sales services are created by the firm. The professional training is technical and regarding the development of the competences of the technicians and the automotive products (Table 2). The automotive products area courses for the after-sales services are designed according to each brand of the large automotive company products and are adapted to the specific characteristics of each brands' products. These specific characteristics are on the products' systems, engines and transmission and technologies of the products. And, the technical competences area for the after-sales services consist in various types of courses as follows:

- Engines: which regard the development of the technical competences of the technicians on the engines of the products and also the development of their capacities for the provision of the after-sales services on these engines.

- Transmission: which regards the systems and components of the transmission in the products, as the gear box and transmission systems. Thus, there are courses for the development of the technical competences and the capacities for the after-sales services on this.
- Electronics: which are courses for the understanding of the electronic systems and components of the products as the airbag systems, the air-conditioning systems and info telematic systems and the development of the capacities for the after-sales services provision.
- Diagnosis: this regards the development of the technical competences of the technicians regarding the diagnosis tool features and operations for problem-solving and also the development of their capacities to use it for the provision of optimal after-sales services.
- Body: courses regarding the body of the product, thus for example the understanding of air and water leakage and also the methodologies for the provisions of the after-sales services to the body of the products.

Finally, after the courses, there is the 'Automotive Technician Accreditation' which are assessments for the accreditation of the technical competences of the technicians about the products and their after-sales services. The results of the assessments provide to the technicians the accreditation proving their competencies.

6. Strategic changes of the information systems architecture and issues

The professional training and the ASS systems and tools used are continuously the subject of evaluations and re-engineering to sustain competitive advantage, customer satisfaction and economic performance. For this purpose, the after-sales services organizations also provide their point of view and satisfaction level concerning them. In general, the key elements to investigate on for the satisfaction level is the perceived ease of use and usefulness of the systems and tools, the computer self-efficacy of the technicians and the usefulness and quality of the professional trainings (Taifi, 2008, a;b). These elements contribute to the continuous re-engineering and development of the information systems architecture of the after-sales services.

In 2007, the systems and tools used for the operations and management of the after-sales services were perceived as useful and easy to use by the dealers of the automotive company (Figure 4) (Taifi, 2008). As stated by a dealer: 'without the IT-tools and systems, we cannot provide the after-sales services to the customers'. However, the objective of the large automotive company is to achieve a complete satisfaction of the dealers about the systems and tools, thus for example, it is also integrating the IT and social interactions awareness of the dealers in dealing with the after-sales services. That is, the awareness of the dealers about the importance of the continuous integration of high-technological tools and systems in their work activities for the after-sales services and also their awareness about the importance of continuous interactions among different small and medium- sized after-sales services organizations ranging from direct to indirect connections in the network for collaborations and operations management, and knowledge sharing and creation about the after-sales services. For example, in 2007, the automotive company created a strategic community (illustrated in Figure 4) in which 50 expert after-sales services small and medium-sized organizations were invited to participate and share their expertise about the automotive products and after-sales services and by this to contribute to the new products and services development and their processes (Taifi, 2007; Taifi and Passiante, 2010; 2011) in an incremental or radical manner (Figure 5).

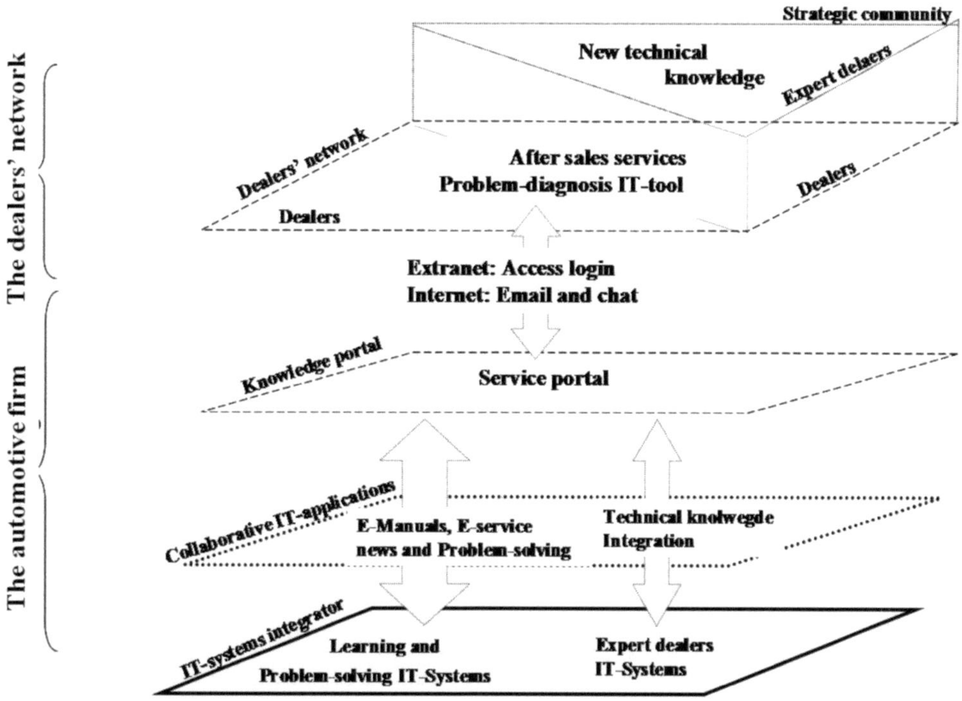

Fig. 4. The information system architecture of the professional training and the after-sales services operations management.

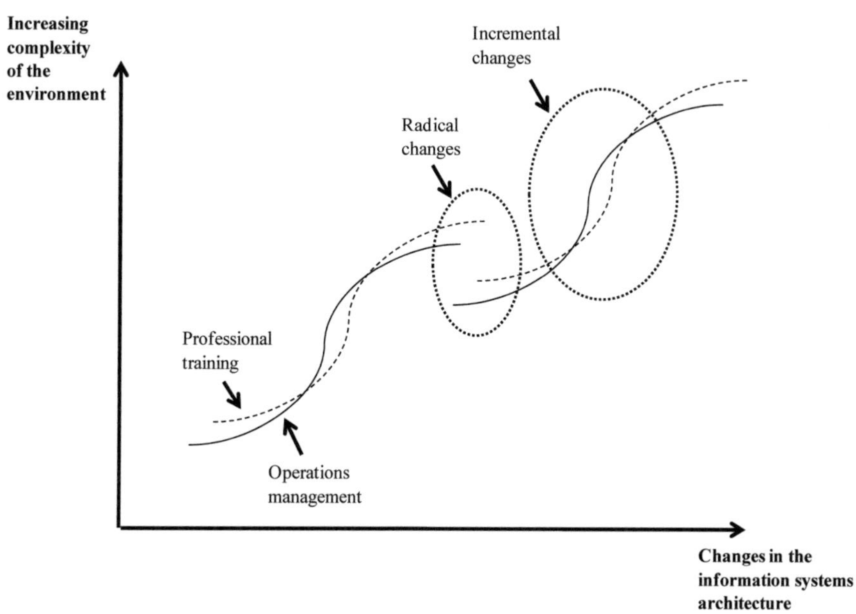

Fig. 5. The environmental complexity and the architecture changes

Also, in 2007, concerning the professional trainings (Figure 4) (Taifi, 2008), their efficiency and quality were considered as satisfying and the expertise of the trainers as well. As stated by a dealer: 'When I go to the professional trainings, I learn many new issues I did not know before'. However, the large automotive company keeps restructuring and reorganizing the content, processes of the trainings including the IT-based ones in an incremental or radical manner (Figure 5) and improves even the skills of the trainers according to the incremental or radical changes in the products and after-sales services activities. Here the development of the capacities of the trainers is also one of the most important factors of success of the professional training and for that, the large automotive firm continuously works on understanding and filling the gaps of the trainers' skills to keep an efficient professional training provision.

The professional training and operations management for the after-sales services are thus subject to changes that are either incremental or radical according to the after-sales services themselves in order to pace with the increasing complexity of the environment (Figure 5). These incremental or radical changes also show the concurrent role that the professional training and the operations management plays. Whenever there are incremental changes in the products or after-sales services, there are incremental changes in the operations management and in the professional training, and when there are radical changes in the products or after-sales services, there are radical changes in the professional training and in the operations management (Figure 6). Besides, each radical change lead to the start of new incremental changes in the professional training and operations management and as the complexity of the environment increases as there are more complex products and services and thus more complex incremental and radical changes in the professional training and operations management (Figure 6).

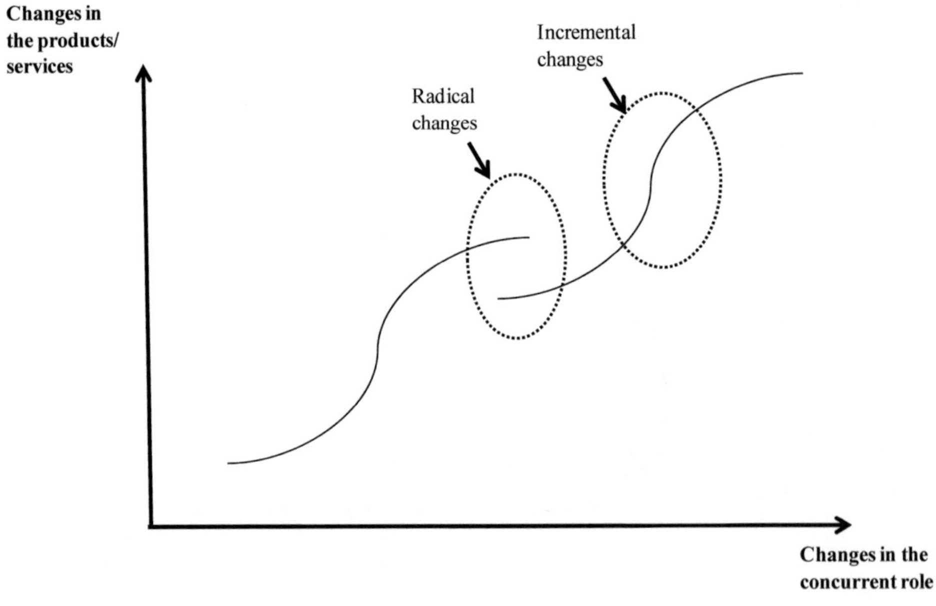

Fig. 6. The types of changes in the concurrent role

7. The concurrent role of professional training and operations management for after-sales services optimization

The processes of interactions taking place among the large automotive firm and the after-sales services organizations are leading to the achievement of the purposes of collaboration. The systems and tools used for the after-sales services and the professional trainings courses regarding the competences of the technicians for the after-sales services and the automotive products are directly connected. As long as there are after-sales services, there are the professional trainings and as long as there is a need for the development of the after-sales services partners competences and capacities, the professional training plays a major role. That is, the more the technological capacities and competences needed to provide the after-sales services, the more likely the professional training will be complex and the less the technological capacities and competences needed to provide the after sales services, the less likely the professional training will be complex. We develop on when and whether the professional training should be complex or simple according to the after-sales services operations management complexity. This shows the concurrent role of professional training and operations management (Figure 7).

Professional Training \ Operations Management	Simple	Complex
Complex	I - Practical experiments for building competences on products' components - Seminars on the characteristics of the products' components of each brand	II - Seminars on technical competences development and on the products' components for the use of the problem-solving tool and system
Simple	IV - Seminars on products components and technical competences development - Web-based courses on the problem-solving tool and system	III - Seminars on technical competences development and practical experiments on the products' components - Courses and seminars for technical competences development for understanding the e-Manuals and e-Service news

Fig. 7. What types of concurrent roles are there among the professional training and the operations management

In order to provide successful professional training to the needed after-sales services, leading to customer satisfaction and high economic performance, there is a need for a complete focus on the technical competences of the technicians and the types of systems and tools used for the after-sales services and the products' components. The more complex the operations management of the after-sales services are, the more complex is the method used for the professional training. Figure 7 shows the simultaneous, concurrent and tight links between operations management and professional training through the methods of professional training used for the operations management.

If the operations management of the after-sales services is simple, the professional training for it ranges from simple to complex contents accordingly. First, in Figure 7, through the development of new products and after-sales services, there is a need for seminars in the different areas whether the technical competences development or the automotive products' components. That is, for example, as soon as there is a new product launched in the market, the technicians are subject to trainings in order to provide the adequate after-sales services or as soon as there is the launch of a new service and its system, the professional training is involved again for the development of the capacities of the technicians on it. When the operations management for the after-sales services is simple, the professional training is simple; Web-based courses and classes are preferred for the diagnosis IT-tool and problem-solving system since they are IT-based and the technicians have the technological capacities to follow the IT-based professional trainings.

Second, when the operations management is simple, the professional training can also be complex and there are two kinds of trainings showing that (cell I of figure 7). That is, for example there can be practical experiments for building competences on product's components in which the technicians can learn practically on products' components. The practical experiments are complex and require major efforts and concentration from both sides since the trainers have to continuously, during the practical experiments, follow the technicians and these latter have to base the practical experiments on the technical knowledge acquired during the seminars. This type of concurrent role of professional training and operations management can be mostly successful if continuously applied as above since both sides are using their knowledge and technical capabilities for the after-sales services. The ones on the characteristics of the products' components of each brand are also complex since they are consisting of more specific and detailed information and knowledge and are related to each product brand. This makes them more complex in comparison to other seminars since they require more precisions and more explanations and preparations for the characteristics are different from one product to the other. These seminars are complementary, necessary and strategic since the technicians in this case have all required knowledge about the products' components and through the seminars acquire new knowledge about the specific characteristics.

In cell II of Figure 7, both the professional training and operations management are complex. Their concurrent role is complex since the links among them are generic and in relation with all the other types of professional training and operations management. For example, the diagnosis IT-tool and problem-solving system are also related to the courses in which the technical competences of the technicians are developed about the products' components. The technicians base their use of the diagnosis IT-tool and problem-solving system on the technical knowledge acquired during the professional training about these tools and systems and about the products' components. The complexity of the concurrent role of professional training and operations management here is that first the technicians are subject to professional training related to products' components and technical competences development. Second, they follow seminars for the optimal use of the IT-based system and tool for the after-sales services. Finally, the integration of the technical competences acquired and the knowledge about the use of the IT-system and tool lead to the building of new capacities and make this cell the most complex type of the concurrent role of the professional training and operations management for the after-sales services.

There is another example about the concurrent role of professional training and operations management for the after-sales services (Cell III of Figure 7) concerning the e-manuals and e-service news used as means and tools for the diffusion of the necessary updated data and information for the after-sales services. These are included in the operations management and are technical support contributing to solving the problems in cars. They are also a support for the professional training or vice versa that is the technicians, in order to understand them, have to follow the courses and practical experiments for the after-sales services and can use them not only for the after-sales services but also as a support for the professional training course. Thus, from here, we can also see the complex concurrent role of the operations management and the professional training.

8. Conclusion and future directions

The goal of this chapter was to explore what the concurrent role of professional training and operations management is. More precisely, to demonstrate it through the presentation of the information system architecture dedicated to the after-sales services. The idea was to connect the professional training and operations management in the automotive industry by presenting the case of this life-long sustainable collaboration among the large automotive company and the after-sales services firms, and their efforts to keep competitive advantage and customer satisfaction within the dynamic and complex environment that this is.

The concurrent role of professional training and operations management have been demonstrated through the presentation of the professional training and operations management separately, the illustration of the information systems architecture that provide an overall more elaborated and clear schema and the derivation of the types of strategic changes in it, and the study of the connections among professional training and operations management through the analysis of the complexity of the different types of links (Figure 8).

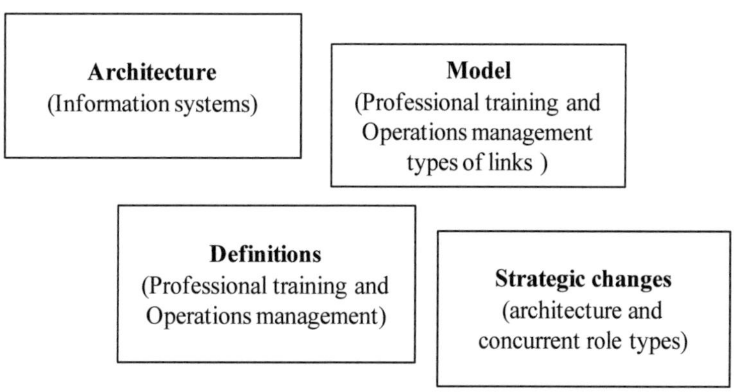

Fig. 8. Research devise and blocks

Most of the time, information system architectures are not only internal to firms but also inter-organizational and this chapter is providing the information systems architecture among two major players in the product and new product development processes and more precisely in the automotive industry that is a complex and dynamic industry. The information system architecture and the main subject of the chapter that is the concurrent role of the professional training and operations management provide IT-based commands and a new theory contributing to the inter-organizational processes of collaboration, the information systems fields of study, the product development and value chain processes of collaboration.

Besides, the four types of concurrent role of professional training and operations management provided in this chapter shows that the link among them can go from simple to complex and give birth to different types of trainings adapted to the operations management of the after-sales services. The simultaneous role of the professional training and operations management is demonstrated through these types of relationships ranging from simple to complex and do show that these have an intertwined role playing a strategic role for customer satisfaction, high economic performance of both the firm and the after-sales services and thus competitive advantage to them. The model contributes to the integration of different fields of research that are competences management, human resources management, and services operations-more precisely the after-sales services.

Moreover, the strategic changes in the concurrent role of professional training and operations management are directly linked to the fields of research on the business processes re-engineering fields, innovation management and reforming the competences development management since the changes in the concurrent role are depending on the changes in the products and services and on the complexity of the environment firms to be continuously innovative thus create new knowledge for that.

Finally, the after-sales services optimization, that is the main purpose of the concurrent role of professional training and operations management, is leading to the customer satisfaction, high economic performance and contributing to the sustainability of competitive advantage (Figure 9). For that, the concurrent role of the professional training and operations management rely on four main success factors that are the people involved in both the professional training and operations management, the strategy followed to design both of them, the processes and systems used that are mainly high-technological based and the structures created for the organization of the professional training, the operations and their management. These four success factors are strategic and contribute to all the areas in the disciplines of professional training and operations management research.

The concurrent role of professional training and operations management is not limited to the relationships between manufacturing and after-sales services firm, thus this subject can be extended to the relationships among suppliers of products' components, designers and manufacturing firms. Also, the book chapter subject can be investigated in other complex industries for the comparison of the strategies, processes and systems and the structures created or other elements constituting the concurrent role of professional training and operations management. It will be interesting to make research on the information and communication technologies used in other industries on the subject or on the relationships showing the concurrent role among different actors in other industries.

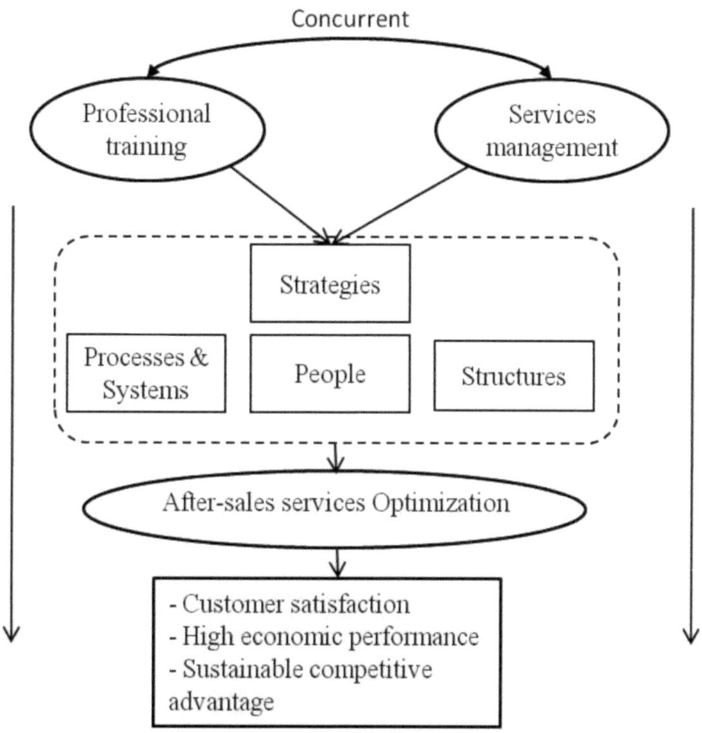

Fig. 9. The concurrent role success factors leading to after-sales services optimization

9. References

Adam, E.E. and Swamidass, P.M., 1989, Assessing Operations Management from a Strategic Perspective, *Journal of Management*, Vol. 15, N. 2, (June 1989), pp (181- 203).

Alexander, W.L., Dayal, S., Dempsey, J.J. and Vander Ark, J.D. (2002), The secret life of factory service centers, *The McKinsey Quarterly*, Vol. 3, pp.106–115.

Allan, M. And Chisholm, CU., (2008), Achieving engineering competencies in the global information society through the integration of on-campus and workplace environments. *Industry and Higher Education*, Vol. 22, N. 3, pp. 145-152.

Ambjörn, N., Miguel-Angel, S. and Lytras, M., (2008), Learning processes and processing learning: from organizational needs to learning designs, *Journal of Knowledge Management*, Vol. 12, N. 6, pp. 5-14

Angell, L.C. and Klassen, R.D. (1999), Integrating environmental issues into the mainstream: an agenda for research in operations management, *Journal of Operations Management*, Vol. 17, Issue 5, pp. 575- 598.

Argyris, C. and Schön, D. 1996, Organizational learning II: Theory, method and practice, Reading, Mass: Addison Wesley.

Aurich, J.C., Fuchs, C. and DeVries, M.F. (2004), An approach to life cycle oriented technical service design, *CIRP Annuals – Manufacturing Technology*, Vol. 53, Issue 1, pp. 151-154.

Baets WRJ, van der Linden G. 2003. Virtual Corporate Universities: A Matrix of Knowledge and Learning for the New Digital Dawn; Norwell, MA: Kluwer Academic Publishers.

Blanchard, B.S. (2004) System Engineering Management, Hoboken, New Jersey: John Wiley & Sons, Inc. ISBN: 0471291765.

Cisternino, V., Campi, E., Corallo, A, Taifi, N. and Zilli, A., 2008, Ontology-based knowledge management systems for the new product development acceleration: Case of a community of designers of automotives, *IEEE proceedings of the 1st KARE workshop in the 4th SITIS*. Nov. 30- Dec. 3, 2008, Bali, Indonesia

Cohen, W. M. (1990), Absorptive Capacity: A New perspective on Learning and Innovation, *Administrative Science Quarterly*, Vol. 35, N. 1, Special Issue: Technology, Organizations and Innovation, pp. 128-152.

Corallo A., Laubacher R.J., Margherita A. and Turrisi G. (2009) "Enhancing Product Development through Knowledge Based Engineering (KBE): a Case Study in the Aerospace Industry", *Journal of Manufacturing Technology Management*, 20(8): 1070-1083.

Corallo, A., Magherita, A. and Pascali, G. (2010a), Digital mock-up to optimize the assembly of a ship fuel system, *Journal of Modelling and Simulation of Systems*, Vol. 1, Issue 1, pp. 4-12.

Corallo, A., Lazoi, M., Taifi, N. and Passiante, G. (2010b), Integrated systems for product design: the move toward outsourcing, *proceedings of the IADIS international conference on information systems*, 18-20 March, Porto, Portugal.

Corallo A., Lazoi M., Margherita A. and Scalvenzi M. (2010c) "Optimizing Competence Management Processes. A Case in the Aerospace", *Business Process Management Journal*, 16(2): 297-314. Online ISSN: 1737-9385

Coughlan, P. and Coghlan, D, 2002, Action research for operations management, *International Journal of Operations & Production Management*, Vol. 22 Issue 2, pp.220 – 240.

Creswell, JW. 2003. Research Design; Thousand Oaks, CA: SAGE Publications.

Kundra, T.K., Tewari, N.K. and Rao, P.N., 1993, Computer aided manufacturing, New Delhi: Tata McGraw-Hill Publishing Company Limited, ISBN: 9780074631034.

Lepak, D.P. and Snell, S.A. (1999), The Human Resource Architecture: Toward a Theory of Human Capital Allocation and Development, *Academy of Management*, Vol. 24, N. 1, pp. 31-48.

Lucertini, M., Nicol, F., Rossetto, S., Telmon, D., Ukovich, W. and Villa, A., 1995, How to evaluate integration between operations management and organizational structures in CIM design, *Control Engineering Practice*, Vol. 3, Issue 1, pp. 51-58.

Maglione, R. and Passiante, G. (2009), Towards 'Stakeholders University' as a dynamic learning network: the Finmeccanica case, in Dynamic learning networks: Models and Cases in Action, Romano, A. and Secundo, G. (eds), Springer.

Morelli, N. (2002), Designing product/service systems: a methodological exploration, *Design Issues*, Vol. 18, N. 3, pp. 3-17.

Nonaka, I. and Takeuchi, H. (1995), The Knowledge Creating Company. New York: Oxford University Press. ISBN 0195092694

Paton, R., Peters, G., Storey, J., Taylor, S. (2005), Handbook of Corporate University Development : Managing Strategic Learning Initiatives in Public and Private Domains, Ashgate Publishing.

Ramaswamy, R. (1996), Design and management of service processes, Reading, Mass, Addison-Wesley Pub Co.

Robert Johnston, (1999) "Service operations management: return to roots", *International Journal of Operations & Production Management*, Vol. 19 Issue: 2, pp.104 - 124.

Roth, A. V. and Menor, L.J. (2009), Insights into service operations management: a research agenda, *Production and operations management*, Vol. 12, Issue 2, pp. 145-164.

Saccani, N., Johansson, P. and Perona, M. (2007), Configuring the after-sales services supply chain: a multiple case study, *International Journal of Production Economics*, Vol. 110, pp.52–69.

Sakao, T. and Shimomura, Y. (2007), Service Engineering: a novel engineering discipline for producers to increase value combining service and product, *Journal of Cleaner Production*, Vol. 15, Issue 6, pp. 590-604.

Salvendy, G., 2001, Handbook of Industrial Engineering: Technology and Operations Management, Canada: John Wiley & Sons, ISBN: 0471330574.

Scalvenzi, M., Corallo, A., Lazoi, M. and Taifi, N. 2008, Efficient IT-based competence management: an aerospace company case-study, *presented at the 10th AISS*. July 3-5, 2008, Ostuni, Italy.

Slack, N., Chambers, S. and Johnston, R. (2007), Operations Management, fifth edition ISBN, 9780273708476. Prentice Hall Financial Times, London.

Taifi, N. (2007) "The Virtual Communities Knowledge Integration in the New Product Development of the Extended Enterprise: Case of a Large Automotive Company", *Paper presented at the 3rd Workshop on Visualizing, Measuring, and Managing Intangibles and Intellectual Capital, Ferrara, Italy,* October 2007.

Taifi, N. (2008a) "The Strategic Orientation of the Managerial Ties: The Emergence of a strategic community from a managerial network", *PhD thesis in eBusiness Management*, eBMS-ISUFI- University of Salento, Lecce, Italy.

Taifi, N. (2008b) "Collaborative Knowledge Networks: Lessons to learn from a Large Automotive Company, *LNCS proceedings of the 3rd IFIP/FIDIS summer school*, Vol. 262, pp. 423-429. Aug. 6-10, 2007, Karlstad, Sweden.

Taifi, N. and Passiante, G. (2010a) "Speeding up products and services development through strategic community creation: case of an automaker after sales services partners, *proceedings of INBAM conference*, June 1-4, 2010, Valencia, Spain.

Taifi, N. and Passiante, G. (2011), Speeding up NPSD through strategic community creation: case of an automaker after-sales services partners, Special Issue: 'Innovation and Creativity in Services', Service Industries Journal (forthcoming).

Taifi, N., Campi, E., Cisternino, V., Zilli, A., Corallo, A. and Passiante, G. (2011), Technology Engineering For NPD Acceleration: Evidences From The Product Design, In: Semantic Web Personalization and Context Awareness: Management of Personal Identities and Social Networking", IGI Global.

Taifi, N., Lazoi, M., Corallo, A., Monaco, M. and Passiante, G. (2012), Integrated systems for product design outsourcing: A case study of an aerospace firm and its partner, International Journal of Innovation and Technology Management (Forthcoming).

Teece, D., Pisano, G. and Shuen, A., 1997, Dynamic Capabilities and Strategic Management, *Strategic Management Journal,* (18:7), pp.509-533.

Voss, C., Tsikriktsis, N. and Frohlich, M. (2002) "Case research in operations management", *International Journal of Operations and Production Management,* Vol. 22 Issue: 2, pp.195 - 219.

Yin RK. 2003. Case Study Research: Design and Methods; Thousand Oaks, CA: SAGE Publications.

5

Human Factors, Ergonomics Model and Application in Automotive Industries: Focus on Job Satisfaction

Siti Zawiah Md Dawal[1], Zubaidah Ismail[2], and Zahari Taha[1]

[1]*Department of Engineering Design and Manufacture, Faculty of Engineering, University of Malaya, 50603 Kuala Lumpur,*

[2]*Department of Civil Engineering, Faculty of Engineering, University of Malaya, 50603 Kuala Lumpur, Malaysia*

1. Introduction

Industrial work design is defined as a specification of work content, method and relationships to satisfy the requirement of the worker and the system. Industrial work design must be developed as an integrated whole, taking into consideration the inter-dependencies among skills, organization and technology. On the other hand, the concept of job satisfaction is typically defined as an individual's attitude about work roles and the relationship to worker motivation. There can be no job satisfaction where there is no motivation.

The most important evidence that indicates the worsening conditions of an organization is the low rate of job satisfaction. Thus job satisfaction is the key to establishing a healthy organizational environment in an organization. Nonetheless, factors related to job satisfaction are relevant in the prevention of employee frustration and low job satisfaction because employees will work harder and perform better if they are satisfied with their jobs.

Many factors affect job satisfaction according to Bowen et al, 1994, DeSantis and Durst, 1996 and Gaesser and Whitbourne, 1985. Although numerous studies on the effect of job satisfaction in industries exist, findings were often specific to the particular investigation, and to date mainly consider individual components of the physical environment (Clegg et al, 1997).

Nadin et al, 2001 stated that many researchers have suggested a number of workplace design strategies to improve the quality of the workplace. In contrast, Oldham, 1996 stated that little attention is given to the actual process of workplace design. Clegg, 1995 suggested that methods should be developed to facilitate this process; and to do so, a more thorough understanding is needed about how various factors affect industrial job satisfaction. Workplace design research can make progress by applying what is already known and by asking a more comprehensive set of research questions (Holman et al, 2002). An approach to the design of workplaces that is human centered is needed. This approach should adequately incorporate factors that contribute to the effectiveness of workplace design.

2. Background materials

Ergonomically, job satisfaction can be affected by job characteristics, job environment, and job organization. The relationships between job satisfaction and these factors can be analyzed statistically using correlations to provide new insights into factors that affect job satisfaction in automotive industries. The Job Diagnostic Survey or JDS by Hackman and Oldham (1974) can be used as a tool to diagnose job characteristics, job environment and job organization.

2.1 Job characteristics

Hackman and Oldham (1974) divided job characteristics into five factors. They were skill variety, task identity, task significance, autonomy, and feedback from job.

Firstly, skill variety was defined as the degree to which a job requires a variety of different activities which involve the use of a number of different skills and talents. Secondly, the degree to which a job requires completion of a "whole" and identifiable piece of work was defined for task identity. Thirdly, task significance was defined as the degree to which a job has a substantial impact on the lives or work of other people, whether in the immediate organization or in the external environment. Fourthly, the degree to which the job provides the employee substantial freedom, independence and discretion in scheduling the work and in determining procedures to complete it was the defined as autonomy. Finally, feedback from job was defined as the degree to which completing work activities required by the job results in the employee obtaining direct and clear information about the effectiveness of his or her performance.

2.2 Job environment

Air temperature, humidity, noise, and light were four environmental factors included that could influence job satisfaction. Parsons (2000) stated that environmental factors such as temperature and humidity can have important effects on psychological parameters such as level of arousal and motivation. To assess thermal comfort, JDS's questionnaire developed adopts the definitions of Parson (2000) as "the condition of mind which expresses satisfaction with the thermal environment". The reference to "mind" indicates that satisfaction is a subjective measure. However, warmth discomfort has been shown to be related to the stickiness caused by un-evaporated perspiration. Consequently, questions regarding thermal comfort addressed satisfaction and comfort.

Noise levels can also affect worker satisfaction. The term comfort is not usually used when assessing the effect of noise on the occupants of the buildings. In practice, again according to Parsons, (2000), annoyance levels are the most useful criterion. Thus, questions regarding noise addressed annoyance and comfort.

Lastly, Parsons (2000) stated that light can cause both discomfort and positive sensations. Questions regarding light addressed satisfaction and the degree of comfort in seeing the work task.

2.3 Job organization

Das (1999) stated that the organization factors such as workers' participation in job related decision self-regulation and worker autonomy can affect job satisfaction. The questionnaire addressed respondents' perception of these factors.

De Jong (1989) classified five factors for job organization and they were job rotation, work method, training, problem solving, and goal setting. Job rotation allows workers to rotate among jobs to increase variety. This technique has been widely used to increase the competence of workers and to reduce monotony (Helander, 1995). The objective of job rotation is to broaden an employee's experience and to train backup staff to allow the company to cope with worker vacations and illnesses, and also with periods of increased production (The Ergonomics Group, 1986).
Work method describes how tasks are being organized (Rouse et al., 1991) and according to Quirk (1999), the methods could include procedures, instructions and documentation that define how manufacturing steps or processes are accomplished. Training is defined as the systematic development of worker skills. Individuals need knowledge and skills to perform adequately on a given task (Stammers and Patrick, 1975). Problem solving describes how the workers handle work related problems by giving them the resources and authority to do so (Ugboro and Obeng, 2001), whereas goal setting is the process of developing, negotiating and formalizing the targets or objectives that an employee is responsible for accomplishing (Umstod et al., 1976).

3. Case study: automotive industries in Malaysia

A survey was done to discover new insights into factors that affect job satisfaction in automotive industries in Malaysia. The primary objective of this survey was to investigate how job satisfaction is affected by job characteristics, job environment and job organization. The methods used in this survey to address the objectives include questionnaire design, observation, measurements, and data collection. The job diagnostic survey (JDS) by Hackman and Oldham (1974) was used as a tool to diagnose job characteristics, job environment and job organization. The questionnaires used consist of a set of multiple-choice items that use a seven-point Likert scale (Rodeghier, 1996). The relationships between job satisfaction and the tested factors were analyzed statistically using correlations.

3.1 The survey

The questionnaires were distributed to the subjects individually. Two automotive manufacturing industries ("Auto1" and "Auto 2") were involved in the survey; 170 male subjects (ages 18 to 40 years) completed the survey, where 120 were from Auto1 and 50 were from Auto2.

3.2 The questionnaires

The questionnaires were organized into four sections covering:

a. General background data: age, gender, years of employment, marital status and education levels.
b. Job characteristics factors: skill variety, task identity, task significance, autonomy and feedback from the work.
c. Environmental factors measured at each workstation: air temperature, humidity, noise in decibels (dB) and light in lux.
d. Job organization: job rotation, work method, training, problem solving and goal setting. This section intends to determine how the respondents felt about the tasks being organized, the type of procedures being used, and the related work being loaded.

Before initiating the questionnaire sessions and as a reference for the analysis, the management was first interviewed and a checklist was made.

3.3 The analysis

The data were analyzed for correlations using the Spearman rank order correlation. To test the reliability of each question in the survey, reliability of factors tested in the survey was quantified using Cronbach's α. This statistic is derived from the average correlations of all items on the scale, which will measure the internal consistency of the test scores. As a rule of thumb, values that are greater than 0.70 will indicate that the questions are reliable.

4. Results and discussion

The results were divided into several sections covering general background data, reliabilities measures, and correlations of job satisfaction with job characteristics, environment and job organization factors.

4.1 General background data

Eighty percent of the 170 respondents in both companies hold "Malaysian Certificate of Education" (SPM) equivalent to "O" levels; while others hold SPM certificate together with other skill certificates.

Respondents in Auto1 were older and more experienced than those in Auto2. The respondents from Auto1 were 23 to 40 years old (mean = 31.3, s.d. = 3.9; 83% ≥ 26 years) with a mean of work experience of 10.6 years (s.d. = 3.8); those from Auto2 were 18 to 27 years old (mean = 22.6, s.d. = 2.1; 10% ≥ 26 years) with a mean of work experience of 2.6 years (s.d. = 1.8). In Auto1, 69% were married, but in Auto2 only 13% were married. Ages were normally distributed but work experience was not.

Work experience for Auto1 was negatively skewed but work experience for Auto2 was positively skewed. At Auto1, 90% of the respondents from Auto 1 had worked for there for more than 5 years, but at Auto2, only 10% had worked there for more than 5 years. These demographic differences between the plants may occur because Auto1 was established before Auto2.

4.2 Reliabilities measure

Cronbach's α was derived from the average correlations between the Likert-scale assessments of all pairs of items (Rodeghier, 1996). Out of twenty-eight reliability measures in both companies, 18 had $\alpha > 0.7$, 9 had $0.6 < \alpha < 0.7$ and one had $\alpha = 0.5$ (Table 1). Therefore, the reliability measures were high for job factors in both companies, especially for skill, task identity, autonomy and feedback ($0.69 < \alpha < 0.88$).

4.3 The correlation coefficient

Job satisfaction was significantly ($p < 0.01$) correlated with several job characteristics, environment and job organization (Figures 1 to 3). Eight factors had strong significant correlation in with job satisfaction Auto 1: skill variety, task identity, autonomy, light, job rotation, work method, training and goal setting. Four factors had strong significant correlation with job satisfaction in Auto 2: skill variety, humidity, job rotation and work method.

Tested Factors	Auto1, n = 120	Auto2, n = 50
Job factors		
Skill	0.77	0.79
Task identity	0.76	0.69
Task significance	0.61	0.63
Autonomy	0.72	0.69
Feedback	0.86	0.88
Environmental factors		
Perception of temperature	0.72	0.69
Perception of humidity	0.64	0.86
Perception of noise	0.76	0.67
Perception of light	0.78	0.64
Job organization factors		
Job rotation	0.83	0.83
Work method	0.89	0.88
Training	0.83	0.50
Problem solving	0.69	0.79
Goal setting	0.90	0.82
Job satisfaction	0.89	0.82

Table 1. Reliability Measures (Cronbach's α).

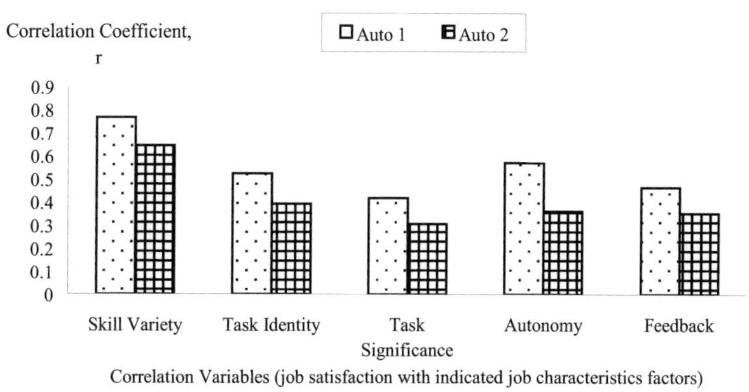

Fig. 1. Correlations of job satisfaction with five job factors

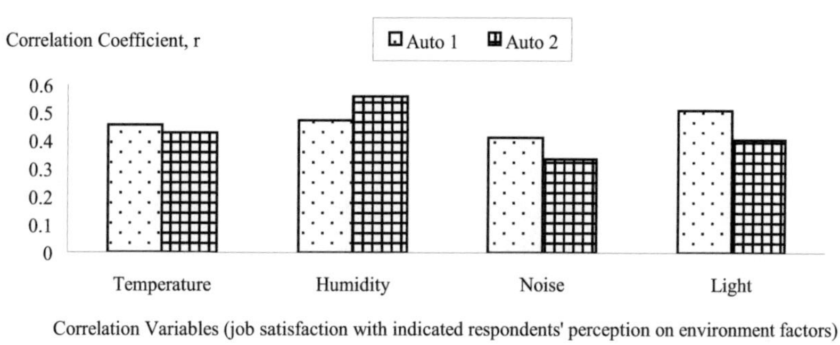

Fig. 2. Correlations of job satisfaction with four environmental factors.

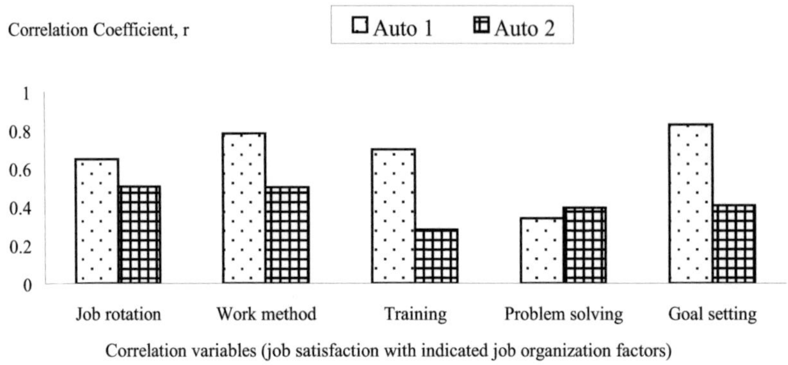

Fig. 3. Correlations of job satisfaction with job organization factors.

4.4 Effects of job characteristics, environmental factors, and job organization on job satisfaction

The study detected significant positive correlations between job satisfaction and several job characteristics. This result was in agreement with those of empirical studies by Hackman and Oldham (1974) and Umstod et al., (1976). In this study, the correlations of job satisfaction with job characteristics were stronger than those observed by Hackman and Oldham (1974), possibly because the present study considered only automotive industries.

One outstanding result was that job satisfaction was significantly correlated with skill in both companies. Generally, more than 80% of respondents agreed that they utilized moderate to very much skill. Based on this finding, they tend seem to appreciate skill variety, and this variety has a greater influence on job satisfaction than other job factors. Hackman and Oldham, (1974 and 1976) stated that skill, task identity and task significance are psychological factors that help workers see their work as meaningful. However, results from this study suggest that skill variety had greater effect on that matter than did other factors.

Significant positive correlations occurred between job satisfaction and perception of all environmental factors as shown in Figure 2. The outstanding correlation for Auto1 was perception of light and for Auto2 was perception of humidity. The results indicated that environment conditions, especially temperature, humidity, noise and light affect job satisfaction in automotive industries. The management of both companies should attempt to optimize temperature, humidity and noise because measurements of these factors are outside the comfortable boundary and respondents are not satisfied with them. Standard environmental conditions (including temperature, humidity, noise, and light) for automotive industries in Malaysia must be revised to maintain workers' health physically and mentally, thereby increasing productivity and job satisfaction as well as performance.

Significant positive correlations were observed between job satisfaction and job organization factors. In Auto1, job satisfaction was strongly correlated with job rotation, work method, training and goal setting showed strong correlations with job satisfaction, but only intermediate correlation with problem solving. In Auto2, job satisfaction showed intermediate correlations with most job organization factors, except for the training factor, for which the correlation was low. The correlations for all factors were higher in Auto1 than Auto2.

The findings indicated that job satisfaction is affected by job rotation, work method, training, goal setting and problem solving. More than 70% of the respondents in both companies were satisfied with the implementation of job rotation, work method, problem solving and goal setting. In contrast, more than 80% of the respondents in Auto1 felt that they have moderate to adequate training and only 55% of the respondents in Auto2 felt the same way. In addition about 30% of workers in Auto2 felt that they have training opportunities, but only 5% respondents in Auto1 felt the same way. Management therefore should emphasize training opportunity, because the result reflected a decrease in job satisfaction with too training much Auto2. Moderate to adequate training will lead to higher job satisfaction (Gaziolu & Tansel, 2002).

4.5 Effects of age, work experience and marital status on job satisfaction

The correlations between job satisfaction, job characteristics and job organization factors are higher in Auto1 than in Auto2 (Figures 1 and 3). One possible explanation is that older, married and more experience workers in Auto1 were more satisfied with their work than the younger, single and less experienced workers in Auto2. Older, married and more experienced workers had higher levels of job satisfaction and are more committed than the younger, single and less experienced men; furthermore, younger, single and less experienced workers may still be deciding on a career and this may interfere with job satisfaction and organizational commitment (Bowen et al. 1994). Benin and Nienstedt (1985) found that job satisfaction influenced marital happiness and that the effects of job satisfaction and fulfillment interacted with the effects of marital happiness in producing overall happiness.

Research on relationships between work satisfaction and marital characteristics is extensive and is primarily found in literature on marital satisfaction, work identity and satisfaction, and dual-career couples (Blair 1998, Ray 1990, Gaesser & Whitbourne 1985). These studies suggested that career and family lives are mutually entangled, and that to understand strain in one domain, information on both facets of an individual's life is necessary (Ludlow & Salvat 2001). Therefore further research should be conducted into this interaction.

Job satisfaction was also correlated with worker age. Studies in five different countries prove that older workers are more satisfied than their younger counterparts (Kaya 1995). The results also supported findings by Janson and Martin (1982) and McCaslin & Mwangi (1994) who found that older employees have higher job satisfaction than younger ones, and those by Lee & Wilbur (1985) which suggested that job satisfaction increases with age. One explanation for such a finding is that older employees are more able to adjust their expectations to the characteristics of their work (DeSantis & Durst 1996). The lack of job satisfaction amongst younger workers may cause them to be more mobile and seek new jobs. If this occurs in Auto2, the plant will experience a shortage of skilled and experienced workers.

Work experience is only one of the many aspects related to length of employment that can be correlated with perceived job satisfaction. Bowen et al. (1994), McCaslin & Mwangi (1994), Manthe (1976), Boltes et al. (1995) and Bertz and Judge (1994) found that overall job satisfaction increased as the years of experience increased.

5. Conclusions

This study found that job satisfaction was significantly correlated with job characteristics, environment, and job organization. The strength of the correlation between job factors and job satisfaction was influenced by age, work experience and marital status whereas job satisfaction was only significantly correlated with environmental and job organization factors. The environmental factors did affect job satisfaction and the strength of the correlation was influenced by the workers' surroundings, depending on the function of the building.

This study established that skill variety had an outstanding effect on job satisfaction in the automotive industries.

6. References

Amrine, H.T, Ritchey, J.A, Moodie, C.L and Kmec, J.F. (1993). Manufacturing Organization and Management, Prentice Hall, Inc.

Bedeian, A.G, Farris, G.R and Kacmar, K.M. (1992). Age, tenure and job satisfaction : A tale of two perspectives. Journal of Vocational Behavior, 40(1), 33 - 48.

Benin, M.H., and Nienstedt, B.C. (1985). Happiness in single and dual-earner families: The families of marital happiness, job satisfaction and life cycle. Journal of Marriage and the Family, 47(4), 975 -984.

Bertz, R.D.Jr. and Judge, T.A. (1994). Person - organization fit and the theory of work adjustment: Implications for satisfaction, tenure and career success. Journal of Vocational Behavior, 44(1), 32-54.

Blair, S.L. (1998). Work roles, domestic roles and marital quality: Perceptions of fairness among dual-eraner couples. Social Justice Research, 11(3), 313-335.

Boltes, B.V, Lippke. L.A, and Gregory, E. (1995). Employee satisfaction in Extension: A Texas study. Journal of Extension [online]. 33(5).

Bowen, C. F., Radhakrishna, R., and Keyser, R. (1994). Job satisfaction and commitment of 4-H agents. Journal of Extension [online]. 32 (1).

Clegg, C.W. (1995) Sociotechnical theory. In N. Nicholson (Ed.), Encyclopaedic dictionary of organizational behaviour. Oxford: Blackwell.

Clegg, C.W., Axtell, C.M., Damodaran, L., Farbey, B., Hull, R., Lloyd-Jones, R..Nicholls, J., Sell, R., Tomlinson, C. (1997) Information technology: A study of performance and the role of human and organizational factors. Ergonomics, 40, 851-871.

Das. B. (1999) Comprehensive Industrial Work Design Model. Human Factors and Ergonomics in Manufacturing, (4), 393 - 411.

De Jong , J.R. (1989) The management of work and effort. In R. Wild (Ed), International handbook of production and operations management, London: Cassell Education Ltd.

DeSantis, Victor and Durst, S. (1996) Comparing Job Satisfaction among Public and Private - Sector Employees. American Review of Publiv Administration, 26 (3), 327-343.

Gaesser, D.L and Whitbourne, S.K. (1985) Work identity and marital adjustment in blue-collar men. Journal of Marriage and the Family, 47(3), 747 -751.

Gazioglu, S and Tansel. A. (2002) Job satisfaction in Britain: Individual and job related Factors. ERC Working paper in Economics. assess on 1.6.04 : http://www.erc. Metu.edu.tr

Hackman J. R and Oldham G.R. (1974) The job diagnostic Survey. Technical Report No. 4, Department of administrative Science Yale University.

Hackman, J.R, OldHam, G.R. (1976) Motivation through the design of work: Test of a theory. Organizational Behavior and Human Performance, 16, 250- 279.

Hamermash, D. (1977) Economic aspects of job satisfaction, in Essays in Labor Market Analysis, edited by Ashenfelter O and Oates W, Toronto: John Wiley & Son.

Helander, M (1995). A Guide to the Ergonomics of Manufacturing. Taylor and Francis.

Holman , D., Chris Clegg, Patrick W. (2002) Navigating the territory of job design. Applied Ergonomics. 33, 197-205.

IES Nomenclature Committee. (1979) Proposed American national standard nomenclature and definitions for illuminating engineering (proposed revision of Z7.1R1973). Journal Illuminating Engineering Society, 9(1), 2-46.

Janson, P & Martin, J.K. (1982) Job satisfaction and age: A test of two views. Social Forces, 60(4), 1089-1102.

Kaya, Ebru. (1995) Job Satisfaction of the Librarians in the Developing Countries. 61st IFLA General Conference.

Lee, R and Wilbur, E. (1985) Age, education, job tenure, salary, job characteristics and job satisfaction: A multivariate analysis. Human Relations, 38, 781 - 791.

Ludlow, L.H and Salvat, R.M.A. (2001) Spillover in the academy: Marriage stability and faculty evaluations. Journal of Personnel Evaluation in Education, 15(2), 111-119.

Manthe, R.D. (1976) A job satisfaction and dissatisfaction study of the West Virginia University Extension Service. Unpublished doctoral dissertation, University of Wisconsin, Madison.

McCaslin, V.L and Mwangi, (1994) J. Job satisfaction of Kenya's Rift Valley Extension agents. Journal of Extension [online],32(3).

Nadin, S.J., Waterson, P.E., Parker, S.K. (2001)Participation in job redesign: An evaluation of the use of a socio technical tool and its impact. Human Factor and Ergonomics in Manufacturing, 11(1) 53-69.

Oldham, G.R. (1996) Job Design. In C.L. Cooper & I.T. Robertson (Eds), International review of industrial and organizational psychology, 11, 33-60. John Wiley & Sons.

O'Reilly, C.A. III and Roberts, K.H. (1975) Individual differences in personality, position in the organization and job satisfaction. Organizational Behavior and Human Performance, 14, 144-150.

Parsons,K.C. (2000)Environmental ergonomics: a review of principles, methods and models. Journal of Applied Ergonomics, 31, 581-594.

Quirk, M. (1999) Manufacturing, Teams, and Improvement: The human art of manufacturing, Prentice Hall, Inc.

Ray, J. Interactional patterns and marital satisfaction among dual-carrier couples. Journal of Independent Social Work, 1990; 4(3), 61-73.

Rodeghier, M. (1996) Survey with Confidence. SPSS Inc.

Rouse, W.B, Cody. W.R and Boff. K.R. (1991) The human factors of System design: Understanding and Enhancing the role of Human Factors Engineering, Human Factor engineering, 87-104.

Stammers, R and J. Patrick. (1975) The Psychology of Training. London: Methuen.

Steadman, R.C. (1979) The assessment of sultriness. Journal of Applied Meteorology, 18(7), 861-884.

The Ergonomics Group. (1986) Health and Environment Laboratories Eastman Kodak Company, "Ergonomics Design For People At Work", Eastman Kodak Company.

Ugboro, Isaiah and Obeng, Kofi. (2001) Managing The Aftermaths of Contracting In Public Transit Organizations: Employee Perception of Job Security, Organizational

Commitment and Trust, Technical Report, North Corolina A&T State University, Greenboro. USA.

Umstot, D.D, Cecil H Bell Jr. and Terence R. Mitchell. (1976) Effect of job enrichment and task goals on satisfaction and productivity. Journal of Applied Psychology, 61 (4), 379 - 394.

Vroom, V. H and Deci, E. L. (1970) Manegement and Motivation, England: Penguin Books Ltd.

6

A Sustainable Service Program for the Automotive Refinishing Industry

Andrea Zavala[1,4], Rafael Moure-Eraso[2,3], Nora Munguía[1] and Luis Velázquez[1,2]
[1]*University of Sonora*
[2]*University of Massachusetts Lowell*
[3]*US Chemical Safety Board*
[4]*Autonomous University of Baja California*

1. Introduction

The automotive body repair and refinishing industry can be considered a subsystem within the complex automotive system. Daily around the world, this industry generates small amounts of pollutants and for that reason, this pollution is not usually considered in the big picture; however, the total environmental impact of a large number of these small generators might be significant. Occupational risks are other sustainability concerns in this industry, exposure to isocyanates and poor work conditions are often present because a lack of safety and health programs in place.

There have been generated a lot of information about the occupational and environmental practices performed in auto body shops from transnational companies such as Ford, GM, Honda, General Motor, Toyota and others where workers performs in safer occupational and environmental settings, but little had been revealed about sustainability practices in small auto body shops from the developing world until researchers from the University of Sonora and the University of Massachusetts Lowell started a collaborative multi-tiered research in Mexico to fill this gap in the knowledge.

This chapter is aimed at encouraging the implementation of a Sustainable Service Program (SSP) in the small auto body shops that holds the exciting potential of preventing, eliminating and/or reducing the environmental and occupational risks associated to this industry while enhancing and strengthening its economic growth.

Evidently, none can expect the small auto body shops has the financial structure for hiring specialists in implementing strategic projects. Often, small auto body shops lack the resources to maintain formal management programs like those described in handbooks. For that reason, the Sustainable Service Program presented in this chapter has been tailored to being implemented at small and medium sized auto body repair where there is a general dearth of adequate managerial and engineering skills for the implementation of ultimate philosophies, techniques, procedures, and rules such as it is done in larger companies. The program was designing with the idea of helping owners in auto body shops who cannot afford expensive consultant fees, but have the support of a higher education institution, or other non-profit organization, through a graduate student or an advanced undergraduate student enrolled in a sustainable development program.

Before beginning a detailed description of the specific stages of the Sustainable Service Program, it is important to explore some important aspects of sustainability on this industry.

In 1987, the Brundtland Commission defined Sustainable Development as "development that meets the needs of the present without compromising the ability of future generations to meet their own needs (WCED; 1987). This concept is conceived as a broad structural approach that takes into account economic, social, and environment aspects (Byrch et al., 2007).

Cleaner production and pollution prevention are concepts used to contribute to sustainable development based on a prevention approach that looks for eliminating risks at the sources. Both concepts are found on literature interchangeably because both are grounded on aiming its efforts at the source of the problem. Bass (2005) documented that the United Nation Environment Program (UNEP) exported the term pollution prevention from North America to Europe in the second half of the decade of 80's; yet, it is very usual to find the terms together such as noun "Cleaner Production and Pollution Prevention".

This approach calls for identifying operations and tasks that potentially could pollute the environment or could pose harm to workers, but where it is created "at the source". On the contrary, the goal of control approach is isolate the worker or environment from the hazard and by this mean avoids the potential damage. For that reason, U.S. EPA (2005) prioritizes prevention over control initiatives.

Professor Ken Geisser (2001) proposes dematerialization and detoxification as pathways of clean production to promote a sustainable material system that reduce the risks to environment and workers. These techniques results important in a cleaner pollution and pollution prevention program.

Pollution prevention programs are usually perceived as a planning process to obtain environmental benefits. According to Bishop (2010) a careful planning is required to ensure the success of pollution prevention activities. Two decades ago, the U.S EPA outlined the major steps needed for planning and implementing a pollution prevention program (U.S. EPA; 1992); this outline has been used for implementing such kind of voluntary programs in the United States of America and world-wide.

A critique to pollution prevention programs is their bias to environmental issues paying inadequate attention to social and economic initiatives such as all related to occupational health and safety. Environmental sustainability is necessary, but it is not enough for sustainable development. A sustainable program conceptually differs from a pollution prevention program on its purpose for potentially integrating economic, environmental, and social dimensions, which include occupational health concerns.

Evidence of prevailing occupational and environmental work practices have been studied since a decade ago (Enander et al., 1998) especially about exposure to chemical hazards and solvents (Woskie et al., 2004 and Enander et al., 2004). Occupational health studies have been conducted mainly in developed countries such as United States, countries within Western Europe and Australia have (Redlich et al., 2001 and Alexanderson et al., 1987).

Workers in automotive body shops daily use materials and products that may contain hazardous components such as isocyanates, solvents, heavy metals among others. The main effects of hazardous exposures are occupational asthma and other lung problems, as well as irritation of the eyes, nose, throat, and skin. (OSHA,2006). According to Di Stefano (2004) and Redlich (2006) spray painters are more likely to develop asthma compared to other occupations. Pronk (2006) and Bello (2004) refers to isocyanates as one of the most common causes of occupational asthma in workers from industrialized countries. The major contributors of exposure to isocyanates group are the polyisocyanates which are inadequately regulated in many work places (Bello, 2004). The isocyanates exposure characterization is often limited to airborne exposure (Liu, 2000). In an epidemiological study conducted in New Haven, U.S. found hexamethylene diisocyanate (HDI) in the blood of auto body shop workers; although there were no records of asthma because this finding, workers reported respiratory problems (Redlich, 2001). The workers on shops faces are also exposed to numerous physical hazards such as burns, cuts from sharp objects, and injury in eyes (CESVI, 2006).

Hazardous wastes in auto body shops are not insignificant (Enander et al., 2003); chemicals not only may harm workers but also they may pollute the environment. A report from the U.S Environmental Protection Agency (2000) estimate this industry emits 70,420 tons per year of volatile organic compounds (VOCs); from those, 11,155 tons per year were generated by small auto body shops. EPA also estimates that more of 150,000 workers in the U.S. auto body shops have been affected by toxic emissions (EPA, 2002).

2. Methods

As previously explained, the sustainable service program has its foundation on a multi-tiered research. The first leg of the study consisted on surveying several Mexican auto body shops to characterizing this industry; researchers gather data about economics, demographics and dynamics trends, occupational and environmental work practices, regulatory framework and current practices, as well as identifying opportunities to improve actual occupational conditions and environmental performance.

The study was conducted in the city of Hermosillo, the Capital of the State of Sonora, Mexico, which has experienced substantial economic growth and industrial development since 1986 when Ford Motor Co. opened a manufacturing plant in the city. The rapid industrial development was accompanied by a fast growth of automobile ownership. According to the 2004 Census, the number of cars on the road in Hermosillo increased from approximately 69,200 in 1989 to 214,000 in 2004, a 3 fold increase (INEGI, 2004) Traffic-related auto accidents also increased by almost 30% from 4,880 in 1999 to 6,275 in 2003 (INEGI,2004). In response to the high rate of auto accidents, the government of Sonora passed a new transit law in January 2006, which enforced third-party liability insurance for all auto operators with residence in Hermosillo. Because this reality; the increased number of cars on the road, accident rates, and the third-party insurance law; it is expected an increment on the job demand and the volume of services in auto body repair shops.

This first phase of the study was based on an inductive reasoning; it consisted on observing specific work practices in 40 Mexican auto body shops; then, generating broader generalizations and theories to create, in the second phase, a Sustainable Service Program framework that responds to the necessities and characteristics of small sized auto body shops.

3. Automotive refinishing industry's profile

The Mexican auto body industry is formed mainly of family-owned business that has little capital to invest in technology; the workers in these shops generally operate in unhealthy working conditions. The average shop is often set up on the back yard of the house, and most of the pre-paint work is done either on the street, the yard, or an open bay area with a bare ceiling/cover for shade without little or no engineering controls for exposure. Improvised enclosure areas are used as spray booths. Such enclosures typically did not have sufficient light and artificial ventilation.

The average shop has been on the market for 13 years and has a workload of 15 cars per month generating a current annual income of US$15,400. The average shop has four full-time employees; the auto mechanics perform only engine repairs, whereas technicians, in addition to body work, typically perform all painting, unless it requires great specialty which it is done by specialized painters. Although technicians spend most of the time on pre-painting preparatory tasks, they preferred to refer to themselves as painters. Nearly all shops participating in the study were located in residential areas.

The labor market relies on personal contacts in Sonora, Mexico to a far greater extent that in the US. The work as well as its management is organized differently as well.

In spite of the fact that tasks in the Sonoran auto body shops are very similar to those described in the literature for the US auto body shops, services are often processed in batches (e.g. mechanical repairs of all cars, then bondo on all, then painting of all at once) rather than in a single car-full repair model.

Work practices are tremendously variable, and written records are rarely maintained and/or inaccessible. It is common practice to hire outside dedicated spray painters to complete jobs as demand arises, especially for specialty jobs. Therefore, painters may work for two or more auto body shops during the same week and in the spare time work on second jobs. Tasks duration as everywhere varied largely depending on the extent of the car's damage.

4. Unsustainability patterns

To ensure that the Sustainable Service Program is connected to the real world; the scheme presented has as a foundation the occupational and pollution prevention findings from this multi-tiered research that have been already published in peer-reviewed international journals (Velazquez et al., 2008 and Munguia et al., 2009).

Key findings of previous tiers suggest the lack of good environmental and occupational and safety practices performed in the auto body shop industry. Today's trends in auto body shops are going in an unsustainable direction, workers work in unhealthy and unsafe settings. In spite of working with well-known hazardous chemicals, notably isocyanates, methylene chloride, welding fumes, and chromium and lead, workers accept these risks as a part of their jobs.

The Sonoran auto body workers usually wear disposable dust masks that offer no protection against solvent vapors and little protection from fine aerosols. Training program on occupational safety and hygiene are needed because the workforce lacks basic education and training on workplace hazards and safer work practices.

Facilities did not have engineering controls and in the best situation, they were inadequate. As a family business, almost all participant shops were located on the back yard of their house, and the pre-paint task was done either on the street, the yard, or an open bay area with a primitive ceiling/cover for shade. Usually, owners improvise an enclosure area as spray booths. These areas do not count with sufficient light, and ventilation.

Environmental practices in the Sonoran auto body shops were found inappropriate for reducing the impact of these small businesses on the environment. Much of the solvent is

emitted to the atmosphere as volatile organic emissions because the lack of enclosed gun cleaning systems. In addition, overspray isocyanates and base coat aerosols may be also released in the surrounding community. Liquid paint waste is typically dumped in the sink or soil, and solid hazardous waste is treated as regular municipal waste. In essence, these shops dump all their pollution in their own backyard.
Housekeeping, handling and storage of materials were found to be significant associated with poor safety conditions. Solvent and isocyanates were commonly stored in inadequate containers which were either not labeled at all, or hand labeled with a marker.
Changing these unsustainability patterns of service in the auto body industry would have implications for many different stakeholders in this process; yet, owners of shops have more accountability. They must assume the responsibility of the negative occupational and environmental effects of their current practices.
The Sustainable Service Program is a management strategy for helping auto body shops to shift unsustainability patterns in this industry through a better understanding of the human health and environmental impacts of materials, processes, and activities.

5. Sustainable service program

The Sustainable Service Program reflects the philosophy of sustainable development understood as a broad structural approach that takes into account economic, social, and environment aspects. The program´s framework is based on a continuous improvement, the plan - do - check - act cycle, which identifies areas of opportunity and then, proposes ideas for solving problems. This iterative process must be repeatedly executed in order to achieve goals. With each succeed iteration, the organization advance toward sustainability.
The Sustainable Service Program is an adaptation of classic Cleaner Production and Pollution Prevention Programs, but it focuses more on the particularities of a small service organization. Tangibility is perhaps the main difference between goods and services production; yet, the production of services requires goods that may impose harm to workers and environment; from this perspective, a Sustainable Service Program does not differ too much from cleaner production and pollution prevention programs because they are concerned with make interventions at the source.
The ultimate goal of a Sustainable Service Program is to prevent, eliminate and/or reduce, at the source, the creations of risks or severe impacts that processes, operations or activities can impose to workers, environment, and society.
Designing a Service Sustainable Program, exhibit in figure 1, involves five stages: management support, planning, implementing, checking, and acting.

Stage 1: Management Support

A typical recommendation when setting a program is to get management support because it is understood that without management support it is very hard to make changes. In succeed organizations, high administration set goals and policies, allocate resources and then, managers at all levels supervise working practices to make sure everything runs according their plan.
Unfortunately, owners in small auto body shops have poor management styles and not to say poor sustainability knowledge. In consequence, gaining management support requires periodical visits to explain them how workers and environment are affected in other auto body shops around the world and how they can reduce these risks and impacts by

implementing such a program. Cases studies from world-wide experiences showing example of good, or even bad, practices are excellent auxiliary materials to create sympathy for sustainability.

In general, once owners are aware of the potential benefits of the sustainable program, they are interested in supporting changes and willing to establish a sustainable service program as a way to reduce its liability and promote sustainability. Additionally, the lack of managerial structure makes easier to start from the scratch which is very difficult in well established companies where organizational structures make decisions from top to down are hard to change.

Sustainability commitment must be written on a formal document called mission statement; yet, it is very likely the lack of missions in auto body shops. Therefore, it is necessary to work in mission statement that will tell customers and groups of interest the inspiration and motivation of the auto body shop for sustainability. This statement will lead future practices not only about sustainability but also about general practices. A basic mission statement can be:

"The name of the shop" is an auto body shop with more than 10 years of service to the community and is committed to promote Sustainable Development by preventing, reducing or eliminating the use of toxic substances that harm the environment or employees through a continuous improvement process.

Like this mission statement can be many other, the importance is to prove a real commitment to take care of workers, environment, and society. Equally important is the definition of sustainability policies that basically are the means to successfully achieve the mission. A policy leads the aim of the goals and objectives and also the procedures to fulfill the goals.

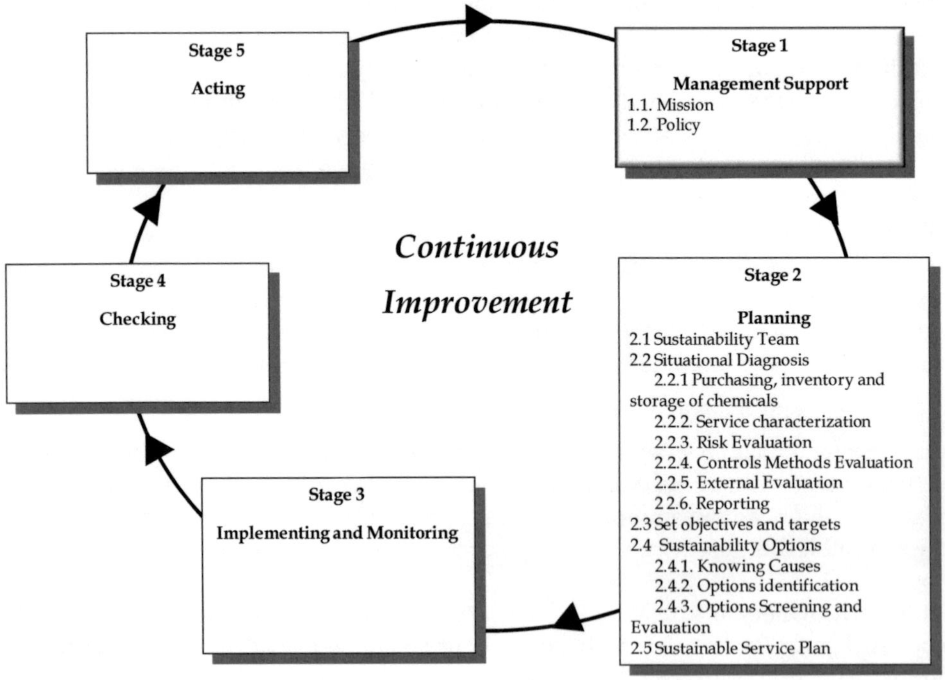

Fig. 1. Sustainable Service Program Scheme

In a typical auto body shop, there are several elements that prevent the establishment of sustainability policies perhaps the main obstacle is its informal organization where owners have the privilege of making all decisions in the shop; however, if the owner has a truly commitment to sustainability he/she will set policies to promote it.

A sustainability policy can be as simple as:

- Whatever possible, it will be bought ecological paint and goods used to repair cars.
- All workers will wear personal protective equipment,
- It is encouraged energy conservation initiatives,
- And so on.

Of course, there is a need for putting emphasis in the enforcement of polices in order to effectively promote sustainability.

Stage 2: Planning

At this point, it is necessary to address the efforts of the team to identify and document the core processes for delivering the service. The product of this stage is a detailed plan regarding future initiatives for improvement environmental and occupational conditions at the shop. The success of the Sustainable Service Program depends on a goon planning; for that reason, this is a critical phase of the program mainly during the first iteration when nothing or almost nothing is known or documented.

Again, the logic to start this phase in a small auto body shops notoriously differ from big organizations because on these organizations high administration set the goals and objectives; then, it is assigned accountability to managers or supervisors to be sure that the goals and objectives are met. In small organization, the creation of the sustainability team is necessary before the establishment of objectives.

The role of the consultant in this stage is to obtain information from the owner and workers and after that, write a formal document which is going to be the written plan. The leader asks the "what" and the "how" and the owner and the workers answer the questions according to their empirical knowledge. Observation of current practices during walkthroughs in the shop is another technique that results useful to identify sustainability opportunities.

The sustainability team is constituted by the owner, the workers, and the graduate student that for her/his sustainability knowledge and skills plays the role of the leader as if he/she were an external consultant. It is a small team whose purpose is to do a diagnosis, set the objectives, look for opportunities, generate alternatives, and evaluate them to hierarchy.

During the first iteration of the program, the leader works along with the owner and the workers overseen the process of creating, maintaining, monitoring, and evaluating the program to be sure it succeed; subsequently, for next iteration, the consultant must left the program on the hands of the owner and workers. If necessary, the consultant intervenes again in specific problematic during the second or following iterations; however, the goal is to eliminate the dependence of the auto body shop's stakeholders by building their capacity.

2.1 Situational Diagnosis

The purpose of the diagnosis is identifying all data that could be helpful in reveal occupational and environmental risks as well as risks to society generated in the auto body shop as a foundation to set goals.

Because at the beginning of the program, during the first iteration of the PDCA cycle, records are not available; it is necessary to conduct a detailed materials accounting and a work practice assessment. Once the first iteration is done and the cycle starts again, data are going to be useful as a preliminary assessment and then, another detailed assessment will be necessary under new conditions.

Understanding how the service is done is indispensable to identify occupational and environmental risks. A service analysis is the first step to performance the service characterization; saying in other words, how inputs are used for producing the service. A material accounting is a complete inventory and assessment of all materials through the process from when they are bought to when they are released or disposal. It is important to be aware that although some materials are easy to track; others, like chemicals, are more difficult because they can be found as constituents in products, raw materials or be present as by-products.

2.2.1 Purchasing, inventory and storage of chemicals

The first target is aimed at identifying general aspects of the processes of purchasing, inventory and storage. The process of buying in shops is very simple; basically, the damage in the automobile determines how much to buy and the owners decide where to buy. Generally, they don't have influence over aspect such quality, packaging or prices.

The automotive refinishing industry usually buy products that contains hazardous chemicals such as isocyanates, solvents, and heavy metals which have the potential to pollute the environment and have adverse effects on workers' health.

Packaging used to contain paints and solvents requires special attention because this is disposal after single use increasing the generation hazardous waste that is hard to reuse or recycle. In addition, the absence of adequate labelling increases the chances of an accident.

Owners must strive to avoid buying toxic and hazardous materials which most of the time are inherent in the materials used to produce the service such as paints and solvents. By toxic, it is understood any substance that pose a harm to humans and environment.

It is desirable the use of minimum inventory level of toxic substances as in a just in time system; in addition, it is necessary to establish a storage procedure for hazardous substances aimed at avoid foreseeable circumstances that may results in potential problems such as spills, fugitive emissions, explosions, fire, or accidents.

Toxic substances must be storage in secure containers and correctly labelled. Compatibility is the criteria for storage toxics substances; chemical with same characteristics are less susceptible to cause dangerous reactions if the container is accidentally torn; yet, the store procedure also has to consider the place when the chemicals are going to be stored to be sure they are going to be manageable, this includes a good housekeeping.

2.2.2 Service characterization

The service is analyzed by using a process flowchart containing symbols to identify the elements of a process; for instance: tasks with rectangles and flows with arrows. In auto body shop there are two core processes: painting and mechanic work; their flowcharts are shown at figure 2 and figure 3.

Once the process has been depicted, the following step is to specify the work activities of workers. Ergonomics considerations are very important to describe the physical arrangement of work stations and tools used by workers to perform their tasks.

Describing work methods is also important to identify occupational risks and also to know waste sources; the leader must ask and observe what is done and how the task is done. The intention is to create a process chart with a clear description of all activities in the core

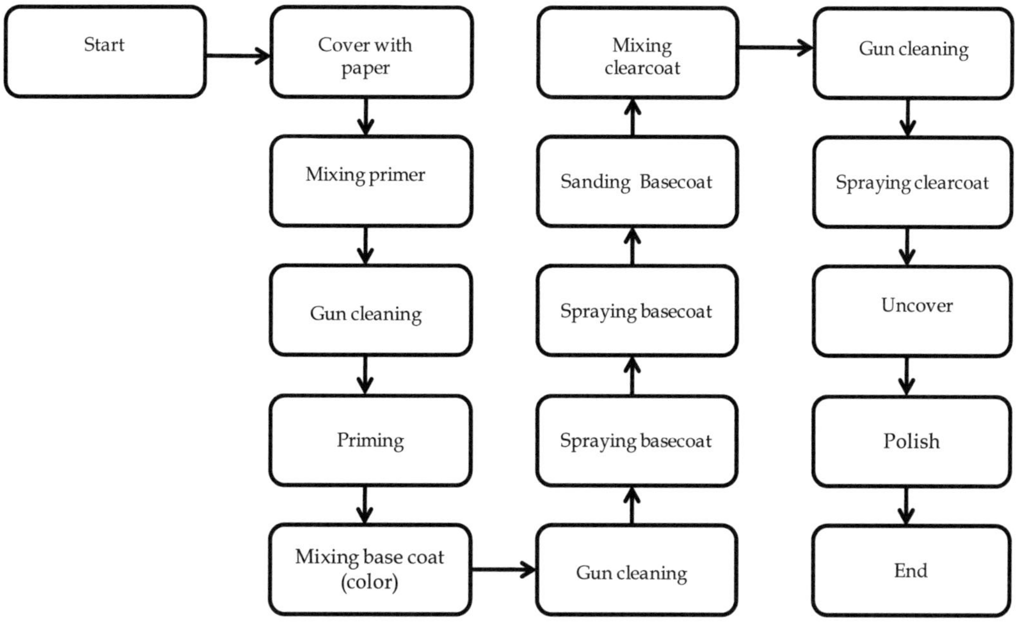

Fig. 2. Painting process flowchart

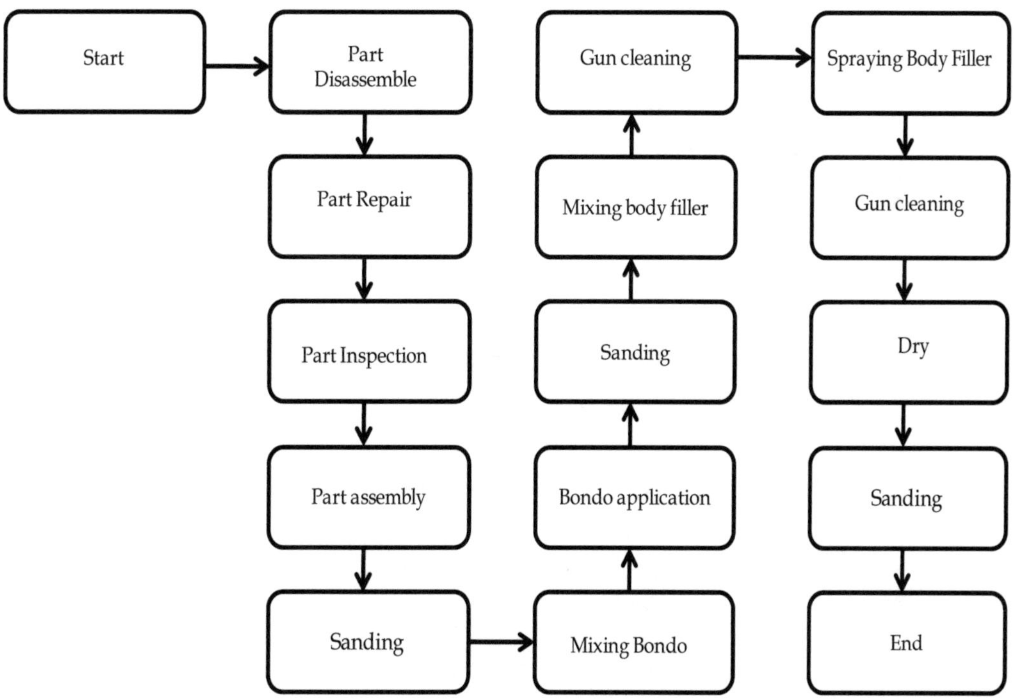

Fig. 3. Mechanic work flowchart

Process Chart

DESCRIPTION	SYMBOL	TIME	DISTANCE	OBS.
Part Disassemble	● ⇨ D □ ▽	40 min		
Part Repair	● ⇨ D □ ▽	25 min		
Inspection	○ ⇨ D ■ ▽	5 min		
Part assembly	● ⇨ D □ ▽	.5 min		
Sandin	● ⇨ D □ ▽	10 min		
Mixing Bondo	● ⇨ D □ ▽	4 min		
Bondo application	● ⇨ D □ ▽	10 min		
Sanding with sandpaper #36	● ⇨ D □ ▽	10 min		
Sanding with sandpaper #80	● ⇨ D □ ▽	30 min		
Drying	● ⇨ D □ ▽	10 min		
Mixing body filler	● ⇨ D □ ▽	5 min		
Gun cleaning	● ⇨ D □ ▽	3 min		
Spraying Body Filler	● ⇨ D □ ▽	5 min		
Gun cleaning	● ⇨ D □ ▽	3 min		
Drying	○ ⇨ ◗ □ ▽	5 min		
Sanding with sandpaper #400 o #1200	● ⇨ D □ ▽	60 min		
Cover with paper	● ⇨ D □ ▽	20 min		
Mixing primer	● ⇨ D □ ▽	5 min		
Gun cleaning	● ⇨ D □ ▽	3 min		
Priming	● ⇨ D □ ▽	5 min		
Gun cleaning	● ⇨ D □ ▽	3 min		
Part transport to the paint area	○ ➡ D □ ▽	5 min		
Mixing base coat (color)	● ⇨ D □ ▽	10 min		
Gun cleaning	● ⇨ D □ ▽	3 min		
Spraying basecoat	● ⇨ D □ ▽	10 min		
Drying	○ ⇨ ◗ □ ▽	60 min		
Spraying basecoat	● ⇨ D □ ▽	10 min		
Sanding basecoat	● ⇨ D □ ▽	60 min		
Mixing clearcoat	● ⇨ D □ ▽	10 min		
Gun cleaning	● ⇨ D □ ▽	3 min		
Sraying clearcoat	● ⇨ D □ ▽	10 min		
Uncover	● ⇨ D □ ▽	5 min		
Drying	○ ⇨ ◗ □ ▽	5 min		
Polish	● ⇨ D □ ▽	30 min		
Final inspection	○ ⇨ D ■ ▽	3 min		

processes with their time requirements. Often activities in auto body shops are non-standardized and non-paced; therefore, characterizing these processes requires studying several services given in different automobiles because the damage varies from one car to another. Figure 4 shows a process chart for core processes in an auto body shop. The characterization of the service concludes studying the interaction of workers with equipment.

2.2.3 Risks evaluation

The service characterization allows a full understanding of the service and consequently, the identification of sources of occupational and environmental risks. The assessment includes not only the identification but also the evaluation of risks. Chemical, mechanical, physical and ergonomics hazards are often found in different magnitude at auto body shops; for the reason, it is necessary to conduct an assessment to estimate the probability to cause harm.

Evidently, chemicals are the biggest concern because they can cause serious adverse health effect or even death. The assessment of chemicals hazards depend on their toxicity, the ability of a substance to produce an unwanted effect, and their hazardous, the probability that chemicals cause poisoning given certain conditions. The most common routes of entry in the shops are inhalation and skin absorption.

Collecting sampling is required for measuring chemicals concentration in the air; results must met specific standards determined to provide a healthy work environment. Once chemicals are discarded, they have the potential to pollute the environment; therefore, it is also necessary to compliance with environmental standards.

Other hazards at auto body shops includes cuts, heat , noise, and bad postures can cause fatigue and musculoskeletal disorders; each of these and other risks must be assess and compare against occupational standards to be sure their magnitude don't represent a threat for workers.

2.2.4 Controls methods Evaluation

This assessment is conducted to identify and evaluate the engineering and administrative controls as well as personal protective equipment on place to protect workers from workplaces hazards.

Engineering controls remove hazards from the work stations or isolate workers from the hazard to avoid damage. They are effective, but have the inconvenient of being expensive. Ventilation, barriers and enclosures are typical examples of engineering controls. Administrative controls are used as complement of engineering controls; they reduce the period of exposition of workers to hazards. It includes: training and education, job rotation, reducing the period of time of exposition to particular risks and other administrative alternatives.

Personal protective equipment is required to provide protection for limited periods of working; often this method is ineffective because workers feel uncomfortable when performing their tasks. Yet, when other methods are not enough to control exposure it is recommendable to wear gloves, respiratory masks, goggles or other necessary equipment.

2.2.5 External evaluation

Service organizations requires interaction with the customers to produce the service; on this context, a Sustainable Service Program cannot not be created, maintained and operated in a vacuum; at the contrary, this must be linked to all stakeholders and other groups of interest even outside of the company.

External factors are those facts over which the auto body shop has not direct control such as the legal framework, clean technology, relations with neighbourhoods and civil protection groups, and so on. The external assessment is often ignored; however, owners need to be aware of external factors because they are constantly changing and might affect the business.

2.2.6 Reporting

The diagnosis has been satisfactorily concluded when the service has been characterized and most of the sustainability risks along the service have been identified and document in a baseline report. The mission, the goal and findings in this report are the basis for establishing objectives and targets.

It is necessary to include in the report a prioritization of occupational and environmental risks based on some criteria such the feasibility to prevent, reduce or eliminate, the potential to harm workers or pollute the environment, risk magnitude, or any other criterion that the team consider useful. Compliance with National Official Standards is critical to prioritize the risks; if there are magnitudes above standards, an objective in the program must be to compliance with all standards.

2.3 Set objectives and targets

As mentioned lines above, the goal of the a Sustainable Service Program is to prevent, eliminate and/or reduce, at the source, the creations of risks or severe impacts that processes, operations or activities can impose to workers, environment, and society. This goal defines a general direction to accomplish the sustainability mission in the long term.

At this point, when diagnosis has revealed areas to focus on, it is possible for the sustainability team the establishment of objectives. Objectives are milestones that serve as specifics guidelines to be met in a short term. Objectives must be clearly stated in the written program; an objective must be understandable, achievable, measurable, and have a specific term to be accomplished. Targets are quantifiable measures for reach objectives; for example:

Objective: "Decrease the generation of hazardous waste this year"

Targets:

- *Reduce hazardous solid waste by 50%*
- *Reduce solvent air emission by 10%*
- *Implement an efficient method of storage for chemicals.*

It is important to avoid operational conflicts by being certain that targets are useful for the accomplishment of objectives; objectives are useful for the accomplishment of the goal, and the goal of the program is consistent with the sustainability mission.

2.4 Sustainability options

2.4.1 Knowing causes

A risk is not eliminate just because it was identified; it is necessary to know the causes that origin it; cause and effect diagrams are helpful to explore causes that result in a single workplace hazard or environmental risk. Preferably, the team must think about all causes and not only the most obvious.

It is necessary to include in the report a prioritization of occupational and environmental risks based on some criteria such the feasibility to be prevented, reduced or eliminated, the

potential to harm workers or pollute the environment, or any other that the team consider useful.
The cause and effect diagram can depict as many causes as the team considers convenient; however, follow the issues in the assessment results a good start. For instance: toxic materials, bad storage, low- efficiency equipment, un-training workers, process, etc. Figure 5 and 6 show the cause and effect diagrams for typical environmental and an occupational risks within an auto body shops.

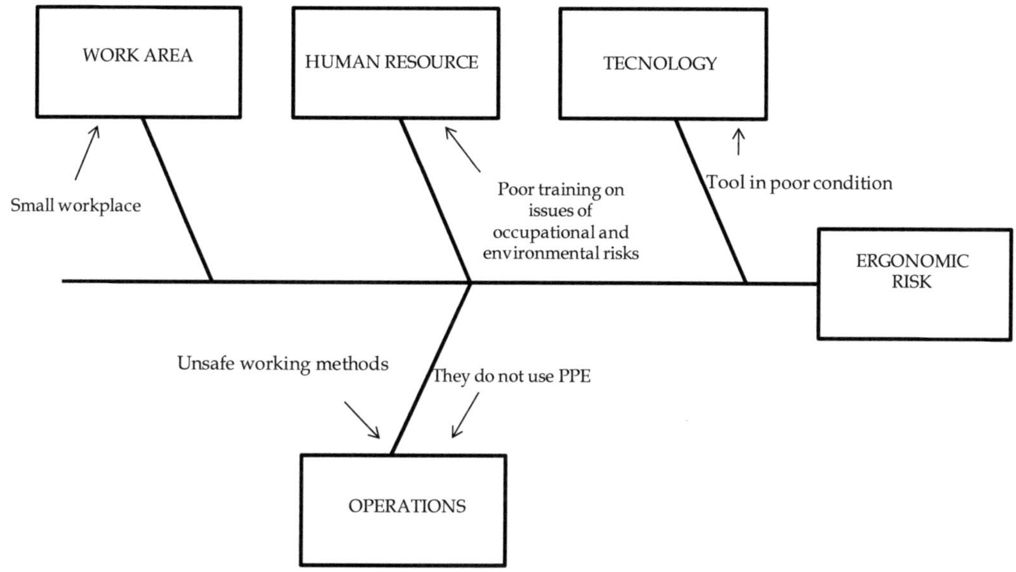

Fig. 5. Cause and effect diagrams for typical occupational risk

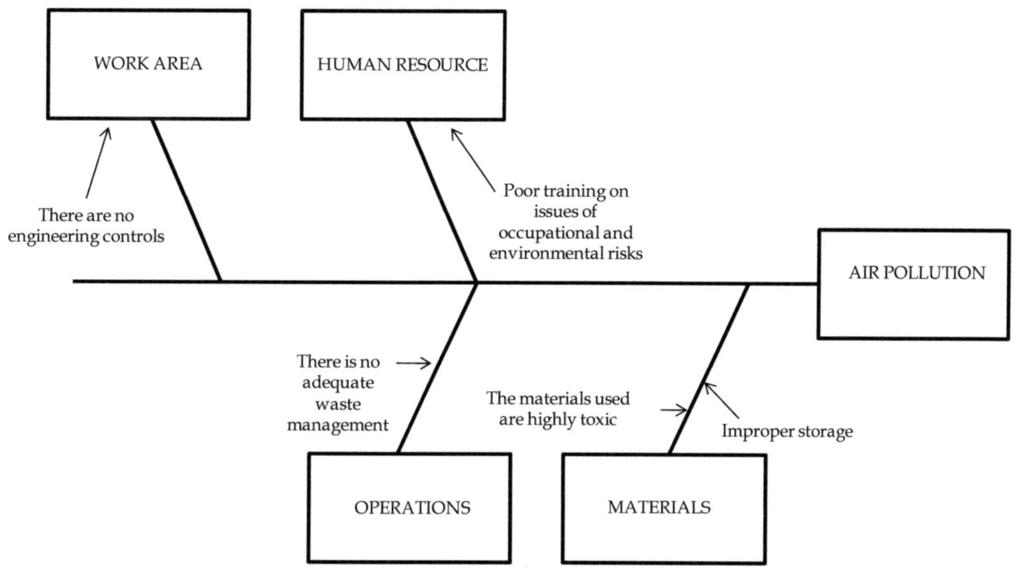

Fig. 6. Cause and effect diagrams for typical environmental risk

2.4.2 Options identification

Once causes that origin the risks are known, the leader must start discussions within the team regarding finding sustainability options aimed at first eliminating and, if this is not possible, reducing their impacts. It is necessary to identify causes for all single risk already identified; however, the generation of sustainability options will be first for the risks that were ranked high in terms of feasibility or importance.
There are several sources for generating option; for instance, brainstorming that allows gathering as many options as possible from the simplest to the most complex. In this technique, creativity and participation are encouraged and no ideas are discarded by any reason, at least, they are not based on sustainable principles; this that options create bigger risks or affect quality. Suppliers are another source for getting good ideas as well as literature on internet, books and others divulgation means.

2.4.3 Options Screening and Evaluation

All identified options must be screened and then narrowed into a prioritized list for each risk. Sustainability options are screening out based on their feasibility, ease of implementation and effectiveness. Prevention is always preferable than control to decide the initial screening.
The options selected must be evaluated for technical, environmental and health, financial, and social feasibility; this is a simple evaluation that consists most of the times only on gathering technical data about options in the market; mainly local market.
Small auto body shops are not technology-intensive shops; technological options range from substitution of raw materials to equipment used to produce the service; inclusive personal protective equipment options are considered in this assessment. For instance, if the option is to switch from a toxic chemical t to another less toxic chemical; then, it is necessary to think about potential obstacles that might affect the implementation of the option such as if the technology option is available at local or regional market, if switching might decrease the quality of the service, if there is evidence on its effectiveness, and so on.
A special attention is also required to assess if the proposed option, chemicals switching, is going to create a new, even worst, environmental or occupational hazard. This is a complicated analysis because the lack of conclusive information on chemicals, even for those chemicals which have been extensively studied such as isocyanates and solvents. Therefore, efforts are addressed to obtain as much information as possible for a better understanding of the option.
If that option is found feasible, an economic assessment is required; often, the option is accepted as long as this is affordable. However, any investment in proposed options, mainly in equipment, requires a justification at least in terms of payback.
The payback period refers to the period of time required to recover an investment; the payback is calculate with the equation: investment/ annual cash inflow; this is a simple and convenient measure of profitability; however, for a better analysis, the payback period method has to be complemented with other method such as the accounting rate of return.
Last but not least, the social assessment provides an opportunity for consulting external stakeholders about the potential affectations derivate when implementing the proposed option. Following the example of the material switching, owners should find what society think about a new paint or if the proposed option is to buy a new compressor for painting, it is necessary to know if noise will bother neighbours. Knowing on advance what is going to

generate conflicts with society enable auto body shops to think twice the implementation of an alternative just taking into account technological and financial aspects.

2.5 Sustainable Service Plan

This is a written plan used as a guide for actions to achieve the goal of the Sustainable Service Program. There is not a single format to present a plan; yet, all items of the planning stage from mission statement to selected options, must to be clearly stated within the plan in order to reduce the occurrence of misperceptions.

This plan provides the basis for the implementation of the options that resulted feasible; however, not all feasible options have to be taken into account in the initial written plan. Despite their feasibility, some options are left to following iterations on an options waiting list. It is important to consider a contingency plan section in order to opportunely react in case of an accident or something doesn't occur as expected.

Stage 3: Implementing and Monitoring

The resources allocation to secure the implementation of the options that were foreseen in the planning stage is the purpose on this stage. Implementation projects describing the labor, technological, and financial resources indispensable to carry out them are done for each of the options. The details about the resources allocation will depend on the auto body shop´s structure, but at least, this has to cover the duration of the activities necessary for the logistics to which it refers.

Monitoring is done to evaluate the efficiency of the options implemented and based on results; some projects could need to adjust them. Efficiency has to be measure not only in engineering or financial terms but also in health and environmental terms. Besides monitoring quality service factors, the program encourages to be annually monitoring the health conditions of workers in order to prevent occupational illness. Recognition of occupational and environmental hazards is also important when monitoring the program.

Sustainability indicators are one of the most important elements in the monitoring stage. They must provide reliable, relevant, and useful information about relevant factors such as waste, water, energy, noise, emissions and so on; therefore, the sustainability team must collect information for interpreting indicators to make decisions about how the projects have been working according to the goals of the program.

One of main challenges in monitoring the implementation of options is the involvement of workers and the main requirement is to prepare workers with the necessary knowledge and skills. Training is vital for assure a good implementation and also for monitoring the implementation because it enables workers not only to increase the productivity of the service but also to reduce their exposure by recognizing occupational and environmental risks on advance to possible impacts.

The Sustainable Service Program does not demand engineering skills to identify eliminate or reduce new hazards that might recently exist. It only requires getting observations of practical situations that are desirable for ensure appropriate interventions, but without being a data intensive activity.

Safety Material Data Sheets are excellent sources of information because they provide data for coping with chemical substances or products such as their physical data, toxicity, health effects and they may also include storage, disposal, labelling and other safety procedures to reduce exposure. Unfortunately, information in safety sheets is hard to understand for worker; consequently, training sessions should include how to understand and use them.

Stage 4: Checking

At this stage, the team review the information collected to find out if data fits with the goal and objectives of the program. The check stage tests the plan for completeness and if there are deviations from the initial plan are found, the team must make decisions to reorient current projects. In some cases checking reveals the need for review objectives rather than particular projects. Preferably, options should be checked on a continuous basis.

Stage 5: Acting

It is particularly desirable that the results of checking server as a feedback for acting. Taking corrective actions is often necessary to keep on the planning track toward reach objectives. Decisions can be from small modifications to cancel the project and select another option from the options waiting list. On the other hand, if objectives have been reached, the team must start the cycle again by strengthen the program and set new objectives to keep reaching sustainability.

6. Conclusions

Automakers have reduced the environmental impacts of motor vehicles by making cars more fuel efficient and conserving natural resources. Unfortunately, cleaner technology has been not sufficient to reach sustainability in this industry. When study the sustainability of the automotive industry, it is a common mistake to pay attention only to the automobile. The automotive system is complex and environmental and social impacts are generated not only in the manufacture and use of car, but also in the services necessaries to keep the car appropriately working.

The service offered in the Mexican auto body shop is clearly not consistent with the accepted precepts of sustainability because it is not addressing the underlying sustainability principles of protect the environment and improving health and safety conditions within shops.

The implementation of a Sustainable Service Program is not sufficient to guarantee sustainability in this industry; but it increases the chances of small auto body shops to develop services strategies to secure long-term economic growth while improving environmental and working conditions.

The goal of preventing, eliminating, and/or reducing, at the source, the creations of risks or severe impacts that processes, operations or activities can impose to workers, environment, and society cannot be achieved without an honest commitment to sustainability from owners and workers in auto body shops.

Even objectives were achieved, it is necessary to sustain success. Continuous improvement ensures that the program does not stop after the first iteration of the cycle. The sustainable service program is a tool to help small service organizations to transit in incremental steps to Sustainable Development.

It has been stated in this chapter that the major accountability for changing unsustainability patterns of services in this industry falls on the shoulders of auto body shops´ owners; unfortunately, they are not in conditions to achieve this goal for themselves.

Around the world, there are examples of positive collaboration between society and universities. Higher education institutions are in an incomparable position of helping small auto body owner to transit to sustainability by implementing the proposed Sustainable Service Program because their mutual interest for pushing for clean production and better

working conditions and because the moral obligation of universities for involving in regional development process by developing clean technology, green and safety process, testing new chemicals, and in general making workplaces safe.
As a part of the research; nowadays, the Sustainable Service Program is being testing in a variety of small auto body shops in different Mexican settings. Hence, it is very likely that some stages and activities in the model would have to be debated and adapted, but the main arguments might provide the means necessary to overcome the barriers confronted by sustainable advocators in their particular situations in specific shops.

7. References

Alexandersson, R., G. Hedenstierna, N. Plato and B. Kolmodin-Hedman: Exposure, lung function, and symptoms in car painters exposed to hexamethylendiisocyanate and biuret modified hexamethylendiisocyanate. *Arch Environ Health* 42:367-373 (1987).

Baas, L., Cleaner production and industrial ecology: dynamic aspects of the introduction and dissemination of new concepts in industrial practice. *Eburon Academic Publishers, Delft* (2005)

Bello. D., Woskie., S., Streicher, R., Liu, Y., et al. Polyisocyanates in Occupational Environments a Critical Review of Exposure Limits and Metric. *American Journal of Industrial Medicine* 46: 480-491 (2004)

Bishop, L., P. Pollution Prevention: Fundamentals and Practice, Boston. Waveland Press, INC 2004, Long Grove, IL., The States Unites of America (2010)

Byrch, C.; Kearins, K.; Milne, M.; and Morgan, R.: "Sustainable "what"? A cognitive approach to understanding sustainable development". *Qualitative Research in Accounting & Management,* : 4(1), 26-52 (2007)

CESVI, (Centro de Educación Vial, México): Manual de prevención de riesgos en las carrocerías, 2da. Edition, P. 46-47

Di Stefano, F., S. Siriruttanapruk, J. McCoach, M. Di Gioacchino and P. S. Burge: Occupational asthma in a highly industrialized region of UK: report from a local surveillance scheme. *Allerg Immunol (Paris)* 36:56-62 (2004).

Enander, R. T., D. M. Gute and H. J. Cohen: The concordance of pollution prevention and occupational health and safety: a perspective on U.S. policy. *Am J Ind Med* 44:312-320 (2003).

Enander, R. T., D. M. Gute and R. Missaghian: Survey of risk reduction and pollution prevention practices in the Rhode Island automotive refinishing industry. *Am Ind Hyg Assoc J* 59:478-489 (1998).

Enander, R. T., H. J. Cohen, D. M. Gute, L. C. Brown, A. M. Desmaris and R. Missaghian: Lead and methylene chloride exposures among automotive repair technicians. *J Occup Environ Hyg* 1:119-125 (2004).

Environmental Protection Agency, (EPA) Emission Inventory Improvement Program, Auto Body Refinishing, Chapter 13, page 4-3,(2000)

EPA (2002) U.S. Environmental Protection Agency, Design for the Environment Projects; EPA 744-F-00-019 (2000), (Internet) Avalaible at http://www.epa.gov/dfe/pubs/tools/dfefactsheet/dfefacts8-02.pdf

Geiser, K. Materials Matter: Toward a Sustainable Materials Policy. Cambridge, MA, MIT Press (2001)

INEGI, SIMBAD (Sistema Municipal de Base de Datos)/accidentes de tránsito, Censo Económico 2004." [Online] Available at www.inegi.gob.mx

INEGI, SIMBAD (Sistema Municipal de Base de Datos)/Vehículos registrados en circulación, Censo Económico 2004." [Online] Available at www.inegi.gob.mx

Liu, Y, L., Sparer J., Woskie, S., Cullen, M., Chung, J., Holm, C., and Redlich, C., Qualitative Assessment of Isocyanate Skin Exposure in Auto Body Shops: A Pilot Study in *American Journal of Industrial Medicine 37:265-274* (2000)

Munguia, N., Zavala, A., Marín, M., Moure-Eraso, R. and Velazquez, L.E: Identifying pollution prevention opportunities in the Mexican auto refinishing industry, Management of Environmental, *Quality: An International Journal,* 21(3) 324-335 (2009)

OSHA (2006) Occupational Safety & Health Administration: Safety and Health Topics: Isocyanates, (Internet) Available atn: http://www.osha.gov/SLTC/isocyanates/index.html

Pronk, A., Tielemans, E., Skarping, G., et al. Inhalation Exposure to Isocyanate of Car Body Repair Workers and Industrial Spray Painters. Annal Occupational Hygiene, Vol. 50, No. 1. pp 1-14 (2006)

Redlich, C. A., D. Bello and A. V. Wisnewski: Isocyanate exposures and health effects. , in *Environmental and occupational medicine,* 4th Edition, Rom, W. N. (ed), pp. Philadelphia: Lippincott-Raven, in press, 2006.

Redlich, C. A., M. H. Stowe, A. V. Wisnewski, E. A. Eisen, M. H. Karol, R. Lemus, et al.: Subclinical immunologic and physiologic responses in hexamethylene diisocyanate-exposed auto body shop workers. *Am J Ind Med* 39:587-597 (2001).

Redlich., C., Stowe., M., Wisnewski et al., Subclinical Immunologic and Physiologic Responses in Hexamethylene Diisocyanate Exposed Auto Body Shop Workers: *American Journal of Industrial Medicine* 39: 587-597 (2001)

U.S EPA. Facility Pollution Prevention Plan Guide. EPA/600/R-92/088. Washinton, DC; (1992) [Online] Available at http://www.epa.gov/agriculture/apol.html

U.S. EPA. Pollution Prevention Framework (P2). EPA-748-B-04-001 (2005) [Online] Available at http://www.epa.gov/p2/pubs/p2policy/framework.htm

Velazquez, L.E., Bello, D., Munguia, N., Zavala, A., Marin, M. and Moure-Eraso, R. "A survey of environmental and occupational work practices in the automotive refinishing industry of a developing country: Sonora, Mexico", *International Journal of Occupational and Environmental Health,* 14 (2) 104-11.(2007)

WCED, The World Commission on Environment and Development, Our Common Future. Oxford Univ. Press: Oxford, NY. (1987) Trends and Developments in Automotive Engineering

Woskie, S. R., J. Sparer, R. J. Gore, M. Stowe, D. Bello, Y. Liu, et al.: Determinants of isocyanate exposures in auto body repair and refinishing shops. *Ann Occup Hyg* 48:393-403 (2004).

7

An Analysis of the Automaker-Systemist Supplier Relationship in an Automotive Industrial Condominium

Mário Sacomano Neto[1] and Sílvio R. I. Pires[2]
[1]*Professor at Methodist University of Piracicaba (UNIMEP)*
[2]*Professor at Methodist University of Piracicaba (UNIMEP)*
Brazil

1. Introduction

Recently Brazil's automotive industry has attained a reasonable performance as a world-class player in the assembly of automobiles. In 2007, 2.97 million units were assembled. This result is 13.9% higher than that achieved in 2006 and represents the best result of the sector, according to ANFAVEA (National Association of Motor Vehicle Manufacturers; Brazil) (2008). In 2007, Brazil ranked as the world's sixth largest vehicle manufacturer, outranking France and Spain. The world's largest producer in 2007 was Japan, followed by the United States, China, Germany and South Korea (ANFAVEA, 2008). Specialists point to the rapid rise of emergent markets among the world's largest vehicle manufacturers, especially the case of China (LUNG, 2000).

Since the mid-1990s, several productive arrangements have been implemented in Brazil's automotive sector, among them the modular consortium and industrial condominiums. These arrangements are characterized by high levels of outsourcing, long-term contracts, integrative agreements, coproduction of components, exchanges of specific resources, information interchange, and support to suppliers. These practices have led to substantial modifications in the relationship and in the measurement of performance among the actors in the supply chain (McCORMACK, LADEIRA & OLIVEIRA, 2008; LEE, KWON & SEVERENCE, 2007; FYNE, VOSS & VÚRCA, 2005).

The relationship standard between automakers and suppliers is a central aspect of the new strategies of the automotive sector and it supports the process of internationalization of automakers and suppliers. Cooperation and partnerships with suppliers are also forms of capturing resources (Gulati, 1999; Gnyawali & Madhavan, 2001) and of minimizing uncertainties (Friedberg & Neville, 1999), which are such prominent characteristics for the insertion of companies into the global market. Automakers use these strategies to implement new plants in emergent markets.

Brazil is an attractive country due to the rapid growth of the automotive market, lower cost production units, accelerated growth of driving rates (LUNG, 2000), and privileged fields for new organizational and labor experiments (Humphrey et al., 2000). However, the vulnerability of these markets requires that automakers adopt adaptive strategies that are able to reach domestic and export markets, allowing for economies of scale and scope (Lung,

2000). To this end, automakers simplify products, reduce the number of platforms, adhere to new forms of labor relations, and, principally, reduce costs through partnerships with suppliers. This fact has modified the relationship standard and the level of integration among these companies.

A primary consequence of this change is the series of responsibilities attributed to auto parts suppliers (Humphrey et al., 2000), especially through the activities introduced by *follow sourcing, global sourcing* and by the modularization of production. Other activities that have been "attributed" or "delegated" to suppliers encompass research and development, quality, new investments, new technologies and supply chain management.

The demands of automakers on first tier suppliers range from design capability and manufacturing excellence to product delivery (Humphrey et al., 2000). These authors highlight three trends in the change of the relationship between automakers and auto parts suppliers: first - greater supplier responsibility for design; second - a trend for the supply of complete functions (systems, subsystems or modules); and third - automakers are standardizing their platforms among their sister companies in the different markets.

This new relationship standard between automakers and suppliers in Brazil's automotive sector motivated the present research, which was conducted by means of interviews with five executives from the areas of production and logistics at the automaker and a director of production at the systemist supplier. The study of the relationship between automaker and systemist constitutes the central theme for an understanding of the strategies and the new configuration of the automotive sector in Brazil. Our efforts focused on gaining an insight of the reflexes of this new relationship standard on production and logistics practices and on measures of performance.

To achieve the proposed objective, this paper discusses the dynamics of the structure and the relations in the context of supply chain management, the *configurations of the world's and Brazil's automotive industry, the research methodology, the companies of this study,* the relationship among companies in the industrial condominium, the impacts on product planning, production, supply and measurement of performance in the chain, and our final conclusions.

2. Supply chain structure and relationships

Nowadays structure and relationships are central elements in the analysis of supply chains (LAMBERT et al., 1998). However, before understanding the structure and the relationships in the chain, one must grasp the core concepts of supply chain management. Supply Chain Management - SCM is originating from the literature about logistics, specifically the issues of purchasing and administration of stocks (TRIENEKENS, 1999). The council of Logistics Management defines logistics as "a part of the supply chain management that plans, implements and efficiently and effectively controls flows, product stocks, services and correlated information, from the point of origin to the point of consumption, with the objective of meeting the clients' needs" (LAMBERT et al., 1998 p.3). The authors point out that logistics has a functional role involving the flows of information and materials in the supply chain.

SCM involves intra- and interorganizational integration and coordination from suppliers to final clients, the integration of many distinct organizations, and the presence of bidirectional flows of products and information. Lastly, SCM seeks to value the client with the appropriate use of resources and also to build competitive advantages in the supply chain.

Pires & Carrretero Diaz (2007, p. 25) emphasizes that the "SC is a network of autonomous or semi-autonomous companies that are effectively responsible for the obtainment, production and release of a given product and/or service to the final client."

Listed below are some of the assumptions of SCM found in the literature about supply chain management: competition among chains and no longer between isolated companies (CHRISTOPHER, 1998); alignment of the competitive strategies among the companies participating in the chain (BAUM & DUTTON, 1996); coordination and planning of the activities and processes among the companies that make up the chain (COOPER et al., 1997); alignment of the business processes and integration of functions in an intra- and intercompany process (COOPER et al., 1997); existence of a bidirectional flow of products (materials and services) and information among the companies belonging to the chain and establishment of cooperative relationships among the companies involved (PRAHINSKI & BENTON, 2004); existence of long-term commitments between suppliers and clients (CHRISTOPHER, 1998); joint investments in research and development and co-design. supplier involvement in the product fabrication process (PIRES & CARRETERO DÍAZ, 2007); electronic data exchange (LAUER, 2000; SANCHES & PERES, 2003; KOUDAL & WELLENER, 2003); trust between clients and suppliers in the chain (SVENSSONS, 2001), among several other issues widely discussed in the literature on the theme.

In this sense, the structure of the chain, understood as the set of relationships upstream and downstream of the chain, and the relationships - cooperative or not, begin to represent essential aspects for the chain's management and, hence, for the improvement of the levels of stocks and services rendered to the client. Lambert et al. (1998) cite three interrelated elements: the structure, the process and the components for the SCM. The structure of the chain involves the types of actors, the vertical structure, the horizontal structure and the horizontal position of the organizations of the chain of suppliers. Business structures are strucrures of activities designed to add value to the end product. The management of the chain's components involves managerial activities in which the business processes are integrated and managed along the chain.

The format of the supply chain and logistics structure can be a competitive advantage (LAMBERT et al. 1998). However, structuring and managing a set of relationships has become an extremely complex strategic issue. Uzzi (1997) reflects on the consequenced of adopting different supply chain configuratios, as illustrated in Figure 1. Each chain is composed of a contractor (the focal company) connected to the first and second tier suppliers. The thick lines represent a higher degree of reciprocity, cooperation, trust, exchange of refined information, etc. The tenuous lines indicate the market relationships (arm's length), with supplier selection criteria based on the best price.

If an organization is deeply inolved in cooperative relationships with a few suppliers and clients, it becomes highly dependent (*overembedded* chain) on these actors, making it difficult for the focal company to adapt to the competitive dynamics and to innovations (UZZI, 1997).

If an organization has market relations solely inside the chain, which the author calls an *underembedded* chain, the business and relationships among companies are conducted based on the criterion of price, with little cooperation, trust and integration, i.e., they are strictly market relations.

The integrated chain, according to Uzzi (1997), would be the most suitable way to struture a supplier chain, for it combines: 1) cooperative relationships with high interdependence and refined exchanges, and 2) market relations with a cost-based criterion. In the integrated

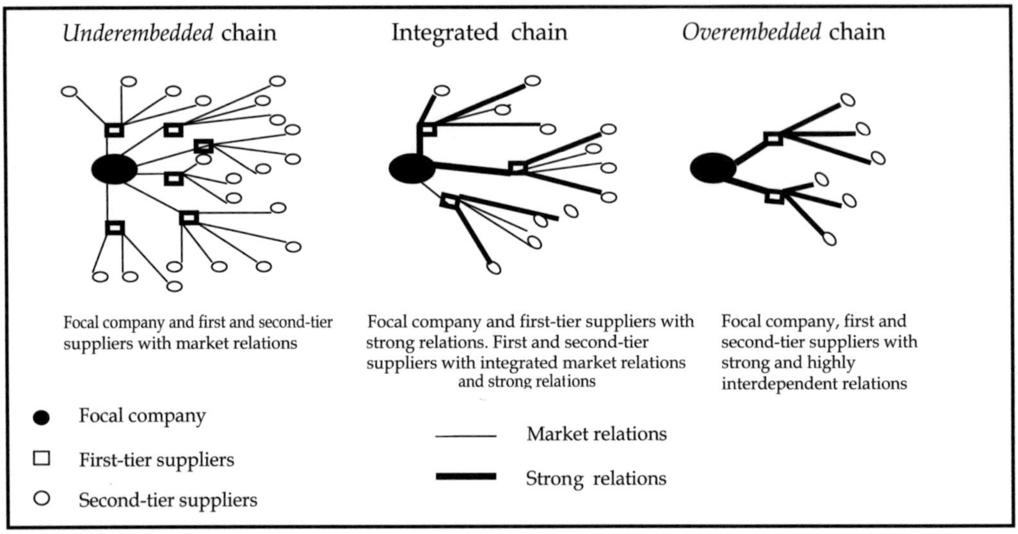

Fig. 1. Types of chain structures and their respective links (Source: UZZI, 1997)

chain there is no exclusive dependence on a few suppliers and there is also the possibility of receiving non-redundant information. As Uzzi (1997) points out, the degree to which relations of cooperation and little cooperation facilitate transactions depends on the quality of the connections, the position and the key companies in the chain. For this reason, understanding the dynamics of the structure of the chain is essential in order to compete.

A contribution concerning relationships in the supply chain was presented by Lambert et al. (1998). The authors mentioned four types of connections in supply chains: managed, monitored, non-managed and indirect connections. Figure 2 illustrated the types of connections in supply chains.

Managed connections are those that occur when the central company integrates its processes with clients and suppliers through collaboration. Monitored connections are forged when a central company monitors and audits the supply chain processes. Non-managed connections occur when the central company does not monitor the participating actors due to the mutual trust existing between the actors. Indirect connections are the ones that influence the central company indirectly in the absence of a relationship with the actor in question. Every supply chain varies according to the diverse types of connections existing in it. The different types of connection can influence the type of information, the mechanisms of performance control, and the forms of production management, among various other aspects.

Several studies have found that more cooperative relations among companies in the chain lead to gains (GHOSH & FEDOROWICZ, 2008; SOOSAY, HYLAND & FERRER, 2008; HADAYA & CASSIVI, 2007).

Supplier relations management is a central process in the model of Lambert et al. (1998). Companies should develop partnerships with key suppliers to underpin the management of manufacturing flow, product development and commercialization (PIRES & CARRETERO DÍAZ, 2007).

The structural and relational dimensions in the chain help one to understand the nature of the relationships among productive actors and to design new supply and distribution channels. Supply chain managers need to map the participating actors, identify the critical connections to be monitored, and establish, or not, cooperative bonds among the actors.

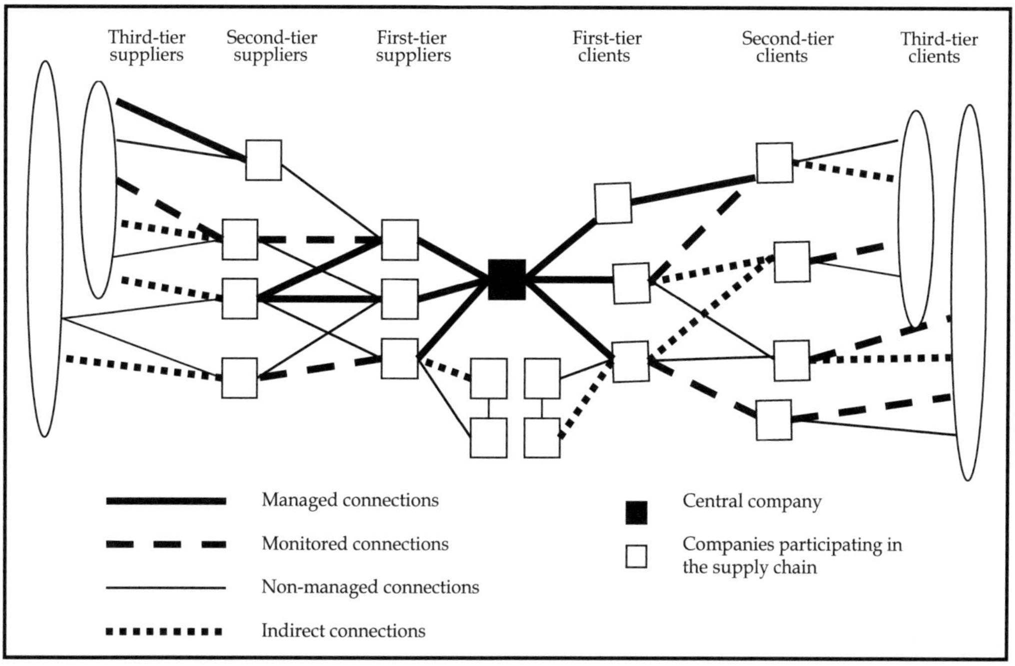

Fig. 2. Types of connections in supply chains (Source: LAMBERT et al. 1998)

As Podolny (1994) asserted, market uncertainties and failures lead organizations to adopt a more cooperative orientation in search of partnerships and long-term relationships. Increasingly, the actors of the automotive sector have been using this orientation as a facilitating element in exchanges and as a principle for selecting exclusive partners. This process will be analyzed within the context of the Brazilian automotive industry.

3. Contemporary configurations of the automotive industry

The last few years have seen an intensification of the internationalization of the automotive industry, a process that represents one of the core strategies of automakers. Internationalization, geographic distribution and international division of labor represent primary themes for studies of the sector, in view of the stabilization of vehicle production and sales in the markets of the triad: United States, Japan and Europe, according to Humphrey et al. (2000). As a result, there has been a significant change in the role of regional markets, as in the case of the Mercosur.

Currently, the sector is expanding its productive structures in a large part of the world's countries. According to Humphrey et al. (2000), the dynamics of the automotive sector is divided principally into three markets: protected autonomous markets (PAMs), integrated peripheral markets (IPMs), and emerging regional markets (ERMs). The first is composed of the countries that protect themselves against outside competition through domestic markets, such as India, China and Malaysia. The second comprises the countries located close to large markets, such as Mexico, the Czech Republic, Hungary and Poland. The third market is composed of countries inserted in emergent blocs, such as Brazil and Argentina, Russia and Turkey. Although these markets represent "new spaces" of action for automakers and auto parts suppliers, the internationalization of the automotive sector is not a homogeneous process.

Although the expansion of the automotive industry has a global character, the realities and peculiarities of each market call into question the existence of single production models, as a result of the legitimation of the best management practices. Volpato (2002) points out that the internationalization of the automotive sector has two extremes: one the one hand, a significant standardization of organizational forms and of decision-making processes originating from headquarter companies and, on the other, localization and adaptation to each regional context. Cultural, social, political and economic differences require different forms of implementation and diffusion of productive systems, leading, according to Boyer et al. (1998), to a process of hybridization. These authors believe that the diffusion of productive systems depends, for its consolidation, on a series of economic, social and historical aspects.

Emerging markets, such as the Brazilian market, are considered attractive due to the following factors: rapid growth of the vehicle market, production units in lower-cost locations, accelerated growth of driving rates (LUNG, 2000), and privileged fields for new organizational and labor experiments (HUMPHREY et al., 2000).

The transformations of the structural foundations of Brazil's automotive sector open up a fast field of research, which involves new production models. The large number of mergers, acquisitions, co-production, consortiums, franchising, strategic alliances, long-term contracts and joint ventures demonstrate the sector's dynamic and complex characteristics. In recent years, several new organizational arrangements have been implemented in Brazil's automotive sector, among them the modular consortium and the industrial condominium. These new arrangements are marked by a high degree of outsourcing, long-term contracts, integrative agreements, component co-production, exchange of specific assets, information transfer, and support to suppliers. These arrangements modify the relationship standard of automakers with auto parts suppliers.

The strategic change in automakers is also related with a greater rationalization in the relations with auto parts suppliers. Economic, technological and market uncertainties lead to the establishment of cooperative agreements with suppliers (KNIGHT, 1998). This fact has given first-tier suppliers high-status positions and, hence, new roles to play in the supply chain of Brazil's automotive industry.

These changes have led to two consequences for the auto parts sector: 1- a significant increase in automaker demands concerning quality, just-in-time deliveries, global sourcing, follow sourcing, product development, co-design, and financial and technological capacitation (CARVALHO et al. 2000); and 2- concentration of the auto parts suppliers in the hands of large international groups and a deep denationalization of the sector.

The introduction of new productive arrangements - the modular consortium and industrial condominium - have placed Brazil's automotive sector on the map in the discussion of industrial models (HUMPHREY et al., 2000). The Brazilian automotive sector is becoming a model for several countries, including the most industrialized nations, where the headquarters of the companies that have manufacturing units in Brazil are located.

4. Methodology of research

This research is classified as exploratory, descriptive, qualitative and case study-based. Table 1 classifies the research methodology.

Subject	Methodology
Objective	**Exploratory and descriptive research** - seeks to understand the relationship between the automaker and the systemist supplier
Approach	**Qualitative** - allows for an understanding of the objective and subjective elements of the relationship
Method	**Case studies** - the automaker and the systemist supplier and their relationship were researched
Data collection technique	**Interviews with semi-structured script** - these were held with 5 executives at the automaker and one at the systemist supplier - these interviews allow for an explanation of the "world view" of the interviewee

Table 1. Classification of the research methodology

The exploratory research approach is suitable when: a) the situations analyzed are contemporary, encompassing and complex; b) the focus falls more strongly on understanding rather than on quantifying the facts; and c) there are several methodological sources to uncover the facts and applicable to situations in which there is no control over the events/behaviors of the facts/people involved in the research (YIN, 1994). This research is exploratory and descriptive inasmuch as it examines the relationship between automaker and systemist and the impacts of this relationship on the forms of production organization and on the measures of supplier performance.

Qualitative research involves meaning, relationships and people for an understanding of the phenomena. According to Chizzotti (1999, p.79), "the qualitative approach starts from the premise that there is a dynamic relationship between the real world and the observer (…). The observer is an integral part of the process of knowledge and interprets the phenomena, giving them a meaning." Qualitative research allows for a deeper analysis of the research object. The study does not allow for generalizations, but makes a more subjective analysis of the relationship between automaker and systemist supplier.

The case studies involved a vehicle manufacturer and a supplier of automotive systems (the systemist). According to Yin (1994), case studies are indicated in three situations: 1) when the case study represents an opportunity to confirm, contest or extend a theory; 2) when it is an extreme and rare case; and 3) when it involves something revealing, a unique opportunity for analyzing an inaccessible phenomenon. The present research is related with situation 3, because it studies very particular and specific aspects of the relationship between the automaker and the systemist within the context of an industrial condominium. That is the main reason for the choice of these companies for this research.

In addition to a review of bibliographic material and observation, data were collected through interviews *in loco* with five executives from the areas of logistics, purchasing and production at the automaker and one executive of the systemist supplier. The semi-structured script for collecting the data involved a study of the supply chain structure, the relationship between companies, performance measurements, logistics, and production planning and control. The semi-structured script allows the interviewee to describe his "world view" and gives greater depth to any given topic. In this research, we attempted to combine closed and objective information with more ample and subjective information.

The interviews and non-participant observation, in which the researcher does not join the observed group, complemented each other in the data collection process. The visit and act of interviewing provide the researcher with a variety of relevant information. The language, stories, behavior, and treatment are some of the aspects observed in the researcher's contact with the interviewees. The case study was conducted during the second semester of 2007 and the first three months of 2008.

5. The companies studied

A set of relevant characteristics and/or information of the studied companies are briefly described below.

5.1 The automaker

The oldest automaker in the country was established in 1957, in the city of São Bernardo do Campo. This plant represents one of the icons of the automotive industry and of Brazilian industrialization. In the end of 2007, this manufacturing unit employed approximately 20,545 people (automaker: 15,000; third parties: 5,545) and more than a thousand engineers in its Engineering Center. The São Bernardo do Campo plant has a production capacity of 1,600 vehicles/day.

In 2002, the São Bernardo plant was restructured to transform it into an industrial condominium. As Pires & Carretero Díaz (2007) explain, in the industrial condominium the suppliers are physically installed next to the automaker. The industrial condominium is composed of 8 companies that supply parts in sequence and in real time to the automaker. Installed in this condominium are the suppliers of tires, wheels, door accessories, cables, chassis components, fuel tank, exhaust pipes, brake and accelerator pedals, instrument panels, and interior car door panels. The modules and systems manufactured by the suppliers are transported and sequenced according to the automaker's production schedule, operating according to the *just-in-sequence* system.

According to data published by Automotive Business (2005), the company invested R$ 2 billion in a highly automated and modern structure. Laser welding, robotized framework islands, automated paint line, the use of palmtops to control production in real time, assembly by a modular system, and car body transporting devices that adapt to the height of the worker are some of the improvements and innovations implemented in the new plant.

5.2 The systemist supplier

The systemist supplier (SS) is installed inside the automaker's plant. The supplier belongs to a German group of the automotive sector, supplying parts and systems for car bodies, chassis and engines (powertrain systems). In the end of 2007, the group to which the systemist belongs had 140 plants in 17 countries and employs 184,000 people. In addition to the automotive sector, the group acts in the steel, elevators, technologies and services sectors. In Brazil, the group has 22 subsidiaries and employs approximately 9,000 people. When the negotiations to set up the industrial condominium began, the automaker asked the SS for two estimates, one corresponding to the plant located outside and the other for the plant located on the automaker's premises. The supplier made a detailed study of its needs to present to the automaker. The SS was thus able to reduce the cost of its part to the automaker by 15% simply by being located on the premises of the condominium. The logistic cost and the synergy between the supplier and the automaker were identified as the principal factor for the cost reduction. Other benefits are energy, water, restaurants, buildings, and security, among other aspects offered by the automaker. "However, the main factor responsible is logistics since, if we were located outside, we would have to produce, transport by truck, unload, sequence the parts and deliver them to the automaker," all of which generates costs, according to the SS manager.

6. Analysing the relationships in the industrial condominium

Intense consolidation with the supplier base in the automotive sector has been observed in several countries, and in Brazil it was no different. The consolidation of the supplier base led to a significant structural change in the automotive chain. The large transnationals began buying up national companies and placing themselves as first-tier suppliers in the chain (SALERNO, et al. 2003). This structural change, whereby the automaker maintains relations with fewer suppliers, affected the relationship standard and the level of integration between suppliers and automakers.

The relationship between automaker and systemist can be classified as highly managed and integrated, according to the classification proposed by Lambert at al. (1998). Integration is a central aspect in the relationship between automaker and systemist. Pires & Carretero Díaz (2007) present different levels of relationships among companies, from the least the the most integrated (commercial relations, non-contractual agreements, licence agreements, alliances, partnerships, joint ventures and vertical integration). Partnerships and joint ventures represent relationships with high levels of integration in the chain. In this classification, the relationship involves a marked level of collaboration, of alignment of objectives, and of integration of processes and information.

The joint venture established between the SS and the automaker indicates a high degree of integration and complementarity of competencies. The industrial manager of the SS stated that "*integration, cost reduction and collective facilitators*" are the main reasons for being in the condominium. This gives the SS access to resources, information and systems that other suppliers do not have: real levels of demand, changes in products and technologies, etc. The industrial manager of the SS also highlighted a technological partnership between the group to which the SS belong and the automaker in Germany for the fabrication of modules similar to those manufactured in Brazil. All the technology the SS uses here in Brazil also comes from the know-how generated jointly in Germany with the automaker of this research. Thus, the relations between the SS and the automaker display a high level of sharing of knowledge relating to component production and development.

Another point cited by the interviewee is the fact that because the SS is right beside the assembly line, "*you can go there and solve any problem of logistics, quality or production.*" The SS and the automaker also have daily meetings to deal with operational questions of logistics and production. The frequency of contact is very high, enabling them to operate with low levels of stocks and a high level of integration for the solution of problems. These contacts create bonds of trust between the companies, rendering the system even more efficient, as indicated by the findings of the study by Morris et al. (2004), who point out that modularization does not refer solely to technology but also to organizational and social relationships between companies. This format of supply chain structure, which privileges a high degree of reciprocity, trust, and exchange of refined information, is worldwide trend in the automotive sector. The new relationship structure facilitates management of the supply chain in several aspects, such as stocks, information flows, and client demands, allowing for coordination and planning of the activities and processes between the companies that make up the chain.

This research confirmed several assumptions set forth in the literature about supply chain management in the relations studied here, such as the establishment of cooperative relationships among the companies involved; long-term commitment between suppliers and clients; joint investments in research and development and in co-design; supplier involvement in the product manufacturing process; and the electronic exchange of data. These elements favor high levels of integration, coordination of work methods, and transfer of added value to the companies. Evidently, one cannot generalize this finding to other links in the chain. The relationships with non-systemist suppliers are not characterized by the same level of integration and complementarity of competencies, as reported by Pires & Carretero Díaz (2007).

The high degree of integration of systemist companies is reflected in the practices and innovations in the condominium. Some innovations involve the use of modules, just-in-sequence deliveries, the Kanban system, EDI (electronic data interchange), cross-docking, logistic consolidator, the poka-yoke system, and the joint venture between the automaker and the SS. Some of these innovations will be described below.

The SS produces in a sequenced form. When the vehicle leaves the automaker's paint shop, it must be assembled. The automaker sends an electronic signal (label) placed at the beginning of the systemists' production line. The label contains the number of the vehicles, the sequential number, and the model of the module to be assembled. The SS copies the label and begins production. The production line tells the operators, by means of lights (poka-yokes), what parts are needed to assemble that module. At the end of the line, the SS attaches the automaker's label to the parts, which are then sent in sequency to the automaker's assembly line. All the systemists of the industrial condominium are responsible for sequencing. This sequencing is also a requirement for some outside suppliers.

One fact clearly illustrates the integration between automaker and SS. At the moment the automaker sends the labels indicating the vehicles to be assembled, the SS has only 1 hour and 30 min to send and sequence the parts according to the automaker's production plan. That is why there is a high investment in preventive maintenance (planned). The industrial manager had the following to say about sequencing and short-term deliveries: "it generates a lot of stress… and the line cannot be stopped." This requires from clients and suppliers a high level of intercompany coordination in the management of stocks, demand and productive capacity.

7. Analyzing three automaker's business processes

The automaker's production is organized into three macro-processes, involving the product planning system, production planning and supply of the assembly line. The product planning activities begin with an alteration in drawings or with a new vehicle design (new designs or modifications of existing designs). This phase is marked by numerous meetings and teams to discuss the new design or vehicle design changes, such as: types of parts, financial aspects, development of tooling and supplier qualification. The initial phase consists of product development and production planning with the key suppliers.

During production planning the automaker selects all the suppliers. For purposes of the architecture of raw materials, the suppliers have a visibility of 6 months of the schedule through releases. Ten weeks before production begins there is still a flexible period (up to the 8th week) until the schedule is frozen in the last two weeks. The suppliers have access to these phases to suggest and introduce changes in the vehicle. After this phase comes the planning phase of how many cars will be produced (monthly, weekly and daily). The entire production is managed by the Manufacturing Information System (MIS), which stores and coordinates the orders from dealers and then determines the models, versions to be manufactured. The system also sends the parts orders to the suppliers to supply the line.

Supply of the production line also has several innovations. The automaker has about 400 outside suppliers. To supply the line, the plant operates by the milk run and Kanban systems with outside suppliers. A logistic consolidator was hired to collect raw materials to supply the plant's production line. The Kanban system is applied for large and expensive parts. Both the milk run and Kanban systems reduce logistics costs and stock, as well as the number of trucks circulating on the plant's premises. The EDI (Electronic Data Interchange) system is being implemented with some suppliers and others via the Internet.

The above phase is followed by the supply operation, which has to be planned from the point of use (packaging, ergonomics, transportation, supply flow, weight, warehouse, installations, information system, type of vehicle, FIFO and LIFO control) in order to avoid unnecessary warehousing. The plant has 22 warehouses to make raw materials available at the moment of assembly. The placement of material at the point of use is called *cross-docking* in the literature (PIRES & CARRETERO DÍAZ, 2007).

All the production of the SS is made to order. There is some safety stock to cover eventual problems, which the automaker's production line for the SS takes advantage of to make a small buffer stock of the most frequently used parts. Due to the need for low stocks, production and supply planning play a significant role in reducing logistics and transport costs.

8. Analysing the supply chain performance measurement process

We found no single system for dealing with supply chain performance measures. The automaker itself has an internal system that includes some supply chain performance measures. It was found that companies have internal systems for measuring performance that extend to supply chains.

The automaker measures performance indicators using KPIs *(Key Performance Indicators)* that encompass all the internal areas of the company, from departments to people. The criteria are based on corporate objectives of highest added value for specific areas to individual sectors, directorates, departments and manufacturing. The performance criteria are indicated by a color code (green, yellow and red) with well-defined limits of control. These numbers are consolidated by the finance area (controller), which determines the results of the indicators, not only financial but also of other areas. The results are similar to those presented by Gulledge & Chavusholu (2007).

Some of the logistics performance measures are detailed down to the supplier level. One of the logistics performance measures is *cripple*, which measures the number of missing units and may indicate supply chain-related issues, e.g., "it indicates if the car is lacking parts and measures failures of the entire supply chain", explained one of the automaker's logistics managers. Indirectly, the KPIs end up reflecting on the suppliers. The reflexes of the KPIs on the systemists are discussed by executives in daily meetings, where preventive and corrective actions are decided.

Logistics has 10 sets of indicators: percentage of lost production, bill of materials, volumes of rejects, cripple (number of missing units/incomplete vehicles), inventory levels, deliveries of replacement parts, overtime (payment of direct or monthly personnel), logistic cost (purchase of materials/supplies), overhead, and inventory volume and fidelity. These sets are specifically for logistics. Each of the supervisors and employees has his own KPI chart and criteria that add value in their area.

According to the interviewees at the automaker, the suppliers are evaluated constantly based on at least three major performance criteria:

- Logistics - operational aspects (packaging, type of delivery, innovations, new systems, and especially faithfulness in fulfilling the program,
- Engineering - development potential (software, technical knowledge),
- Quality - a system that ensures the quality of the process (internal process, means of control, calibration of tooling, maintenance).

For new projects, the purchasing area indicates the suppliers that are qualified to supply. The choice of suppliers involves identifying the best price and selecting three suppliers, whereupon the proposal is submitted to the other areas. The area of quality may veto the choice, claiming that the supplier is unqualified. If an area vetoes a supplier, the supplier will need to have a very efficient action plan.

The automaker also has a supplier evaluation system which is multifunctional, involving logistics, engineering, quality, finance and commercial. The forums for defining suppliers are also multifunctional. Both systems work somewhat like external supplier performance measure systems.

The suppliers also have internal performance measure systems that extend to external measures. The SS has several performance measures, such as physical sales, customer complaints, field failures, client assembly line stoppage, average failure time, waste and scrap control, material blocked by suppliers, product audits, and maintenance control. The criteria related directly to the automaker1s supply chain are customer complaints and client assembly line stoppage. According to the interviewee, the SS has much stricter internal performance measures than those used by the automaker to measure supplier performance. What the automaker really controls is line stoppage and SS quality.

9. Final comments

The implementation of the industrial condominium brought major changes in the automaker's relationship with auto parts suppliers, particularly with the systemists. As Doran (2004) explains, in the modular system there is a transfer of added value from the automaker to first-tier suppliers, especially to modulists. Morris et al. (2004) stated that the modular system gives rise to mutual development between automaker and suppliers in accepting work methods, standard procedures, rules, documents and methods of communication. This characteristic renders the system mode interactive than impository, "in other words, modularization does not refer solely to technology but also to organizational and social relationships between companies" (Morris et al., 2004, p. 130).

Simplification of the supply system was an issue brought up by the automaker's interviewees. In the words of the automaker's logistics manager, "if we did not have the systems, we might have 25,000 items rather than 40,000 to manage, so the systemist manages a very large parcel of items" with high added value. The modular design is used in Mercedes-Benz's "Smart" design. While a typical Mercedes-Benz car requires 100 suppliers, the "Smart" model uses 25 modulist suppliers. The benefits to the automaker are reduced risks, investments and costs, according to Doran (2004). For the modulists there is an increase in responsibilities and a high degree of involvement in product and process development. Pires (1998) adds the factor of opportunity for the supplier to develop new competencies by becoming a modulist, as was the case of modular consortium suppliers.

The items supplied by the SS are items with high added value for the automaker and present vehicle safety characteristics. The SS is physically next to the automaker and presents high levels of efficiency, as indicated during the interviews. All product and process development, production and supply are carried out in an integrated way between the SS and the automaker. This finding is consistent with that of Doran (2004), who points out that there is a considerable transfer of added value from the automaker to the modulists, who must have a culture of quality, supply items at low cost, have research and development capabilities, achieve global presence and the capacity to develop modular solutions for automakers.

The automaker consolidated its supplier base into a small number of partners, going from multiple suppliers to single suppliers. However, the systemists must present high levels of performance, for if the systemist allows the assembly to stop, the cost is very high (the value of the cost of the lost cars). Therefore, the SS has invested heavily in maintenance, daily meetings and several performance measures to ensure its supply to the automaker. In the words of one of the automaker's managers, "It is much easier to measure performance issues with the systemists, since they cause us infinitely fewer problems, not least because we scrutinize them more closely because they are in the plant. Sometimes it is much more troublesome to solve a problem of a glove compartment screw than that of a systemist."

Considering the types of connections presented by Lambert et al. (1998), the relations between the SS and the automaker are marked by a high degree of monitoring. According to the SS manager, each day production line leaders go to the automaker to check if there is any problem. "These relations produce good results... ...that is the advantage of being in here... ...our employee goes directly to the person who receives our module, so this communication is very intensive." The connections with the SS are managed and strongly monitored by the automaker. This finding is in line with Mchung, Humphreys & Mclvor (2003), who point out that cooperation is greater the greater the participation in the cost of the product supplied in the end product.

This research contributes to show how strategic bonds in the supply chain receive higher investments in innovations and monitoring. This fact indicates that the traditional relationships are limited when it comes to generating collective gains. Partnerships and cooperation have proved to be important elements in the configuration and formation of supply chains in the automotive sector. Relationships of this type imply collective gains for the actors with high positions in the structure of the chain. This position of prominence is, in large part, occupied by first-tier suppliers and in large part by transnationals.

One point noted here is that the higher the strategic value and the asset specificity of the item supplied the greater the possibilities of expanding partnerships in the search for mutual gains. At the same time, the auto parts suppliers assume greater responsibilities and the automakers make greater demands.

This study has limitations, the first of which refers to sampling. The case study at only two companies cannot lead to generalizations to other automaker-supplier relationships. Another limitation is the depth of the analysis of the relationship between the automaker and the SS. Since these relations involve intense information interchange and are assured through contracts, many of them are confidential and could not be revealed. The greatest difficulty was the field data collection, due to the novelty of the theme.

Far from finalizing the discussion on this subject, further researches are necessary in order to contribute to the rich discussion concerning the partnership relations between automakers and auto parts suppliers, and the local and global configuration of supply chains in the automotive sector. This is even more necessary due to the increasing importance of the Brazilian plants of subsidiary multinationals in the global context of the automotive industry.

10. References

ANFAVEA (2008), ANFAVEA (National Association of Motor Vehicle Manufacturers) Letter, Available in: www.anfavea.com.br. Access in: January 20th, 2008.

AUTOMOTIVE BUSINESS (2005), *Nova Anchieta, uma revolução dentro da fábrica pioneira.* Available in: www.automotivebusiness.com.br, Access in: February 1st, 2005.

CARVALHO, R. Q. QUEIROZ, S. R. R., HUMPHREY, J., CONSONI, F. L., COSTA, I., & FONSECA, R. R. (2000) *Globalização e reestruturação da cadeia produtiva na indústria automotiva: qual é o papel do Mercosul?* Partnership IPEA-DPCT/IG/UNICAMP. Final Report. FUNCAMP, 252p.

CHIZZOTTI, A. (1995) *Pesquisa em ciências humanas e sociais,* Second Edition. São Paulo, Cortez.

COOPER, M, LAMBERT, D., & PAGH, J. (1997), Supply chain management: more than a new name for logistics. *The International Journal of Logistics Management,* Vol. 8, No. 1, pp. 1-13.

BAUM, J. A. C.; DUTTON, J. E. (1996), The embeddedness of strategy. *Advances in Strategic Management,* Vol.13. pp. 363-388.

BOYER, R. CHARRON, E., JURGENS, U., TOLLIDAY, S. (1998), *Between imitation and innovation*: the transfer and hybridization of productive systems in the international automobile industry. New York, Oxford University Press.

CHRISTOPHER, M. (1998) Logistics and Supply Chain Management, London, Prentice Hall.

DORAN, D. (2004), Rethinking the supply chain: an automotive perspective. *Supply Chain Management: an International Journal,* Vol. 9, No.1, pp.102-109.

FYNES, B.; VOSS, C.; BÚRCA, S. (2005), The impact of supply chain relationship dynamics on manufacturing performance. *International Journal of Operations & Production Management,* Vol. 25, No. 1, pp. 6-19.

FRIEDBERG, E.; NEVILLE, J.P. (1999), Inside partnership Trust, opportunism and cooperation in the European automobile industry. In: Grandori, Anna (1999), *Interfirm Networks: Organization and Industrial Competitiveness,* Routledge Studies in Business Organization and Networks, 11.

GNYAWALI, D.; MADHAVAN, R. (2001), Cooperative networks and competitive dynamics: a structural embeddedness perspective. *Academy of Management Review,* Vol. 26, No.3, pp. 431-445.

GHOSH, A.; FEDOROWICZ, J. (2008), The role of trust in supply chain governance. *Business Process Management Journal;* Vol. 14, No. 4.

GULATI, R. (1999) Where do interorganizational networks come from? *American Journal of Sociology,* Vol.104, No.5, pp. 1439-93.

GULLEDGE, T.; CHAVUSHOLU, T. (2008), Automating the construction of supply chain key performance indicators. *Industrial Management & Data Systems,* Vol. 108, No. 6, pp.750 - 774.

HADAYA, P.; CASSIVI, L. (2007), The role of joint collaboration planning actions in a demand-driven supply chain. *Industrial Management & Data Systems,* Vol. 107, No.7, pp. 954-978.

HUMPHREY, J.; LECLER, Y.; SALERNO, M.S. (2000), *Global strategies and local realities: the auto industry in emerging markets.* London, Macmillan.

KNIGHT, J. (1998), The bases of cooperation: social norms and rule of law. *Journal of Institutional and Theoretical Economics,* Vol.154, No. 4, pp. 754-63.

KOUDAL P.; WELLENER, P. (2003), Digital loyalty networks: continuously connecting automakers with their customers and suppliers. *Strategy and Leadership,* Volume 31, Number 6, pp. 4-11(8). pp.4-11.

LAMBERT, D. M.; COOPER, M.C.; PAGH, J.D. (1998), Supply chain Management: implementation issues and research opportunities. *The international Journal of Logistics management,* Vol. 9, No.2, pp. 1-19.

LAUER, T. W. (2000), Side effects of mandatory EDI order processing in the automotive supply chain. *Business Process Management Journal,* Vol. 6, No.5, pp. 366–375.

LEE, C. W.; KWON, I. G.; SEVERANCE, D. (2007), Relationship between supply chain performance and degree of linkage among supplier, internal integration, and customer. *Supply Chain Management,* Vol.12, No. 6., pp. 444 - 452.

LUNG, Y. (2000), *Is the rise of emerging countries as automobile producers an irreversible phenomenon?* In: HUMPHREY, J.; LECLER, Y.; SALERNO; M.S. Global strategies and local realities: the auto industry in emerging markets. London, Macmillan.

MCCORMACK, K.; LADEIRA, M. B.; OLIVEIRA, M. P. V. (2008) Supply Chain Maturity and Performance in Brazil. *Supply Chain Management,* Vol.13, p. 272-282.

MCHUNG, M; HUMPHREYS, P; MCLVOR, R. (2003) Buyer-supplier relationships and organizational health. *The Journal of Supply Chain Management,* Vol.39, No.2, pp.15-25.

MORRIS, D., DONNELLY, T., AND DONNELLY, T. (2004) Suppliers parks in the automotive industry. *Supply Chain Management: An International Journal,* Vol.9, No.2, pp.129 - 133.

PIRES, S., CARRETERO DIAZ, L.E. (2007) *Gestión de la Cadena de Suministros,* McGraw Hill, Madrid, 258p.

PIRES, S. (1998), Managerial implications of the modular consortium model in a Brazilian automotive plant. *International Journal of Operations & Production Management,* Vol. 18, No. 3, pp. 221-32.

PODOLNY, J. M. (1994), Market uncertainty and the social character of economic exchange. *Administrative Science Quarterly,* Vol.39, pp. pp.458-83.

PRAHINSKI, C., BENTON, W. C. (2004), Supplier evaluation: communication strategies to improve supplier performance. *Journal of Operation Management,* Vol. 22, pp.39 - 62.

SALERNO, M. S., MARX, R., ZILBOVICIUS, M. (2003), A nova configuração da cadeia de fornecimento na indústria automobilística no Brasil. *Revista de Administração* -USP , São Paulo, Vol. 38, No. 3, pp. 192-204.

SÁNCHES, A. M., PÉRES, M. P. (2003), The use of EDI for interorganizational co-operation and co-ordination in the supply chain. *Integrated Manufacturing Systems,* Vol. 14, No.8, pp. 642 - 651.

SOOSAY, C. A.; HYLAND, P. W.; FERRER, M. (2008), Supply chain collaboration: capabilities for continuous innovation. *Supply Chain Management,* Vol. 13, No.2, 160 - 169.

SVENSSON, G. (2001), Perceived trust towards suppliers and customers in supply chain of Swedish automotive industry. *International Journal of Physical Distribution & Logistics Management,* Vol. 31, No.9, pp.647-62.

TRIENEKENS, J. (1999), *Management of Process in chains: a research framework.* Thesis. Wageningen University.

UZZI, B. (1997), Social structure and competition in inter-firm networks: the paradox of embeddedness. *Administrative Science Quartely,* Vol. 42, pp.35-67.

VOLPATO, G. (2002), *Carmakers internationalization strategies: an overview.* La Lettre du Gerpisa, No.158, pp. 2-5.

YIN, R. K. (1994), *Case study research.* California: Sage Publications.

Authors:

Mário Sacomano Neto has a Doctoral degree in Production Engineering (UFSCar); He was a Visiting Scholar in University of Chicago (USA). He has a Master's degree in Production Engineering (USP/EESC) and a Bachelor's Degree in Business Administration (PUC). Nowadays, he is an associate professor of Operations, Supply Chain and Network Analysis in Business Administration Mastership at Methodist University of Piracicaba (UNIMEP). Email: msacomano@unimep.br; pmsn@terra.com.br.

Sílvio R. I. Pires has a B.Sc. (UFSCar), and a M.Sc. and a Doctoral degree (EESC-USP) in Production Engineering. He is a full professor of Operations and Supply Chain Management at the Methodist University of Piracicaba (UNIMEP). Formerly he was a Production Planning and Control manager in a large heavy equipments company in Brazil, a visiting research-professor at IMD in Switzerland, and a visiting professor at the Instituto de Empresa Business School in Spain. He also conducted several projects to large companies operating in Brazil. He has 25 years of practical experience in Operations Management. E-mail: sripires@unimep.br

8

Strategic Priorities and Lean Manufacturing Practices in Automotive Suppliers. Ten Years After.

Juan A. Marin-Garcia[1] and Tomas Bonavia[2]
[1]*Universidad Politecnica of Valencia,*
[2]*University of Valencia*
Spain

1. Introduction

Currently, automobile manufacturers have transformed their philosophy of production in favour of the lean manufacturing paradigm. By doing so, they hope to improve efficiency and to obtain better results in the markets in which they operate. This transformation must occur not only in their plants, but it seems important that their suppliers should also modify their production systems in line with the lean manufacturing philosophy (Liker & Wu, 2000; Morris et al., 2006; Oliver & Delbridge, 2002). The effects of this wave will probably result that one integrated supply chain can be built.

Supply companies can see benefits from the implantation of lean manufacturing practices in certain of their strategic priorities, be they quality, lead-times or costs.

On different matters, there appears to be sufficient empirical and theoretical evidence to affirm that human resource management practices play a very important role in the successful implantation of lean manufacturing and above all in its maintenance thanks to the creation of a continuous improvement culture that supports the other lean manufacturing practices (Garcia-Sabater & Marin-Garcia, 2010).

The objective of this chapter is to present an architecture of lean management practices, indicating the implantation sequence recommended for suppliers and the necessities that can help to resolve each one of the practices. We will also describe how, over the past decade, the necessities as much as the practices of lean manufacturing have evolved using some Spanish automotive suppliers as examples. Finally we will analyse in detail the evolution of continuous improvement programs to support lean manufacturing in 11 companies that had started their deployment in the year 2000 (Marin-Garcia et al., 2009).

2. Lean manufacturing

The environment in which most industrial companies find themselves is characterized by the rivalry with competitors, the speed of change and the instability of demand. The majority of their markets are mature and their customers demand quality products that fit their specifics needs, one of which is the demand for quicker and more regular deliveries

(Marin-Garcia et al., 2009; Peng et al., 2008; Devaraj et al., 2004; Ketokivi & Schroeder, 2004). Everything seems to indicate that this tendency will only increase in the future. With this in mind, it is advisable that companies position themselves and decide their strategic operation priorities (Urgal González & García Vázquez, 2005; Ketokivi & Schroeder, 2004; Martín Peña & Díaz Garrido, 2007). Currently a consensus exists regarding two large blocks of strategic priorities in the area of production. Companies whose principal corporate strategy is the emphasis on costs will see as priority the efficient management of operations (reduce costs, investments and inventory) (González Benito & Suárez González, 2007; Avella et al., 2001; Ketokivi & Schroeder, 2004; Hayes & Wheelwright, 1984). Those companies that emphasise differentiation will see the quality operations area as priority (error free products and quality as perceived by the customer); or flexibility (range of the product line, modification of production volume or mix and design modifications) or delivery (production cycle time, delivery speed and on time delivery) (Lewis & Boyer, 2002; González Benito & Suárez González, 2007; Avella et al., 2001; Devaraj et al., 2004; Ketokivi & Schroeder, 2004; Hayes & Wheelwright, 1984; Lewis & Boyer, 2002).

According to some authors it is difficult for a given company to satisfy all these priorities at the same time, given that there exists incompatibilities (trade off) between them (Avella et al., 2001; Devaraj et al., 2004; Skinner, 1969). However, there are also investigations where a sequential or accumulative model is proposed (González Benito & Suárez González, 2007). According to this model the companies focus on a few priorities at any given moment, but once they are satisfied they move onto others without losing the developed abilities. In this way they accumulate abilities that allow them, over time, to simultaneously satisfy a wide range of priorities.

Currently, to combat competitive pressures, it is necessary to complement the efforts towards continuous improvement in production and quality being made since the eighties (White & Prybutok, 2001; Suzaki, 1993; Vazquez-Bustelo & Avella, 2006). To this end, it is necessary to identify problems before their consequences are seen, to analyse solutions for the suppression of unnecessary activities; and to reduce fabrication time, change over time and batch size (Garcia-Sabater & Marin-Garcia, 2010; Marin-Garcia et al., 2008; Marin-Garcia et al., 2009). These activities are the basis of a group of practices that shape systems of advanced manufacturing. These systems have received many names, amongst others: lean manufacturing, just in time, total quality management or world class manufacturing. There exist many similarities in these concepts (Prado Prado, 2002; White & Prybutok, 2001; Marin-Garcia & Carneiro, 2010). To sum up, the objective of these practices is the systematic elimination of all types of waste (Callen et al., 2000), considering as waste anything beyond the minimum needed in terms of equipment, materials, components, space or worker time to give added value to the products (Suzaki, 1993).

To put lean manufacturing systems into place, it is proposed a group of practices related to operations management (production planning and control, materials flow, maintenance system, quality system...), the relationships with clients and suppliers, product design or human resource management (participative management, worker involvement...). Amongst the most common we can find the following (see Table 1)(Shah & Ward, 2007; Dabhilkar & Ahlstrom, 2007; Jorgensen et al., 2008; Gurumurthy & Kodali, 2008; Carrasqueira & Machado, 2008; Marin-Garcia et al., 2006; Marin-Garcia & Carneiro, 2010; Treville & Antonakis, 2006; Doolen & Hacker, 2005; Marin-Garcia & Conci, 2009; White & Prybutok, 2001; Prado Prado, 2002; White et al., 1999):

Practice	Definition
Visual Management (VM)	The whole workplace is set-up with signs, labels, color-coded markings, etc. such that anyone unfamiliar with the process can, in a matter of minutes, know what is going on, understand the process, and know what is being done correctly and what is out of place.
Continuous Improvement (CI)	Incremental improvement of products, processes, or services over time, with the goal of reducing waste to improve workplace functionality, customer service, or product performance.
Total Quality Management (TQM)	Establish quality as the top priority of the organization's business objectives. This includes involvement in the quality effort by all functions and employees and implementation of statistical quality control methods for defect prevention is an integral part of the program.
Process Standardization (SOP)	Systematize how a part is processed, and includes man-machine interactions and studies of human motion. All tasks are organized in the best known sequence and by using the most effective combination of people, materials, methods and machines.
Reduction of Set up time (SMED)	This practice would attempt to reduce the time and costs involved in changing from the tooling, layout, etc. required to produce one product to that required to produce other products.
Total productive Maintenance (TPM)	Attempt to establish and refine routine preventive maintenance and replacement programs, getting machine operator actively participating in the minor machine maintenance functions in order to eliminate losses tied to equipment maintenance or, in other words, keep equipment producing only good product, as fast as possible with no unplanned downtime.
Just in Time (JIT)	To make one piece at a time correctly all the time, without unplanned interruptions and without lengthy queue times. Everything that is needed to process the part is within easy reach, and no part is allowed to go to the next operation until the previous operation has been completed. Ensures that suppliers deliver the right quantity at the right time in the right place.
Supplier relationship (SR)	Develop suppliers so they can be more involved in the production process of the firm. The objective is to improve quality, flexibility and levels of service form suppliers by increasing the quantity of orders, reducing the number of suppliers and developing a long term relationship based on trust and providing regular feedback to suppliers about their performance.

Customer relationship (CR)	Focus on a firm's customers and their needs with closer customer-company relationship, customer needs surveys, operation integrated with those of customers and delivery quality information.
Design integrated with manufacturing (DFMA)	Designing products in such a way that they are easy to manufacture, choosing the right material, the machines and processes used to work the raw material, and the assembly of the product.
High Involvement Work Practices (HIWP): empowerment, training and rewards	Consistently advocates a bottom-up approach to management. Jobs at the lowest level are thought to be best designed when individuals or teams do a complete part of an organization's work process, such as making an entire product or providing a complete service. Ensuring that people are fully trained and truly empowered and installing "enlightened" and realistic performance measures, evaluation, and reward systems

Table 1. Common lean manufacturing practices.

In diverse papers it has been stated that the application of these practices has beneficial effects for the given company. These effects are greater if large groups of practices are implanted, and not in an isolated way, taking advantage of a synergetic effect between them (White & Prybutok, 2001). Thanks to lean manufacturing companies can improve their productivity as much in terms of workforce as machinery, and can reduce time between receipt of order and completion of production, can improve internal and external quality and reduce inventory levels and unit costs (Callen et al., 2000; White & Prybutok, 2001; Marin-Garcia et al., 2009). All of this allows an improvement in competition.

The majority of experiences related to lean manufacturing have taken place in companies that make elevated quantities of the same product in repetitive processes (production lines). Amongst these the automobile industry, and its auxiliary companies, and the consumer electronics industry stand out. However, jobs that justify the benefits of these systems also exist in other sectors, as much in process companies (food, chemicals, pharmaceuticals, detergents...) as other types of company (textile, industrial machinery, metallic components, compressors, hydraulic valves, electrical consumer goods, plastics...)(Schonberger, 1996), even in those companies that make highly differentiated products of which few of each model are made (James-moore & Gibbons, 1997; White & Prybutok, 2001). Nevertheless, we have to take into account that the use of these practices is most seen in companies with repetitive configurations (line or process) than in those with non repetitive configurations (projects or workshops) (White & Prybutok, 2001). As well, the results obtained by companies are relatively better with repetitive configurations, where complex and standard consumer products are made. However, other types of processes can also be improved with these techniques, although at a lower level (Lee, 1996).

On the other hand, it appears to be demonstrated that plants with less than 250 employees use these systems less (Schonberger, 1996; White & Prybutok, 2001). For smaller companies it is better to undertake a sequential implantation of those tools available to them, starting with the easiest and cheapest. Perhaps even for bigger companies this is the most efficient tactic for launching lean manufacturing.

Taking into account the fact that the implantation of these practices is a gradual process, it is vital to reflect if it is possible to find the most appropriate order in which the practices should be put into action in the auxiliary automobile industry (Marin-Garcia et al., 2010). A general proposal exists by Monden (1998) and, in this paper, we will present an adaption to the particular necessities of the sector upon which we are focusing. We will begin with describing how, over the last decade, the necessities as much as the practices of lean manufacturing have evolved in a Spanish automotive supplier sample. We will also analyse the evolution over these 10 years of those continuous improvement programs implanted to give support to lean manufacturing in 11 companies within the sector. From these experiences a proposal for a lean manufacturing practice architecture will be proposed, indicating the implantation procedure most appropriate for automobile supplier companies.

3. Evolution of the use of practices and priorities

The grade of deployment of lean manufacturing practices has been analysed via a survey taken by Spanish companies belonging to the cluster of suppliers to the automotive industry. The majority of these companies are small and medium sized (Table 2) and there is no a significant difference between the samples from the years 2000 and 2010, although over this decade there had been a process of concentration that gave origin to mergers, acquisitions, and closures, generating an increase in the number of large companies in the sector.

Workers	2000	2010
less than 50	39%	24%
between 50 and 249	48%	56%
250 or more	13%	21%
N	31	33

Table 2. Distribution of companies by size.

The data that we are going to present in this section were obtained via a questionnaire within which companies were asked about the level of deployment of the different practices with a range of answers between 0, nothing, up to 5, very much (Marin-Garcia & Carneiro, 2010).

In Figure 1 we show the evolution of strategic priorities in the area of production. Over these 10 years the relative importance of priorities has changed, although the principal priorities in 2010 continue to be lead-times and the reduction of defects. However, in these 10 years, the necessity to fight competitors on the basis of cost has moved from third to eighth place, and to achieve an increased workforce productivity has changed from penultimate to fourth place. We can appreciate how the sensation of having to attend to all competitive factors has increased over the decade, and that current companies score the 11 dimensions with a grade of importance greater than 4, while in the year 2000 there were 6 dimensions with an importance of below 4. Without a doubt this has increased the complexity of operations management in having to try to simultaneously improve diverse indicators that, in some cases, can have some grade of incompatibility. Lastly, to highlight that the increment of strategic priorities is statistically significant in 5 of these (to increase the flexibility to modify products, attend to variable demand, reduce leadtimes, to integrate production decisions with the company strategy and to increase workforce productivity).

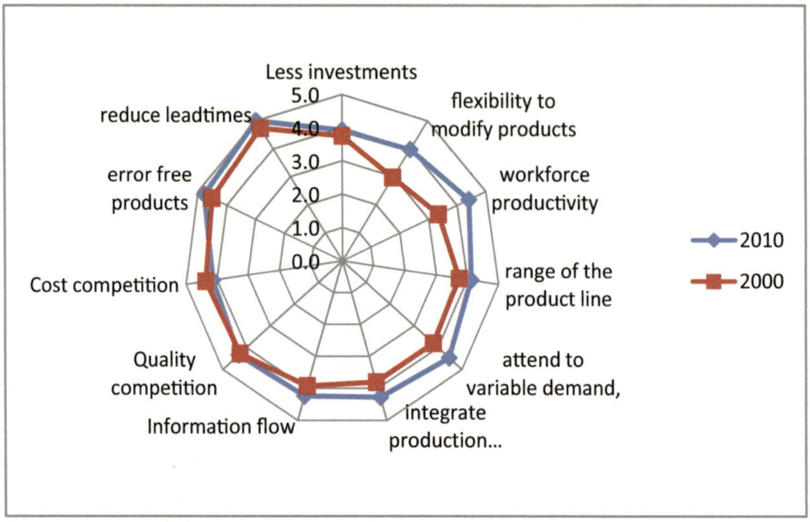

Fig. 1. Evolution of competitive priorities in operations management.

In Figure 2 we show the evolution of deployment of lean manufacturing practices. The differences are statistically significant in all except preventive maintenance, supplier relationship and JIT with suppliers. TPM and supplier relationship were at a high level of deployment in 2000 and have remained amongst the most developed in 2010. The third was one of the least used in 2000 and continues to be one of the most complicated to implant in 2010. In part because it requires the prior deployment of other tools than still have not achieved an adequate level of development in the sector, and in part because second level suppliers are smaller companies with fewer resources that encounter many difficulties when implanting and maintaining lean manufacturing in their companies. However, over these 10 years the grade of implantation of practices such as SOP, SMED, customer relationship, internal JIT and empowerment has increased greatly, allowing that the sector has passed from an initial stage to a moderate deployment of lean manufacturing practices.

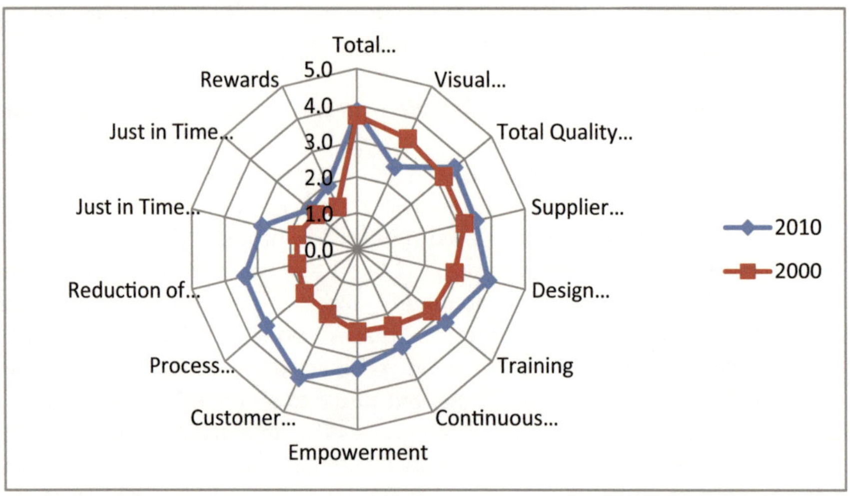

Fig. 2. Grade of use of lean manufacturing practices.

The only practice that has fallen back in its grade of use is visual management. Perhaps this is due, as we will see in the next section, to the fact that in the year 2000 companies had just launched these practices (those which all companies began with) and, with time, the practice has diminished in use for a lack of discipline in maintenance. It could also be due the ever more frequent use of computers in the collection and analysis of data while the use of informative screens (touch screen or not) has not been popularized in production lines; which means that the information that previously was distributed on paper (and even hand written) is now transmitted in electronic format, without having achieved the visual impact of traditional procedures.

To complete the analysis of the evolution of the last decade, in the next section we are going to comment upon how the process of deployment has been in 11 companies that have in common having taken part in a development programme for automotive suppliers around the year 2000 (Marin-Garcia et al., 2009).

4. Evolution of continuous improvement programs to support the implantation of lean manufacturing

In this section we resume the qualitative investigation undertaken in 11 companies from the previous sample. These 11 companies participated during 2000-2001 in a series of Kaizen events led by external consultants (lean managers of the main client) with the objective of deploying lean manufacturing in the company. These companies, belong to different industries and manufacture various products, among which are soundproofing, metal stamping, welded parts, nuts and bolts, plastics (injection and moulded), mechanical sets and electrical products.

All the companies received the same intervention, summarised in the four steps described below:

Step 1. Selecting the line or process to be observed in the plant.

Step 2. Initial diagnosis of the situation of the line selected. This diagnostic period usually lasted 2 days.

Step 3. Development of the Kaizen-Blitz activities and action. A workshop dynamic of 4-5 complete days duration was used, under the guidance of expert consultants. Groups of 5 to 14 people participated in the workshops, half of whom were workers. These tools ranged from 5S tools, visual factory and re-design of layouts for the less developed plants in lean manufacturing, to Kanban or TPM techniques for those in which some others tools had already been introduced. At the end of the week, the group had developed the chosen improvements and had proposed an immediate action plan for further improvements that would require the approval of the management. Finally, a date was agreed for follow-up on the evolution of the indicators of productive efficiency. This process was repeated two or three times in each company during a 9-month period until the objectives specified in the initial diagnosis was fulfilled. In other words, two or three Kaizen events were carried out in each company.

Step 4. Drafting a report to reflect the summary of the activities, to be added to the research database.

The main results obtained in the eleven cases analysed are summarised by a notable improvement in the efficiency of the machines (approximately 18%), mainly obtained due to a radical improvement in the changeover time (reductions of almost 60% of the original time); improvement in the quality rate of nearly 5%; reduction of inventory levels by almost 40% and an increase in productivity between 9% and 60%. Along with this, we also detected important improvements in the use of the space in the plant, a reduction in the number of containers and the distance travelled by products (Marin-Garcia et al., 2009).

We interviewed the production directors of these companies with the objective of learning how they valued the workshops undertaken ten years ago, what was the deployment process of lean manufacturing after that experience, what difficulties they found and how they overcame them (Fendt & Sachs, 2008; Charmaz, 2006).

The majority of the interviewees do not doubt that the experiment was a success. To value it in this way is not only based on the positive evolution of the Key Performance Indicators (KPIs) such as FTT, OEE, DTD or productivity (see below), they also take into account the impulse needed for the deployment of lean manufacturing, or the knowledge that it allowed them to attain. In this sense, the involvement of the consultants was valued, the practical experience they had, and the transfer of real solutions that had been tried in similar situation. For many of the interviewees, these workshops from 10 years ago showed them "all I know about lean manufacturing". However, not all the opinions are favourable. In a few companies it is considered that "it isn't worth anything", "the customer came to sniff around our processes and to impose a cost reduction, with hardly any help in achieving this end". It is interesting to observe that the assessment of success or failure of the workshops did not depend on whether the company had begun or not the path towards lean manufacturing before the arrival of the external consultants. Although it is possible that the action of the consultants was not exactly equal in all the companies, it appears to be more probable to think that the reaction from the companies can be seen as culturally conditioned (there are companies where they do not like it when outsiders come to tell them how to do things, or that try to introduce methodologies that clash with company or holding group politics), or for reasons of commercial friction far from the Kaizen events.

With respect to when the companies began the deployment of lean manufacturing, the majority undertook it around 2000. One company had started with lean manufacturing implantations around 1995. Another company began in 1999 with 5S, SMED and TPM. Amongst the others, some had undertaken Kaizen events after the continuous improvement approach, but without a methodology of lean manufacturing deployment perspective. Other had not undertaken anything more than have started up a suggestions system. Therefore, for the majority, the first real contact with a lean manufacturing deployment was the Kaizen events. The evolution over the ten previous years differed in each of the companies. However, two groups can be seen.

The first of these, the most common, is the gradual loss of impulse once the Kaizen events are over. The attained achievements and the initiated dynamics gradually degraded and, after 12-24 months, the situation with respect to lean manufacturing was very similar to that of the year 2000. Perhaps not all of the tools lost their effect. For example, it has been stated that some maintenance of 5S and SMED has been seen. But in general terms the system remains at 15-20% of what it could have achieved if the implantation had been continued. The motives for this were principally the lack of management support. In some cases because "they didn't believe in the system" or "the management support was like a theatre, the client wanted us to do it so we did it". In others, due to the fact that the growth in business overwhelmed capacity and "to attend to urgent matters robbed us of time we were able to dedicate to important matters". Another common cause for the fall off in the system was due to the companies not being able to give the necessary resources for the system to work. One of the resources was money for small investments. But the principle resources lacking, in the opinion of the managers interviewed, was the ability to dedicate the time of someone who took command the lean manufacturing deployment or the ability to free up workers from the production line so they could dedicate some time to working on the pilot production line in lean manufacturing tools. This difficulty is still current in the year 2010 in some companies. Lastly, another cause for the interruption in lean manufacturing deployment was the wear and tear that it generates in those who keep the systems moving. These people have to be convincing management and workers alike, training, following, paying attention to possible improvement methods... and this task is never done. Something which can begin as an interesting challenge ends up becoming "a pain when the necessary support and resources are not available".

The second group is characterized by companies who continue with lean manufacturing system deployment, and some of the first groups that one, two or three years after they stop it (which is to say 4-5 years after the first implantation) decide to look again at, and restart, the implantation of lean manufacturing. In these cases, the principal driver of the new initiative comes from changes in management personnel. All the companies in this group coincide in that the success of the continued implantation is based in various things. Perhaps the principal is the explicit support of management. Another, very important, is to achieve a change in culture to highlight a philosophy of continuous improvement where the maintenance of improvements is seen as important as putting them into place. In this sense, standardization is a key part in sustaining the system. This cultural change has been brought about by training and "preaching the example" by management. The third of the key things seems to be "most focused" which is to say all the actions are focused to achieve something, and it is available a system of indicators (KPIs) to confirm, in time, whether everything is going according to plan, and in the case of problems that can guide as to which corrective actions are necessary. Lastly, those polled agreed that the existence of a "lean champion", with either full time or part time commitment to the role, is crucial to make sure all functions as it should.

5. Proposal for the lean manufacturing implantation process

Starting with the experience of the companies interviewed, the implantation process should begin with the breaking down of competitive priorities into KPIs that allow us to measure how the company is evolving. In the auxiliary automotive sector it is common to find these indicators (Maskell, 1995; Giffi et al., 1990; Dal et al., 2000; Suzaki, 1993):

- Production: Manpower productivity.
- Quality: FTT (First-Time-Through); customer returns/warranty; rejection/rework
- Cost: Buying cost/unit produced; cost of logistics; Dock-To-Dock (DTD), Overall Equipment Effectiveness (OEE), Build To Schedule (BTS).
- Delivery: Delay in delivery, lead time.
- Safety: Accidents.
- Morale: Employee satisfaction surveys, number of suggestions, absenteeism, turnover.

When the company has chosen its priority indicators it is advisable to undertake a prior diagnostic and the drawing up of a Value Stream Map (VSM) (Tapping et al., 2002; Rother & Shook, 1998). In this way, the current state can be documented and a better focus towards that which most interests the company can be considered. With the data from the diagnostic the most suitable pilot area can be chosen, along with the group of action to be undertaken.

Perhaps workers can be involved in the diagnostic, with this helping to start the implantation process.
In general, it is possible to draw up an itinerary for the recommended implantation order of the tools. Although we must take into account that the sequence proposed can need to be altered in an actual implantation, in function of the analysis of the diagnostic undertaken by both the project team and the external experts collaborating in the implantation.
The next stage following the diagnostic would be to raise awareness and to involve all personnel in the process of continuous improvement. Often the deployment of some 5S followed by visual management can be a good start in the pilot area if it is combined with the use of human resource management practices (training, empowerment and rewards), in such a way so as to achieve worker commitment and so the worker takes on board and even brings about the necessary changes in the company (Lee, 1996; Lee, 1996; Martínez Sánchez et al., 2001; Lawler III et al., 2001).
Following this, if the company has automated processes, it is convenient to undertake the implantation of SMED and TPM. The next stage, for those companies that need it, would be line balancing and cellular manufacturing.
Standardization of processes is advisable between each of the processes thus far commented upon, to maintain the advances achieved. Afterwards JIT and Kanban systems can be looked at.
In parallel, there are other practices that can be gradually incorporated, enough to satisfy the competitive necessities. We refer to integrated design, TQM, client relationships and supplier relationships.
The Figure 3 represents the stages thus far stated. The tools on top act as support and should be present in all implantations. Those on the right complement other system tools although it can be said that they are not necessary in all companies, or do not have an exact moment to be placed into action (they have fewer precedence restrictions than other practices represented in the figure).

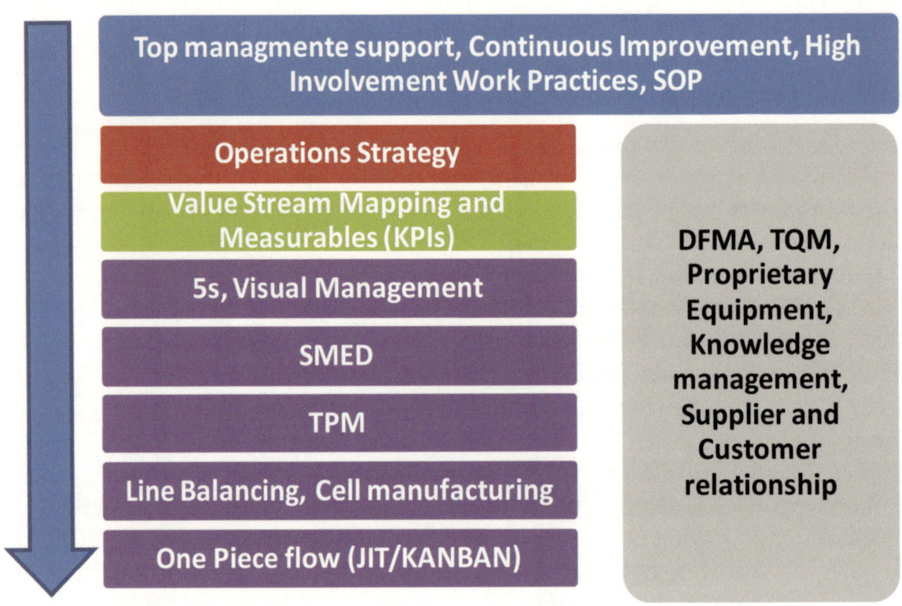

Fig. 3. Implementation process

6. Conclusion

In this paper we have analysed the different practices of lean manufacturing, the evolution of its grade of use in the auxiliary automotive industry between 2000 and 2010 and how this evolution has been experienced in some companies.

Starting from the experience of a group of companies, a success lean manufacturing implantation process should have the following steps:

1. Explicit support from upper management: implantation requires continuous effort from the whole company. Much can be gained from implantation, but it is necessary to maintain constant striving towards continuous improvement. Towards this end it is advisable that all personnel are clear that the upper management unconditionally support the project and provide the necessary resources.
2. The establishment of a project team to lead the implantation. Heading this group it is convenient to have a lean manufacturing "champion" or leader. The objectives of this team are usually, amongst others: spread good practice throughout the company, provide training on tools and techniques, and establish implantation objectives and to supervise the advancement. Probably the support of an industry cluster association would be the key in giving support to these teams.
3. Choosing a methodology that guides and structures the implantation project.
4. Selection of pilot projects and the progressive deployment of the implantation.

The order in which practices are implanted suggested by us in the implantation process section allows a progressive construction of a solid base for lean manufacturing. First phase practices tend to be easier to implant, but we must advise that even the simplest practice is complicated to maintain, thus meaning a change in attitudes and collective conduct is necessary. Support, supervision and constant reminder from upper management is required so that the gains obtained from the implantation are maintained over time, and so that we do not return at the beginning.

7. References

Avella, L., Fernandez, E., & Vazquez, C. J. (2001). Analysis of manufacturing strategy as an explanatory factor of competitiveness in the large Spanish industrial firm. *International Journal of Production Economics*, Vol. 72, No. 2, pp. 139-157.

Callen, J., Fader, C., & Kirnksky, I. (2000). Just-in-time: A cross-sectional plant analysis. *International Journal o Production Economics,* No. 63, pp. 277-301.

Carrasqueira, M. & Machado, V. C. (2008). Strategic logistics: Re-designing companies in accordance with Lean Principles. *International Journal of Management Scienceand Engineering Management*, Vol. 3, No. 4, pp. 294-302.

Charmaz, K. (2006). *Constructing grounded theory. A practical guide through qualitative analysis,* SAGE, 10 0-7619-7353-2, London.

Dabhilkar, M. & Ahlstrom, P. (2007). The Impact of Lean Production Practices and Continuous Improvement Behavior on Plant Operating Perfomance, *Preceedings of 8th International CINet Conference*, Gothenburg

Dal, B., Tugwell, P., & Greatbanks, R. (2000). Overall equipment effectiveness as a measure of operational improvement - A practical analysis. *International Journal of Operations & Production Management*, Vol. 20, No. 12, pp. 1488.

Devaraj, S., Hollingworth, D. G., & Schroeder, R. G. (2004). Generic manufacturing strategies and plant performance. *Journal of Operations Management*, Vol. 22, No. 3, pp. 313-333.

Doolen, T. L. & Hacker, M. E. (2005). A Review of Lean Assessment in Organizations: An Exploratory Study of Lean Practices by Electronics Manufacturers. *International Journal of Manufacturing Systems*, Vol. 24, No. 1, pp. 55-67.

Fendt, J. & Sachs, W. (2008). Grounded Theory Method in Management Research: Users' Perspectives. *Organizational Research Methods*, Vol. 11, No. 3, pp. 430-455.

Garcia-Sabater, J. J. & Marin-Garcia, J. A. (2010). Can we still talk about continuous improvement? Rethinking enablers and inhibitors for successful implementation. *International Journal of Technology Management*, Vol. In Press.

Giffi, C., Roth, A., & Seal, G. (1990). *Competing in worl-class manufacturing*, Irwin, 1-55623-401-5, Homewood.

González Benito, J. & Suárez González, I. (2007). El alineamiento de la estrategia competitiva, la estrategia de producción, las capacidades productivas y los resultados empresariales, pp. 325-334, International Conference on Industrial Engineering & Industrial Management - CIO, Madrid.

Gurumurthy, A. & Kodali, R. (2008). A multi-criteria decision-making model for the justification of lean manufacturing systems. *International Journal of Management Scienceand Engineering Management*, Vol. 3, No. 4, pp. 100-118.

Hayes, R. H. & Wheelwright, S. C. (1984). *Restoring Our Competitive Edge: Competing Through Manufacturing.*, John Wiley & Sons, New York.

James-moore, S. M. & Gibbons, A. (1997). Is Lean Manufacture Universally Relevant - An Investigative Methodology. *International Journal of Operations & Production Management*, Vol. 17, No. 9-10, pp. 899+.

Jorgensen, F., Laugen, B., & Vujovic, S. (2008). Organizing for Continuous Improvement, *Preceedings of 9th International CINet Conference*, Valencia

Ketokivi, M. A. & Schroeder, R. G. (2004). Strategic, structural contingency and institutional explanations in the adoption of innovative manufacturing practices. *Journal of Operations Management*, Vol. 22, No. 1, pp. 63-89.

Lawler III, E. E., Mohrman, S., & Benson, G. (2001). *Organizing for high performance: employee involvement, TQM, reengineering, and knowledge management in the fortune 1000. The CEO report*, Jossey-Bass, 0-7879-4397-5, San Francisco.

Lee, C. Y. (1996). The applicability of just-in-time manufacturing to small manufacturing firms: An analysis. *International Journal of Management*, Vol. 13, No. 2, pp. 249-259.

Lewis, M. W. & Boyer, K. K. (2002). Factors impacting AMT implementation: an integrative and controlled study. *Journal of Engineering and Technology Management*, Vol. 19, No. 2, pp. 111-130.

Liker, J. K. & Wu, Y.-C. (2000). Japanese automakers, U.S. Suppliers and supply-chain superiority. *MIT Sloan Management Review*, Vol. 42, No. 1, pp. 81.

Marin-Garcia, J. A. & Carneiro, P. (2010). Desarrollo y validación de un modelo multidimensional de la producción ajustada. *Intangible Capital*, Vol. 6, No. 1, pp. 78-127.

Marin-Garcia, J. A. & Carneiro, P. (2010). Questionnaire validation to measure the application degree of alternative tools to mass production. *International Journal of Management Science and Engineering Management*, Vol. 5, No. In press.

Marin-Garcia, J. A. & Conci, G. (2009). Exploratory study of high involvement work practices: Identification of the dimensions and proposal of questionnaire to measure the degree of use in the company. *Intangible Capital*, Vol. 5, No. 3, pp. 278-300.

Marin-Garcia, J. A., Garcia-Sabater, J. J., & Bonavia, T. (2009). The impact of Kaizen Events on improving the performance of automotive components' first-tier suppliers. *International Journal of Automotive Technology and Management*, Vol. 9, No. 4, pp. 362-376.

Marin-Garcia, J. A., Pardo del Val, M., & Bonavía Martín, T. (2006). The Impact of Training and *ad hoc* Teams in Industrial Settings. *International Journal of Management Science and Engineering Management*, Vol. 1, No. 2, pp. 137-147.

Marin-Garcia, J. A., Pardo del Val, M., & Bonavía Martín, T. (2008). Longitudinal study of the results of continuous improvement in an industrial company. *Team Performance Management*, Vol. 14, No. 1/2, pp. 56-69.

Marin-Garcia, J. A., Pardo del Val, M., & Bonavía Martín, T. (2009). Los sistemas productivos, el aprendizaje interno y los resultados del área de producción baldosas cerámicas. *CIT- Revista de Información Tecnológica*, Vol. 20, No. 1, pp. 39-52.

Marin-Garcia, J. A., Perello-Marin, M. R., & Garcia-Sabater, J. J. (2010). Desarrollo de una metodología para identificar dependencia de camino en gestión de operaciones. *Working Papers on Operations Management*, Vol. 1, No. 1, pp. 37-40.

Martín Peña, M. L. & Díaz Garrido, E. (2007). Impacto de la estrategia de producción en la ventaja competitiva y en los resultados operativos, pp. 367-377, International Conference on Industrial Engineering & Industrial Management - CIO, Madrid.

Martínez Sánchez, A., Pérez Pérez, M., & Urbina Pérez, O. (2001). Flexibilidad organizativa y relación entre JIT y calidad total. *Alta Dirección*, Vol. 35, No. 210, pp. 74-84.

Maskell, B. H. (1995). *Sistemas de datos de industrias de primer nivel mundial*, TGP-Hoshin, 84-87022-15-4, Madrid.

Monden, Y. (1998). *Toyota Production System: An integrated approach to Just in Time*, Engineering and Management Press, 978-0898061802.

Morris, M., Bessant, J., & Barnes, J. (2006). Using learning networks to enable industrial development - Case studies from South Africa. *International Journal of Operations & Production Management*, Vol. 26, No. 5-6, pp. 532-557.

Oliver, N. & Delbridge, R. (2002). The characteristics of high performing supply chains. *International Journal of Technology Management*, Vol. 23, No. 1-3, pp. 60-73.

Peng, D., Schroeder, R. G., & Shah, R. (2008). Linking routines to operations capabilities: A new perspective. *Journal of Operations Management*, Vol. 26, pp. 730-748.

Prado Prado, J. C. (2002). JIT (justo a tiempo), TQM (calidad total), BPR (reingeniería),...¿Distintos enfoques para incrementar la competitividad? *Esic Market*, No. 112, pp. 141-151.

Rother, M. & Shook, J. (1998). *Learning to see. Value stream mapping to create value and eliminate muda.*, Lean Enterprise Institute, 0-9667843-0-8, Massachusetts.

Schonberger, R. J. (1996). *World Class Manufacturing: the next decade*, Free Press, 0-684-82303-9, New York.

Shah, R. & Ward, P. T. (2007). Defining and developing measures of lean production. *Journal of Operations Management*, Vol. 25, No. 4, pp. 785-805.

Skinner, W. (1969). Manufacturing. Missing link in corporate strategy. *Harvard Business Review*, No. May-June, pp. 136-145.

Suzaki, K. (1993). *The new Shop floor management: empoweing people for continuous improvement*, Free Press, 0-02-932265-0, New York.

Tapping, D., Luyster, T., & Shuker, T. (2002). *Value Stream management eight steps to planning, mapping, and sustaining lean improvements*, Productivity Press, 1-56327-245-8, New York.

Treville, S. d. & Antonakis, J. (2006). Could lean production job design be intrinsically motivating? Contextual, configurational, and levels-of-analysis issues. *Journal of Operations Management*, Vol. 24, No. 2, pp. 99-123.

Urgal González, B. & García Vázquez, J. M. (2005). Análisis estratégico de las decisiones de producción estructurales desde un enfoque basado en las capacidades de producción. *Revista Europea de Dirección y Economía de la Empresa*, Vol. 14, No. 4, pp. 101-120.

Vazquez-Bustelo, D. & Avella, L. (2006). Agile manufacturing: Industrial case studies in Spain. *Technovation*, Vol. 26, pp. 1147-1161.

White, R. E., Pearson, J. N., & Wilson, J. R. (1999). JIT manufacturing: A survey of implementations in small and large U.S. manufacturers. *Management Science*, Vol. 45, No. 1, pp. 1-16.

White, R. E. & Prybutok, V. (2001). The relationship between JIT practices and type of production system. *Omega*, Vol. 29, No. 2, pp. 113-124

9

Identifying and Prioritizing Ecodesign Key Factors for the Automotive Industry

Miriam Borchardt, Miguel Afonso Sellitto, Giancarlo Medeiros Pereira, Leonel Augusto Calliari Poltosi and Luciana Paulo Gomes
UNISINOS - Vale do Rio dos Sinos University
Brazil

1. Introduction

One of the key causes that most contribute to the environmental degradation that threatens the planet is the increasing production and consumption of goods and services. Some of the factors that contribute to that are: (a) the lifestyle of some societies; (b) the development of emerging countries; (c) the ageing of population in developed countries; (d) the inequalities among regions of the planet; and (e) the ever smaller life cycle of products (Maxwell et al., 2006).

The balance between environmental "cost" and functional "income" is essential for sustainable development, resulting that environmental issues must now be merged into "classical" product development processes (Luttropp & Lagerstedt, 2006). Concepts such as ecodesign, cleaner production, design for (the) environment, recycling projects and development of sustainable products promote a re-design at techniques, like conceptualization, design and manufacturing of goods (Byggeth et al., 2007).

Ecodesign is a concept that integrates multifaceted aspects of design and environmental considerations aiming to create sustainable solutions that satisfy human needs and desires. The product is a part of life-style and design, as well as ecodesign, relate to more than the rational function of a product or service (Karlsson & Luttropp, 2006).

There are several motivations for implementing ecodesign besides the environmental aspects, e.g. cost savings, competitive advantage, image of the company, quality improvement, legal requirements. Large companies consider the implementation of ecodesign as a way to preserve the environment as well the competitiveness and the image of the organization. Nevertheless, small and medium enterprises still need to be convinced of the advantages and possibilities of ecodesign (Vercalsteren, 2001). A priori, SMES rarely integrate the analysis of environmental restrictions to their field of knowledge (Pochat et al., 2007).

Another difficulty presented for companies in general, and SMES in particular, refers to the ecodesign tools. Most require application by experts (Pochat et al., 2007; Rao, 2004). Moreover, many tools for ecodesign fail because they do not focus on the design, but seek retrospective analysis based on existing products (Lofthouse, 2006). Indeed, ecodesign, as a process, must be integrated into the design and management processes of the company. Not only appropriated tools for ecodesign are needed, but also tools that can help designers to link then to their conventional tools (Pochat et al., 2007). A lot of different requirements for ecodesign are proposed in literature. Main of them regards materials, components, processes and products characteristics, use of energy, storage and distribution, packaging and waste (Wimmer et al., 2005; Luttropp & Lagerstedt, 2006; Fiksel, 1996).

Among others, the automotive electronics industry hosts ecodesign initiatives in response to the regulations and to the innovation's demand verified in this industry (Ferrão & Amaral, 2006; Mathieux et al., 2001).

Aiming to contribute to increase knowledge on ecodesign practices and management, the first part of this chapter highlights some of the key factors that influence the adoption and implementation of ecodesign practices in manufacturing companies. The discussion focuses particularly on a case study which illustrates how ecodesign is being incorporated into the design of products manufactured by a mid-sized automotive electronics supplier in Brazil.

An analysis of the performance of ecodesign is also contributive in this subject. Authors such as Cabezas et al. (2005) and Svensson et al. (2006) have been working on the development of performance indicators associated to ecodesign; they highlight, however, there is no common sense to that matter. Despite of how frequent the environmental performance is present in literature, it has not been found a shape of guide lines or an objective method that might generate an instrument for measuring the application or performance for ecodesign practices. Such instrument would avoid all efforts towards ecodesign to result contradictory and ineffective and could, as well, guide the organizations giving priority to resources where environmental gains are more meaningful.

For the prioritization of resources and actions related to ecodesign, supported by papers that discuss evaluation and performance in environmental aspects, it is understood to be relevant the identification of the degree of importance of each key factor of ecodesign for companies of a particular industry and how much each company fulfils each requirement.

This investigation also aims to prioritize resources and actions of ecodesign. Supported by Hermann et al. (2007), which speak on measurement of performance on environmental aspects, the authors consider relevant to identify the degree of importance of each ecodesign construct for companies in a particular industry and to evaluate the degree of application of ecodesign constructs.

Considering the context presented, the main objective of the second part of this chapter is to assess the performance of ecodesign in a chemical company that supplies the automotive industry. Secondary objectives were: (a) to identify latent constructs and indicators that explain the ecodesign performance of the operation; (b) to assess the relative importance of ecodesign constructs (practices), supported by the Analytic Hierarchy Process (AHP); (c) to assess the degree of application of ecodesign constructs (practices); (d) to evaluate the gaps between importance and application of ecodesign constructs. For doing so, it was developed a method to evaluate the performance in ecodesign. The method was developed taking into account that the application in other industries is feasible.

After this introduction, the chapter presents: theoretical background about ecodesign implementation, practices and discussion about the reasons for adoption; theoretical background about environmental performance measurement; research methodology, findings, discussions and contribution for the first and the second objectives; and conclusions and suggestions for continuity. Limitations of the research are those related to the research method, that is, the results are valid for the case, nor for the entire industry, but the method can be replicated elsewhere, if applicable.

2. Ecodesign

2.1 Concepts, implementation and practices

Kazazian (2005) focuses on eco-conception, which is the process of applying the concepts of ecodesign. With this approach, the environment is considered to be equal in importance to factors such as technical feasibility, cost control, and market demand. Eco-conception can lead to three different levels of eco-design intervention when designing a product: (a) optimization for environmental impact reduction, (b) more intensive development efforts, such as modifying the product, and (c) "radical" intervention, such as substitution of different products or services (Kazazian, 2005).

Boks (2006) stresses the importance of product designers, emphasizing their unique position and ability to influence environmental strategies. Designers can have a key impact when they enlarge the focus of their efforts, giving the environment a prominent position in defining the parameters of product development.

Karlsson & Luttropp (2006) note that ecodesign incorporates priorities related to sustainability into the overall business scenario. The "eco" in ecodesign can refer to both economics (reflecting a business orientation) and ecology (reflecting the importance of environmental aspects) (Figure 1).

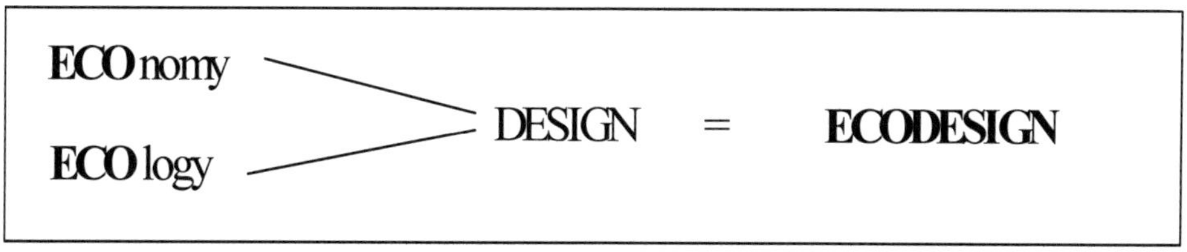

Fig. 1. The linguistic map of the word ecodesign (Karlsson & Luttropp, 2006).

2.1.1 Potential of a company for the application of ecodesign

Regarding the potential of a company for the application of ecodesign, and consequently its insertion on products development routine, the organization must evaluate internal facts, external facts and the product (Vercalsteren, 2001). Internal factors are: (a) company motivation; (b) innovation, considering the ability of the company into influencing the specifications of the product; (c) competitiveness, once a company that is leader of a specific sector in the market has more chances of re-sketching the products, the smaller companies can consider ecodesign as an opportunity to increase its participation in the market; and, (d) sector, considering that if there already are equivalent initiatives in the sector, the company can learn from these experiences. External factors are: (a) regulation; (b) clients and market, where it is necessary to evaluate whether the market will accept or not the green products; and, (c) suppliers, once it is essential their willing in cooperate. As per the product, it must have the potential for a redesign based under the environmental ponderings (Vercalsteren, 2001).

2.1.2 Practices for ecodesign

Recognized the potential of a company for the application of ecodesign, it is necessary the identification of the key factors that constitute ecodesign. In order to do so, we evaluated propositions from Fiksel (1996), Wimmer et al. (2005) and Luttropp & Lagersted (2006). The synthesis of the proposed practices is presented on Table 1.

First level (key factors)	Second level (items)
Materials: choice and use	(i) ability to use raw material closer to their natural state, (ii) ability to avoid mixtures of non-compatible materials, (iii) ability to eliminate the use of toxic, hazardous and carcinogenic substances, (iv) ability to not use raw materials that generate hazardous waste (Class I); (v) ability to use recycled and / or renewable materials, and (vi) ability to reduce atmospheric emissions caused by the use of volatile organic compounds.
Product components: selection and choice	(i) ability to recover components or to use components recovered, (ii) ability to facilitate access to components, (iii) ability to identify materials and components, and (iv) ability to determine the degree of recycling of each material and component.
Product/Process characteristics	(i) ability to develop products with simpler forms and that reduce the use or consumption of raw materials, (ii) the ability to design products with longer lifetime (iii) capacity to design multifunctional products, (iv) capacity to perform upgrades to the product, and (v) ability to develop a product with a "design" that complies with the world trends
Use of energy	(i) ability to use energy from renewable resources, (ii) ability to use devices for reduction of power consumption during use of the product, (iii) ability to reduce power consumption during the production of the product, and (iv) ability to reduce power consumption during product storage.
Products distribution	(i) ability to plan the logistics of distribution, (ii) ability to favor suppliers / distributors located closer, (iii) ability to minimize inventory in all the stages of the product lifetime, and (iv) ability to use modes of transport more energy efficient.
Packaging and documentation	(i) ability to reduce weight and complexity of packaging, (ii) ability to use electronic documentation, (iii) ability to use packaging that can be reused, (iv) ability to use packages produced from reused materials, and (v) ability to use refillable products.
Waste	(i) ability to minimize waste generated in the production process, (ii) ability to minimize waste generated during the use of the product, (iii) ability to reuse the waste generated, (iv) ability to ensure acceptable limits of emissions, and (v) ability to eliminate the presence of hazardous waste (Class I).

Source: adapted from Wimmer *et al.* (2005); Luttropp & Lagerstedt (2006); Fiksel (1996).

Table 1. Syntheses of practices proposed for ecodesign

2.1.3 Ecodesign tools

Over the past decade or so, a wide range of ecodesign tools have been developed in order to support the application of the ecodesign practices. In many cases, tools have grown out of pilot projects and partnerships between private companies and academic research centers. Pochat et al. (2007) identified more than 150 ecodesign tools. More tools have been created as interest in ecodesign increases.

Despite the plethora of tools available, ecodesign is not always promptly adopted by manufacturing companies. Several authors note that industry designers often find the tools difficult to use (Lofthouse, 2006; Pochat et al., 2007; Luttropp & Lagerstedt, 2006; Byggeth & Hochschorner, 2006; Byggeth et al., 2007). According to Lofthouse (2006), tools often fail to be adopted "because they do not focus on design, but instead are aimed at strategic management or retrospective analysis of existing products." The author notes that what designers actually need is "specific information on areas such as materials and construction techniques to help them become more easily involved in ecodesign projects." The environmental information associated with ecodesign tools is often very general. In most instances, tools do not provide the detailed and specific information that designers find necessary when working on design projects.

Pochat et al. (2007) note that effective use of ecodesign tools generally requires input from experts. This can create difficulties for many companies, especially small and mid-sized enterprises, in which often lack the resources required to bring in expert assistance.

Moreover, the amount of information available about both materials and product environmental aspects has increased substantially in recent years. This has made ecodesign tools even more difficult and cumbersome to use, and requires them to be updated frequently (Luttropp & Lagerstedt, 2006).

Several authors mention ecodesign checklists. These checklists typically include lists of questions relating to the potential environmental impacts of products. Pochat et al. (2007) see the ecodesign checklist as a qualitative tool that is useful primarily for identifying key environmental issues associated with the life cycle of products. According to Lofthouse (2006), many designers view ecodesign checklists as too general to be useful. In addition, the checklists often are perceived as including too many requirements. Byggeth & Hochschorner (2006) note that ecodesign checklists often require the user to make trade-offs among a variety of different aspects and issues without sufficient direction on which options are the most preferable from the standpoint of promoting sustainability. The checklist user typically must evaluate whether the solutions offered "are good, indifferent, bad or irrelevant."

A number of different ecodesign checklists exist, many of which have been developed by designers and engineers. Despite their potential drawbacks, using these checklists can help implementers record their ecodesign activities and work more cooperatively with other teams (Côté et al., 2006).

2.2 Environmental practices in the automotive industry

Regulation clearly can play an important role in promoting ecodesign. Much of the relevant literature that was reviewed concentrated on regulation in the European Union (EU), which has implemented some important environmental regulatory directives affecting the automotive and electronics industries. These studies include the end-of-life vehicles (ELV) directive, the waste electrical and electronic equipment (WEEE) directive, and the restriction of hazardous substances (RoHS) directive. In addition, the EU has finalized a framework directive for reducing the environmental impacts of energy-using products through ecodesign (Park & Tahara, 2008; Pochat et al., 2007).

The automotive industry operates in a highly competitive market, with worldwide sales and distribution of products. The tolerance for product flaws is low, especially in the case of vehicle safety features. These factors can operate as constraints on the adoption of ecodesign practices by companies in the industry.

2.2.1 Negative environmental impacts

In terms of natural resources, the "environmental balance" for vehicles has always been negative. According to Kazazian (2005), production of a vehicle typically requires displacing fifteen tons of raw material (about ten times the weight of the final product). The production phase also uses large amounts of water. For example, about forty thousand litters of water are required to manufacture a car. During their useful life, vehicles consume fuel and lubricating oils, most often in the form of non-renewable fossil-based resources. Some of the fuel and oil products leak into the environment as contaminants. In addition, each vehicle uses several tires, many of which are not recycled. Moreover, vehicles emit significant quantities of air pollutants, including carbon dioxide (a major greenhouse gas) and sulphur dioxide (which contributes to acid rain).

Vehicles can also be difficult to recycle at the end of their life cycle. They typically contain a variety of different materials (including plastics and metals, as well as electrical and electronic components) that may be costly and challenging to separate.

2.2.2 Efforts to green the automotive industry

These negative impacts, related to the environmental balance for vehicles, reinforce the perception that automobiles and other vehicles are not designed with an emphasis on preserving the environment and promoting sustainability. Partly in response to these perceptions and concerns, car makers are working to make the industry more environmentally friendly.

In recent years, the automotive industry has developed high-performance and hybrid engines. Car makers are using more parts manufactured with recycled composite materials. In addition, more vehicles now run on renewable bio-fuels and use high-durability synthetic lubricating oils.

As noted in the following sections, the automotive industry is also seeking to restrict the use of hazardous substances and to increase the quantity of packaging and materials that are recycled and reused. These issues are particularly relevant to automotive manufacturers that sell products in the European Union. The EU's RoHS directive bans the use of certain hazardous materials as constituents in specified types of electronic equipment (Donnelly et al., 2006).

2.2.3 Restrictions on the use of hazardous materials

Many automotive car assemblers now provide their suppliers with lists identifying hazardous materials that are subject to restriction of use pursuant to applicable laws or standards. Typically, "white lists" identify materials that can be used. "Gray lists" indicate materials that can potentially be used if certain conditions are met or there is sufficient reason to do so. "Black lists" identify materials that are prohibited (Luttropp & Lagerstedt, 2006; Tingström & Karlsson, 2006).

As part of product development, companies that supply automotive assemblers generally must produce statements confirming that they are in compliance with any applicable restrictions on the use of hazardous substances. If they cannot do so, they may be able to request a temporary waiver from the assembler. In connection with such a request, the supplier generally must describe the reasons for the deviation and present a plan of action for meeting the restrictions in the future.

Suppliers to automotive assemblers must also register their products into the International Material Data System (IMDS), a database that contains information (including chemical

composition) on all materials used in the manufacture of cars. The supplier's registration can then be checked against the automotive assemblers' gray and black lists to determine whether there are any deviations.
The company investigated on the first part of the research develops and manufactures products for vehicle assembly. These products are subject to hazardous-materials restrictions and are registered on the IMDS.

2.2.4 Reducing and reusing packaging

The process of assembling an automotive product involves a large number of different items, and the assembly line requires a high degree of standardization. As a result, any reusable forms of packaging that are adopted also generally must be standardized. Boxes typically have identifying information that allows their supplier to be traced. In addition, pallets typically must meet standards that have been established for size dimensions and maximum weights.
The study company involved in the first part of this research is an approved supplier to automotive assemblers. The company employs reusable forms of packaging, even though doing so adds extra costs in terms of administration and transportation.

2.2.5 Conflicts between ecodesign practices and automotive safety requirements

In the automotive industry, parts that are related to safety must be disposed of if they fail. Under the applicable automotive assembler standards, such parts cannot be repaired and re-sold on the market. They may, however, be dismantled and recycled.
This disposal requirement conflicts with the principles of ecodesign. However, the integrity of the automotive product clearly must be safeguarded. In this instance, the automotive industry has indicated that it values accident prevention over the ecodesign principles related to component reuse.

2.3 Assessment of performance in ecodesign

Tingström & Karlsson (2006) highlight the ecodesign´s multidisciplinary, affirming this is not a linear and repetitive process, for it must be tested or measured the effect of the product on the environment by using models. They also point out that in environmental practices and strategies the execution of the plans must be measured by measuring systems that hold the complexity of the object. Sellitto et al. (2010) present the importance of performance measurement systems in several managerial strategies, including those regarding environmental issues. It is seen in Borchardt et al. (2009) the application of AHP (Analytic Hierarchy Process) in the integration of environmental goals in ecodesign.
It has been observed in the researched literature that there are no clear distinctions among performance measurement and performance evaluation terms. For this research, it was adopted the definition proposed by Sellittto et al. (2006): one should talk about performance evaluation when based on assessment of categorical variates and one should mention performance measurement when based on measurement of quantitative variates.
A system for measurement or for performance evaluation must: (a) avoid under-optimize the place; (b) unfold strategic goals up to operational levels; (c) help with full understanding of goals and conflicts structure, strategy trade-offs; and (d) consider aspects of the organizational culture (Bititci, 1995). The usage of several variates in performance measurement remits to multicriteria decision. As per French (1986), it is hardly ever found a

model to be clear and uniformly structured in a multicriteria decision. Deepened discussions about the theory of decision based on multicriterial focus are found in French (1986).

The evaluation of performance requires a model for measurement and communications, which is obtained by mental construction. The most abstract construction is the theoretical term that holds aspects of a definition wide enough, structured upon constructs and concepts. The other constructs are also of abstract construction, deliberately created to answer a scientific purpose, however closer to reality. The concept, at last, it is not the phenomena yet, but it can already communicate its implications. Its dimensions are represented by numerical values - the indicators - that might be combined and summed quantitatively in indexes, according hierarchical theoretical schemes that help represent the intangible reality (Voss et al., 2002).

The structure of performance, in this paper known as ecodesign performance, can be organized in a tree-like structure, illustrated in Figure 2. The tree-like shape can be pondered by methods of decision support, such as AHP (Analytic Hierarchy Process).

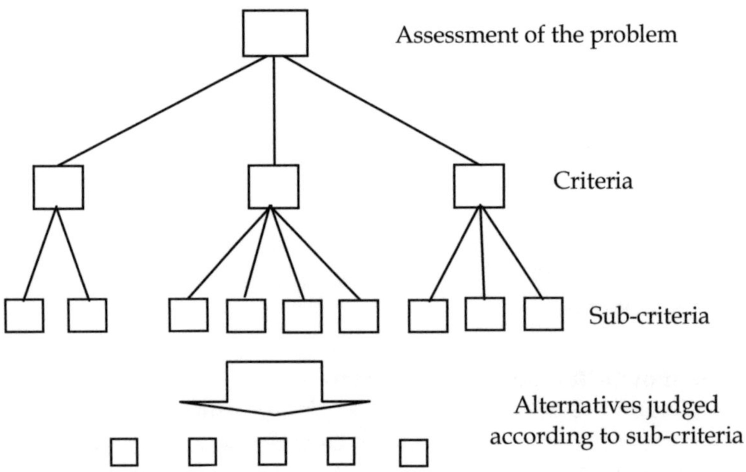

Fig. 2. Structure of hierarchic decision (adapted from Forman & Selly, 2001).

According to Forman & Selly (2001), the AHP forces the decision makers to consider perceptions, experience, intuitions and uncertainties in a rational manner, generating scales of priorities or weights. It is a methodology of compensatory decision, once weak alternatives to an objective can have strong performance in other objectives. The AHP operates in three steps: (a) description of a complex situation of interest under the shape of hierarchic concepts, shaped by criteria and sub-criteria up to the point when, as per decision makers, the assessment of the problem has been enough described; (b) comparing two by two the influence of the criteria and sub-criteria on higher hierarchic levels; and (c) computing the results. The options with preference on pared base comparison, used on AHP, are presented on Table 2. Saaty (1991) recommends the determination of the CRs, the reasons of consistency on assessments, which must be smaller than 0.10. Although the recommendation, we stress that the lower the CR is the better the decision will be, so it is worth seeking lower values for the variate by eventually reviewing judgements.

if a_i related to a_j =	then c_{ij} =	if a_i related to a_j =	then c_{ij} =
equals	1	equals	1
a little more important	3	a little less important	1/3
a lot more important	5	a lot less important	1/5
strongly more important	7	strongly less important	1/7
absolutely more important	9	absolutely less important	1/9

Source: Saaty, 1991, p.22 and 23.

Table 2. Preferential options based on pared comparison

3. 1st Part – Ecodesign implementation at manufacturing company

3.1. Research methodology for the 1st part of the research

The research discussed in this part of the chapter involved a case study of an automotive supplier. The case study methodology allows researchers to examine a subject in depth without separating the subject from its contextual environment (Voss et al., 2002)

Authors have recognized three main types of case studies: exploratory, descriptive, and explanatory. An exploratory case study seeks information and suggests hypotheses for further studies. A descriptive case study investigates associations between the variables defined in exploratory studies. Finally, an explanatory case study presents plausible explanations for associations established in descriptive studies (Yin, 2001).

It has been suggested that a case study can contribute to theoretical research in at least five ways: first by providing, for subsequent studies, a deep and specific description of an object; second by interpreting some regularities as evidence of more generic and not yet verified theoretical postulates; third by heuristic: a situation is deliberately constructed to test an idea; fourth by doing a plausible search based on the theory proposed by the heuristic method; and fifth by the crucial case, which supports or refutes the theory (Easterby, 1975).

3.1.1 Characteristics of the case study

The case study described here is exploratory; we have gathered information and hypotheses for future studies. The contribution this case study makes to theory is of the first type: a thorough description of a specific subject. It is also inductive, as the first in a potential series of studies that could lead to a grounded theory of motivation for ecodesign implementation. This case study was guided by the following questions:

a. Why the company decided to adopt ecodesign practices?
b. How are ecodesign practices being incorporated into routine product design at the study company?

Ultimately, the goal of the case study described here was to provide insights, at the exploratory level, about the elements that induce organizations to adopt ecodesign practices and about the ways in which ecodesign practices can be incorporated into organizations' product design procedures.

3.1.2 Data collection

Much of the information for this case study was collected via five semi-structured interviews with managers in the company's research and development (R&D) department, managers in product design, and the manager of the company's environmental management system. In order to further develop data, we also relied on direct observation and document analysis.

3.2 Results and discussion for ecodesign implementation analysis

The research described here was carried out at a company that supplies electronic components to the automotive industry. The study company operates in Rio Grande do Sul, a state of Brazil and can be classified as mid-sized. The company has obtained certification to both ISO 9001 and ISO 14001.

3.2.1 Products made by the study company

The company produces on-board electronic components for vehicles. Some of the items it supplies were developed to meet individual customer specifications, while others are standardized products.

The first product category consists mainly of electrical relays for switching and voltage converters; these items affect automotive safety since they directly influence the basic function of vehicles. The latter product group includes standardized components used for entertainment applications, such as on-board video and audio systems for buses.

3.2.2 Relationships with vehicle assemblers

The study company supplies its products directly to assemblers of trucks and buses. Some of the company's personnel have in-depth knowledge regarding the design of the vehicles that use its components. As a result, there are confidentiality agreements between the study company and its key employees and between the study company and the assemblers it supplies.

The company has developed a complex business-to-business relationship with its customers. The company must meet applicable regulatory requirements and also depends on customers' approval in order to make changes to its products. The study company has little autonomy in making such decisions.

Since the products manufactured often involve special safety and security features, the company is not allowed to reuse parts, since doing so could compromise functional reliability. However, raw materials (such as plastics, metals, and other materials) can be recycled since they are routed to the primary supplier for inclusion in the overall process of manufacture.

3.2.3 Company environmental management policy

For the past nine years, the study company's environmental management policy has included provisions that are intended to address problems related to resource scarcity. Key issues covered in the company's environmental management policy include (a) energy consumption, (b) materials consumption, and (c) waste handling and treatment.

When automotive assemblers go through the process of qualifying suppliers, they primarily evaluate characteristics such as the supplier's ability to deliver products reliably. Suppliers also must be able to meet all relevant environmental requirements, such as those pertaining to restrictions on the use of hazardous substances. However, using techniques that exceed the applicable environmental protection requirements does not constitute a preferential factor for a given supplier.

3.2.4 Motivations for adopting ecodesign

When asked about their motivation for adopting ecodesign practices in strategic planning, respondents at the study company said that the main drivers involved reducing costs, which had the effect of increasing the company's profit margin and providing it with more flexibility. In the study company's view, cost reduction could be facilitated by dematerializing (using the smallest possible amount of raw material) and by lowering expenditures related to the treatment of waste.

The study company sees implementation of ecodesign as a way to formalize eco-concepts in the new-product development process, allowing for better control of results and continuous improvement.

3.2.5 Ecodesign implementation process

Because the scope of ecodesign is broad, the company formed a multidisciplinary group to handle the study, planning, and strategic deployment of ecodesign techniques. Top management at the company organized a working group that included people with expertise in a range of relevant areas, such as trade, development, product quality, logistics, and industrialization.

The working group focused on activities related to the development of products and processes. The steps they followed in implementing ecodesign are outlined in the following sections.

Study phase

Members of the working group read the relevant literature and made contact with other companies that had already implemented ecodesign methods. Personnel throughout the whole company received training on the basic principles of ecodesign, and staff members' suggestions were collected.

At this stage of the process, the company also analyzed customer demands, along with internal company rules and the requirements of applicable standards such as ISO 9001, ISO/TS 16949 (a quality management system for the automotive industry), and ISO 14001.

Planning

The ecodesign implementation project was framed using the company's projects management methodology, with timelines and financial guidelines established. Regular meetings were held for critical and risk analyses.

Formulation of primary guidelines

The company prepared primary guidelines (IMP - Integrated Management Procedure) that incorporated ecodesign practices and guidance on the development of products and industrial processes.

Formulation of secondary guidelines (operating procedures)

The actual operating procedures for application of ecodesign were deployed via engineering specifications. These procedures involved a high degree of detail and were implemented through checklists, as recommended by Donnelly et al. (2006) for "knowledge management in ecodesign." The company frequently reviews and updates its checklists, allowing new contributions to be recorded and preserving the knowledge gained for future use.

Table 3 offers sample checklists of items to be considered in electric-electronic design and mechanical design of products, along with ecodesign-related recommendations. The checklists consider aspects such as materials recovery, energy efficiency, product simplification, separation of materials, and use of specific manufacturing components, including plastics, metals, and printed circuit boards. These parts are used in various phases of the product design process, including detailing and meeting critical analysis.

The development team suggested extending the principles of ecodesign to software development. Ecodesign principles can be applied to extend the useful life of installed software by providing the ability to receive updates, making the product multifunctional, and preventing downtime with software maintenance routines and remote systems The company encountered some difficulties in the course of implementing ecodesign practices. In particular, when assessing ecodesign concepts and seeking to apply checklists, it lacked technical information on environmental impacts.

For example, in a case where the project team was trying to choose among alternatives for the surface treatment of metals, it was hard to make a choice due to the lack of information indicating the environmental impacts.The team also believes that ecodesign implementation could be expanded to include the company's suppliers. The members agreed that suppliers could be educated about ecodesign and encouraged to adopt proactive attitudes regarding the environmental impact of manufacturing. It was understood, by the group, that sustainability can be achieved only with the engagement of the whole production chain.

Ecodesign item	Checklist to Electric-Electronic Design of the Product	Checklist to Mechanical Design of the Product
1. Material recovery	Give priority to constituents who may have recoverable raw material: for example electrolytic capacitors have recyclable aluminum; tantalum capacitors have not.	Try using plastics and thermoplastics instead of termofixes; do not unite incompatible plastic materials that would make the separation impossible therefore recycling impossible.
2. Components recovery	As standards in the automotive industry, electronic items cannot be repaired at risk of compromising the reliability.	Metal trimmings should be used for smaller parts manufacturing
3. Ease of access to components	Allow repairs during the production line and during the use of the vehicle.	Ease of assembly of the product with minimal fixing components.
4. Simplicity aimed projects	Developing projects with as few electronic components as possible to not compromise the MTBF (Mean Time Between Fails) of the product; occupy less area of the printed circuit board.	Using forms that allow a maximized use of the metal sheet; plastic boxes that allow multiple applications. Using modular cabinets.
5. Reducing the use of raw material	Using SMT (Surface Mountain Technology) components: small electronic components, fixed directly on the printed circuit card, without the use of terminal and connectors. Use the thickest PCB (printed circuit board) possible.	Using aluminized metal sheets, which exempt anti-corrosive treatment preliminary. Using the thickest sheet metal possible avoids screws and painting process.
6. Severability	Using electro-electronic products with fixing elements allowing easy separation of the parties. Identify the requirements of the RoHS on PCB.	Identify all plastic parties with the code of recycling; using adhesives that do not prevent the separation of not compatible parts in terms of recycling.

Ecodesign item	Checklist to Electric-Electronic Design of the Product	Checklist to Mechanical Design of the Product
7. No use of contaminant materials	No use of welding material with lead alloys (lead free solder)	Do not use mechanical materials with contaminants.
8. Recovery and reuse of waste	Waste of paper, copper and aluminum must be separated for subsequent recycling.	Remains of the process of plastic injection should be recycled; all metallic material must be separated for subsequent forwarding to recycling.
9. Waste incineration	All components must meet the regulatory ROHS, with no emission of toxic waste in the incineration process.	All components must meet the regulatory RoHS, with no emission of toxic waste in the incineration process.
10. Reduction of the use of energy in production	Using only one side component PCBs simplifies the solder process and saves energy	Avoid using ultrasound, laser, and other kinds of modern production tools.
11. Employment of devices for reducing energy consumption	Using intelligent electronic circuits that save energy while on stand-by. Using as low speed microprocessors as possible to avoid high energy consumption. Decrease backlight LCD (liquid cristal display) intensity during the night to save energy. Using energy dissipated in equipment for electrical testing of power as heating for stages of the manufacturing process (cure of painting oven, for example).	Design the mechanical parts as light as possible to save fluel during the vehicle's life cycle.
12. Reduction of the use of energy in the distribution	Optimize the process of transport of raw materials and the distribution of the final product.	Optimize the process of transport of raw materials and the distribution of the final product. Package as compact as possible to save transport volume in the transport.
13. Use of renewable energy.	Not applicable.	Not applicable.
14. Multifunctional products	Developing printed circuit board that meet more than one use by mounting options.	Developing plastics and metal cabinets that meet more than one use by assembly options.
15. Specific use of recycled materials	Use of recycled welding material, copper cables, etc.	Using plastic and metal with a high content of recycled material.
16. Use of renewable materials	Use of printed circuit boards made of cellulose.	Use of packaging made of cellulose.

Ecodesign item	Checklist to Electric-Electronic Design of the Product	Checklist to Mechanical Design of the Product
17. Products with higher durability	Implement protection devices to prevent damage to the product in the event of overload or short circuit.	Plastic and/or metal cabinets with index of protection consistent with the application and UV (ultra-violet) resistant, corrosion, temperature and vibration.
18. Packaging recovery	Returnable packaging, reuse of the packaging of raw materials as pads for the packaging of the final products.	Returnable packaging, reuse of the packaging of raw materials as pads for the packaging of the final products.
19. No use of hazardous substances	Mounting boards using solder free of lead. Using only ROHS components.	Answering the RoHS standards.
20. Use of substances with water basis	Using flux to solder type "no clean," that is, with a water-based solvent	Use of paints and adhesives with a water-based solvent.
21. Use of biodegradable products	Not applicable to automotive industry.	Not applicable to automotive industry.
22. Accident prevention	In the event of electrical failure, the product should take the vehicle to a safe state of operation.	In the event of mechanical failure, the product should take the vehicle to a safe state of operation.

Table 3. Checklist to electro-electronic design and mechanical design

a. Training

Employees at all levels of the company were provided with information on the operating procedures involved in applying ecodesign techniques. This training was adapted to the employee's particular involvement with ecodesign implementation.

b. Implementation

This step marked the point at which the company began using ecodesign procedures on both new and ongoing projects, as well as in activities related to improvement of existing products.

c. Maintaining improvement

As knowledge management procedures require, the ecodesign checklists established by the company are continuously updated whenever new information is developed. New insights and experience arising from the application of ecodesign techniques are also incorporated into the company's critical analysis mechanisms.

d. Consideration of Life-Cycle Assessment

After studying the commercial software available for life-cycle assessment (LCA), the working group decided not to adopt the technique. As Chehebe (2002) has noted, LCA results are considered reliable only when the database used for analysis is compatible with the actual conditions at the application site. Thus, in order to be effective for the study company, an LCA database would have to accurately reflect factors such as the availability of raw materials, the cost of transport, and the matrix of energy generation as they exist in Brazil. When a trustworthy LCA database is not available, companies typically will not adopt life-cycle assessment methods.

3.2.6 Shortage of technical information

The company encountered some difficulties in the course of implementing ecodesign practices. In particular, when assessing ecodesign concepts and seeking to apply checklists, it lacked technical information on environmental impacts.

For example, in a case where the project team was trying to choose among alternatives for the surface treatment of metals, it was hard to make a choice due to the lack of information indicating the environmental impacts.

3.2.7 Results of ecodesign implementation at the study company

The study company is still in the process of measuring the results of its ecodesign implementation effort. In addition, products developed entirely under the company's ecodesign system are still undergoing approval by customers.

However, the company has already recognized a positive change in its R&D team's degree of involvement with new materials, new technologies, and environmental issues generally in the design of products. Moreover, the company has observed the following results (short-term, medium-term and long-term) as a consequence of using ecodesign practices:

a. reductions in product costs resulting from dematerialization (medium-term);
b. reduction in the number of products offered by the company as a result of increases in product multifunctionality (long-term);
c. improvement in knowledge management through systematically recording in checklists the development of practices learned (short- term);
d. decrease in the number of raw material items in stock (medium-term);
e. decrease in the number of test sets and assembly devices used in the manufacturing process as a result of streamlining the life cycle of these items (long-term);
f. reduction in the need for investment in the industrial process as a result of the less extensive and diverse set of devices now required (long-term);
g. reduction in environmental management costs, especially with respect to waste (medium-term); and
h. reduction in transport costs for raw materials and semi-ready products (short term).

3.2.8 Prospects for future expansion of ecodesign

As the process of ecodesign continues to be incorporated into the study company's management system, the respondents interviewed reported their optimism about the eventual long-term results. They hope they can effectively transmit their experiences with ecodesign to their suppliers, thereby broadening the range of small and medium-sized businesses that use ecodesign principles as guidelines in the development of products.

4. 2nd Part – Assessing ecodesign implementation dimensions

4.1 Research methodology for the 2nd part of the research

In this part of the chapter a method to evaluate the performance in ecodesign is presented. To exemplify and improve the method, the same has been applied to a company on the chemical sector that supplies the automotive industry.

4.1.1 Characteristics of the research developed on the 2nd part

This second part of the study was guided by the following question: how to assess the ecodesign performance of an industrial operation.

The main objective was to assess the ecodesign performance of a manufacturing operation. Secondary objectives were: (a) to identify latent key words and indicators that explain the key factors of the ecodesign performance of the operation; (b) to assess the relative importance of ecodesign constructs (practices), supported by the Analytic Hierarchy Process (AHP); (c) to assess the degree of application of ecodesign constructs (practices); (d) to evaluate the gaps between importance and application of ecodesign constructs. For doing so, it was developed a method to evaluate the performance in ecodesign.

In the intention of keeping coherence on the terminology used in this chapter, it has been adopted: ecodesign is the top term; ecodesign practices are the constructs; the elements part of the ecodesign practices are the items of application (also known as concepts).

4.1.2 Data collection and method of work

The stages of development of this research were: (a) the construction of a tree-like structure able of representing the top end ecodesign and its constructs, (b) the weighing of the structure using the AHP method, suitable for chemical company, (c) the split of the ecodesign constructs into items of application, and the preparation of a questionnaire to identify the degree in which every item is reached, (d) the comparison of the performance obtained for each item of a particular construct with the degree of importance assigned for that construct.

The tree-like structure for ecodesign, unfolded in constructs has been built in focus group meetings. Four researches that act in ecodesign co-related areas and two managers, one from an automotive company and another from a chemical company that supplies the automotive industry, both with expertise in environmental management have participated. They all fully know about productive processes, products employment and logistic processes. The procedures of the focus group followed the Thietart et al (2001) recommendations. The same focus group, guided by the researchers, weighing the ecodesign constructs by AHP method. The researchers split the ecodesign constructs into items of application and prepared a questionnaire; the same was validated and tested with the members of focus group. The questionnaire was answered by four engineers from the company.

4.2 Results and discussion for the assessment of ecodesign implementation dimensions in the automotive industry

4.2.1 Characteristics of the company

The company has six manufacture units in the country; the study took place in a large unit located at the South region in Brazil. The main products are adhesives and laminated for the shoe making industry, as well as furniture and automotive industries.

The following characteristics were identified in the company: (a) a history of environmental concern since the late 1980s; (b) strategic positioning and focusing on developing innovative products and solutions and new technologies; and (c) cost reduction in developing new products or in the redesign of existing ones.

The company provides products and services for the automotive industry, furniture industry and footwear industry, especially adhesives and laminates. Besides these points related to the company, aligned with Vercalsteren (2001) point of view, the company had expressed interest in ecodesign.

4.2.2 Three-like structure for ecodesign

The first line (the criteria) of the tree-like structure for ecodesign, unfolded in constructs, is presented on Figure 3. The requirements proposed by Fiksel (1996), Luttropp & Lagersted (2006) and Wimmer et al. (2005) and the expertise of the group members served as base for the development of this part of the research.

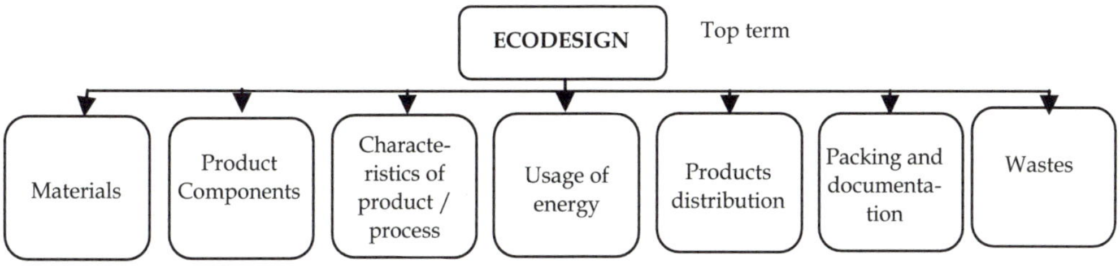

Fig. 3. Tree-like structure representative of ecodesign

4.2.3 Weighing the ecodesign tree-like structure and unfolding the constructs

This section consisted on the weighing of a tree-like structure using AHP. This weighing was based on the criteria presented on Table 2 at the company of study. The authors of this paper mediated the sections.

Table 4 illustrates the matrix of ecodesign construct preferences using AHP for the company studied. The computing of matrix preference data shows the relative importance of each ecodesign construct. For the company in study it was obtained: Materials with 12% of relative importance; Product components with 3%; Characteristics of the product and process 34%; Usage of energy 3%; Distribution of products with 8%, Packaging and documentation with 11% and Wastes with 29% of relative importance for the ecodesign. The CR index was of 0.064, what indicates the preferences of the decision makers have an acceptable degree of rationality.

Company	materials: choice and employment	product components	characteristics of the product	usage of energy	distribution of the products	packing and documentation	wastes
materials: choice and employment	1	3	1/3	5	3	1	1/3
product components	1/3	1	1/7	1	1/3	1/5	1/9
characteristics of the product	3	7	1	7	3	3	3
usage of energy	1/5	1	1/7	1	1/5	1/7	1/9
distribution of the products	1/3	3	1/3	5	1	1	1/5
packing and documentation	1	5	1/3	7	1	1	1/5
wastes	3	9	1/3	9	5	5	1

Table 4. Matrix of ecodesign construct preferences

The next step of the research consisted in unfolding the constructs into application items (concepts) of ecodesign, elaborating an evaluation instrument that allows identifying the degree of performance of each item. The instrument has 32 evaluation questions and each question refers to an application item. The evaluation items and its respective constructs can be identified on Table 5. For the answers, it was used a Likert scale from 1 to 5, where 1 represents the case where the item is not present or is never reached, and 5 is equivalent to the case where the item is completely met. NA (not applicable) indicates that the item is not applicable in its presence; in this case, this item is not considered by the company in the calculation of the degree of construct application. The degree of performance of each evaluated item will be determined in a consensual manner among the participants of the company.

Construct	Query (evaluation items)	Application
Materials	1) ability to use raw material closer to its natural state	1
	2) ability to avoid mixtures of non-compatible materials aiming recycling or reusing materials	5
	3) ability to eliminate or not use toxic, hazardous or carcinogenic substances	3
	4) ability to eliminate or not use raw material that generate Class 1 residuous – hazardous	2
	5) ability to use recycled and/or renewable materials	5
	6) ability to limit atmospheric emissions originated by the use of volatile organic compounds	2
Product components	7) ability to recover or use recovered components	2
	8) ability to easy the access of components	NA
	9) ability to identify materials and components to help in later recycling or reuse	2
	10) ability to determine the degree of recycling of a material or component	1
Characte-ristics of the product / process	11) ability to elaborate products with simpler shapes and that reduce the employment or use of raw material	4
	12) ability to project products with longer lifetime	5
	13) ability to project multifunctional products	4
	14) ability to generate an upgrade on the product	4
Usage of energy	15) ability to use energy generated by renewable resources	2
	16) ability to employ energy reduction batches during the use of the product	3
	17) ability to reduce employment of energy during the production of the product	3
	18) ability to reduce employment of energy during storage of the product	NA
Distribution of products / storage	19) ability to plan in a wise manner and €optimize the distribution logistics	5
	20) ability to privilege suppliers and distributors closer located	1
	21) ability to minimize raw material storage, during productive process, finished product and product for reuse	3
	22) ability to use more efficient transport model in energetic terms	1

Construct	Query (evaluation items)	Application
Packing and documen-tation	23) ability to reduce packages weight and complexity	2
	24) ability to use packages that can be reused	2
	25) ability to use packages produced from reused raw material (eg.: recycled paper)	3
	26) ability to use electronic documentation	3
	27) ability to use refilled products	3
Wastes	28) ability to minimize wastings generated in the productive process	4
	29) ability to minimize wastings generated during usage of the product	3
	30) ability to reuse generated wastings	4
	31) ability to assure acceptable emission limits	3
	32) ability to eliminate Class 1 wastings presence - hazardous	2

Table 5. Application of evaluation items referent to ecodesign constructs

4.2.4 Analysis of the results

The degree of the application of each construct is obtained by the average of values attributed to each query related to the evaluated construct. The average grade of each construct is converted into percent points. In doing so, when all percent points related to the degree of application of all constructs are summed, one obtains an ecodesign performance index for the company. Table 6 indicates the degree of implementation of each ecodesign construct and the level of total implementation.

Construct	Weight of construct	Degree of application
Materials	12%	7.2pp
Products components	3%	1.0pp
Characteristics of the product / process	34%	28.9pp
Usage of energy	3%	1.6pp
Distribution of products / storage	8%	4.0pp
Packing and documentation	11%	5.7pp
Wastes	29%	18.6pp
Total	**100%**	**67.0pp**
CR	0,064	

Table 6. Degree of ecodesign implementation

The analysis of the results indicates the company answers 67% of what could be reached related to ecodesign. The construct Characteristic of the product/process appears with higher relative importance; this construct is answered in 85% out of total (28.9pp out of possible 34%). The second construct of importance is Wastes; this construct is answered in 60%. The third construct of importance is Materials which is answered in 60%. Packaging and documentation is the fourth important construct, answering 51.8% of constructs' demands. The next is Distribution of products/storage, Usage of energy and Products components with 50%, 53.3% e 33.3%. Fig. 4 presents the values (% to the weight and pp to the application) for each ecodesign constructs.

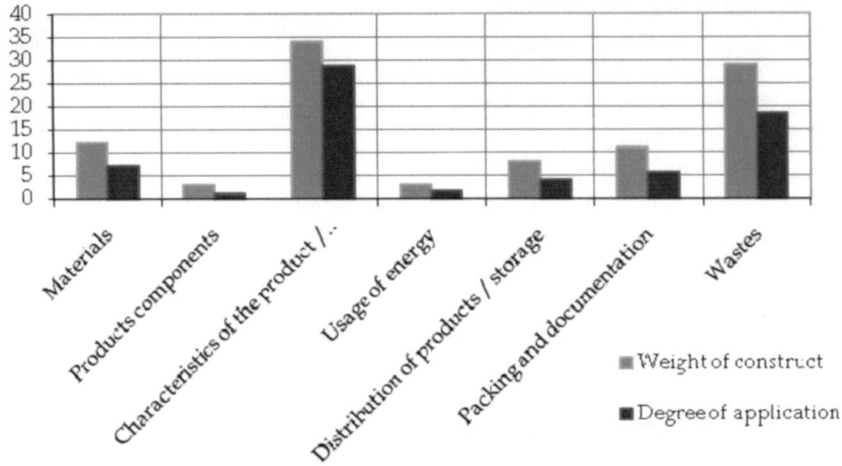

Fig. 4. Weight of ecodesign constructs and degree of application

If the company decides to simply increase the overall presence of ecodesign, the gaps should be reduced by managerial actions. Those actions may require new productive resources. Once resources are finite and constrained, before allocating more resources to developing new products, it might be better to relocate the current resources.

For such relocation, the two-dimensional graphical analysis [Importance - Performance] can be used. As far as research has gone, the earliest reference to the method was Slack (1993). It is considered that the importance of the elements or success key factors (constructs) and their assessed performance (presence) should be similar. According to this idea, in a graph [Importance - Presence], it would be ideal that the constructs be distributed along a diagonal: constructs with more importance would have more presence.

If, for a construct, the represented point by importance x presence intersection is below the diagonal, this may mean that improvement actions in relation to this construct shall be taken. If this point is far above, it means there is an excess in the construct performance in relation to its importance. In this case, resources might be reallocated to constructs located far bellow the diagonal (urgent acting zone).

Figure 5 presents the analysis. The two-dimensional space consists of the area for urgent action (priority much higher than presence); area of improvement (priority higher than presence); appropriate area (priority and presence balanced) and zone over (priority lower than presence). Characteristics, Materials and Energy constructs are balanced and should not be modified. Components and Wastes should be the target of improvement actions. Packaging and distribution are in the bordering areas. Constructs in the urgent area and the overflow area were not observed. Therefore, this may mean that there is no construct that removes productive resources for allocation to other constructs.

There are no "urgent" or constructs in zone "excess". At first, constructs that are in the excess performance area or that need urgent acting and therefore justify the reallocation of the resources of the organization were not observed.

Under the managerial view, according to the perception of the participants from the company, the relative importance of ecodesign constructs shows the strategy and the actions of the company. It is primarily focused in the characteristics of the product and process, bound directly with management and wastes. The innovative characteristics of the products make their differences. The non generation of wastes is prioritized; once not possible, it is aimed the maximization of wastes usage on the productive process of the company itself.

Energy was one of the constructs with the smallest degree of importance; the usage and the respective costs with energy in the company are considered low when compared to the others; yet, there is the concern to have the user of the company product (usually shoes maker companies and furniture industry) to reduce the consume of energy in drying of adhesives, and in doing so focusing in the reduction of the usage of energy when using the product. Components also present a low relative importance; it is observed few presence of this construct due to the product technical characteristics.

A criticism that can be made to the method relates to the subjectivity of the focus group. To avoid skewness on constructs appraisal it can be made up ad hoc, by experts with no interest in operation, but with knowledge on it. Another criticism concerns the AHP method. The method allows some inconsistency: the 10% limit was arbitrarily set by its proposers. Other methods can reach less then 10%. Another point is the lack of coherence in the results when the alternatives change, by criteria that is entered or already exist. The use of AHP is research delimitation. If this criticism is unsurpassed, the essential contribution of the case becomes the method and not the specific outcome of the case.

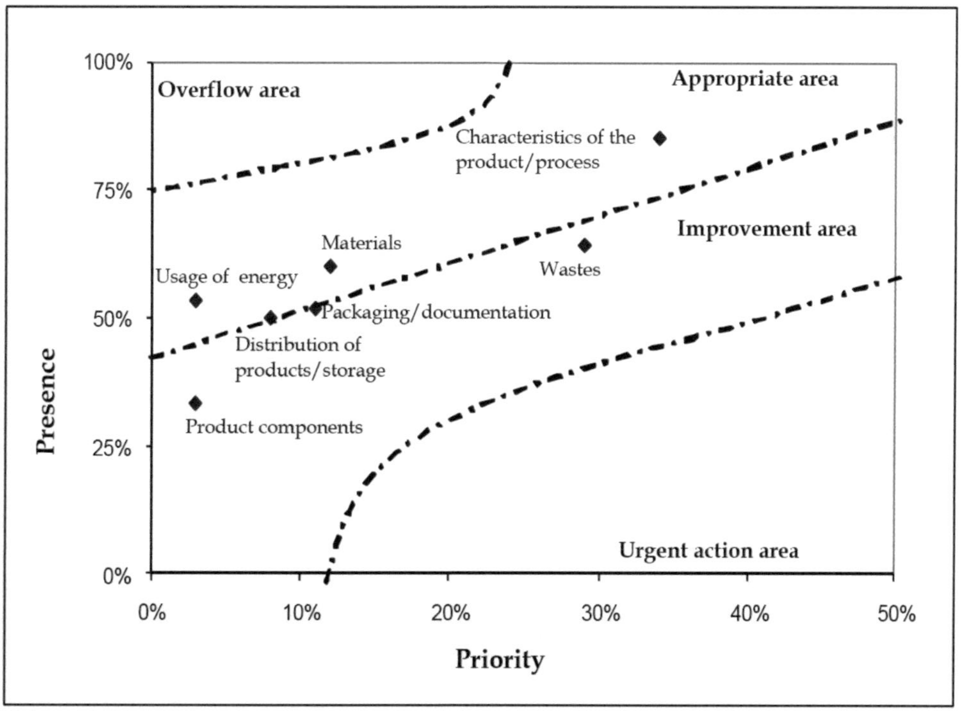

Fig. 5. Analysis [Priority x Presence] of the constructs of ecodesign

7. Conclusion

The main objective of the first part of this article was to analyze the process of deployment of ecodesign in a company belonging to the automotive electronics industry in order to identify the elements that justified the motivation for the employment of this technique. The study was exploratory and does not allow generalizations about the process of implementation of ecodesign in automotive electronics industry. The repetition of the cases will allow that.

The theoretical reference approached the concept of ecodesign, the critical factors for success and difficulties of implementation.

Among the elements capable of sustaining the implementation of ecodesign in the company studied, there is the prospect of cutting costs, because the technique is based on dematerialization and the reduction of waste and its subsequent treatment.

The introduction of the ecodesign practices in the process of the company was guaranteed by the adoption of existing routines of project management. A multifunctional team with coordinator was constituted, and the scope, the schedule and the risk analysis were accompanied by the high management, similar to other products and processes developments in the company.

Another essential element to ensure the implementation of ecodesign in the studied company was the commitment of top management. Indeed, this point can be demonstrated in the strategies of the company - performance indicators related to ecodesign were inserted in the Balanced Scorecard. It is also highlighted the training in ecodesign to all employees of the company, according to each one's involvement with the process of implementation of ecodesign.

The assumptions of ecodesign were implemented through checklists for the procedures of development of product and process. These checklists should be regarded as "living documents", i.e. each new event or experience of an aspect of ecodesign should be added as information to the correspondent checklist.

It was also noticed that the industry of automotive electronics has peculiarities which must be adapted in the adoption of the technique. Problems of a technical nature were addressed in the implementation, such as the absence of information about the local reality in the analysis of the Life Cycle Assessment of products. The lack of information about the environmental impact of each alternative available in the design phase was also another problem identified. It suggests fronts for the development of future work in the area.

As a continuity of this research is proposed to examine, in the medium and long term, the parameters about development of products affected by ecodesign and their impact on indicators of the organization. Similarly, it is suggested to extend the study to other organizations of various sizes and segments, in order to determine a method of deployment of ecodesign that is adaptable to various organizational realities.

In relation to the second part of the chapter, the main objective was to assess the ecodesign performance on a chemical company. The constructs that compose and explain the ecodesign were established in the theoretical framework, based on the set of practical proposals by Fiksel (1996), Wimmer et al. (2005) and Luttropp & Lagersted (2006). The group of managers who participated in the survey felt that the scope is adequate, it is not necessary to include other practices. The importance of the constructs was obtained from a tree-like structure representing the ecodesign and it was weighted by AHP. An assessment tool was used to determine the degree of presence of each construct. Finally, the overall performance of the company in relation to the practice of ecodesign was determined. It is understood that this work contributes to further development of performance indicators associated with the ecodesign, complementing the exposed by Cabezas et al. (2005) and Svensson et al. (2006). The key performance indicators could be related to items included in ecodesign practice with a greater degree of importance.

The research findings from the perspective of organizations will enable decision making, involving resources and priority for actions related to ecodesign and focusing on the constructs that most contribute to it. Under the scientific perspective, the contribution is focused on developing a method applicable to other organizations, including other industries.

It is understood that a study could be continued by widely assessing the performance of an industry, or even establishing performance indicators tied to each construct.

8. References

Bititci, U. (1995). Modelling of performance measurement systems in manufacturing enterprises. International Journal of Production Economics, Vol. 42, No. 2, pp. 137-147, ISSN 0925-5273.

Boks, C. (2006). The soft side of ecodesign. *Journal of Cleaner Production*. Vol. 14, No. 15 - 16, pp. 1346 - 1356, ISSN 0959-6526 .

Borchardt, M.; Pereira, G.; Sellitto, M. (2009). Assessment of ecodesign applications using the Analytic Hierarchy Process: a case study in three furniture companies. *Chemical Engineering Transactions*. Vol. 18, No.1, pp.177-182, ISBN 978-88-95608-04-4.

Byggeth, E.; Hoschschorner, I. (2006). Handling trade-offs in ecodesign tools for sustainable product development and procurement. *Journal of Cleaner Production*. Vol.14, No. 15 - 16, pp. 1420 - 1430, ISSN 0959-6526.

Byggeth, S. ; Broman, G. ; Robèrt, K. (2007). A Method for Sustainable Product Development based on a Modular System of Guiding Questions. *Journal of Cleaner Production*. Vol. 15, No. 1, pp. 1 - 11, ISSN 0959-6526.

Cabezas, H.; Pawlowski, C.; Mayer, A.; Hoagland, T. (2005). Sustainable systems theory: ecological and other aspects. *Journal of Cleaner Production*. Vol. 13, No. 5, pp. 455 - 467, ISSN 0959-6526.

Chehebe, J. (2002). *Análise do Ciclo de Vida de Produtos – Ferramenta Gerencial da ISO 14000*. Ed. Qualitymark, ISBN: 85-7303-169-7, Rio de Janeiro.

Côté, R. ; Booth, A.; Louis, B. Eco-efficiency and SMEs in Nova Scotia, Canada. *Journal of Cleaner Production*. Vol. 14, No. 6-7, pp. 542 - 550 ISSN 0959-6526.

Donnelly, K.; Beckett-Furnell, Z.; Traeger, S.; Okrasinski, T. (2006). Ecodesign implemented through a product-based environmental management system. *Journal of Cleaner Production*. Vol. 14, No. 15 - 16, pp. 1357-1367, ISSN 0959-6526.

Easterby, E. (1975). *Case study in social sciences*. Chapman Hall, London. ISBN 9780201026016

Ferrão, P.; Amaral, J. (2006). Design for recycling in the automobile industry: new approaches and new tools. *Journal of Engineering Design*. Vol. 17, No. 5, pp. 447-462, ISSN 1466-1837.

Fiksel, J. (1996). Design for Environment. Mc Graw Hill, ISBN : 0071605568 / 9780071605564, New York.

Forman, E. ; Selly, M. (2001). *Decisions by objectives*. Expert Choice, Inc. Avaiable in http://www.expertchoice.com, accessed in may/2009.

French, S. (1986). *Decision Theory: an introduction to the mathematics of rationality*. Ed. Ellis Horwood, ISBN-10: 0470203080 and ISBN-13: 978-0470203088, West Sussex (UK).

Hermann, B., Kroeze, C., Jawjit, W. (2007). Assessing environmental performance by combining life cycle assessment, multi-criteria analysis and environmental performance indicator, *Journal of Cleaner Production*, Vol. 15, No.18, pp.1787-96, ISSN 0959-6526.

Karlsson, R., Luttropp, C. (2006). Ecodesign: What´s happening? An Overview of the subject area of ecodesign and the papers in this Special Issue. *Journal of Cleaner Production*. Vol.14, No. 15-16, pp. 1291 - 1298, ISSN 0959-6526.

Kazazian, T. (2005). *Haverá a Idade das Coisas Leves – Design e desenvolvimento sustentável*. Ed. SENAC, ISBN 8573594365, São Paulo.

Lofthouse, V. (2006). Ecodesign tools for designers: defining the requirements. *Journal of Cleaner Production*. Vol. 14, No. 15-16, pp. 1386 - 1395, ISSN 0959-6526.

Luttropp, C.; Lagerstedt, J. (2006). Ecodesign and the Ten Golden Rules: generic advice for merging environmental aspects into product development. *Journal of Cleaner Production*. Vol. 14, No. 15-16, pp. 1396 - 1408, ISSN 0959-6526.

Mathieux, F.; Rebitzer, G.; Ferrendier, S.; Simon, M.; Froelich, D. (2001). Ecodesign in the European Electr(on)ics Industry - An analysis of the current practices based on cases studies. *The Journal of Sustainable Product Design*. Vol. 1, No. 4, pp. 233-245, ISSN 1367-6679.

Maxwell, D., Sheate, W., Vorst, R. (2006). Functional and systems aspects of the sustainable product and service development approach for industry. *Journal of Cleaner Production*. Vol. 14, No. 17, pp. 1466 - 1479, ISSN 0959-6526.

Park, P.; Tahara, K. (2008). Quantifying producer and consumer-based eco-efficiencies for the identification of key ecodesign issues. *Journal of Cleaner Production*. Vol. 16, pp. 95 -104, ISSN 0959-6526.

Pochat, S.; Bertoluci, G.; Froelich, D. (2007). Integrating ecodesign by conducting changes in SMES. *Journal of Cleaner Production*. Vol. 15, No.7 , pp. 671 - 680, ISSN 0959-6526.

Rao, P. (2004). Greening production: a South-Asian East experience. *International Journal of Operations & Production Management*. Vol. 24, No. 3, pp. 289 - 320, ISSN: 0144-3577.

Saaty, T. (1991). Método de Análise Hierárquica. Makron Books, São Paulo. ISBN 9780201026016

Sellitto, M.; Borchardt, M.; Pereira, G. (2006). Avaliação multicriterial de desempenho: um estudo de caso na indústria de transporte coletivo de passageiros. *Gestão & Produção*, Vol. 13, No. 2, pp. 339-352, ISSN 0104-530X. (in portuguese)

Sellitto, M.; Borchardt, M.; Pereira, G. (2010). Modelagem para avaliação de desempenho ambiental em operações de manufatura. *Gestão & Produção*, Vol. 17, No.1, pp. 95-109, ISSN 0104-530X. (in portuguese)

Slack, N. (1993). The manufacturing advantage : achieving competitive manufacturing operations. Ed. Mercury, ISBN 8522432600, London.

Svensson, N.; Roth L.; Eklund, M.; Märtensson, A. (2006). Environmental relevance and use of energy indicators in environmental management and research. *Journal of Cleaner Production*. Vol.14, No. 2, pp. 134 - 145, ISSN 0959-6526.

Thietart, R. et al. (2001). *Doing management research – a comprehensive guide*. Sage Publications, ISBN 0 7619 6516 3, London.

Tingström, J.; Karlsson, R. (2006). The relationship between environmental analyses and the dialogue process in product development. *Journal of Cleaner Production*. Vol. 14, No. 15 - 16, pp. 1409 - 1419, ISSN 0959-6526.

Vercalsteren, A. (2001). Integrating the ecodesign concept in small and medium-size enterprises: Experiences in the Flemish Region of Belgium. *Enviromental Management and Health*. Vol. 12, No. 4, pp. 347 - 355, ISSN 0956-6163.

Voss, C.; Tsikriktsis, N.; Frohlich, M. (2002). Case Research in Operations Management. *International Journal of Operations & Production Management*, Vol.22, No. 2, pp.195-219, ISSN 0144-3577.

Wimmer, W.; Lee, K.; Jeong, T.; Hong, J. (2005). Ecodesign in twelve steps: provinding systematic guidance for considering environmental aspects and stakeholder requirements in product design and development. International Conference on Engineering Design; ICED; Melbourne.

Yin, R. (2001). *Estudo de caso: planejamento e método. Bookman*, ISBN 853630462 , Porto Alegre, 2001.

10

Context Analysis for Situation Assessment in Automotive Applications

L. Ciardelli[1], A. Beoldo[2] and C. Regazzoni[1]
[1]*Department of Biophysical and Electronic Engineering, University of Genoa*
[2]*TechnoAware s.r.l*
Italy

1. Introduction

In the last few years, the application of ICT technologies in automotive field has taken an increasing role in improving both the safety and the driving comfort. In this context, systems capable of determining the traffic situation and/or driver behaviour through the analysis of signals from multiple sensors (e.g. radar, cameras, etc...) are the subject of active research in both industrial and academic sectors (Trivedi et al., 2007); (Yu et al., 2009); (Schneider et al., 2008). These systems, unlike autonomous vehicles or automated control systems are more acceptable for near future real applications since they try to improve driver's sensing capabilities and to help the decision process in a non-intrusive way. According to this statement, car manufacturers are starting to introduce driving support systems in luxury vehicles as, for instance, parking sensors.

The extraction of contextual information through the analysis of video streams captured by cameras can therefore have implications in many applications focused both on prevention of incidents and on provision of useful messages to drivers as traffic flows analysis, dangerous behaviour detection, traffic laws infringements (e.g. speed limits), etc.... For these applications the analysis of what happens inside and outside a car is a relevant source of information.

A framework is proposed, integrating a dual camera network and a set of heterogeneous sensors communicating through a *CAN-bus* for the extraction of context data from on-board cameras mounted on vehicles and of other data as steering angle, speed, brakes, etc.... A camera is oriented so as to frame the portion of road in front of the vehicle while the other one is positioned inside the vehicle and pointed on the driver. As a matter of fact, the joint analysis of on board/off board car context can be used to derive considerations on driver's behaviour and then to detect possible dangerous situations (sleep, dangerous lane changes, etc.) or a driving style which does not respect traffic regulation.

Three blocks are considered: a) internal processing, b) external processing and c) vehicle status data coming from the *CAN-bus*. The most significant data that can be extracted from a camera monitoring the driver are the gaze direction, the position of face, frequency of blinking eyes and mouth movement. As well, the camera looking outside the vehicle allows detecting the position of the vehicle on the road and the lane changes. Moreover, the analysis of the road type (highway, urban road, etc.) and of traffic can provide relevant information to evaluate the possible risks of the driving behaviour. Finally, the data collected through the *CAN-bus* allow a more robust interpretation of internal context providing information concerning the steering angle, the speed of the vehicle, etc… which can be considered in order to evaluate the behaviour of the driver and consequently of the vehicle with respect to the surrounding environment.

In this work, the architecture of the proposed system will be introduced, then the different blocks will be discussed in detail as well the impact of the proposed solutions on driver's behaviour. Moreover, promising results will be presented concerning the single processing frameworks and the integrated framework where data alignment and data association techniques will be applied to provide a comprehensive description of the driving context towards situation assessment. Finally, the impact of potential improvements of the proposed system and the introduction of a statistical representation of interactions among the driver, the vehicle and the surrounding environment will be discussed.

2. System architecture

2.1 Physical architecture

The physical architecture of the system (see Fig. 1) is composed as follows:

- A processing unit directly mounted on the vehicle allowing real time storing and processing of the acquired signals;
  - A standard camera looking outside the vehicle and positioned as to frame the portion of the road in front of the vehicle. Such sensor allows, for instance, capturing video sequences concerning other vehicles which are occupying the lanes;

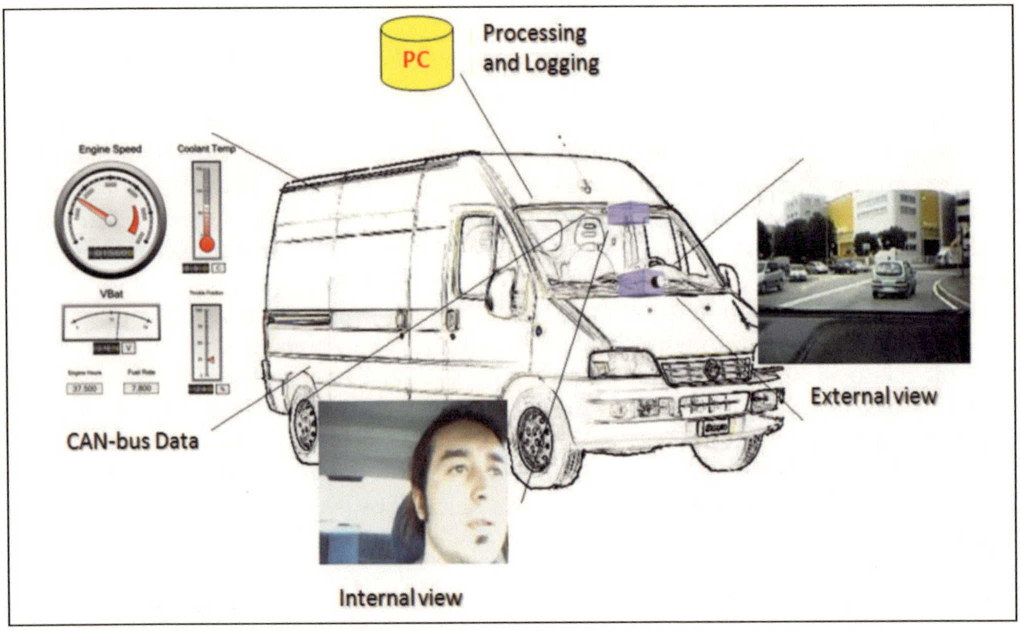

Fig. 1. System physical architecture

- A standard camera looking inside the vehicle and positioned as to frame the face of the driver. Such sensor allows, for instance, capturing video sequences in order to analyze the level of attention of the driver through the analysis of the facial traits;

- A Controller Area Network (CAN-bus) which communicates data concerning the internal state of the car (e.g. steering angle, acceleration) to the processing unit. Data are sent over an Ethernet network in an asynchronous mode as UDP datagram.

2.2 Logical architecture

The logical architecture of the system is presented in Fig. 2.

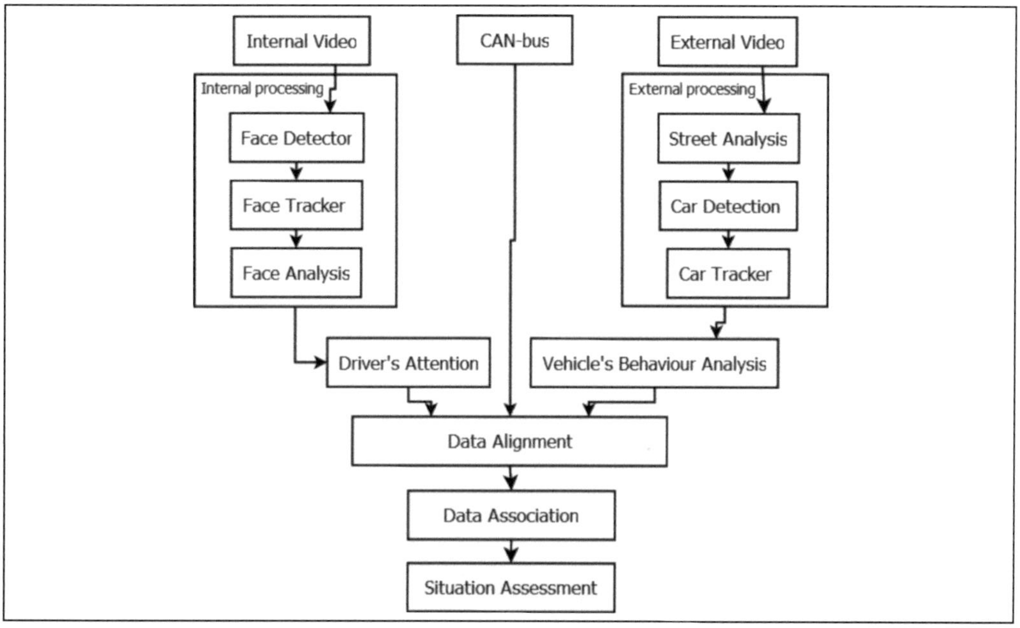

Fig. 2. System logical architecture

Three blocks have been considered, as described previously in the introduction, for the processing of the data coming from the camera looking inside the vehicle, the camera looking outside the vehicle and the CAN-bus (such blocks will be discussed in detail in the following paragraphs).

Then, the output of each block provides different information as the level of attention of the driver, the safe/not safe behaviour of the vehicle with respect to the correct driving patterns (also taking into account what happens in the surrounding environment) and the parameters related to the internal state of the car (speed, status of the lights, etc…).

Finally, data fusion techniques allow aligning and associating the collected data towards the assessment of potentially dangerous situations/behaviours which could affect not only the "intelligent vehicle" but also other cars. The preliminary studies concerning the application of bio-inspired data fusion models for understanding and, eventually, predicting anomalous events will be discussed in the following.

3. Internal video processing and driver's attention analysis

3.1 Related work

The most significant data that can be extracted from a camera monitoring the driver are the gaze direction, the position of the face, the eyes' blinking frequency and the mouth state.

In the state of art several works can be found regarding face pose estimation that is the feature we will mainly focus on for the purposes of the internal processing block. Such feature has been considered particularly relevant in the context of Human-Computer Interaction (HCI) or automotive applications.
A significant body of research has been carried out in the last few years. Various approaches can be categorized, according to the different proposed methodologies, by pointing out the condition of usage, the assumptions and the obtainable performances.
In (Asteriadis et al., 2009) gaze is estimated by analyzing the motion of some relevant features in the eyes and mouth area. This last method, however, does not take into account possible illumination changes since it is designed for indoor Human-Computer Interaction applications.
(Whitehill & Movellan, 2008) describe a method based on frame by frame head pose tracking robust to illumination changes and different facial appearance. Two classifier trained with the GentleBoost (Friedman et al., 2000) approach are used and the output are integrated using linear regression to estimate pose angles.
Other works try to cope with the difficult environmental conditions of the automotive applications (i.e. frequent and relevant illumination changes) by detecting eyes using infrared cameras (Bergasa et al., 2006).
These approaches are usually more robust and they can operate also with very low illumination even if not cost-effective (the cost of an infrared camera is much higher than a standard camera).
Recently, researchers (Yang et al., 2002); (Murphy-Chutorian & Trivedi, 2009) have presented a complete and up-to date survey of head pose estimation algorithms in order to fix the most common issues and to provide adequate solutions.
More in detail, head pose estimation methods can be not exhaustively classified as follows:

- Appearance Template Methods: a comparison among an image of the head and a set of models is performed in order to find the most similar;
- Detector Array Methods: a set of detector, each one specific for a pose discrete value, is trained;
- Non-linear Regression Methods: non linear regression tools are exploited to map the images and the related features as head pose values;
- Geometric Methods: Eyes, nose and mouth positions are exploited to determine the pose;
- Tracking Methods: the pose is evaluated according to the movement observed over consecutive frames;
- Hybrid Methods: two or more of the previously cited methods are combined to overcome drawbacks and limitations of each single method;

3.2 Face detection, tracking and analysis

The proposed internal processing system exploits a tracking based approach for driver's gaze detection to the aim of obtaining a better ratio between accuracy of the method and computational speed. The aim of this work is not to present in deep the technical details of the algorithms but to provide a general description of the method applied for the internal processing block. A more detailed description of the proposed algorithm can be found in (Ciardelli et al., 2010).
In particular, three processing steps have been considered:

- Face detection (face, eyes, mouth and nose);
- Face tracking;
- Face analysis and angle of view calculation.

Firstly, an initialization step is performed for face detection. For each trait the Viola-Jones detector is applied. Secondly, the tracking algorithm enables localizing the position of the face in the video frame and evaluating the relative position of every facial trait like the nose, the mouth and the eyes. For each trait, an instance of Kanade-Lucas-Tomasi (KLT) feature tracker algorithm has been used. Lastly, for each video frame the pose of the face is evaluated in order to extract the angle of view and other relevant information (see Fig. 3).

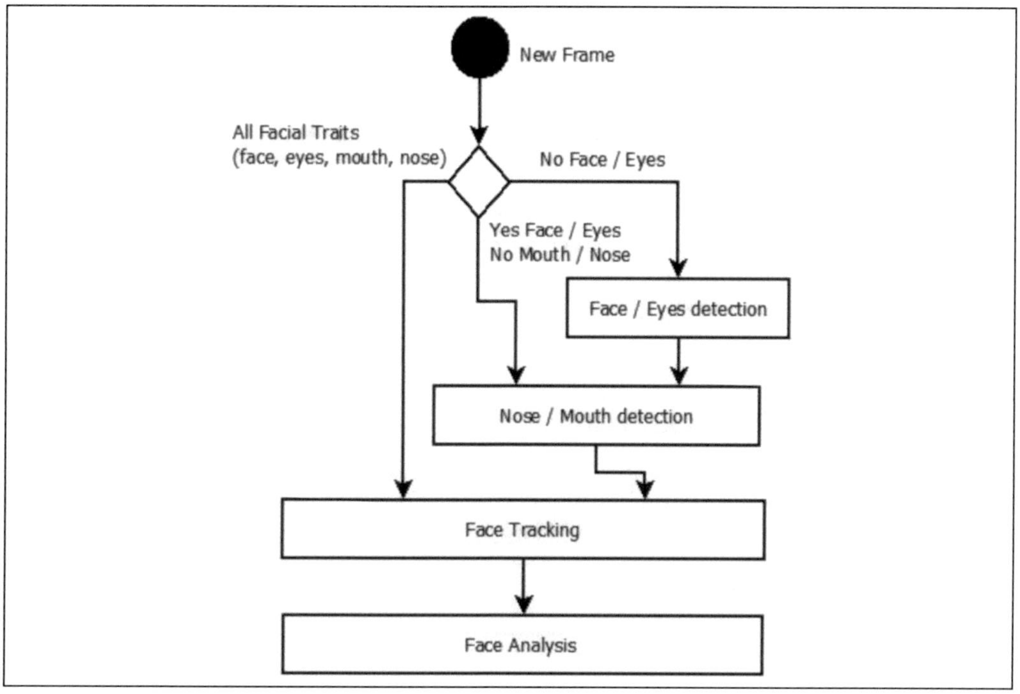

Fig. 3. Internal processing algorithm structure

Face analysis has been focused on the evaluation of the driver's view angle which is one of the most important information that is needed to assess his/her state. The information concerning the angle of view can be disassembled in yaw (rotation with respect to horizontal plane), roll (longitudinal rotation related to movement) and pitch (vertical rotation) angles as shown in Fig. 4.

As a general rule, we have assumed (having been demonstrated in a large testing phase) that the information obtained by the analysis of the yaw component can provide sufficient knowledge about the direction of the driver's gaze. More in detail, we can consider that values of the yaw angle near to 0 correspond to the situation of driver looking straight ahead (i.e. driver is looking at the street and his/her level of attention is adequate) while values far from 0 correspond to the case of driver looking in other directions rather than street one (i.e. a possible dangerous situation can happen because the driver is absent-minded).

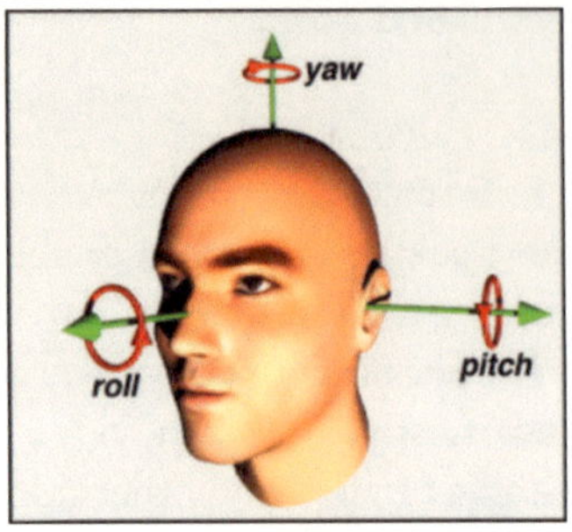

Fig. 4. Yaw, roll and pitch angles

3.3 Driver's attention and experimental results

The aim of the proposed experiments is not to demonstrate the effectiveness of the proposed face detection, tracking and analysis method which has been already proven in other works but to discuss the capability of the proposed system to properly assess the driver's attention in order to provide information useful for the analysis of the driving context.

According to this statement, to calculate the driver's attention, we decided to analyze the angle of yaw extracted from the camera framing the internal context of the car. A time interval μ has been fixed and the follwing formula has been applied:

$$\text{att} = \sum_{t=i}^{i+\mu} w(d_t) + q(|y|) \tag{1}$$

where d_t can take values 0 or 1 depending on whether or not there is the face detection (0 if there is detection), $w(x)$ is the weight function for the non-detection event, $|y|$ is the modulus of the yaw angle and $q(x)$ the corresponding weight function. For each frame, the value of attention *att* thus obtained is compared with two thresholds η_0 and η_1 in order to assess the level of attention (low, medium, high).

A lot of experiments have been performed using a standard camera at 320x240 of resolution. The standard camera, installed on the vehicle as described in the previous paragraphs, has been used to analyze a driver during a thirty minutes drive aiming at identifying the level of attention.

In Fig. 5 some shots are presented showing the capability of the system of correctly recognizing the attention of the driver.

In the top left sub-figure, the exceeding rotation of the head with respect to the camera axis leads to a blank frame (due to a malfunctioning of the detection and tracking algorithms) which corresponds to a "low attention" message.

In the top right one, as well as in the previous frame, the system recognizes a "low attention" situation according to the value of the *att* factor which is lower than threshold η_0.

Finally, bottom left and bottom right images show respectively an average and a high attention situation being the values of *att* respectively within η_0 and η_1 and over η_1.

Table 1 shows the experimental result obtained by the driver's attention analysis. The percentage of frame with errors is obtained comparing algorithm results with observations.

Actually, a more significant percentage of errors occur in the case of low attention because it is more difficult according to the proposed method to correctly detect this case. However, such performance could be improved modifying the thresholds. In this case (i.e. increase of the threshold) the capability of correctly recognize a low attention situation should improve even if the percentage of false alarms (i.e. of incorrectly detected low attention situations) should increase reducing the capability of preventing dangerous events which usually happen while the driver's level of attention is not adequate.

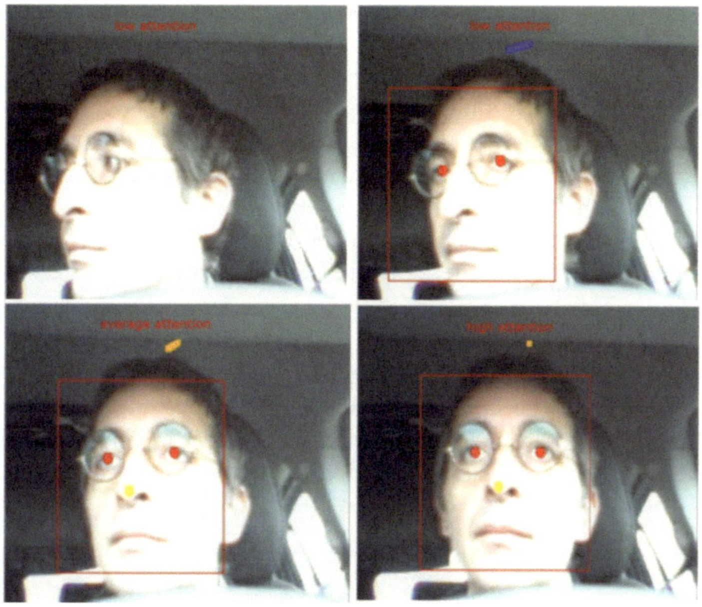

Fig. 5. Examples of driver's attention assessment

Low attention	7,1%
Average attention	5,3%
High attention	4,1%

Table 1. Percentage of frames with errors

4. External processing

4.1 Related work

The analysis of the road type (highway, urban road, etc.) and of traffic represent an important task to provide relevant information to evaluate the possible risks of the driving behavior.

In the literature, several works can be found addressing the problem of lane detection and vehicle's tracking. Concerning the first problem, in (McCall & Trivedi, 2006), a survey of lane detection algorithms is proposed where the key element of these algorithms are outlined. In (Nieto et al., 2008) a geometric model derived from perspective distortion is used to construct a road model and filter out extracted lines that are not consistent. Another widely used technique to postprocess of the output of the road marking extraction is the Hough transform as shown for example in (Voisin et al., 2005).

Among the different potential applications of vehicle's tracking, in (Chen) a security system for detection and tracking of stolen vehicles is discussed. A 360 degrees single PAL camera-based system is presented in (Yu et al., 2009), where authors provide both the driver's face pose and eye status and the driver's viewing scene basing on a machine learning algorithm for object tracking.

In (Wang et al., 2008) a road detection and tracking method based on a condensation particle filter for real-time video-based navigation applications is presented. The problem is also addressed using different approaches in other works. A real-time traffic surveillance system for the detection, recognition, and tracking of multiple vehicles in roadway images is shown in (Taj & Song, 2010). In this approach, moving vehicles can be automatically separated from the image sequences by a moving object segmentation method. Finally, in (Chung-Cheng et al., 2010) a contour initialization and tracking algorithm is presented to track multiple motorcycles and vehicles at any position on the roadway being not constrained by lane boundaries or vehicle size. Such method exploits dynamic models to predict the horizontal and vertical positions of vehicle contours.

4.2 Lane detection and vehicle(s)'s tracking

The logical framework of the lane detection module is presented in Fig. 6. A detailed description of the steps that have been implemented in order to detect the number of traffic lanes and the position of the vehicle with respect to the road is out of the scope of this work and has been already discussed in (Beoldo et al., 2009).

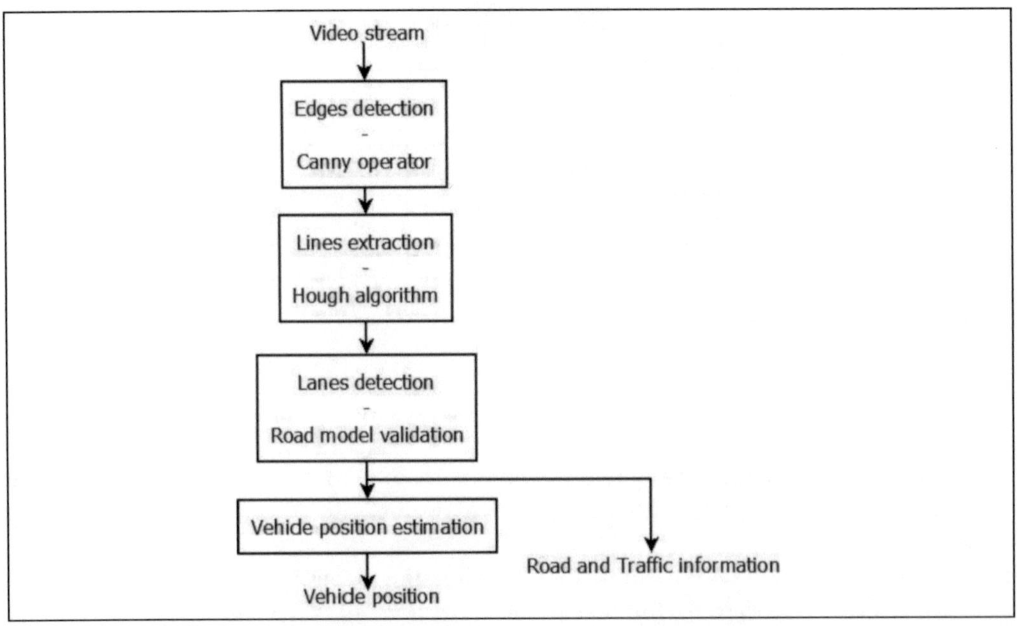

Fig. 6. Lane detection module logical framework

According to the proposed framework, the following steps have been applied to extract road context information from a video sequence:

1. Edges extraction using Canny operator (Fig. 7 - top left);
2. Lines detection using Hough algorithm (Fig. 7 – top right)

3. Lanes detection and road model validation:
   a. The two lines that belong to the lane where the vehicle is driving on are located;
   b. Attention is focused on an area within the triangle formed by the extracted lines. A frame per frame statistical analysis of the pixels belonging to the road is performed to create a model of the road.
   c. All pixels in the image below the point of intersection between the two lines identified at step 3 are considered and each pixel is compared with the model of the road looking for those that are more similar to the model (Fig. 7 - bottom left).
4. Evaluation of whether the road has one or two lanes and which is the position of the vehicle with respect to them (Fig. 7 - bottom right).

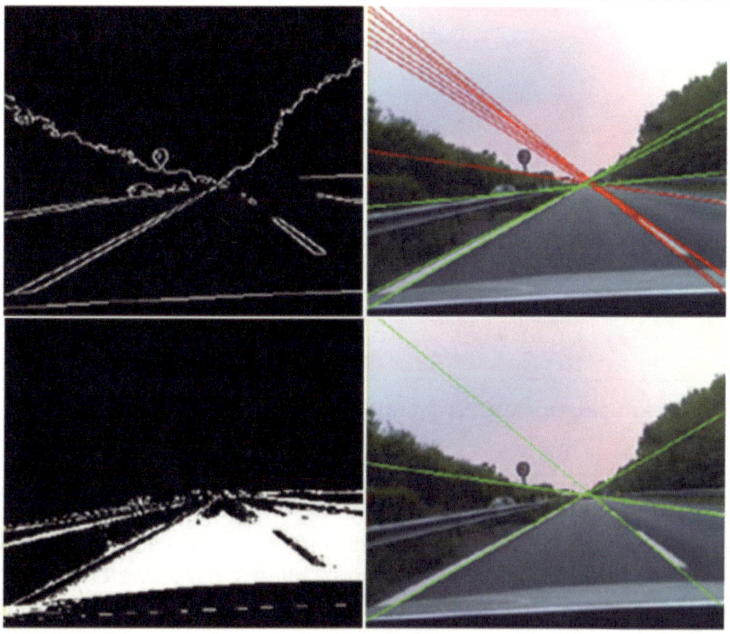

Fig. 7. Lanes detection and vehicle's position estimation: an example

Towards the development on an efficient intelligent system that enables improvements in the cars' safety, the extraction of the most accurate information concerning the space around the vehicle is needed. In such space, the targets to be considered are represented by fixed objects as buildings and trees and/or moving objects, mainly represented by all other vehicles (motorcycles, cars, trucks, etc...).

The main focus of the proposed system is represented by the detection and tracking of vehicles acting in the smart vehicle's surrounding space with particular regard to vehicles in front.

The two main consecutive steps which characterize such a detection and tracking system are:

- Generation of hypotheses (where the vehicle/object to be detected and tracked could be placed in the image);
- Hypotheses testing (previous hypotheses verification concerning the presence of vehicles/objects within the image).

As a matter of fact, the implementation of a solution robust enough to deal with the strict requirements of the proposed application is not easy. In particular, such a system must guarantee, at the same time, a few missed alarms (i.e. the number of missed vehicle/object detections) and a few false alarms (i.e. the number of wrongly detected vehicles/objects).
To this aim a feature-based tracking method is proposed where a Kanade-Lucas-Tomasi (KLT) feature tracking is used in a particle filter framework to predict local object motion (Dore et al., 2009). In particular, such a multitarget tracking algorithm exploits a sparse distributed shape model to handle partial occlusions where the state vector is composed by a set of points of interest (i.e. corners) enabling to jointly describe position and shape of the target.
An instance of the results obtained with the cited algorthm is presented in Fig. 8

Fig. 8. Vehicle's tracking algorithm: an example

4.3 CAN-bus

The Controller Area Network, also known as CAN-bus, is a vehicle bus standard designed to allow microcontrollers and devices to communicate with each other within a vehicle without a host computer. The CAN-bus interface allows extracting context data related to the vehicle's internal state. The data are sent asynchronously via an internal Ethernet network as UDP packets. A not exhaustive list of the data made available by the CAN-bus is provided in the following:

- Light: it indicates activation of the lights of the vehicle;
- Lateral acceleration (positive value corresponds to the left);
- Longitudinal acceleration;
- Parking brake;
- Speed;
- Steering angle (positive value corresponds to the left).

These and other data are made available and properly used according to the different type of application.
Fig. 9 shows an example where the video stream coming from the camera positioned in order to frame the external context and the temporal evolution (graph) of three different data coming from the CAN-bus are considered. The data shown in this example are the speed in metres per second, the acceleration in metres per square second and the steering angle in degrees. For each video frame, the displayed graphs are instantly updated according to the new available data.

In the proposed example the vehicle is moving straight ahead (steering angle equal to zero as shown in the bottom left part of the figure) and is approaching a turn. According to this, the vehicle is in a deceleration phase (see speed Module in the top right of the figure) and the graph of the longitudinal acceleration is negative (see bottom right part of the figure).

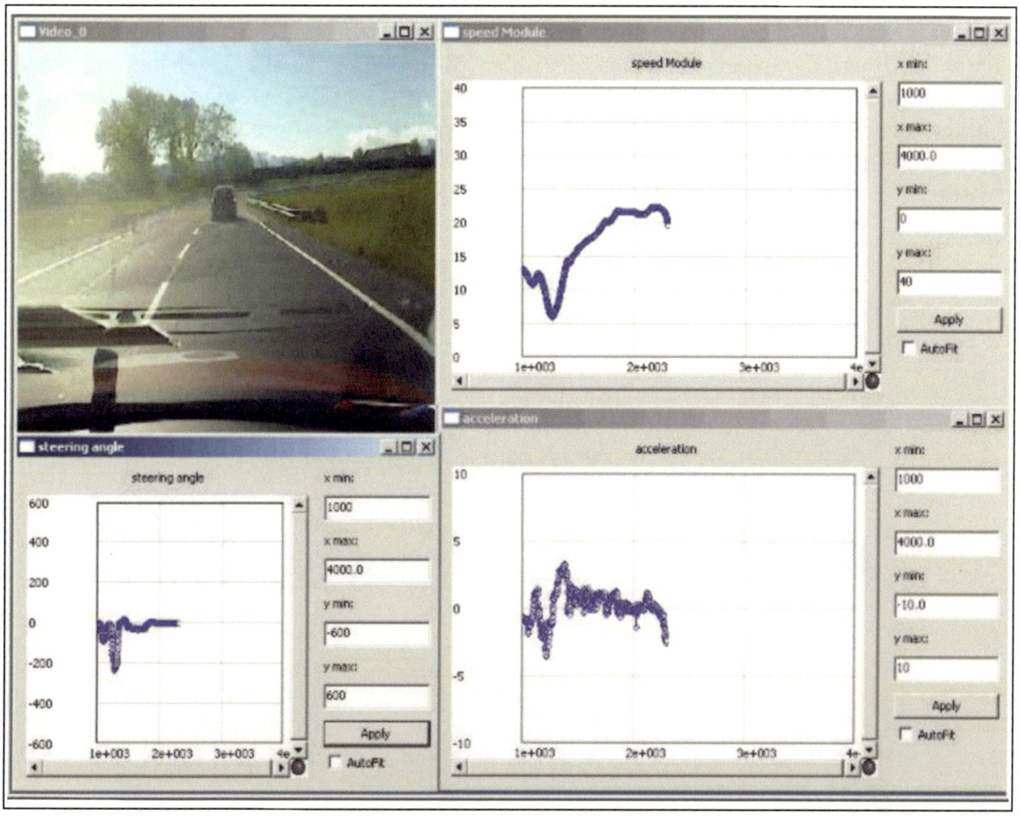

Fig. 9. Video/CAN-bus data visualization

It is important to note that, in our experiments, the video data are stored at 25 frames per second and that each video sequence lasts 5 minutes. Moreover, the video capturing-recording application works also as UDP receiver for the CAN-bus data so that the current frame is used as a reference for the synchronization of video and CAN-bus data. However, the CAN-bus data are sent asynchronously so it may also happen to not receive data for a few frames.

4.4 Vehicle(s)'s behaviour analysis and experimental results

The analysis of all the available information concerning the vehicle and the environment allows to generate alarms when a potentially dangerous situation happens. In particular, we have focused the attention on the analysis of the correlation between the distance with respect to the vehicle in front of the smart car (provided by the further processing of the information obtained from the detection and tracking modules) and the speed and the acceleration obtained via the CAN-bus. After a careful analysis it has been decided to define the following formula in order to establish whether or not to report an alarm:

$$danger = \begin{cases} \text{true} \quad \text{if } dist(x_t) < \varepsilon_n, \ dist(x_{t-1}) > dist(x_t) \text{ and } a(t) > 0 \\ \text{false} \quad \text{otherwise} \end{cases} \tag{2}$$

where $dist(x_t)$ is the function that calculates the distance between the camera and the vehicle which is in the forn of the smart car, ε_n is the threshold below which there may be danger and $a(t)$ is the value of the longitudinal acceleration at frame t.

Figure 10 shows the experimental results obtained applying the proposed method. Three different distances have been considered: a) near (distance below the ε_n threshold), b) average (distance within the ε_n and a ε_a threshold, properly fixed according to the different applications (i.e. highway, street, heavy traffic, etc...)) and c) far (distance over the ε_a threshold).

Fig. 10. Dangerous behaviour analysis: an example

In Fig. 10, the top right image shows a far distance situation. The green rectangle symbolizes the low level of danger. In the top left image an average distance situation is presented characterized by a yellow rectangle. Finally in the bottom images two different near distance situation are presented. In the left one, the system recognizes a near distance potentially dangerous situation and a message is displayed ("Attention"). In the right one, the system recognizes a near distance but not dangerous situation so that no message is displayed. The difference between the two cases resides in the data coming from the CAN-bus. In the first case, an increasing value of longitudinal acceleration is detected (potentially leading to a crash) while in the second one the value of longitudinal acceleration of the car is decreasing.

Future experiments will allow showing in the same GUI both the information coming from the video-sensors and from the CAN-bus.

5. Bio-inspired model for interaction analysis

Parallel activities have been carried out in order to study an approach based on a "bio-inspired" model for the analysis of driver's behavior and to detect possible dangerous situations. In (Dore et al., 2010) has been presented a general framework capable of predicting certain behaviors by studying interaction patterns between humans and the outside world. Such framework takes inspiration from the work of the neurophysiologist A. Damasio (Damasio, 2000).

According to Damasio, the common shared model for describing the behaviour of a bio-inspired (cognitive) system is the so-called Cognitive Cycle which is composed by four main characteristics:

- Sensing: the system has to continuously acquire knowledge about the interacting objects and about its own internal status, sensing is a passive interaction component;
- Analysis: the perceived raw data need an analysis phase to represent them and extract interesting filtered information;
- Decision: the intelligence of the system is expressed by the ability to decide for the proper action, given a basic knowledge, experience and sensed data;
- Action: the system tries to influence its interacting entities to maximize the functional of its objective; action is an active interaction component in relation to decision.

The learning phase is continuous and involves all the stages (within certain limits) of the cognitive cycle. According to the cognitive paradigm for the representation, organization, learning from experience and usage of knowledge, a bio-inspired system allows an entity predicting the near future and reacting in a proactive manner to interacting users' actions.

Damasio states that the brain representation of objects or feelings, both internal and external to the body, can be defined as *proto-self* and *core self*. *Proto-self* and *core self* are respectively voted for the self-monitoring and the control of the internal state of a person and for the relationship with the external world.

Thus, we can define as *proto* state $X_p(t)$ the vector of values acquired by "sensors" related to the internal state of a system and as *core* state $X_c(t)$ the vector of values acquired by "sensors" related to the external world. As well, a change in the *proto* state is defined as *proto* event while a change in the *core* state is defined as *core* event. To learn interactions between the internal and external context, the Autobiographical Memory (AM) algorithm, has been exploited (Dore et. al, 2010). In the proposed model, AM is the structure responsible for the representation of cause/effect relationships between state changes (events) occurring in the external world and in the internal system.

Such relationships are stored in the AM as triplets of events $\{ \varepsilon_{P^-}, \varepsilon_C, \varepsilon_{P^+} \}$ or $\{ \varepsilon_{C^-}, \varepsilon_P, \varepsilon_{C^+} \}$. This collection of relations between an entity (e.g. the system, a human subject, etc...) and the environment can be used to obtain a non-parametric estimation of the probability density functions (PDFs) $p(\varepsilon_{P^-}, \varepsilon_C, \varepsilon_{P^+})$ and $p(\varepsilon_{C^-}, \varepsilon_P, \varepsilon_{C^+})$. The PDFs describe the cause-effect relationships between the *proto* and the *core* events and allow to obtain a prediction of the future behavior of the interacting entities given a couple of *proto* and *core* events.

In the proposed automotive application, preliminary studies have been carried out focusing on the vehicle's behaviour analysis. In such a context, we have considered as *core* events all the data acquired from the sensor framing the external context (i.e. the position of vehicle with respect to the traffic lanes, the position of other vehicles, etc...) and as *proto* events the data collected via the CAN-bus.

Fig. 11 shows the logical architecture of the system based on the above described cognitive approach and designed to analyze the behavior of the driver.

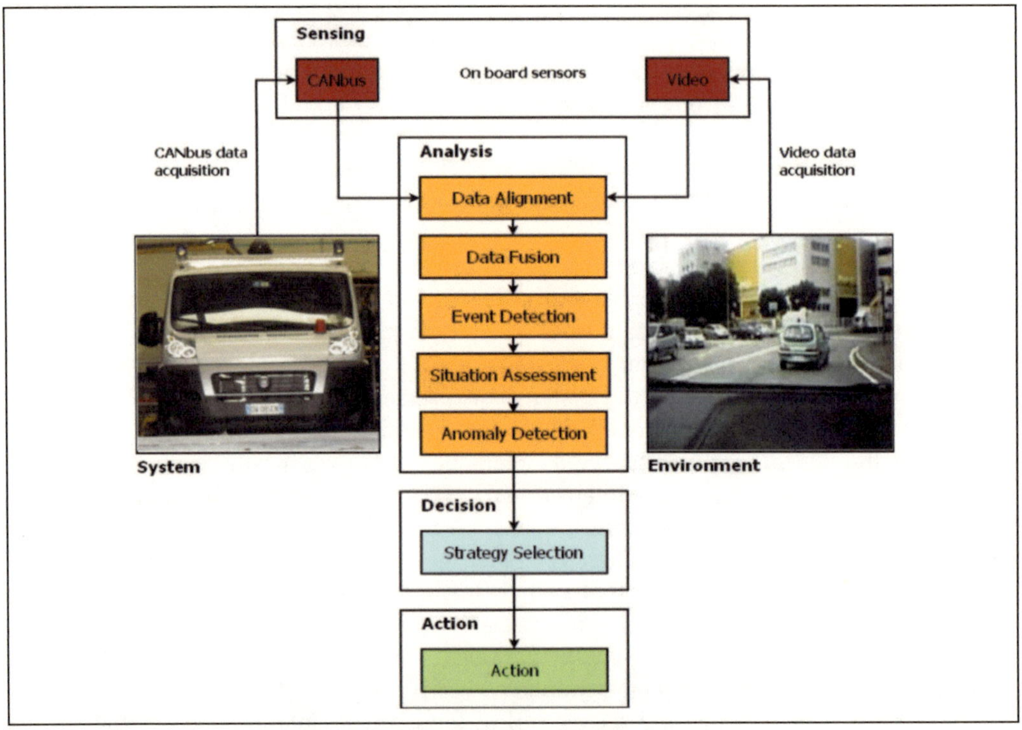

Fig. 11. Cognitive-based logical architecture of the system

Such framework can be divided into a sensing phase corresponding to sensor data (video and CAN-bus) acquisition an analysis phase corresponding to the processing of the available data according to the AM algorithm in order to define causal relationship between internal/external events and to identify abnormal situations. Then, in the decision stage the most suitable strategy to be applied in the incoming situation is selected according to previously acquired experience (properly stored in the AM). Finally, in the action phase, the system interacts with the outside world according to the strategy identified in the previous module.

It is important to highlight that the effectiveness of the proposed approach is strictly related to the amount of data available in the AM. According to this statement, the development of a simulator capable of resembling the behaviours observed in a real scenario is crucial. Actually, first steps have been performed in order to setup a simulation platform capable of providing a large set of training data resembling the scenario (a track) where a lot of real tests have been performed.

6. Conclusions and future work

In this work, a video-based system for context analysis and situation assessment in automotive applications has been presented. Two different solutions have been analyzed involving the internal and the external context of a vehicle. Promising results have been shown concerning both the driver's attention evaluation and the vehicle's dangerous behaviour assessment.

Future steps will deal with the implementation of a cognitive based framework for the joint analysis of internal and external events towards the prediction of incoming dangerous situations and the definition of a proper proactive reaction strategy. A bio-inspired model will be applied to define causal relationship between internal and external events and a simulation platform will be developed to provide a large set of training data.

7. References

Asteriadis, S.; Tzouveli, P.; Karpouzis, K. & Kollias, S. (2009). Estimation of behavioral user state based on eye gaze and head pose-application in an e-learning environment. *Multimedia Tools and Applications*, Vol. 41, No. 3, February 2009, pp. 469-493, ISSN:1380-7501

Beoldo, A.; Dore, A. & Regazzoni, C.S. (2009). Extraction of Contextual Information for Automotive Applications. *Proceedings of the 16th International Conference on Image Processing, ICIP 2009*, pp. 1153 - 1156, Cairo, Egypt, November 2009, ISBN: 978-1-4244-5653-6

Bergasa, L. M.; Nuevo, J.; Sotelo, M.; Barea, R. & Lopez Guillen, M. L. (2006). Real-time system for monitoring driver vigilance. *IEEE Transactions on Intelligent Transportation Systems*, Vol. 7, No. 1, March 2006, pp. 63-77, ISSN : 1524-9050

Chen, L.; Syue, K. & Tseng, Y. (2010). A vehicular surveillance and sensing system for car security and tracking applications. *Proceedings of the 9th ACM/IEEE International Conference on Information Processing in Sensor Networks*, pp. 426-427, Stockholm, April 2010, ISBN:978-1-60558-988-6, ACM New York, NY, USA

Chung-Cheng, C.; Min-Yu, K. & Chun-Yi, W. (2010). Automatic Traffic Surveillance System for Vision-Based Vehicle Recognition and Tracking. *Journal of Information Science and Engineering*, pp. 611-629, Vol. 26, No. 2, ISSN 1016-2364

Ciardelli, L.; Beoldo, A.; Pasini, F. & Regazzoni, C. S. (2010). Driver's Behavior Assessment by On-board/Off-board Video Context Analysis, *Proceedings of the Twenty Third International Conference on Industrial, Engineering & Other Applications of Applied Intelligent Systems, IEA/AIE 2010*, pp. II-450 – II-459, ISBN: 3-642-13024-0, Cordoba, Spain, June 2010, Springer-Verlag, Berlin Heidelberg

Damasio, A. (2000). *The Feeling of What Happens: Body and Emotion in the Making of Consciousness*. Harvest, ISBN 0-434-00773-0, Fort Washington, PA

Dore, A.; Beoldo, A. & Regazzoni, C.S. (2009). Multitarget Tracking with a Corner-based Particle Filter, *Proceedings of the International Workshop on Visual Surveillance, VS 2009, in conjunction with ICCV 2009*, pp. 1251-1258, Kyoto, Japan, 27 September - 4 October 2009, ISBN: 978-1-4244-4442-7

Dore A., Cattoni, A.F. & Regazzoni, C.S. (2010). Interaction Modeling and Prediction in Smart Spaces: a Bio-Inspired Approach Based on Autobiographical Memory. *IEEE Transactions on Systems, Man, and Cybernetics, Part A: Systems and Humans*, in press

Friedman, J.; Hastie, T. & Tibshirani, R. (2000). Additive logistic regression: a statistical view of boosting. *The Annals of Statisics*, Vol. 28, No. 2, pp. 337–407, doi:10.1214/aos/1016218223

McCall, J.C. & Trivedi, M.M. (2006). Video-based lane estimation and tracking for driver assistance: Survey, system, and evaluation. *IEEE Transaction on Intelligent*

Transportation Systems, Vol. 7, No. 1, pp. 20–37, March 2006, ISSN : 1524-9050Murphy-Chutorian, E. & Trivedi, M. (2009). Head pose estimation in computer vision: A survey. *IEEE Transaction on Pattern Analysis and Machine Intelligence,* Vol. 31, No. 4, April 2009, pp. 607-626, ISSN: 0162-8828

Nieto, M.; Salgado, L.; Jaureguizar, F. & Arrospide, J. (2008). Robust multiple lane road modeling based on perspective analysis. *Proceedings of 15th IEEE International Conference on Image Processing, ICIP 2008,* pp. 2396–2399, San Diego, CA, USA, October 2008, ISBN: 978-1-4244-1765-0

Schneider, J.; Wilde, A. & Naab, K. (2008). Probabilistic approach for modeling and identifying driving situations. *Proceedings of the IEEE Intelligent Vehicles Symposium,* ISBN: 978-1-4244-2568-6, Eindhoven, June 2008, pp. 343–348

Tai, J. & Song, K. (2010). Image tracking of motorcycles and vehicles on urban roads and its application to traffic monitoring and enforcement. *Journal of the Chinese Institute of Engineers,* Vol. 33, No. 6, in Press.

Trivedi, M.; Gandhi, T. & McCall, J. (2007). Looking-in and looking-out of a vehicle: Computer-vision-based enhanced vehicle safety. *IEEE Transactions on Intelligent Transportation Systems,* Vol. 8, No. 1, March 2007, pp. 108–120, ISSN: 1524-9050

Voisin, V.; Avila, M.; Emile, B.; Begot, S. & Bardet, J.-C. (2005). Road markings detection and tracking usingn hough transform and kalman filter. *Proceedings of Advanced Concepts for Intelligent Vision Systems,* pp. 76–83, LNCS, vol. 3708/2005 Springer, Heidelberg, ISSN: 0302-9743

Wang, Y.; Bai, L. & Fairhurst, M. (2008). Robust Road Modeling and Tracking Using Condensation. *IEEE Transactions on Intelligent Transportation Systems,* Vol. 9, No. 4, pp. 570–579, December 2008, ISSN: 1524-9050

Whitehill, J. & Movellan, J. R. (2008). A discriminative approach to frame by frame head pose tracking. *Proceedings of the 8th International Conference on Face and Gesture Recognition,* pp. 1 – 7, Amsterdam, September 2008, ISBN: 978-1-4244-2153-4

Yang, M.; Kriegman, D. J. & Ahuja, N. (2002). Detecting Faces in Images: A Survey. *IEEE Transactions on Pattern Analysis and Machine Intelligence,* Vol. 24, No. 1, pp. 34-58, January 2002, doi:10.1109/34.982883

Yu, G.; Xiao, X. & Bai, J. (2009). Analysis of vehicle surroundings and driver status from video stream based on a single PAL camera. *Proceedings of the 9th International Conference on Electronic Measurement and Instruments, ICEMI 2009,* pp. 4-363 – 4-367, ISBN: 9781424438631, Beijing, China, August 2009, IEEE

11

New Concept in Automotive Manufacturing: A System-based Manufacturing

Mohammad A. Omar
Clemson University-International Centre for Automotive Research CU-ICAR
USA

1. Introduction

The automotive industry has been going through a continuous process of adjustment due to the changes in its operating environment. Such factors; the govremental in addition to the National Standards Setting Bodies NSB's regulations, for example the Corporate Average Fuel Economy CAFE standards controls the OEMs fleet fuel economy average, leading to the introduction of smaller vehicles or the use of light weight materials (low density) in the vehicle structures. In addition to the new environmental regulatons that have led to changes in the material usage, the levels of production emissions, and the expended energy. Additionally the NSB's have different focus in different countries so for OEMs operating in different markets, they would have to respond to different regulations; for example the NSB's in Eruope such as the DIN (German Institute for Standardization) and the CEN (The French Creative Environmental Network) have a recent focus on safety systems and standards in automobiles, while the american NSB's such as American National Standards Institute ANSI focuses on the final vehicle testing protocoles. All these regulations have a direct effect on the automotive manufatuirng; to provide specific exmaple; the automotive OEMs have shifted their paint from the typical solventborne into waterborne paints, due to the Volatile Organic Compounds VOCs emissions. This shift led to additional production steps, such as the flash off zone which is necessary to control the amount of water evaporation from the paint once it is applied on the vehicle shell. Also, the waterborne paint requires tigher control over the spray booth air conditioning requirements, whih have led to more energy usage in the paint area. Another effect on the manufacturing came from the use of the Tailor Welded Blanks, Coils and Tubes TW B/C/T technology which is introduced to allow designers to custom mix different steel grades or panel thicknesses for some body panels to meet the different functional requirements (load-bearing vs. Non-load bearing) across the panel; a good example of the TWB is the door inner panel which has a stiffness requirement at the hinge area, hence thicker or stronger steel is used in that area while the rest of the panel is non-load bearing structure, which means that a thinner or weaker steel grade can be used. The TWB technolgy enabled the designers more freedom to met fucntionalities and save weight at the same time, however this technology have added several steps such as laser welding different pieces to from the new blank, which also require stacking and de-stacking steps; more importantly if the OEMs don't have enough epxertise in laser welding then such blanks will need to be shipped to the subcontrator location; adding more time delays and cost implications. Ultimately, each OEM will have to conduct a feaisbility study to assess the TWB technology benefits and challenges, before impelmenting such appraoch. Other impacts on manufacturing include; the role of the ergonomics on the design of the final assembly area and the usage of the different fixture and power-tool hardware to help the line workers in thier frequent activities. Also the ergonomic have a played a major role in the design of the conveyor systems.

Other external factors include, the customer demand trends, which have changed to be more diverse in trems of product type and its power-train propulsion. These days, the OEMs will have to depart from the economy of scope perspective and focus on more economies of scope in terms of their product portfolio. To provide a quantitative example; the Japanese OEMs produced 85 different models in 1960's, which is further increased into 400 models in the 1980's; at the same time the production volume per model kept decreasing, hence the OEM is producing the same umber of vehicles but with a higher product mix; meaning that the automotive manufacturing is shifting from high volume, low mix production strategy into a high volume, high mix, which have major impacts on the factory layout; shifting from the product type layout strategy where the processes and machinery are ordered and sequneced based on a known, repeatable product type with the goal of decreasing the product lead time; into the process and cell based layouts. The process and cell based layouts are better suited for higher product mix because it was designed to increase the manufactuirng flexibility to deal with varying product. Additionally, the change in customer demand is not only limited to vehicles with different body style, size, or platform, but also in terms of the product propulsion or power-train system; internal combustion engine (gasoline or diesel), or internal combusion engine assisted with a recchargeable battery system (hybrid), or a full electric vehicle. These variations in power-train complicates the manufacturing final assembly process due to the different power-train mariage steps required for each type, in addition to the different steps needed to assemble each of these sub-assemblies and of course the associated saftey considerations when dealing with fully charged battery packs.

The manufacturing operating cost is another challenge affecting the automotive OEMs. The operating cost is changing in terms of the raw material cost, which can be qunatified into vehicle structural materials mainly Steel and Aluminium in addition other materials including the trim material and the chemicals such as paint, wax, adheisves and sealants. Another issue with the operating cost is the relative cost of energy between the different countries; for example the electric energy consumption in Italy costs around 30 cents for each kWhr, compared with 10 cents in Germany, and 5 cents in South Africa. The energy cost not only affect the direct manufacturing energy expenditures but also affect the cost of raw materials, because of the intensive material extraction energy requirements; for example the to extract and process 1 kg of wrought Aluminium almost 60 kWhr are expended. Also, the labor wage cost is highly relative between the different countries; to illustrate with an example, the labor wage is South Africa is around $5/hr compared with more than $30/hr in the United States.

The emerging of new markets and more distributed production and supply networks have also challenged the automotive industry and affected their production strategies. This factor is further exagerated with the penetration of new Original Equipment Manufacturers OEMs into established markets such as the Korean and Chinese OEMs. These factors have impacted the vehicle manufacturing because, the new markets introduced new challenges in identifying a supplier base if the production is established in such markets, and maintaining the labor force for long time periods while other OEMs are trying to recriut the same labor force. To illustrate this point with an example, the avergae age for a team-leader working for an automotive OEM in Japan is around 35-40 years, while that for the same OEM in China is around 30 years. Further complications are the result of govermental monitoring regulations and Intellectual Property IP issues.

2. The current state of automotive manufacturing

The automotive manufacturing systems in its three perspectives; the procedural side, the static/structural (machinery) side, and the transformational (material conversion processes) side, have been affected due to the above changes and challenges. To provide specific examples; the oil crisis in the 1970s have led to dramatic loss in customer proudct usage, hence halted the automotive production lines. In 1980's, the entering of the East Asian OEMs with their manufacturing practices into the established markets of Europe and North America, forced the industry to change and adjust its operational side. Such changes included more seious adoption of the lean manufacturing and the value engineering practices into the automotive manufacturing systems; within the American and the European OEMs. This is apparent in the emergence of the company-wide specific production systems such as the Ford Prdouction System FPS, the Mercedes Benz Production System MPS which are built on the same pillars as the Toyota Production System TPS. In addition to changes in the forecasting and the Material Requirement Planning MRP practices and the departure from the mass-production practices; or push systems into the pull based systems, where the customer demand in terms of quantity and frequency decides on the production output. Additional adjsutments were in the static side, where several OEMs started to follow a celluar based production layouts to accomodate the assembly and the fabrication of the vehicle sub-assemblies.

However, such practices did not affect the transformational side in any meaningful way; so the actual manufacturing processes from stamping to body-weld and then painting stayed the same since the 1970's. Such processes not only control the production lead time and cost but also the research and development efforts required to develop the different jigs and fixtures in addition to the stamping dies, which cost the automotive OEM around $5 million per die and around 2 years for the development and the approval processes. Additionally, the actual fabrication processes governs and control the overall flexibility in terms of product volume and scope. Furthermore, the current and coming challenges of added environmental regulations, the wide variations in customer demand in terms of product type (vehicle platforms) and volume can only be met through adjusting the manufacturing processes into more dedicated yet flexible platforms. Additional motivation to adjust and change the existing transformational processes, is due to the increase in demand for light weight vehicles that features hybridized body materials from Aluminum, Magnesium and Adavnced High Strength Steels AHSS. Forming and fabricating such new materials onto the current production lines introduces several technical challanges due to these materials intrinsic propoerties. For example, forming Aluminum using mechanical or hydraulic presses is not trivial due to its narrower forming window and its higher springback levels when compared with steel; this have forced several OEMs to design new body structures based on the space-frame design not the standard uni-body platform. However, the space-

frame is based on extrusions and hydroformed components focring the OEMs to rely on external suppliers for such components in addition the space-frame designs can't be accomodated on mass production basis due to the high level of manual work content involved. Furthermore, with a space-frame platform, it is more difficult for OEMs to selectively incorporate other lower cost materials to provide improved functionalities at lower cost.
So this chapter is intended to highlight some of the potential transformational changes that can be incorporated to change the current manufacturing practices into more streamlined and dedicated platforms that not only consolidate the number of components but also the number of processes involved in making the vehicle body structures. So this chapter is mainly focused on the transformational processes utilized in the automotive assembly plants.

3. The automotive manufacturing transformational processes

The current automotive manufacturing processes can be mainly perceived from the assembly operations which lead to the construction of the complete vehicle structure; and the power-train operations which are responsible for fabricating the power-train and drive-line components. The assembly processes starts with the stamping and forming of the vehicle body sheet panels into the vehicle different structures, using a variety of mechanical and hydraulic presses and dies. The stamped body-parts are then joined to form the vehicle main sub-assemblies that include the under-body sub-assembly, the side members (left and right) sub-assembly also, the fractional sub-assemblies that include; the roof, the cowl, the upper and lower backs, and the engine compartment sub-assembly. Other stamped componenets are called the vehicle closures namely; the doors, trunk, fenders, and the hood. The joinng porcess features an army of robots that join the ~500 stamped components to form the vehicle body shape using a around 5000 spot welds, several meters (~2 m) of Metal Inert Gas MIG welds, and around 10-20 meters of adhesive bonding. Each of these joining technologies is applied based on each body location functionality (stiffness) and styling requirement requirements (surface finish). The spot welded are applied using electordes with different tip material and diameters, and using different combination of welding current and time to accomodate the different material types and thicknesses; even though the spot welding is a very effective joining process it still requires a two-side access for each joint which limits its application for certian locations within the vehicle. On the other hand, the MIG welding can be applied from one side, but results in a larger heating foot-print leading to a larger Heat Affected Zone HAZ around the weld. Additionally the MIG welding process requires a shielding gas environment to protect the welding pool from any contamination or oxidations, which might ultimately lead to a weaker weld. The adhesive bonding has been typically applied to join the inner and outer panels of doors and hoods, after it is hemmed. The adheisve bonding is gaining more acceptances within the automotive industry due to its added advantages, such as its ability to join dis-similar materials whicl alow the OEMs to join Aluminum and Steel without worrying about the differences in thermal characteristics (thermal expansion coefficient, melting point, etc) or the galvanic corrosion issue. At the same time, the adheisve bonded joints are stiffer than the spot welded ones, due to the fact that the joint is made up of folded material, which increases its moment of inertia hence increasing its stiffness. On the other hand, the adhesive bonding should be selected carefully to ensure its compatability with the production conditions such as the curing oven temperatures, and the chemicals in the immersion paint tanks such as the E-Coat and the cleaning tanks. Additionally, the adheisve material should be checked for its comptability with the vehicle service life conditions, to avoid any toxic emissions or degradation in the adhesive performance.

Following the joining process in the body-shop area, the vehicle structure which is called the Body in White BiW at this stage, starts the painting process, which conditions, cleans, and convert the BiW surfaces to provide it with a corrosion resistance finish and prepare it for the sub-sequent spray painting steps. The spary paint covers the vehicle structure with three to five coats composed of the primer, the top-coat or base-coat and finally the clear coat. These coats provide not only a corrosion resistance, and a chip resistance layer but also the vehicles' final asthetics. The final assembly area then installs the different trim parts into the vehicle shell and joins the shell with the power-train, to complete the vehicle build. The final assembly area features mainly manual work assisted with power-tools.

The power-train manufactuirng sequence is composed of a variety of casting, forging processes to form the engine cylinder blocks and head, the connecting rods, the cam and crank shafts. These processes are then followed by multiple machining steps to remove excess material, drill functional holes, and create the required surface roughness. The final assembly of the engine and transmission components is done manually with the aid of different fixtures and fault-proof jigs. The power-train plants rely mainly on in-house components, however the assembly plants receives more material content from the different suppliers. It is important to mention that the power-train plant and the assembly plant are sequenced to follow the same production takt-time, hence at the end of the day each produced engine will meet with a specific body-shell in the final assembly area to output a complete vehicle following one unified takt-time.

Following text tries to highlight the main challenges and shortcomings of the current automotive assembly manufacturing processes and steps;

- Having the stamping process right at the begining with a rigid die shape that can not be changed, locks the shape of the vehicle shell early on in the manufacturing process, hence it does not allow the OEM to operate based on *make to order* principles. The only thing that the OEM can change is the final color of the vehicle shell in addition to some of the trim options because such operations are done late in the process.
- The stamping process yields around 400-500 parts and pieces that need to be joined and assembled which results in large number of processing steps in addition to even larger number of non-value added efforts in stacking, de-staging, staging, and transporting tasks. In addition stamping this large number of pieces requires more dies and adds more complexity to the sequencing and the allocation between/of the different presses and press lines.
- The reliance on the stamping to form the body shell, limits the OEMs material choices and the final shapes and geometries. For example the formability of light weight materials such as Magnesium and Aluminium limits the bending radius to the panel thickness ratio, which can be achieved using a press based forming. So, Aluminium can only be stamped into flat or semi-flat panels such as the hood or the roof. Furthermore the stamping process does not allow the designers to manipulate the cross sectional shape freely to help compensate for the Aluminium lower stiffness from that of Steel.

- The stamping dies not only lock the vehicle shell early in the manufacturing process, but also through the product life, for example a typical vehicle model life is around 4-5 years. However if the vehicle model life is shortened due to low vehicle acceptance or other economical conditions, then the OEM will incur heavy losses due to the high development cost already paid for each major die. At the same time some "face lift" re-design can be done but it will be limited to only one aspect of the vehicle overall geometry.
- Each one of the manufacturing processes within the assembly plant is different in its drivers and sensitivities. For example the stamping process is driven by the material (sheet metal) and the machinery (presses), while the joining process is heavily dependent on the machinery (robotic welders), the painting process is dependent on the paint material and the booth conditions and controls. While the final assembly area is heavily dependent on manual work so it's mainly controlled by the labour productivity and attitude (absenteeism). Hence these processes are not integrated but merely set in a serial fashion to apply different values to the vehicle semi-finished components as it travels through the production line. This adds greater complexity to the control and the monitoring schemes used to synchronize it. In addition this renders the overall production system sensitive to variety of market factors. The Toyota Production System tried to resolve the laakc of integration between the different production stages using the Andon system to highlight any problem areas within the production line, along with the Kanban system to synchronize the one-piece flow between the different stations. However, these systems are effective for known product type and quantity hence it need to be adjusted and changed to add more flexibility to the production sequence.
- The current joining process is composed of around 5000 spot welds per vehicle, which adds more lead time in addition it adds more investment in machinery, because applying 5000 spot welds within a typical takt time of 60 seconds means more robotic welders. Additionally, the high frequency of the welding process translates into more intensive maintenance efforts, especially for dressing and changing the electrode tips for each welding guns. Furthermore, the reliance on the resistance welding schemes limits the materials that can be joined together; for example the direct joining of the Aluminium and Steel panels leads to galvanic corrosion, also the fusion welding is not applicable for plastic parts. Even though, more and more adhesive bonding is applied within the automotive industry, it is done on the expense of Metal Inert Gas MIG welding and the mechanical fastening techniques not the spot welding.
- The tack welding step assembles the automobile shell parts together to form the basic vehicle shape, and only then a series of spot welding processes create the permanent joints and the final fit and dimensions of the panels' relative positions. To create the first shape in tack welding, a fixture with the specific vehicle body shape should be used to hold the different panels in the exact relative positions needed, for each of the body main sub-assemblies. If several vehicle models are built on the same line, this will require multiplicity of these fixtures which add cost and precious manufacturing time to replace, transport, and store these fixtures. Different "flexible" and /or "intelligent" fixture designs have been developed by several of the automotive OEMs such as the *Robogate* system developed by the Fiat Motor Company, the Intelligent Body Assembly

System *IBAS* developed by Nissan, the *Preciflex* system developed by Renault, and the *Intelligent Jig Fixture* introduced by the Toyota Motor Company. However, these flexible fixturing solutions are not yet perfect, for example the Robogate system has a large footprint that limits its applicability and adoption, while the remaining fixturing solutions with the exception of the *Preciflex*, still require transporting the un-used pallet or fixture if it is not in use.

- A more integrated solution to the fixturing problem is to incorporate fixturing and clamping reference points and features through the panel shape and geometry, which is created in the stamping process. These clamping features ensure that the panels will only fit one way, the right way.
- The painting process starts by immersing the vehicle shell into cleaning, conditioning and phosphate tanks, in addition to the electro-coat bath. Due to the complicated geometries of the vehicle shell and the large number of holes and gaps in it, even the immersion process can't guarantee a full coverage of the shell surfaces and fill its gaps and crevices. Even though new dip systems are based on rotating the BiW inside the tank to ensure better coverage, such systems are limited by some physical limits, for example the E-Coat coverage is limited by the Faraday's cage effect; hence the soluble paint does not reach inside the tubes and extrusions more distance than its external diameter.
- The current paining process consumes around 60 to 70% of the total energy within an automotive assembly plant due to the number and the nature of the processes involved. The air conditioning inside the spray booths is the major consumer of electric energy while the curing ovens have the lowest efficiency (around 10 to 20%) consuming the majority of the fossil fuel requirements. Other major energy expenditures are in the water treatment, because the water is used in capturing the over sprayed paint in the under-booth area, then it passes through a scrubber system to separate the water to go to the treatment facility and the over sprayed paint to accumulate as sludge.
- The final assembly processes are heavily dependent on manual operations which are not only difficult to integrate into the overall production control system but also, require further tooling and fixturing solutions to extend the operators reach and facilitate their operations. This adds further ergonomic considerations and training and liability issues.

The mentioning of the above challenges are meant to show the current manufacturing lines' limitations and major shortcomings, so new manufacturing sequence and processes are proposed to solve some of the above issues. Additionally listing the main limitations of current production lines can serve as a starting point for any new manufacturing systems.

Also, from the above challenges, one can conclude that the current, automotive manufacturing main stations are not well integrated, on the contrary each station work might complicate the sub-sequent ones' operation; for example the complex geometries and holes formed in the stamping process not only limits the paint coverage in the E-Coat baths but also complicates the robotic programming to apply the spray paint layers. Additionally, the different geometries created early on in the manufacturing sequence limit the robotic welders' flexibility and accessibility. The robotic welders' flexibility is limited by the fact that each body style will require a different fixture to hold the different panels in space so that the tack welding process can be applied to determine the vehicle body shape. So the proposed manufacturing system will re-consider the sequence of the manufacturing processes to incorporate the ease of manufacturing perspective.

The following section will address new concepts to replace some of the current automotive manufacturing processes to yield a more integrated and dedicated manufacturing sequence that is flexible, with higher value addition per unit time.

4. New concepts in automotive manufacturing

4.1 New manufacturing system, decision making

The first step in proposing a new manufacturing process, sequence and ultimately a system is to recognize the improvements required and decide on metrics of success so that any proposed system can be benchmarked against the current one. One can extract such objectives and metrics from the discussion in section two to be in following aspects; increasing the flexibility of the production lines, reducing the overall production cost, and increasing the production stability and quality levels. These metrics can be further translated into; (a) reduction in the production lead time, which helps in increasing the production flexibility and reduce its overall cost. (b) Reduce the number of processes and manufacturing operations needed to complete the vehicle build, (c) reduce the manufacturing process complexity level, (d) reduce the re-work and scrap levels to reduce the cost and non-value added work. (e) Increase the production standardization level and facilitate equipment re-configuration for different body styles.

The second step is to prioritize these improvements, by using systematic decision making tools. Such tools include the Quality Function Deployment QFD matrices and the Analytical Hierarchy Process AHP. The QFD is a systematic decision making tool that lists the improvements required as "customer demand" for example, short production lead time, along with a numerical rank to show their priorities. Also, the QFD describes the technical implications of each of these demands on the production line and more importantly on the product; also it explores the positive and negative interactions between these technical implications to help finalize a relative rank of each of the improvements. Figure 1 shows a standard QFD template filled with the current case study specifics, where the above mentioned customer demand are listed on the left hand side and the technical implications or the manufacturing system functional requirements are listed in the corresponding columns. Such technical requirements include; the reduction in parts count, the reduction in setup time, the number of materials selected, the panels' intricate shapes and geometries, consolidation of processes, etc. Also, the inter-relationships between these technical requirements are listed in the top of the QFD matrix. For example, the reduction in number of components has a strong relationship or effect on the reduction in setup time, this is indicated as a score of 2, which means that any reduction in the number of parts leads to major reduction in the setup time. However, the reduction in variability in panels' dimensions has a negative impact on the technical requirement of avoiding intricate shapes and geometries that is indicated as a score of -1.

On the other hand the AHP process ranks the different objectives or customer demands based on their performance in achieving the sought goals and objectives. Additionally the AHP is based on straightforward computation that can accommodate qualitative and quantitative metrics and criteria; in the qualitative sense, it decomposes an unstructured problem into a systematic decision hierarchy. It then uses a quantitative ranking through numerical numbers and weights in which a pair-wise comparison is used to determine the local and global priority weights and the overall ranking of the alternatives. It is worth mentioning that both QFD and AHP have provided the same results showing that the part consolidation and the use of modular sub-systems are the highest rank metrics.

4.2 Process perspective

Previous sub-section indicated that a manufacturing system capable of using modular structures and a common product platform will have the highest potential of meeting the automotive OEMs main objectives of cutting cost, lead-time, and increase the flexibility levels. So the proposed manufacturing system should accommodate these two metrics through process selection and sequence.

To reduce the parts count, one can think of consolidating the body panels by re-designing their stamping dies; for example the body side outer is typically composed of 3 -5 pieces including the A, B, C pillars, the quarter panel, and the fender. Combining these panels into one piece can reduce the parts count, however it impacts the product functionality and design. To illustrate with more words, this consolidation approach can limit the freedom of designers to selectively select different materials for their different locations within the same body panel; for example the fender can be made out of plastic to reduce the vehicle weight, and the B-pillar to be made out of two panels with different thicknesses to serve the different load bearing requirement from each location. So from this discussion, the mere changing of the stamping dies to consolidate the body panels into lesser count is not trivial and can lead to loss of design freedom, in addition this approach alone will lead not only to lesser utilization of the generated engineering scrap but also to higher scrap and offal rates, which in turn translates into higher manufacturing cost. One should also consider the current technologies dependant on the parts count such as the Tailor Welded Blank TWB approach, where different panels in grade or thickness are joined together using laser welding then stamped to form a tailored body panel. In addition the consolidation of parts count based on changing stamping dies add another challenge to the OEM, that is the addition of a more expensive die to develop and validate. Also, at the same time, the body shape is still locked early in the production cycle and the complicated geometries and features are still created before the painting or welding activities.

Based on above discussion, one might consider departing from forming the body panels using the press and die based systems, into more flexible fabrication techniques that allow for more flexibility from product and process perspectives. Such forming alternatives include super-plastic forming, which not only allows for fabricating large and complex shaped panels in single process step, but also it allows for the use of light weight materials such as Aluminium and Magnesium. The super-plastic forming can result into higher elongation from those achievable in typical press stamping, and can reach up to 800% total elongation. The super-plastic forming is achieved under, high temperature with low forming force and low strain rate; however not all materials can be formed using this technique due to the fact that the super-plastic forming is dependent on the material grain boundary sliding which is only achievable for materials with very fine grain size. Additionally, the super-plastic forming cycle is in the order of hours when compared with the press-stamping, which can achieve the final shape in few seconds. But one should consider all the time expended in the non-value added steps, such stacking and de-stacking the different panels and pieces that can be replaced with one super-plastic formed panel. To provide a more descriptive example, one can super-plastic form the whole under-body as one piece that can be made out of Magnesium or Aluminium, while stamping the under-body constitute forming many different Steel pieces that need to be transported, setup and then joined together; so when comparing the two under-body manufacturing methods, the super-plastic forming might offer more advantages and flexibility but at higher cycle time.

Another possible forming technique that can be applied for some of the body panels is based on folding the sheet metal using slit and smiles created along the fold line. Even though this forming approach is not suitable for all body panels due to the surface finish requirements, it can eliminate the need for any forming dies in addition it can result in greater flexibility in changing the body style on the fly, because the slits location, number, and shape can be changed through a computer controlled laser cutting machine. Additionally this approach can reduce the engineering scrap and offal by optimizing the cutting process; also this forming approach can be done using different material types. Furthermore, this forming method helps in reducing the part count through integrating the different folds to create different shapes from one cut sheet. It also facilitates transporting the flat cut panels and reduces the efforts needed for stacking and de-stacking.

Industrial Origami Incorporated *IOI* is pioneering the forming through folding technology, a demonstration of their technology is displayed in figure 2, which shows an instrument panel formed from one piece.

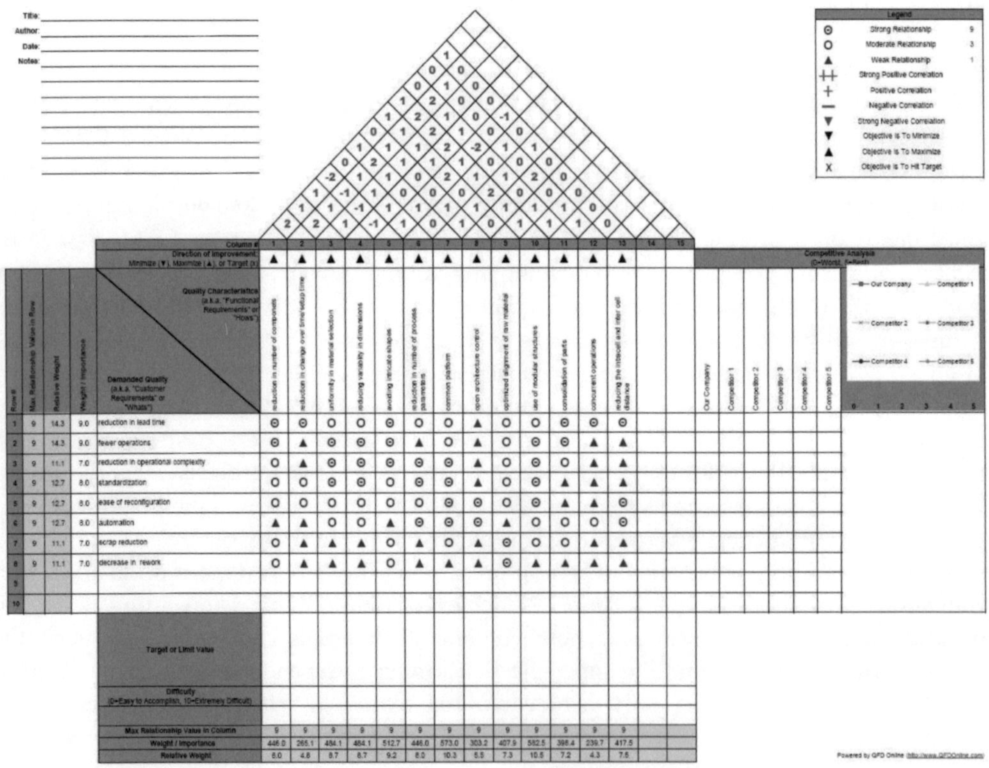

Fig. 1. QFD matrix

Using the Industrial Origami *IOI* technique and the super-plastic forming can also present a new potential for facilitating the automobile production by changing its sequence. In more words, an OEM can move the conditioning phase of the painting process to be before the folding step; because it is easier and more efficient to immerse coat flat sheets than complicated geometries; also this will improve the phosphate and E-Coat coverage over the panels. Additionally, the slits and smiles compose referencing and clamping features within the stamping so no additional fixtures are needed to fixate the panels' relative positions. This reduces the overall investment in fixtures and mounts and the production overall lead time, also it improves the flexibility of the joining process and the better utilization of the space floor; because, some of the proposed flexible fixturing systems as the *Robogate* has a large foot-print as mentioned previously. Additional benefits of the folding-based forming are due to the fact that the panels can be shipped and transported in and out of the plant in flat shape and also reduce the number of joining processes needed to create the different sub-assemblies, which can also be said about the super-plastic formed pieces because it integrates several smaller panels into one.

So, from the above discussion, one can layout a new production sequence to start with conditioning stage of the Steel and Aluminium sheets through immersing it in the cleaning, phosphating baths. Then the forming of the body panels is achieved through super-plastic forming, folding, and press-based forming. The press based forming will be mainly for the exterior panels; such as door outer skins, hood, roof, body-side outer that also incorporate the fenders. The use of super-plastic forming should consolidate several panels to compensate for its slow cycle time, such as the under-body, which should integrate the floor pan, the trunk pan, etc. Any considerations about the effect of the applied coatings on the formability of the material should be analyzed from the frictional interface conditions between the die and the sheet metal. Also, the coating integrity should be tested to ensure that it was not degraded because of the forming process. Even though this arrangement require the use of different forming technologies with different machinery and equipment requirements and expertise, it ultimately focuses on streamlining the overall production by reducing the stamped panels count while at the same time provide several modular sub-assemblies that can be shared between the different vehicle models.

For the new joining step, it starts by folding the interior components and fixating them relative to each other using the embedded clamping points and features. Then join them to the under-body panels to form a basic module that can be used for different body styles. Then, the body side outer can welded to this module. The body side outer along with the other closure panels (doors, hood, etc) will be the changing panels between the different vehicle models. The design of these panels will incorporate also clamping features to help locate the panels' relative positions in the welding process to avoid the reliance on a dedicated fixturing systems or solutions. After the welding is completed, the BiW will go straight into the new paint area which features spray booths only to provide a single coat of anti-corrosion paint formulation to replace the E-Coat layer, in addition to the sealants and the under-body wax applications, then run the BIW through a curing oven to cure the sealants and the paint all at once, which saves energy and space, in addition to reducing the number of processes. The sprayed paint should also be applied for the internal panels and its colour should be neutral. The use of a curing oven is necessary and can't be easily replaced in the automotive industry because some of the currently used steel grades require baking to add more strength to the steel at its fabrication stages; such grades include the Bake-Hardenable BH grades. The BH steel microstructure contains small amounts of carbon in solid solution which when heated comes out of the solution to increase the steel strength and dent resistance. Such steels are developed so that the final steel strength and dent resistance meet the functional requirement without having to use a strong steel grade in the stamping stage, which requires more tonnage. The final paint coats, including the top-coat that include the colour pigments, the UV absorbing pigments, and the metallic flakes, in addition to the clear coat, are proposed to be applied in the dealership per the customer demand for colour and finish. This proposal try to further postpone applying the final colour so that the dealerships have greater flexibility in manipulating their car inventory per customer purchasing trends. Additionally, moving the top and clear coats painting outside the automotive manufacturing plant, hence reducing its overall energy consumption, its overall floor space requirements and also, reduce the process count leading to a shorter lead production times.

For the new final assembly area, it should also use modular trim components that can be easily shared between the different vehicle models; also such modules should be designed in terms of its size and shape to fit into the vehicle without the need to remove the doors. This eliminates the need to remove the doors and have them move uselessly around the plant to meet with the vehicle shell at the end of the assembly line.

The proposed manufacturing sequence and processes are selected based on the main two objectives extracted from the QFD and the AHP results that is reducing the number of parts and using more modules. The coming sub-section will discuss the changes and modifications needed to incorporate the suggested transformational changes and processes.

4.3 Structural and operational issues

The difference in cycle time between the proposed forming techniques should be first addressed to ensure that the proposed production line functions based on an average takt time. To address this issue, one can use the Value Stream Mapping VSM technique to decide on the adequate buffer levels that will ensure the slow forming methods (super-plastic forming) are not bottle necks, using the minimum Work In Process WIP levels.

One of the tools used in VSM is the implementation of a supermarket at the end of the press stamping operation so that the faster process (press stamping) is producing according to the quantity and type of panels pulled from the supermarket, as controlled and specified by the production and the withdrawal Kanban cards. Additionally, one can enrich the press-stamping with more work content; such as assembling the doors inners and outers and the hood inners and outers. The enrichment of the work content has the added advantage of creating lesser number of stations that are easier to control and can accommodate a cellura base layout. The cellular layout offers better communication schemes and enables people and equipment sharing within the cell leading to better resource utilization and higher flexibility levels to accommodate different product types and volumes. For the super-plastic forming multiple stations might be established to help keep up with short cycle times, in addition to a buffer to further compensate for the cycle time difference.

The above discussed plan not only enables the OEMs to use the super-plastic forming and other forming technologies for some sub-assemblies, but also it establishes lesser work-stations that can produce more specialized complete sub-assemblies, that is each work-station adds more value content to the vehicle in continuous fashion without the need for the stacking, de-stacking, and transporting items between smaller stations.

The new joining line will receive bigger sub-assemblies that can be joined directly to the vehicle under-body; this due to the fact that the cells that used to provide the welding line with the different assembled parts such as; floor assembly, rear-floor assembly, engine compartment, roof cowl, are all shifted to be done in the stamping step. Also, joining certain reinforcements can also be done in the stamping process by using the techniques of In-Die joining. So, a shorter body-weld line is achieved.

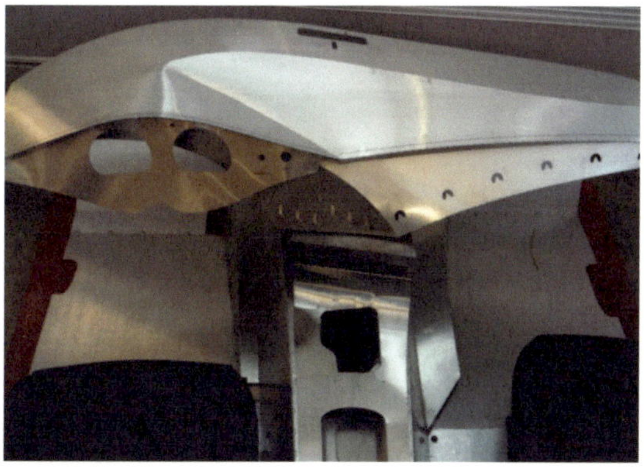

Fig. 2. Folded Instrument Panel

The new paint line layout is much simplified and is composed of; a cleaning station where the BiW are first cleaned from any accumulated sludge from the weld line using paint friendly detergent formulations. Also, this station can accommodate the sealant and under-body wax applications steps; then the BiW travels through a single spray paint booth to receive an anti-corrosion coat, and then pass through the curing oven. The new paint line has much smaller foot-print because it doesn't have the selectivity bank area that is typically used to change the sequence of the production vehicles so different colour batches can be created; in addition to pulling vehicles for paint defects repair. Also, the immersion process is already shifted and only one curing oven and one spray booth are used.

5. Discussion and challenges

To apply the proposed processes and layout into current vehicle designs and OEM facilities, require a collaborative effort between the different manufacturing stations' engineers. For example, paint engineers should work with the stamping engineers to decide on the best way of applying the phosphate coats on the metal sheets, in addition to analyzing the effect of the stamping process on the paint adhesion and integrity. Furthermore, the welding engineers should work with the designers to help embed locating and clamping features into the panel designs to help alleviate the need for locating fixtures and jigs.
Additional efforts are needed from the material scientists to develop new, cheaper Aluminium alloys that can be blow formed in super-plastic forming. Also, equipment supplier should be involved to help in developing faster super-plastic forming machines that can meet the production requirement of the new production line. Additionally, discrete event simulations should be done to optimize the material and content flow through the production shop floor and regular VSM activities should be carried out to ensure the continuous improvement of the production line operation.

6. References

Clarke, C. (2005). *Automotive Production Systems and Standardisation,* Physica Verlag, 3-7908-1578-0, Heidelberg

Hitomi, K. (1996). *Manufacturing Systems Engineering,* Taylor and Francis, 0-484-0323-x, London

Omar, M.; Mears, L., Kiggans, R., & Kurfess, T. (2009). Organizational Learning in Automotive Manufacturing: A Strategic Choice. *Journal of Intelligent Manufacturing,* ISSN 0956-5515, DOI 10.1007/s10845-009-0330-6

Part 3

Industrial Machinery and Tools

12

Tomography Visualization Methods for Monitoring Gases in the Automotive Systems

Krzysztof Polakowski
Warsaw University of Technology
Poland

1. Introduction

Taking into consideration growing ecology and safety demands, there are a need for more accurate and cheaper ways to monitor workflows of gases, which can be used in automotive applications. Such researches come into prominence in case of more commonly applied gas systems, that use hydrocarbons as well as hydrogen, which in case of running down of oil resources, are predicated as the most probable fuel for intern combustion engines or fuel cells used in vehicles. Use of tomography can be an alternative (in relation to currently used) method for such monitoring methods. The ultrasonic flow measurement is one of the most promising. The main advantage of this method is the fact that it is based on contactless, non-invasive flow measurement which does not cause any pressure or other physics-chemical changes in observed environment.

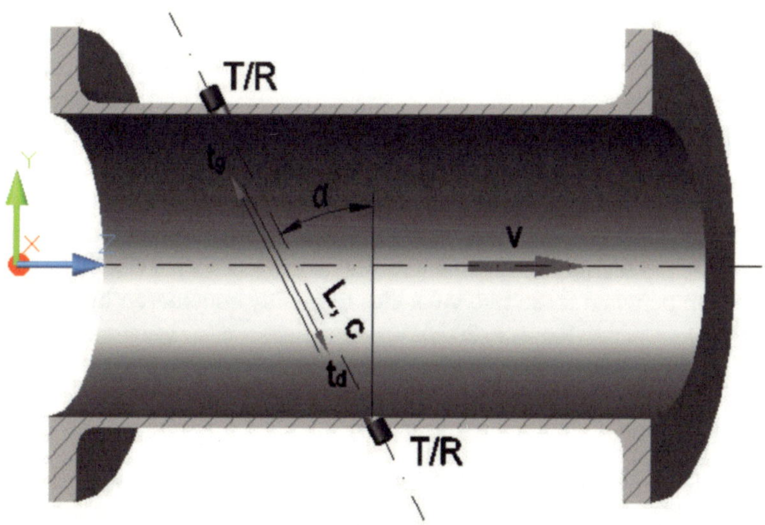

Fig. 1. A scheme of the ultrasonic transit-time flowmeter

The measurement principle of an ultrasonic transit-time flowmeter involves at least one pair of ultrasonic transducers: transmitter (T) and receiver (R). Transmitting transducers are triggered to send an ultrasonic pulse, one upstream and one downstream to the receiving transducers. Due to the fact that the distance between the opposite arranged transducers L is known, the travel time of the pulses can be estimated if the temperature is also known. The measurement principle of an ultrasonic transit-time flowmeter utilizes an ultrasonic travel time measurement for the calculation of both velocity v of the flowing medium and the speed of sound c (Kupnik, 2008):

$$v = \frac{L}{2\sin(\alpha)} \frac{t_g - t_d}{t_g t_d} \tag{1}$$

and

$$c = \frac{L}{2} \frac{t_g + t_d}{t_g t_d} \tag{2}$$

In general, the measure of the matter that moves in a given time through a given transport cross-section is termed flow rate. The matter can be in solid, liquid or gaseous form. I should distinguish between volumetric flow rate Q_v (quantity of a flow in cubic metre per unit time) and mass flow rate Q_m (quantity of a flow in kilograms per unit time) (Kupnik, 2008):

$$Q_v = \frac{dV}{dt} = v_A A \tag{3}$$

and

$$Q_m = \frac{dm}{dt} = \rho Q_v = \rho v_A A \tag{4}$$

where V is the volume, m is the mass, A is the cross-section of the transport way, v_A is the averaged velocity over cross-sectional area of the transport way and ρ is the density.
A fundamental problem in acoustic flow measurement is the fact that the distribution of the velocity in the measurement pipe in each case is not known exactly, because an ultrasonic transit-time flowmeter always determines the averaged velocity along the sound path (v_s), i.e. it integrates the velocity profile over the volume of the sound beam. Exact knowledge of the velocity profile is essential to convert the line averaged path velocity v_s to the velocity v_A. The connection between these two velocities is usually considered by a correction factor k_v, also termed meter factor (Kupnik, 2008):

$$k_v = \frac{v_A}{v_S} \tag{5}$$

For a given axially symmetric flow profile v(r), where r denotes the radius and D denoted the internal diameter of the pipe, can write (Kupnik, 2008):

$$k_v = \frac{v_A}{v_S} = \frac{4}{D} \frac{\int_0^R v(r)\, r\, dr}{\int_0^R v(r)\, dr} \tag{6}$$

Equation (6) shows clearly that meter factor k_v.depends directly on the flow profile $v(r)$. Each deviation of the flow profile from the assumed one, which is used in Equation (6), forcibly leads to uncertainties of the flowmeter. Finding an appropriate model equation for the flow profile $v(r)$ for each flow measurement problem is essential (Kupnik, 2008).
Multipath systems with different sets of emitters and receivers are used in order to enlarge the data and accuracy of measurement (Mandard at al., 2008). In considered system equally spaced sensors around the pipe's perimeter in perpendicular surface to pipe's axis were placed. In the system the transmitters one by one generates ultrasound impulses, which with different delays reach all receivers. The time duration of these waves is the basis for calculation of the average value of measured flow's speed (Roger & Baker, 2005).
If I assume that the transmitters emit ultrasound waves of not too high frequencies, than the angle of spreading groups of waves is big enough to allow them to reach all of the receivers placed in pipe's perimeter (Opieliński at al., 2006).
The speed profile image (a magnitude of speed vectors distribution on the surface perpendicular to flow direction) we can achieve with an aid of ultrasonic tomography method.

2. The inverse problem solution

New generations of computer's tomography systems often based on algebraic algorithms of image reconstruction (ART) from projections. The tomography image is constructed with the help of algorithm, which digitize the diagnosed area to the quadratic cells, which geometric centres are handled like pixels of the image.
Velocity profile image construction in a plane of receivers means estimation of a set of unknown values, which we can denote as $f(x, y)$.
I can get integrated values of speed on a base of time measurement of ultrasonic impulses travelling like rays from transmitters to the receivers (Fig. 2). These values, according to projection method, are named the ray-sum measured with i-th ray s_i. I have assumed that in each cell of the rectangular or square grid, the function $f(x, y)$ is constant. The relationship between the f_j's and s_i's may be expressed as:

$$\sum_{j=1}^{n} w_{ij} f_j = s_j, \quad i = 1,2,...,m \tag{7}$$

where: m is total number of rays (main and auxiliary ones) in all projections, n number of cells crossed by rays, w_{ij} is the weighting factor that represents the contribution of the j-th cell to the i-th ray integral.
The iterative method, based on the "method of projections" as first proposed by Kaczmarz (Kak & Slaney, 1999) is very useful for solving this problem. In this method a grid representation with n^2 cells gives image of n degrees of freedom. Therefore, an image, represented by (f_1, f_2, ..., f_n) may be considered to be a single point in an n dimensional space. In this space each of equations (8), represents a hyperplane. When a unique solution of these equations exists (for $m \geq n$), the intersection of all these hyperplanes is a single point giving that solution.
The equation (7) we can write in an expanded form:

$$\begin{aligned}
&w_{11}f_1 + w_{12}f_2 + w_{13}f_3 + \cdots + +w_{1n}f_n = s_1 \\
&w_{21}f_1 + w_{22}f_2 + w_{23}f_3 + \cdots + +w_{2n}f_n = s_2 \\
&\vdots \\
&w_{m1}f_1 + w_{m2}f_2 + w_{m3}f_3 + \cdots + +w_{mn}f_n = s_m
\end{aligned} \tag{8}$$

In order to solve the system of equations (8) we have used least squares method which gives an excellent results particularly in case when m is bigger than n (over-determined system of equations).

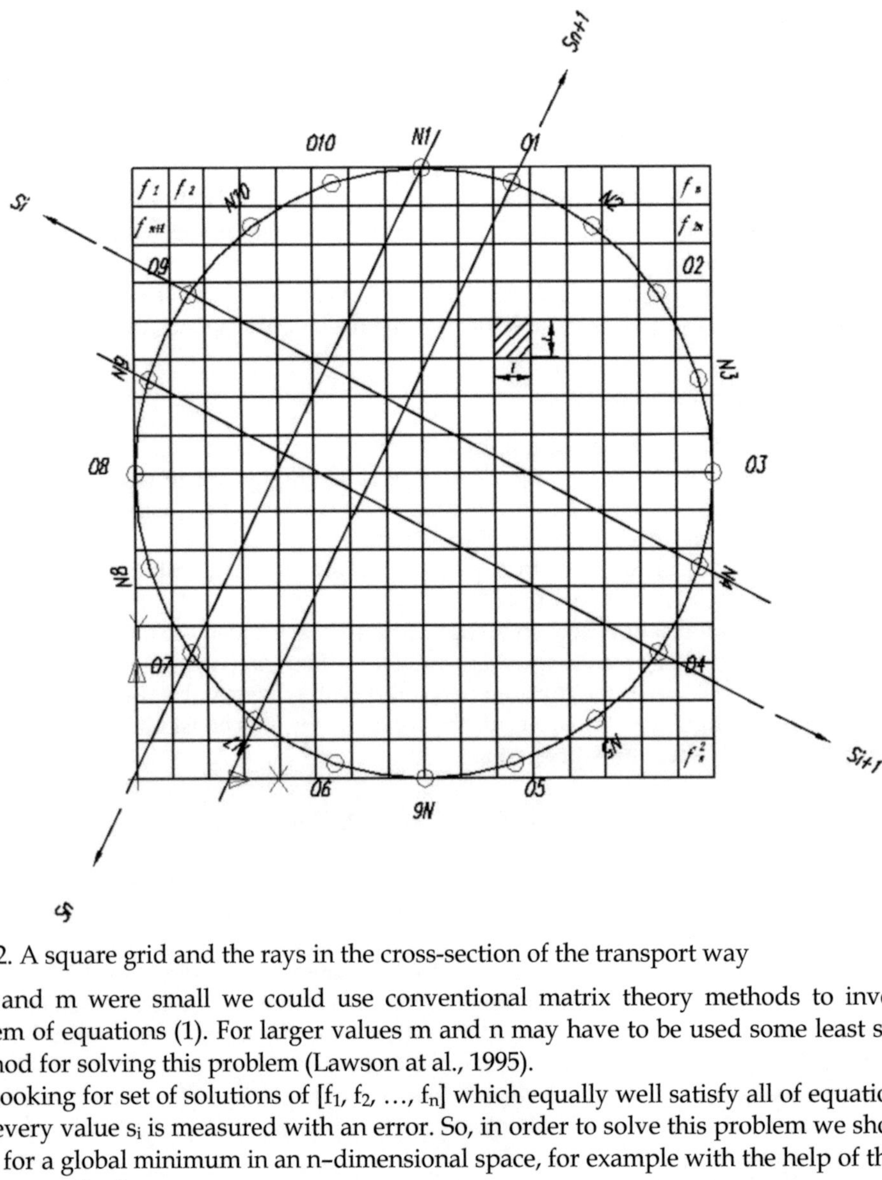

Fig. 2. A square grid and the rays in the cross-section of the transport way

If n and m were small we could use conventional matrix theory methods to invert the system of equations (1). For larger values m and n may have to be used some least squares method for solving this problem (Lawson at al., 1995).

I'm looking for set of solutions of [f_1, f_2, …, f_n] which equally well satisfy all of equations (8), but every value s_i is measured with an error. So, in order to solve this problem we should to look for a global minimum in an n–dimensional space, for example with the help of the least squares method.

The factors w_{ij} of the cells, which are not crossed by the rays are equal to 0. When a j-th cell is crossed by i-th ray this factor could be calculated according to very simple relation (9):

$$w_{ij} = \frac{\sqrt{2}}{lp_{ij}} \tag{9}$$

Problem of image construction in case of the ultrasonic flowmeters very often leads to the over determined algebraic set of equations (2), which in matrix form can be expressed:

$$\boldsymbol{Wf} = \boldsymbol{s} \tag{10}$$

where: $\boldsymbol{W}$ is the matrix of dimensions $m \times n$ and $m > n$, $\boldsymbol{s}=[s_1,s_2,\ldots,s_m]$ – right hand side vector, and $\boldsymbol{f}=[f_1,f_2,\ldots f_n]^T$ – the solution vector. One of the ways of the solution of the problem (8) is to find the vector $\boldsymbol{f}^*$, which minimize Euclidean norm of residual vector $\mathbf{r}$ for the known matrix $\boldsymbol{W}$ and vector $\boldsymbol{s}$. It means: $\|r\|_2=min\|s-Wf\|_2$, $\|f^*\|_2=min\|f\|2$,
where - the last minimum is taken for all vectors $\boldsymbol{f}$ which fulfil the previous relation. This is Linear Least Squares Problem (LSP) (Lawson at al., 1995).
In order to calculate the solution to Problem LS and analyzing the effect of data errors as they influence to it, we will use the Singular Value Decomposition (SVD) theorem. This theorem says that for any arbitrary matrix $\boldsymbol{W}\in R_{m\times n}(m \geq n)$ of pseudo-rank k exists an m by m orthogonal matrix $\boldsymbol{U}\in R_{m\times n}$ an n by n orthogonal matrix $\boldsymbol{V}\in R_{m\times n}$ and $\boldsymbol{D}\in R_{m\times n}$ such that:

$$\boldsymbol{W} = \boldsymbol{UDV}^T \tag{11}$$

where:

$$\boldsymbol{D} = \begin{bmatrix} d_1 & 0 & 0 & 0 \\ 0 & d_2 & 0 & 0 \\ . & . & . & . \\ 0 & 0 & 0 & d_n \end{bmatrix} \in R_{m\times n} \tag{12}$$
$$d_1 \geq d_2 \geq d_k > d_{k+1} = d_{k+2} = \cdots d_n = 0$$

and k is the pseudo-rank of matrix $\boldsymbol{W}$.
The diagonal entries $d_i$of the diagonal matrix $\boldsymbol{D}$ are known as singular values of matrix $\boldsymbol{W}$ and relation (11) as a Singular Value Decomposition (SVD), (Polakowski at al., 2007, a).
This decomposition is closely related to the eigenvalue-eigenvector decomposition of the symmetric nonnegative matrix $\boldsymbol{W}^T\boldsymbol{W}$, and diagonal terms of matrix $\boldsymbol{D}$ are the squared eigenvalues of $\boldsymbol{W}^T\boldsymbol{W}$. Columns of matrix $\boldsymbol{V}$ are the orthonormal eigenvectors of $\boldsymbol{W}^T\boldsymbol{W}$, and columns of matrix $\boldsymbol{U}$ are eigenvectors of $\boldsymbol{WW}^T$.
Knowing the Singular Value Decomposition (11) one can easily to find the solution of LSP:

$$\boldsymbol{f}^* = \boldsymbol{W}^+\boldsymbol{s} \tag{13}$$

where: $\boldsymbol{W}^+=\boldsymbol{VD}^+\boldsymbol{U}^T$ is the pseudoinverse of matrix $\boldsymbol{W}$ (or inverse matrix in Moore - Penrose sense) and $\boldsymbol{D}^+ = diag\left(\frac{1}{d_1},\ldots,\frac{1}{d_k},0,\ldots,0\right) \in R_{m\times n}$.

For a square nonsingular matrix $\boldsymbol{W}$, the pseudoinverse of matrix $\boldsymbol{W}$ is the inverse of matrix $\boldsymbol{W}$.

$$\boldsymbol{W}^+ = \boldsymbol{W}^{-1} \tag{14}$$

In case of ill-conditioned problem (in tomography usually we have to deal with ill-conditioned problems) the solution can be achieved in the following way. Suppose the singular value decomposition is computed for the matrix $\boldsymbol{W}$:

$$\boldsymbol{W} = \boldsymbol{U}\begin{bmatrix}\boldsymbol{D}\\0\end{bmatrix}\boldsymbol{V}^T \tag{15}$$

One can compute:

$$g = \boldsymbol{U}^T \boldsymbol{s} \tag{16}$$

and consider the least squares problem:

$$\begin{bmatrix}\boldsymbol{D}\\0\end{bmatrix} q \cong g \tag{17}$$

where $\boldsymbol{q}$ is related to $\boldsymbol{f}$ by the orthogonal linear transformation:

$$\boldsymbol{f} = \boldsymbol{V}\boldsymbol{q} \tag{18}$$

Problem (18) is equivalent to the problem $\boldsymbol{Wf} \cong \boldsymbol{s}$ in the sense of general orthogonal transformations of least squares problems. Since $\boldsymbol{D}$ is diagonal ($\boldsymbol{D}=diag\{d_1,d_2,\ldots,d_n\}$) the effect of each component of $\boldsymbol{q}$ upon the residual norm is immediately obvious. Introducing a component q_j with the value:

$$q_j = \frac{g_j}{d_j} \tag{19}$$

reduces the sum of squares of residuals by the amount q_j^2.

Assume the singular values are ordered so $d_k \geq d_{k+1}$, k=1,2,…,n. It is then natural to consider "candidate" solutions for problem (18) of the form:

$$\boldsymbol{q}^{(k)} = \begin{bmatrix} q_1 \\ \vdots \\ q_k \\ 0 \\ \vdots \\ 0 \end{bmatrix} k = 0,1,\ldots,n \tag{20}$$

where q_j is given by Eq. (19). The candidate solution vector $\boldsymbol{q}^{(k)}$ is the pseudoinverse solution (i.e., the minimal length solution) of problem (18) under the assumption that the singular values d_j for $j \geq k$ are regarded as being zero.

From the candidate solution vectors $\boldsymbol{q}^{(k)}$ one obtains candidate solution vectors $\boldsymbol{f}^{(k)}$ for the problem $\boldsymbol{Wf} \cong \boldsymbol{s}$ as:

$$\boldsymbol{f}^{(k)} = \boldsymbol{V}\boldsymbol{q}^{(k)} = \sum_{j=1}^{k} q_j V^j, \qquad k = 0,\ldots,n \tag{21}$$

where V^j denotes the j-th column vector of $\boldsymbol{V}$.

Note that:

$$\left\|r^{(k)}\right\|^2 = \left\|q^{(k)}\right\|^2 = \sum_{j=1}^{k} q_j^2 = \sum_{j=1}^{k} \left(\frac{q_j}{d_j}\right)^2 \tag{22}$$

hence $\|f^{(k)}\|$ is a nondecreasing function of k.
The squared residual norm associated with $f^{(k)}$ is given by:

$$\left\|r^{(k)}\right\|^2 = \left\|s - Wf^{(k)}\right\|^2 = \sum_{j=k+1}^{m} g_j^2 \tag{23}$$

Inspections of the columns of the matrix V associated with small singular values are a very effective technique for identifying the set of columns of W that are nearly linearly dependent.
Suppose that the matrix W is ill-conditioned; then some of the later singular values are significantly smaller than the earlier ones. In such a case some of the later q_j values may be undesirably large. One hopes to locate an index k such that all coefficients q_j for $j \leq k$ are acceptably small, all singular values d_j for $j \leq k$ are acceptably large, and the residual norm $r^{(k)}$ is acceptably small. If such an index k exists, then one can take the candidate vector $f^{(k)}$ as an acceptable solution vector. (Polakowski at al., 2007, b).
In order to choose a preferred value of index k one can use a graph of residual norm versus solutions norm $\|r^{(k)}\| = f(\|f^{(k)}\|)$. For ill-conditioned problems we have „L" shaped curve. Using such the graph it is easy to determine the optimal value of index k.
Using the method called FOCUSS (FOCal Underdetermined System Solver) we could solve a system of underdetermined algebraic set of equations (fewer measurements than the unknowns which are a common case for industrial tomography), (Gorodinitsky at al., 1995).
It let us consider the following constrained optimization problem
Minimize

$$J_p(f) = p \mid fj \mid \tag{24}$$

subject to

$$Wf = s$$

where f_j are non zero values of the matrix f and $J_p(f)$ (often called the diversity measure) is some measure of sparsity of signals and it can take form, called the generalized p-norm :

$$J_p(f) = sign(p) \sum_{j=1}^{n} p \mid f_j \mid^p \tag{25}$$

where $p \leq 1$ and is selected by the user.
To minimize the generalized p norm diversity measure $J_p(f)$ in (21), subject to the equality constraint $Wf = s$ we define the Lagrangian $L(f,\lambda)$ as:

$$L(f,\lambda) = J_p(f) + \lambda(s - Wf) \tag{26}$$

where $\lambda \in R^n$ is a vector of Lagrange multipliers and $D_{|f|} \in R^{(n \times n)}$ is a diagonal matrix with the entries $d_j = |f_j|^{2-p}$.

Solving the above equations by simple mathematical operations, we obtain that

$$\lambda_* = |p|\left(WD_{|f|}(f_*)W^T\right) - 1s \tag{27}$$

$$f_* = |p|^{-1} D_{|f|}(f_*) = D_{|f|}(f_*)W^T(WD\,|f_*|\,W^T)^{-1}s \tag{28}$$

The equation (24) is not in a convenient form for computation since the desired vector f_* is implicitly in the right side of the equation f_*. However, it suggests that an iterative algorithm for estimation of the optimal vector f_* is given as:

$$f(k+1) = D_{|f|}(k)W^T(WD_{|f|}(k)W^T)^{-1}s \tag{29}$$

where

$$D_{|f|}(k) = diag\{|f_1(k)|^{2-p}, |f_2(k)|^{2-p}, |f_n(k)|^{2-p}\} \tag{30}$$

The above algorithm, called the generalized FOCUSS algorithm can be expressed in a more compact form:

$$f(k+1) = \tilde{D}_{|f|}(k)[W\tilde{D}\,|f|\,(k)]^+ s \tag{31}$$

where the superscript $(\,.\,)^+$ denotes the Moore-Penrose pseudo-inverse and

$$\tilde{D}_{|f|}(k) = D_{|f|}^{1/2}(k) = diag\{|f_1|^{1-p/2}(k), |f_2|^{1-p/2}(k), |f_n|^{1-p/2}(k)\} \tag{32}$$

It's should be noted that the matrix $D_{|f|}$ exists for all f and even for a negative p. For p=2, the matrix $D_{|f|}$ =J and the FOCUSS algorithm simplifies to the standard LS or the minimum 2-norm solution:

$$f_* = W^T(WW^T)^{-1}s \tag{33}$$

3. Results

The system of equations describing the tomographic flowmeter was solved with the aid of Linear Least Squares method for overdetermined algebraic set of equations. With the method called FOCUSS (FOCal Underdetermined System Solver) I have solved a system of underdetermined algebraic set of equations.

Condition number of the resulting rectangular matrix was high enough so that the classical Kaczmarz's algorithm was not able to produce correct results. That's why I have to take into account pseudo rank deficiency of the matrix coefficients. I have considered all possible candidate solutions, when k was changing from 1 till the full pseudo - rank equal to 1000 (Polakowski at al., 2008, a).

The images and their relief plots constructed on the basis of the candidate solutions are presented in Figure: 4b, 6b and 7b. The images and their relief plots constructed on the basis of FOCUSS are presented in Figure: 4c, 5b, c, d, 6c and 7c.

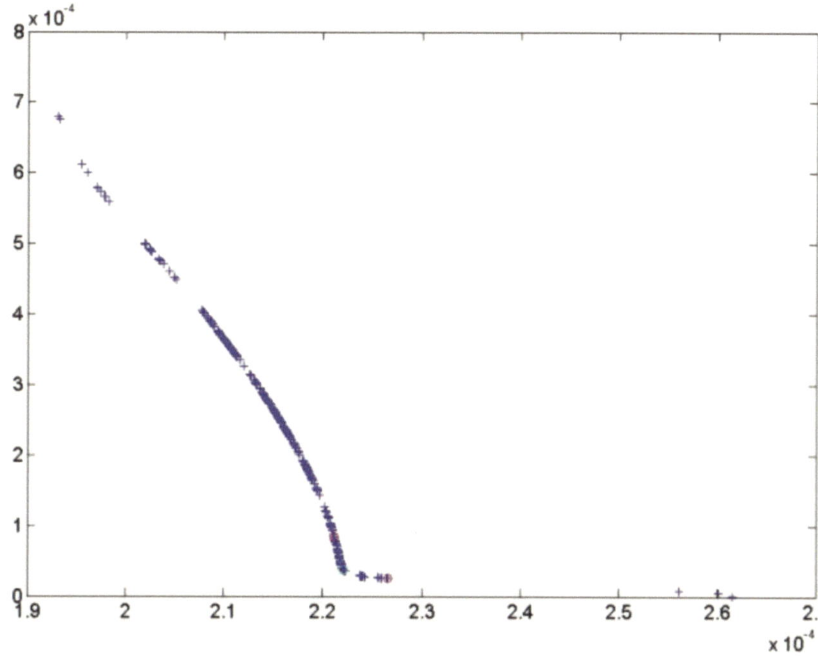

Fig. 3. Diagram $\|r^{(k)}\|=f(\|f^{(k)}\|)$of the residual vector norm versus the solution vector norm for the cross shaped object

	The resolution of grid 16x16, number of the rays 256	The resolution of grid 16x16, number of the rays 624
a)	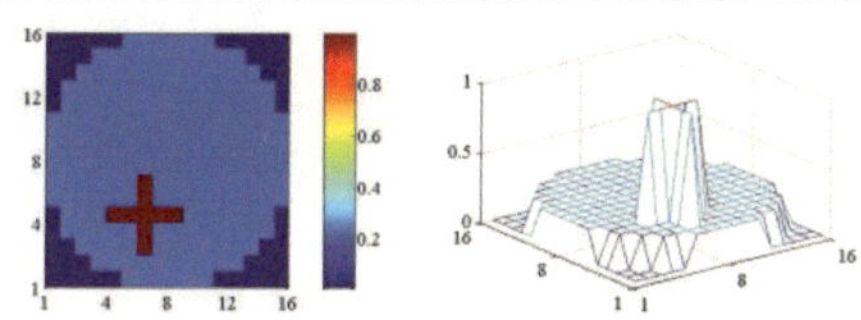	
b)	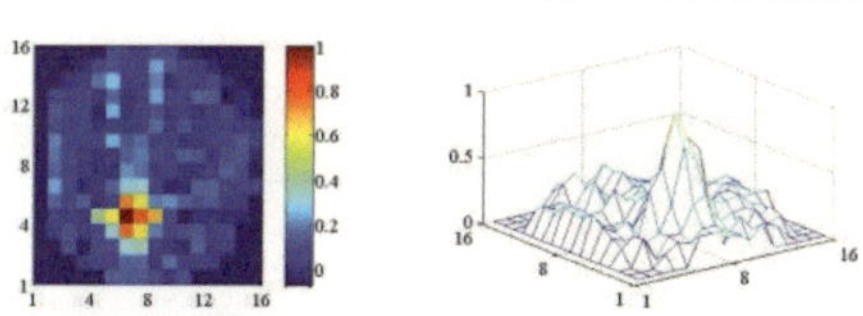	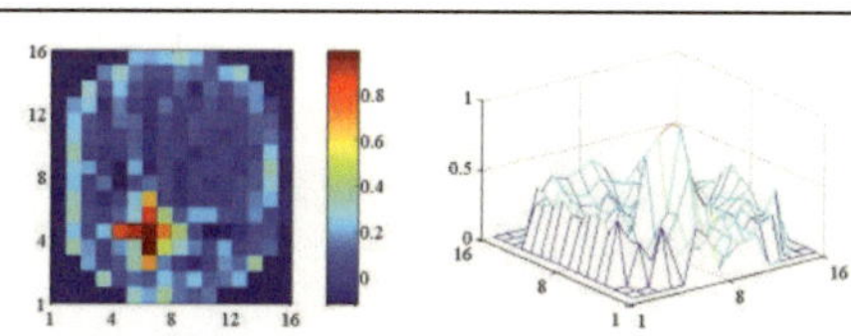
c)	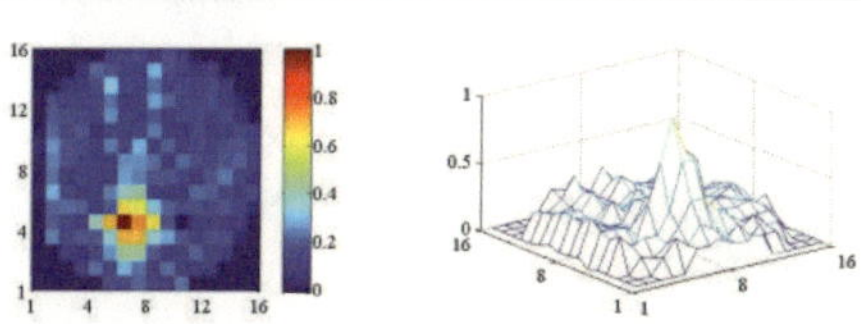	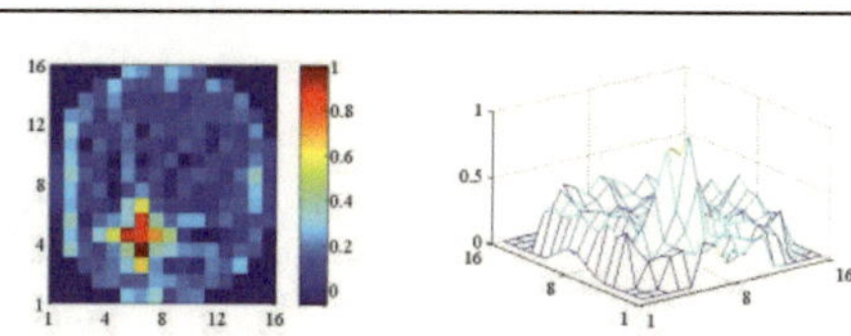

Fig. 4. The changes of images and the relief plots of a cross shaped object in dependence of number of the rays a) reconstructed object, b) reconstruction with the aid of Linear Least Squares Method , c) reconstruction with the aid of FOCUSS

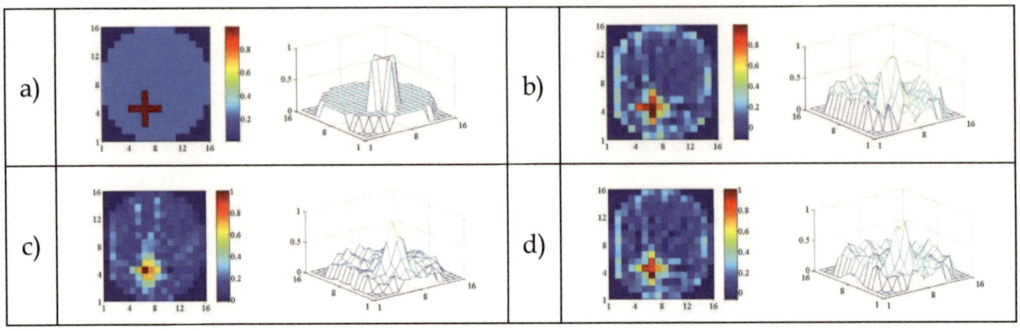

Fig. 5. Images and the relief plots of a cross shaped object reconstructed with the aid of FOCUSS in dependence of regularity index a) reconstructed object, b) regularity index 0.02 , c) regularity index 0.2, d). regularity index 50.

When I compare Figure: 5b, c and d we can also see, that the images have not been improved with the bigger regularization parameter, when the resolution of the grid was not to high.

Inspecting those images we can observe the influence of the resolution of the square greed and number of the rays on the object forming inside the region. The influence of resolution and number of the rays on improving the image we can clearly see on Figure 4, 5, 6 and 7.

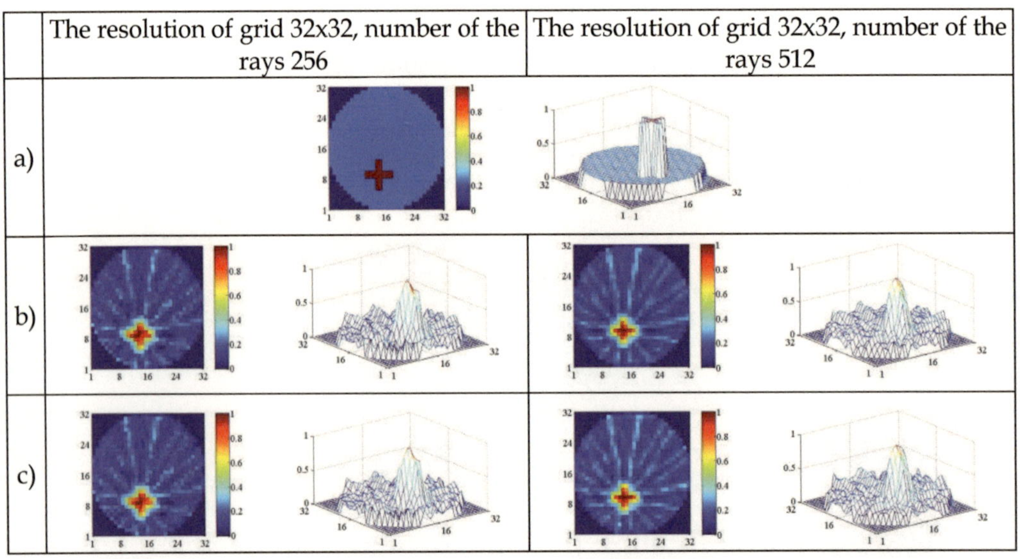

Fig. 6. The changes of images and the relief plots of a cross shaped object in dependence of number of the rays a) reconstructed object, b) reconstruction with the aid of Linear Least Squares Method , c) reconstruction with the aid of FOCUSS

It is worth to mention, that shown in Figures: 4÷7 achieved results were constructed for unpolluted synthetic data and the images were not filtered in order to check the behaviour of the image construction algorithm.

Theoretical and experimental researches carried out in this work prove that by increasing the number of radiuses which cross the pipe we increase the number of rows in the coefficient matrix *W*. It causes the results improvement, but at the same algorithm's

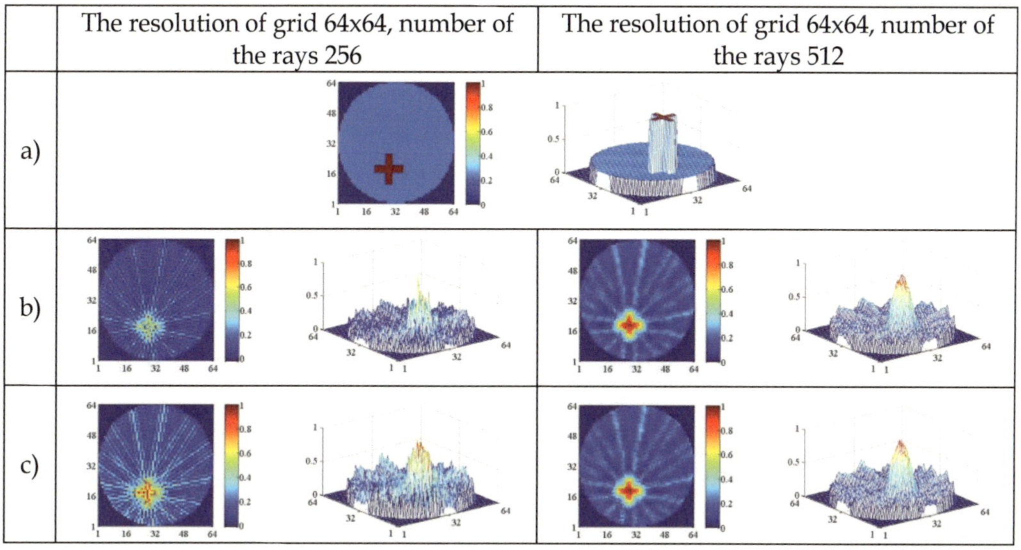

Fig. 7. The changes of images and the relief plots of a cross shaped object in dependence of number of the rays a) reconstructed object , b) reconstruction with the aid of Linear Least Squares Method , c) reconstruction with the aid of FOCUSS

execution time. Thus the number of rays should be selected in reasonable and considered way, thinking about image's quality and algorithm's execution time. By increasing the size of the resolution of the square greed (increase number of pixel) we cause a rise of the number of columns in coefficient matrix **W**. On the other hand, with higher resolution of the square greed, reconstructed object could have more details and it is more similar to the real object. We should notice that with higher resolution of the square greed, the number of rays has to be increased proportionally.

After checking the behaviour of described above the image construction algorithms we tried to receive a velocity profile of the flow. For receive a velocity profile computations were made in area of a pipe with 0,20 m diameter. In the model transmitters (N=32) and receivers (O=48) were evenly distributed around cross-sectional area of the pipe as shown in Fig. 8.

In analyzed model the transmitters one by one generates ultrasound impulses, which with different delays reach all receivers (Polakowski at al., 2008, b).

This work contains examples of simulation computations of the complex shape modelling the flow with complicated 3D shape (Fig. 9). Chosen methods made it possible to obtain tomographic images that accurately map tested shape (Fig. 11).

The tested area with modelled object was divided into 5 surfaces (Fig. 10). In each surface were made 32 projections with help of 32 x 48 rays between 32 transmitters and 48 receivers in each surface. In all surfaces were made calculations, which gave tomography images of calculated area. On figure 9 are shown only 9 from all achieved results with their relief plots.

From these 2,5 D results we can quite accurately reconstruct the whole 3D modelled flow shape.

The system of equations describing that tomographic imaging was solved with the aid of Linear Least Squares Method. Condition number of the resulting rectangular matrix was high enough so that the classical Kaczmarz's algorithm was not able to produce correctly results (Polakowski at al., 2008, b). That is why I have to take into account pseudo rank deficiency of the matrix coefficients. I have considered all possible candidate solutions according Eq. (21), when k was changing from 1 till the full pseudo-rank.

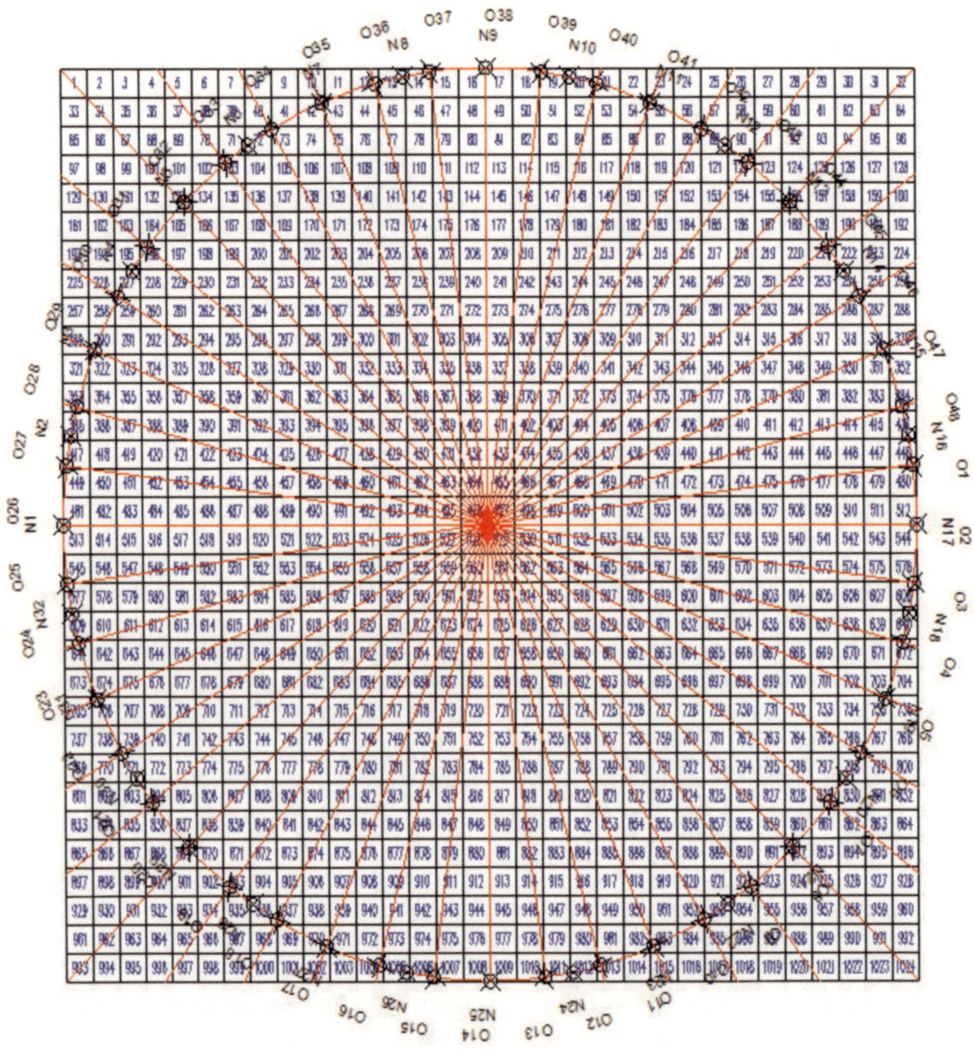

Fig. 8. Modelled area divided with 32x32 pixels and evenly distributed transducers: 32 transmitters N x 48 receivers O

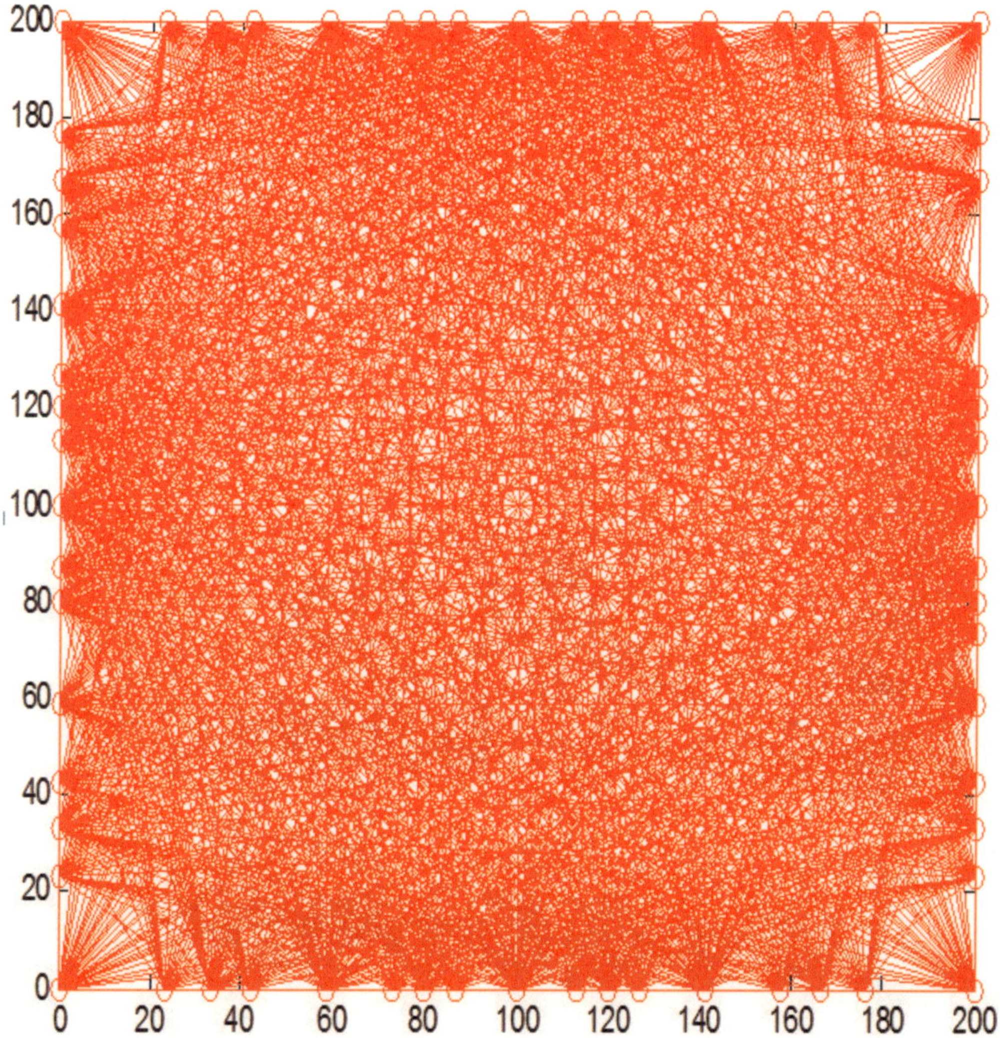

Fig. 9. All possible 1536 rays in modelled cross-sectional area from 32 projections between 32 transmitters and 48 receivers

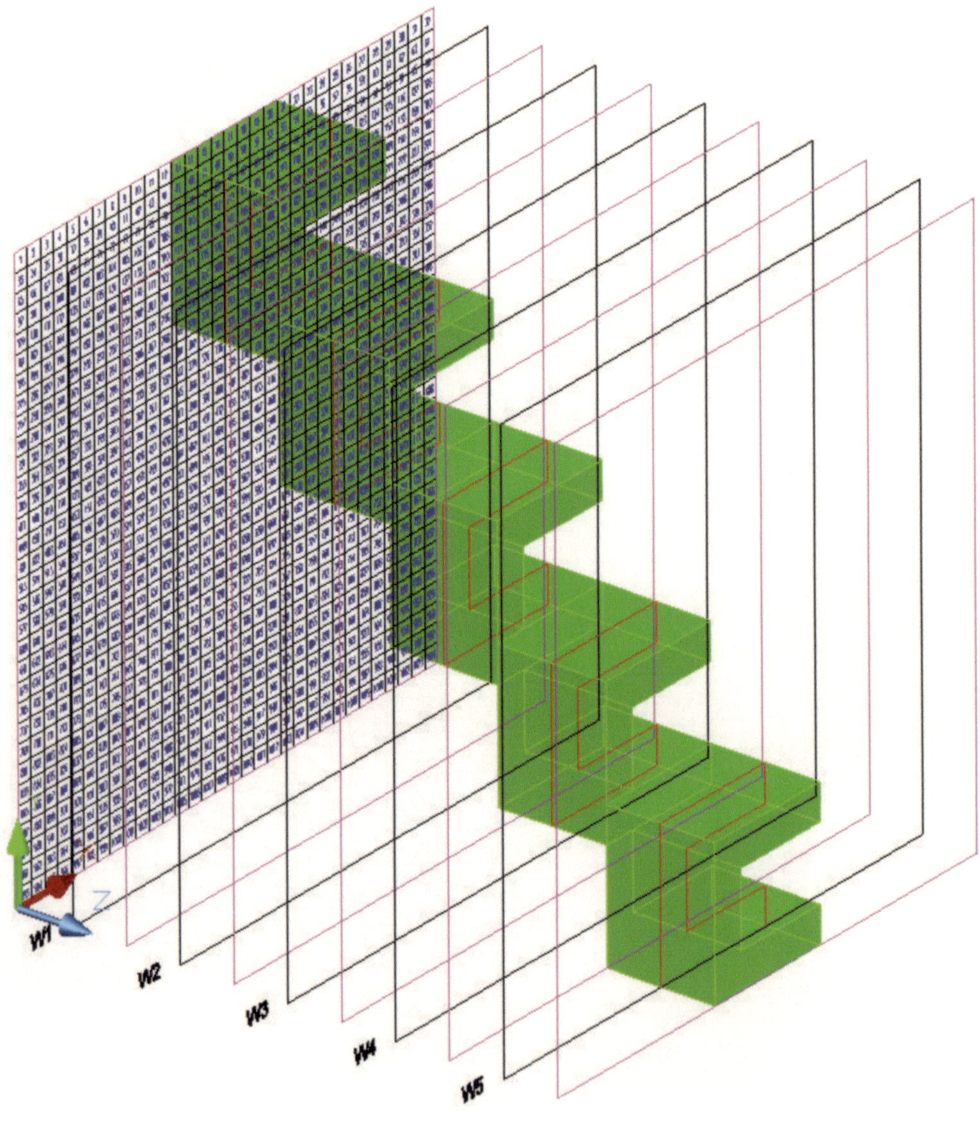

Fig. 10. Model of the complicated flow shape and its 2,5D visualization

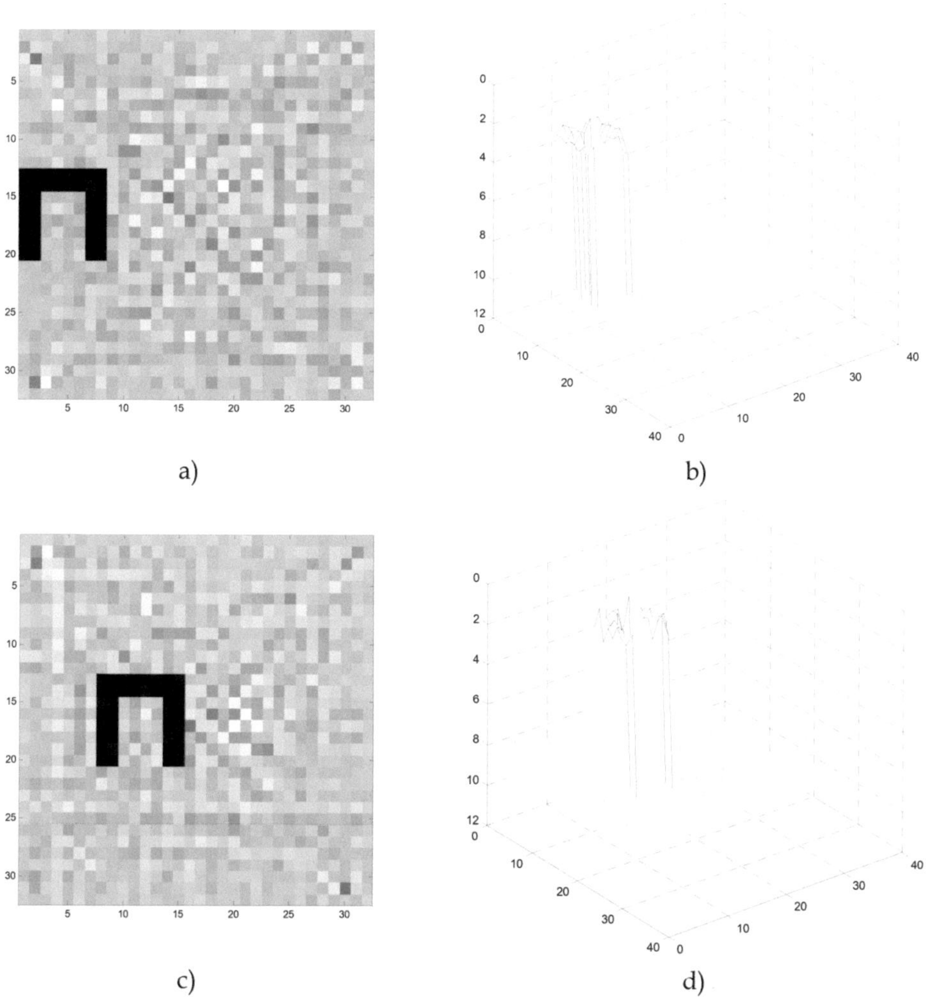

Fig. 11. Two examples of obtained tomography images with their relief plots in W1 (a, b) and W2 (c, d) surfaces used for reconstruction of the flow in 2,5 D

I also have performed calculations for noise polluted data. The noise was generated according to algorithm where the changes in rays flow were achieved through changing the position of transmitters and receivers according to Eq. (33), where in case of noise a random number l_{los} with weight w was added to n_y coordinate l_{los} was within the <0, 1> scope and was calculated by random numbers generator.

$$n_y = n_y + w(l_{los} - 0{,}5)n_y \tag{33}$$

After that the value was reduced by 0,5 in order to get positive or negative values.
In this case even data with high noise haven't caused big image deformation. It is an essential fact, because real data consists of noise from measurement errors.
The obtained tomography images (Fig. 11) confirm that chosen method gave us images that accurately map tested shape.

3. Conclusion

The calculations with use of the modelling in 2,5D space, give a chance to get results which reflect the phenomenon in the analyzed 3D area, quite accurately (Fig. 9). The obtained results are satisfying and further work should be given the answer for the question, if the proposed method will finding practical application in automotive applications. The simplification of calculations with assuring the sufficient accuracy in making tomography images of analyzed physical phenomenon should succeed in faster obtaining of results. This issue is important because the contemporary tomography is expected to bring real time tomography images of dynamically changing environment.

4. References

Gorodinitsky I.F., George J.S and Rao B.D., (1995), *Neuromagnetic source imaging with FOCUSS: a recursive weighted minimum norm algorithm,* Clinical Neurophysiology, vol. 95, pp. 231-251

Kak A., C., Slaney M. (1999) *Principles of Computerized Tomographic Imaging,* IEEE Press, ISBN: 0-87942-198-3

Kupnik M. (2008) *Ultrasonic Transit-time Gas Flowmeter for Automotive Applications,* VDM Verlag Dr Műller, ISBN: 978-3-639-00789-3

Lawson C. L., Hanson R. J. (1995) Solving Least Squares Problems", *Classics in Applied Mathematics 15,* SIAM

Opieliński K., Gudra T. (2006) *Recognition of external object features in gas media using ultrasound transmission tomography,* Ultrasonics, 44, pp.1069-1076.

Mandard E., Kouame' D., Battault R., Remenieras J. P., Patat F. (2008) *Methodology for Developing a High-Precision Ultrasound Flow Meter and Fluid Velocity Profile Reconstruction,* IEEE Transactions on Ultrasonic, Ferroelectrics and Frequency Control, vol. 55, no.1, pp. 161-171

Polakowski K., Sikora J., Filipowicz F.S. (2007) *SVD for image construction in ultrasound tomography,* The International Conference on "Computer as a Tool" EUROCON, Warsaw, pp. 276-281, a

Polakowski K., Sikora J. (2007) *Visualization and image analysis problems in multipath ultrasonic tomography,* 5th World Congress on Industrial Process Tomography WCIPT5, Bergen, pp. 941-948, b

Polakowski K., Sikora J., Filipowicz F.S. (2007) *Idea of 3D Imaging Based on 2,5D Tomography Reconstruction Approach;* 16th International Conference on Systems Science ICSS'07, Wrocław, vol. 3, pp. 206-211, c

Polakowski K., Sikora J., Filipowicz S.F. (2008) *Computer Methods in Monitoring of Flow Processes in Car Systems,* ZKwE, Poznań, April 14-16, pp. 185-186, a

Polakowski K., Sikora J., Filipowicz S.F., Rymarczyk T. (2008) *Tomography Technology Application for Workflows of Gases Monitoring in The Automotive Systems,* Przegląd Elektrotechniczny, R. LXXXIV, 12/2008, pp. 227-229, b

Roger C. Baker (2005) *Flow Measurement Handbook,* Cambridge University Press, pp. 312-351

13

FlexLean - Flexible Automation for Automotive Body Assembly

Sven Soetebier[1], Nicolas Mauser[1], Fabrice Legeleux[2] and Sönke Kock[3]
[1]ABB Corporate Research Center
[2]ABB R&D BIW
[3]ABB Corporate Research
[1]Germany
[2]China
[3]Sweden

1. Introduction

The automotive industry has pioneered the large scale use of robots. Long production runs of identical car bodies were the ideal field of application for early industrial robots, and spot welding lines with hundreds of robots have become a familiar sight. However, a lot of the production equipment is still based on hard automation. Today's market is increasingly putting automobile manufacturers under pressure to offer customers more choice of products and variants with decreasing life cycle, while at the same time demanding lower production costs. To fulfill these apparently contradictory requirements, a single line must be able to produce a mix of different models, and must "learn" to make new models without calling for a total re-design of its equipment - and preferably without even stopping production ("rolling launches").

To responded to these demands it is necessary to make the automotive body assembly more adaptable, easier to install and more economic on space. This can be achieved with the presented concept (FlexLean) that introduces modular and highly flexible solutions based on modular, standardized components consequently in all levels of the automation (Negre & Legeleux, 2006).

First of all the line is based on standardized and freely configurable cell modules. The complete line is made up of these highly flexible modules which are connected by a material handling track motion and standard communication interfaces only. The cells again achieve their high flexibility through a rigorous utilization of robot technology, not only for handling and welding with robots but also for clamping and fixturing systems. Key elements are here programmable flexible positioners, called FlexPLPs. These positioners can have depending on requirements different types of kinematics with one up to four axes and replace the tooling equipment used today for locating, handling and fixing car bodies or other parts. A special Deltapod kinematic is described in more detail that combines high accuracy and stiffness with an excellent compactness and light weight. To fulfill the wide range of requirements on the flexible positioners for the totally different tasks in the car body assembly they are built based on standardized components. A concept will be proposed how different types of positioners again can be designed automatically based on the standard components matching exactly the requirements of the process like workspace, load, stiffness, accuracy. Finally such a flexible automation system needs a control concept that allows the engineering and programming of the cells with minimal effort for the line builder and customer. The proposed control solution profits directly from the concept of the modular and standardized cells. This allows also a very modular control concept based on standardized control modules. Going this way consequently to the end, it will be shown that it is even possible to replace the classical programming by only simply configuring the desired behaviour of the components up to the sequence of a whole cell.

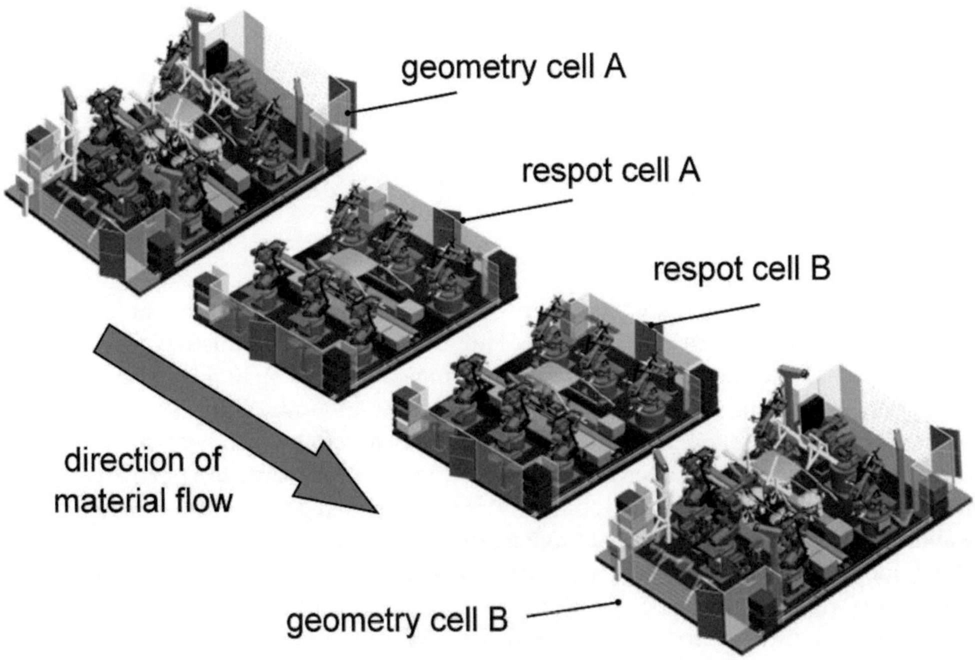

Fig. 1. Assembly line based on standardized modular cells

2. The FlexLean concept

2.1 Flexible and modular assembly line

With FlexLean an automotive body assembly line is composed out of freely configurable and standardized cells (see Fig. 1). Each of these cells is a modular robot cell where all equipment from the robots up to the controllers and cabling are pre-mounted on a platform. Every cell has its own control and operational system responsible for all operations in the cell from robot movements, part handling and transport of the car body up to the whole production cycle. A cell is connected to its neighbouring cells only by a standardized communication interface for handling the handover of parts from on cell to the next and by a handling track motion for the car body. In this way every cell is a standalone system that can, if necessary, be replaced very easily by another cell module at any time or new cells can be introduced in the line to adapt to extended requirements, new processes or a new type of car model.

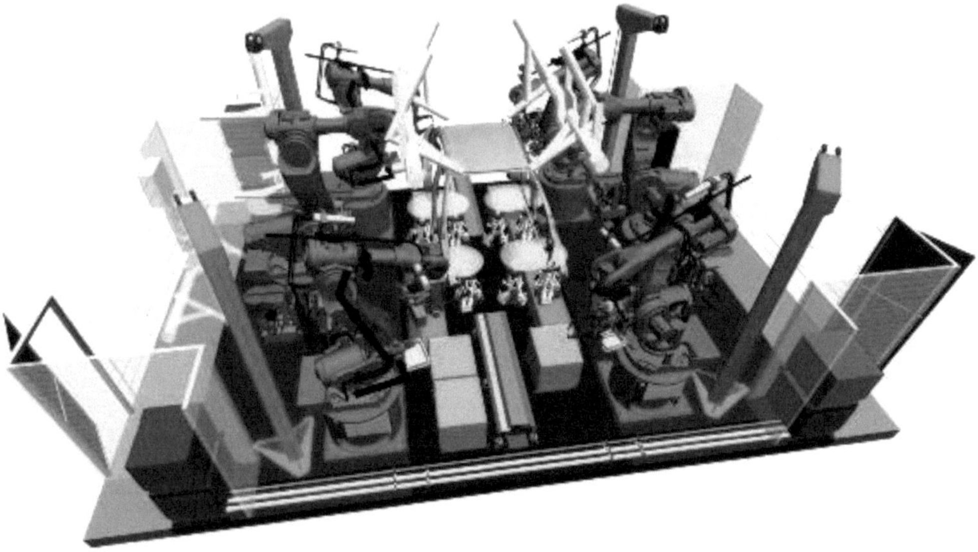

Fig. 2. FlexLean standardized cell with 6 robots, track motion and positioners for multi-model car body assembly

Fig. 2 shows a schematic layout of such a standardized cell. It can have a different number of robots (typically 2 to 6) that can be equipped with a choice of predefined process packages (like spot welding, material handling, sealing ...). Further it can have e.g. a different number of manual loading stations. For the part handling of the car body in the cell and between the cells a robot like servo controlled track is used enabling for precise and high speed part transfer. Further key elements of the standard cells are the programmable flexible positioners that replaces fixed tooling equipment used today holding and fixing car bodies on the track motion as well as for holding and handling of parts in the cell .

2.2 Flexible positioning and gripping

A car body is made out of 300 to 500 parts (Wemhöner, 2005). Robotized spot welding is the most common process to reliably join the parts; to secure the geometry of the car, every part has to be held in place by a fixture prior to welding. The accuracy and stiffness of fixturing of the shaped metal sheets defines the final quality of the car body geometry.

To achieve this, the cars are built often on top of skids which are transported by a conveyor through the factory (see Fig. 3). These skids have some drawbacks: first of all they need to be returned to the beginning of the line after finishing the car. Since they are not flexible and assigned to one car model only, they need to be exchanged when changing to another car model. This requires heavy skid handling equipment in the line and a lot of place for storing the different types of skids. The effort for this explodes with the number of car models produced in one line.

In FlexLean the traditional conveyors have been replaced by the already mentioned track that transports the car body from on cell to the next only. The proposed flexible and lean approach is to avoid car model specific tooling altogether and replace them by fixtures with programmable geometry that can adapt to any car body. This is achieved by the flexible programmable positioners (FlexPLP) replacing every pole carrying a locator by a positioner.

Fig. 3. A model specific skid (pallet) with fixtured car underbody on roller table

Thus instead of the skids a number of FlexPLP is mounted on the track carrying the car body from one cell to the next (see Fig. 4 in the background). In the cell again fixed mounted flexible positioners take over the car body from the track motion allowing the track motion to get the next car body from the previous cell (see Fig. 4 in the foreground).

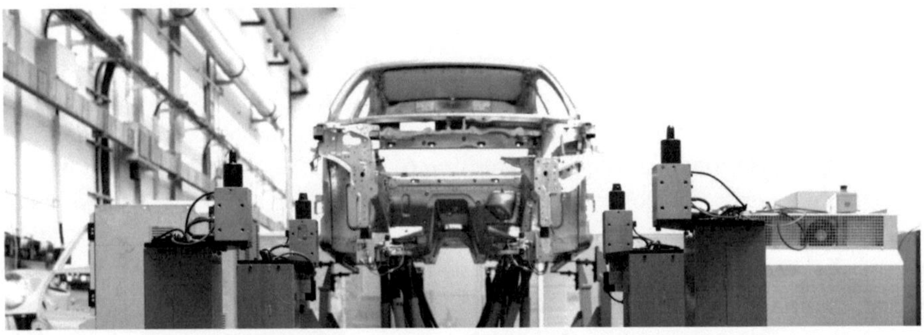

Fig. 4. Car body on track motion carried by flexible positioners (background) and flexible positioners taking over the car body in the next cell (foreground)

As for the handling of underbodies/car bodies it is also necessary to find a new solution for the part handling in the cell. Today mostly geometric grippers are used, which can already have a modular structure but are restricted to a fixed geometric shape for only one part. To achieve the required flexibility to adapt to different sizes of parts for different car models, it is possible in the same way as for the skids to replace the fixed locators off the geometric grippers with programmable positioners. Fig. 8 shows examples for such a type of highly flexible geometric gripper with different types of flexible positioners that will be mounted on robots. Since all them use robot technology they can be programmed and controlled like the conventional robots integrated in each car body assembly cell.

3. Flexible positioners

3.1 Conventional Cartesian positioners

The obvious approach to build a 3-axis positioner for carrying the pin locators and clamping tools typically used for positioning and fixing car body parts is based on a Cartesian arrangement of single linear axis modules.

Such a design involves low design and engineering complexity on one hand which make them quite easy to use for stationary positioner arrangements. Furthermore they allow for robust and dust protected designs which can be easily scaled for different workspace sizes on each axis separately. But on the other hand this concept involves quite high masses and inertia for a certain level of stiffness which implies directly less suitability for mobile application e.g. the positioners are mounted on the track motion based part transfer system or mounted on a flexible gripper which is attached to the robot. Furthermore the TCP of Cartesian positioner axes can hardly reach out of the footprint without a tremendous design effort which is also required for positioner applications on mobile servo shuttles in order to allow for a short part transfer time.

3.2 Positioners based on parallel kinematic machines

After a screening of machine concepts that would fulfill the requirements, especially the small footprint and the fact that the TCP will have to move outside the footprint, posed a challenge. It became obvious that the combination of requirements called for a parallel kinematic machine (PKM) concept, but none of the known kinematics became an obvious candidate.

3.2.1 The 3 dof challenge of parallel kinematic machines

The 3-UPU machine - based on a characteristic kinematic chain using an universal joint U, and one prismatic joint P afterwards and then another universal joint U - was first introduced by Tsai (Tsai, 1996). It looks strikingly simple and promises to be a very lean and low cost machine for pure translational motions compared to the well known Hexapod, as it uses only 3 instead of 6 variable length struts and blocks the unwanted rotational degrees of freedom by using universal joints. Many researchers have built 3-UPUs and reported interesting results, but the main problem remains that the struts experience very high rotational moments, which results almost unavoidably in a low stiffness, or in a very bulky design. Surprisingly, this fact has not been widely published in terms of real measurements. After prototyping a simple 3-UPU machine, this option was ruled out, and it became clear that 6 legs are needed to avoid torsional moments and get high stiffness. Using Hexapods for translational motions, however, is not an option either in applications that are sensitive to cost. The need for extra motors, cables, power amplifiers and control prohibits the use of Hexapods in purely translational applications in most cases. Other machines for 3-axis translational motions use fixed strut length and linear motions that would either violate the footprint constraint (by moving the foot points horizontally) or the requirement to have a workspace area bigger than the footprint (by moving the foot points vertically up).
A Delta machine (Clavel, 1988) would come closest, but still could not be made compact and stiff enough. So the only feasible machine would be a combination of Delta (for 3-axis motion) and Hexapod (Gough & Whitehall, 1962) (for stiffness and footprint). Such a machine would use parallelograms in a Delta configuration that could be changed in length by a single motor and pivot around 2 axes. Unfortunately, such parallelograms did not exist so far, and it was not obvious how to design them.
A parallelogram that can be extended and retracted by a single motor can be made out of two cylinders with ball screws and a mechanical coupling, like gears or belts. If such a parallelogram is required to pivot around two axes, the distance between the cylinders changes and things become more complicated. Nevertheless, several solutions were developed by the authors, and the most compact one was chosen (see Fig. 5). The coupling

uses a sequence of bevel gears and a synchronization belt to accommodate for the parallelogram's pivoting motion (patent pending). Four universal joints provide the required degrees of freedom.

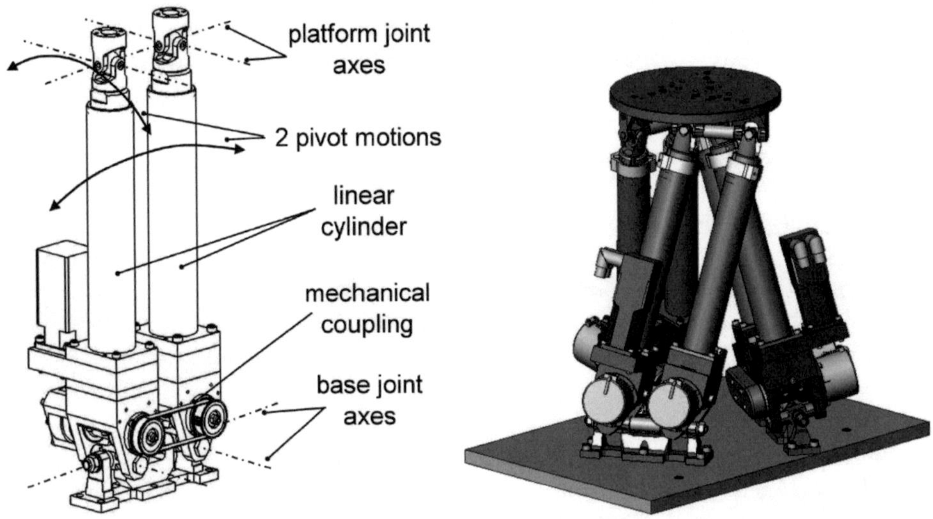

Fig. 5. FlexPLP based on pivotable extensible parallelogram actuator module (left) and its application on a twisted Deltapod mechanism (right)

3.2.2 The Deltapod positioner kinematics

With the parallelogram actuator required machine element in place, a variety of PKMs can be synthesized. According to the above requirements, the Deltapod was constructed, which is based on the Delta geometry with equilateral triangles defining the position of the alignment of the joints on the base and on the moving platform.

However - due to the small footprint and the offset length of the cylinders, a symmetrical Delta configuration exhibits an insufficient stiffness in horizontal direction, and would be badly-conditioned. The solution here was to twist the base joint positions points around the center axis of symmetry, and to twist the upper platform joints in the opposite direction, while at the same time reducing the diameters of the principal circles that define the geometry (see Fig. 6).

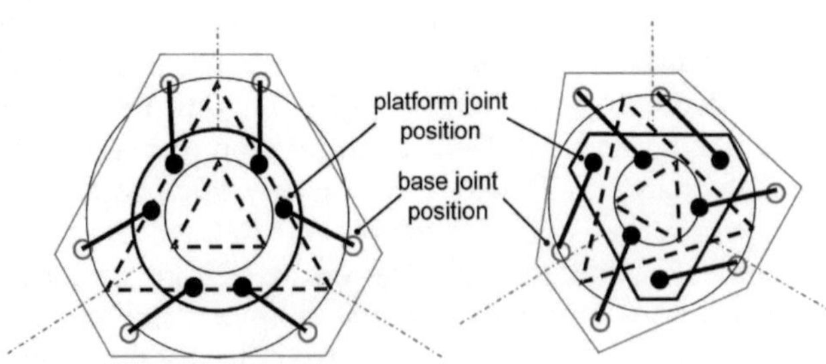

Fig. 6. Compacting footprint by twisting joint axes positions from the classic delta (left) to the twisted delta (right) configuration

The result of this operation is that the machine is compacted, while the condition number remains almost identical - hence stiffness and velocity relations are not affected. The only drawback of this operation is that a twist torque on the plat-form is created due to loss of symmetry. It turns out, though, that this can be easily compensated by anti-rotational measures of the legs - the resulting torsional moments in the legs are much lower than in the 3-UPU design. As a result, a very compact and strong PKM is achieved, and it was possible to meet all of the requirements for the flexile underbody fixturing application.

3.2.3 T-pod4- positioner kinematics

Parallel kinematic machines offer an inherent modularity. It is therefore natural that for any new PKM concept, a variety of derivatives exist.

In the given case, we derive the T-pod based on the developed parallelogram module, by aligning two parallelogram planes such that the upper joints form (nearly) a single line, hereby blocking two rotations, and by placing the third parallelogram plane perpendicular, (see Fig. 7). The name was chosen because the joint locations on the movable platform resemble a T-shape.

The advantage of this machine is the reduced footprint in one direction, so that it can be place in very narrow spaces, but in particular the possibility to remove the synchronization and add an extra motor to the perpendicular parallelogram, so that an optional tilt motion around the symmetry axis can be introduced leading to the T-pod4 configuration. This can be particular useful when fixturing buckled car body parts. The T-pod4 is somewhat related to the Kanuk (Rolland, 1999), even though the drive mechanism is different.

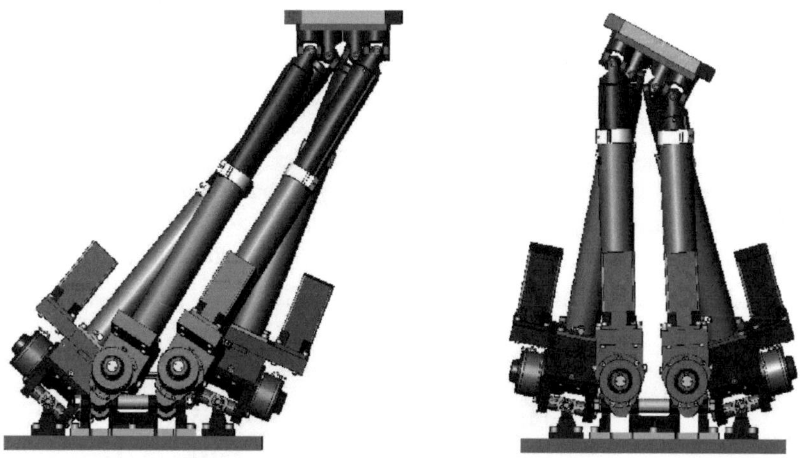

Fig. 7. A T-pod with 3 d.o.f. (left) and a T-pod4 with 4 d.o.f.(right)

3.2.4 Flexible grippers based on positioners

Combining e.g. four of those T-POD positioners to a common backbone attached to a powerful handling robot a flexible programmable gripper can be achieved which provides an excellent payload compared to its own mass (see Fig. 8).

By use of flexible grippers the part logistics within highly flexible car body assembly lines which is another big issue become addressed. With flexible grippers' space and cycle time consuming tool changing of different grippers as well as additional grippers themselves can be abandoned. Different from typical tool (gripper) changing a flexible gripper will be reconfigured according to the part geometry of the successive car model during the transfer motion of the handling robot without affecting the primary processing time.

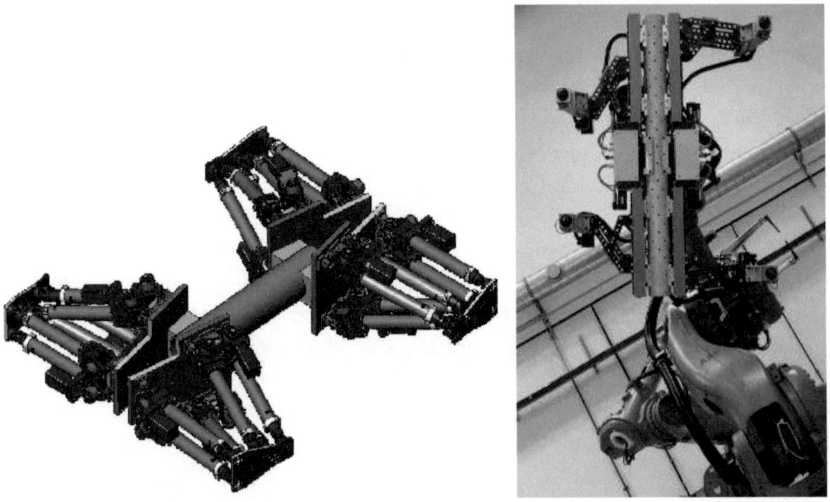

Fig. 8. Flexible Gripper Examples T-pod based (left) and hybrid linear module based attached to a handling robot (right)

4. Engineering of flexible postioners

4.1 Application specific positioner requirements

For the concept for a flexible multi-model car body assembly line presented in this chapter, positioners are used for different applications: These are stationary flexible fixtures, mobile flexible fixtures on the shuttle and mobile flexible grippers. Each application involves different requirements which need to be reflected by an adequate design and configuration of the positioner mechanism. Both the serial Cartesian as well the parallelogram actuator basing parallel mechanisms can be optimized and dimensioned towards these requirements.

	Flexible Fixtures stationary/mobile	Flexible Gripper mobile
Repeatability	± 0.05 mm	± 0.1 mm
Speed	100 - 250 mm	100 mm
Reach per axis	200 - 400 mm	100 - 200 mm
Stiffness	> 1 N/µm	> 1 N/µm
Mass	50 - 130 kg	20 - 40 kg
Payload Dynamic	30-50 kg	5 - 30 kg
Payload Static	150 kg	55 - 100 kg

Table 1. Examples for positioner mechanism related requirements with respect to application

Choosing the optimal mechanism dimensions for the best performance is still a challenging task because variations of the geometric parameters lead to greater variations of the machine performance than for conventional serial structures. Furthermore there are a lot of performance criteria which have to be taken into account. Another problematic issue is to find global values for criteria which depend on the pose of the working platform of the mechanism. The most important requirements on PKM especially but not only for tooling applications are workspace, accuracy, stiffness, velocity and lifetime. But these keywords must be looked at more precisely and put into mathematical expressions. The characteristic functions allow establishing performance criteria and help to recognize the degree of fulfillment of a PKM structure with regard to the requirements (Tsai, 1999).

Typically all requirements and developed characteristic functions assigned to PKM structures are generally not constant or isotropic, but depend on the location or pose (position and orientation) of the working platform in the plane or space. Isotropic behaviour is strongly desired, but is a subsequent task of selection, often in coherence with optimizing procedures. Almost all performance criteria depend on the position and orientation of the TCP of the PKM. However, in an optimization design process we have to compare and assess different parameterized kinematic structures using global criteria characterizing the structural behaviour inside of the workspace (Krefft et al., 2005) (Gosselin, 1998).

Since these complex calculations are mainly reserved to mechanism experts it is hard to use the inherent modularity of PKM to full potential in view of the requirements give by each single tooling application.

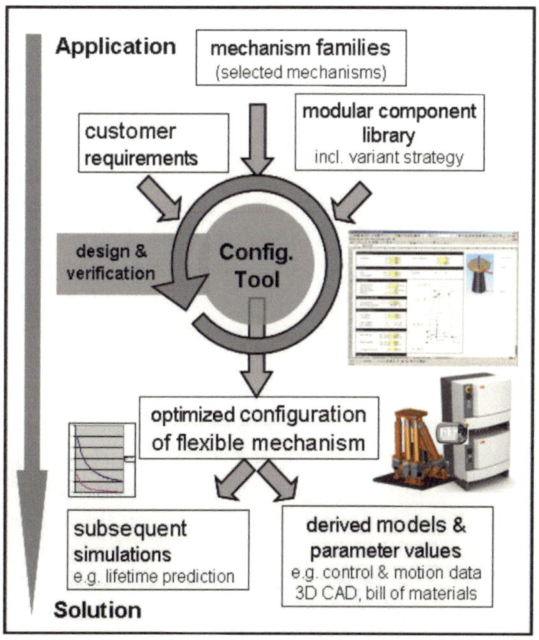

Fig. 9. Schematic architecture of mechanism engineering tool

4.2 Tool concept for efficient positioner engineering

One promising approach is a software tool based methodology that enables an efficient engineering of optimized modular mechanisms for tooling applications based on flexible positioners and grippers. In general it should support the specifics of parallel mechanisms but it is preferably also extendable to serial and hybrid kinematic mechanisms.

According to the concept of the engineering tool the user has to provide all available application related information where the tooling is required for, but the user does not need to provide mechanism design expertise (see Fig. 9). The tool is mapping application specific requirements to an adequate layout of mechanism subsystems and their optimized configuration via defined characteristics functions.

A selection and optimization sub routine ensures to find the best possible mechanism solution in terms of technical requirements but also cost based on the entered information of the specific application.. It considers a variant strategy with a predefined number of standard components and variants provided by libraries for proved positioner mechanism families and their subunits and components.

A set of interfaces provide mechanism related model data for control (motions files and user interface parameters) and for sub-sequent simulations with advanced tools for line concept simulation, offline programming.

These features enable not only an efficient design and planning positioners but also manufacturing, assembly and commissioning of these mechanisms. In order to guarantee for a high uptime of the positioner units during their operation one important element will be to derive some predictions of required maintenance cycles or even lifetime estimates depending on the specific tasks the positioner units are supposed to perform.

5. Flexible engineering of cell control

5.1 Flexible programming and engineering of cells

Full advantage of the FlexLean concept can only be taken if the corresponding manufacturing process control is supporting the concept's flexibility and modularity. Therefore the program of the programmable logic controller (PLC) has to fulfill certain requirements as modular structure, well-defined interfaces, encapsulation and functional flexibility. One PLC program per cell is managing the manufacturing process inside the cell. Each manufacturing device as FlexPLPs, track motion, robots, fixtures, etc. has its own control code module within the PLC program handling all device functionalities. These control modules are interconnected via well-defined standard interfaces allowing high flexibility in module interactions. The working sequence of a device (e.g., handling or welding of car parts) is controlled and executed only by its control module. Working sequences of several devices are synchronized via the module interfaces.

Instead of fixed and pre-programmed working sequences limiting the device functionalities to a few procedures, the working sequences on their part are split into a set of basic standard working steps which can be set in an arbitrary order. For example the taking of a car part by a robot from a fixture and subsequently placing it on the car body for welding is organized in steps as "Approach to fixture", "Pick-part", "Approach to track", "Put-Part on car body on track", "Wait-for Welding", "Back-Approach from track". The single working step is a predefined and independent subsequence without need for synchronization with other device sequences during its execution. Synchronization takes place only between working steps. The working step concept allows high flexibility in coping with a wide range of different working sequences. The control module of each device has a control routine for each possible working step which is executed according to the complete working sequence.

5.2 Configuration instead of programming

One of the most innovative aspects of the engineering concept is that the working sequence is no longer hard coded in the control program but only configured. The customer can configure the desired working sequences for each devices and product type via a wizard on the human-machine-interface (HMI) just by selecting and configuring working steps from a list (see Fig. 10) without knowledge of PLC-programming.

The time-consuming process of programming, compiling, transferring to the PLC and debugging the generated code is no longer necessary. The control program contains all necessary control functionalities to handle all possible working steps. The corresponding control code for a specific working sequence is not generated and then transferred to the PLC but exists already as pre-programmed and fully tested control module on the PLC. The control module of each device reads the customer-configured working sequence and executes it working step by working step. These sequences of working steps are executed for each device by a sub-control module called sequencer.

The customer can synchronize several sequencer with a working step called "Wait", for which a devices and a working step number are parameterized. Once the sequencer is executing this step, it is waiting until the sequencer of the specified device has reached the corresponding working step number.

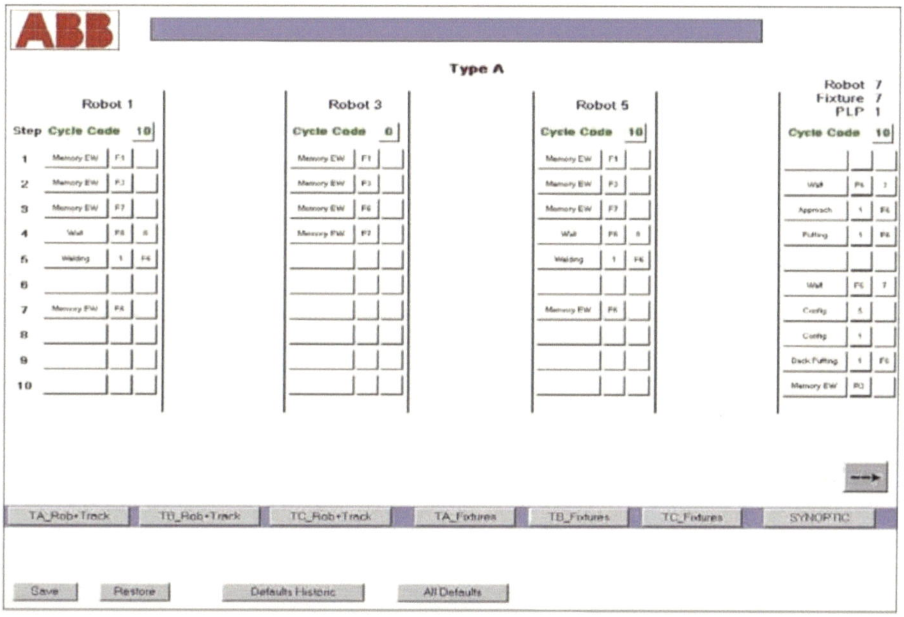

Fig. 10. Configuration wizard for working sequences

Start conditions are checked before executing a working step to ensure that no constraints are violated by starting this step. End conditions have to be fulfilled before declaring a working step as finished and launching the next step in the sequence.

Thus the configuration data contains not only the working sequences but also information about safety constraints for working steps and devices. As for example a robot is only allowed to pick a part from a fixture if certain fixture clamps are open or closed. For each product type different constraints and settings can be configured by the customer. Complete flexibility is available via the configuration of settings for part sensors, fixture clamps, robot moves, safety settings, robot interlocking, working steps and sequences for all devices, etc. Even the cell layout can be configured. If robots, fixtures or operators are present within the

FlexLean cell can be set via parameters on the HMI and thus the corresponding control module is activated for execution on the PLC.

This configuration concept meets the flexibility of the whole FlexLean solution. Starting from basic manufacturing device functionalities, specific working sequences for different product types can be configured by the customer and executed directly after downloading the configuration data to the PLC.

6. Summary

When car model specific and geometrically fixed toolings for fixtures and grippers will be replaced by flexible tooling using programmable and lean positioners, new assembly concepts can be introduced. Thus the FlexLean concept allows designing highly flexible automotive body assembly lines for a changing mix of different models in the same line. The need for redesigning when introducing new models is now replaced by a simple reconfiguration of the cell control. This allows model change on the fly.

7. References

Chablat, D.; Wenger, P. (2003) Architecture Optimisation of a 3-dof Translational Parallel Mechanism for Machining Applications, the Orthoglide. In: *IEEE Trans. on Rob. and Autom.*, Vol. 19(3), pp. 403-410.

Clavel, R. (1988). DELTA, a fast robot with parallel geometry, *18th Int. Symp. on Industrial Robot*, pp. 91-100, Lausanne, 26-28 April.

Gosselin, C. M. (1998) On the Design of Efficient Parallel Mechanisms. In: *Computational Methods in Mechanical Systems*, Springer-Verlag, Berlin Heidelberg New York, pp. 68-96.

Gough, V.E.; Whitehall, S.G. (1962). Universal tyre test machine, *Proceedings of the FISITA Ninth International Technical Congress*, pp. 117-137, May 1962.

Krefft, M.; Kerle, H.; Hesselbach, J. (2005) *The Assessment of parallel Mechanisms – it's not only kinematics*, Production Engineering, Vol. XII No.1, pp. 173-8.

Krefft. M.; Hesselbach J. (2006). The dynamic optimization of PKM, *Advances in Robot Kinematics ARK*, pp. 339-348, June 2006.

Negre, B.; Legeleux, F. (2006). *FlexLean – robots challenge low cost labor*, ABB Review, 4/2006

Ottaviano, E.; M. Ceccarelli, M. (2002). Optimum Design of Parallel Manipulatorsfor Workspace and Singularity Performances, *Proc. of the Workshop on Fundamental Issues and Future Research Directions for Parallel Mechanisms*, pp. 98-105, Québec (Canada).

Rolland, L. (1999). The Manta and the Kanuk: Novel 4-DOF Parallel Mechanisms for Industrial Handling, *Proc. ASME Dynamic Systems and Control Division*, pp. 831-844, IMECE'99 Conference, Vol. 67, Nashville, USA, Nov. 14-19.

Tsai, L.-W.: *Robot Analysis - The Mechanics of Serial and Parallel Manipulators*, John Wiley, 1999

Tsai, L-W (1996). Kinematics of a three-dof platform with three extensible limbs, *ARK*, pp. 401-410, Portoroz-Bernadin, 22-26 June.

Wemhöner, N. (2005). *Flexibilitätsoptimierung zur Auslastungssteigerung im Automobilrohbau*, PD thesis, RWTH Aachen, 2005.

Wenger P. et Chablat D. (2000). Kinematic analysis of a new parallel machinetool: the Orthoglide, *Advances in Robot Kinematics ARK*, pp. 305-314, Piran, 25- 29 June 2000.

Part 4

Design Developments

14

Sustainable Design of Automotive Components through Jute Fiber Composites: An Integrated Approach

Cristiano Alves, Arlindo Silva, Luis Reis, Paulo Ferrão and Manuel Freitas
Instituto Superior Técnico, Universidade Técnica de Lisboa
Portugal

1. Introduction

Nowadays, the world faces unprecedented challenges in the social, environmental and economic dimensions. Industrial design has an important contribution shown in all of these dimensions with solutions that provide positive answers. In particular, due to its relevance, the automotive industry confronts a moment of crisis, and based on the ecodesign of products it has been transforming the challenges in opportunities.

In the broad sense of Bertalanffy´s (1969) theory, earth and its inhabitants form a single live body in constant evolutive dynamics, in which the human activity influences and is influenced by the environment. Since natural and artificial systems work in holistic relationship, synergy between and self-organization of living organisms maintains the global equilibrium of the systems.

According to Uexküll (1982) there are different and tangible environments which depend on the environmental perception of each species. Thus, the activity and influence of each species in nature are directly related to biological scope. In this sense, the human being developed the ability to create goods and tools to ensure its survival in a hostile and unknown world, beginning a new age based, exclusively, on material consumption from local environment, in which societies perceived and recognized themselves as an integral part of nature. Thus, the consumption rate was consistent with the capacity of the natural systems to absorb waste and generate new natural resources. It did not imply large changes in the environment since it was constrained by biological limits of societies to reach so far lands.

After the industrial revolution, landmark of human domain over nature, this equilibrium between natural and human (artificial) systems was changed drastically. The industrialized society expanded its natural limits, accelerating the consumption of natural resources hence the environmental impacts, unprecedented in human history.

The human activity has reached a scale that resulted in severe perturbation of the nature (Schumacher, 1989) and as a consequence environmental issues constitute a central theme in international policy debates since the 60s. In this context, a group of researchers founded the Rome Club, which pointed out the limits for population and economic growth based on the finiteness of natural resources, through the well-known report "The limits to growth" (Meadows et al., 1972). From that moment, policy changes were perceived in some countries and sustainability became a central point among policy makers, managers, environmental movements and scientific debates. In the 70s sustainable development emerged as a concept, providing a new view point for environmental issues from purely qualitative to quantitative approaches, recognizing technological, economic and environmental constrains for that growth.

In the 80s, environmental impacts were assigned to the industrial systems and their technological determinism. Nature was considered as a development constraint and costly to global production, reducing the competitiveness of companies. Then, to emphasize the natural resources finitude and the necessity to reduce the consumption, the United Nations established the World Commission on Environment and Development (WCED), which produced the so called Brundtland Report "Our common future" (United Nations, 1987). In the 90s, the globalization brought a new perspective on environmental issues, recognizing them as a global as well as the common interest of nations to establish policies for sustainable development.

At the global conference Rio-92, sustainable development was defined as "the model of development which aims the environmental sustainability through the rational use of natural resources in order to meet the current generation needs without compromise future generation needs". It was established the "Agenda 21" (United Nations, 1992), suggesting targets for sustainable policies of nations, and after that societies have noted that the environmental depletion implies direct impacts on global economic development, as demonstrated by the results in recent decades. In order to foster the evaluation of the Rio-92 proposals, by 2002, the UN organized the World Summit on Sustainable Development - Johannesburg (Rio+10), and it confirmed the poor advance until then. Indeed, only 40 countries had achieved their environmental targets, 24% of the forests around the world had disappeared only in the 90s, and fossil fuel consumption (and CO2 emissions) had increased about 10% from 1992 to 2002 (United Nations, 2002). It was made clear that the current development model emphasis economic activities as more important than environment, leading to the depletion of natural resources that are essential to satisfy society's needs, even to maintain the current economic system.

According to the UN - Commission for Sustainable Development (United Nations, 1996), sustainability has four dimensions: social, economic, environmental and institutional. It is concerned with the reduction of raw materials usage, and should focus on quality instead of quantity, promoting a better quality of life for everyone, even for future generations and also improving economic and social standards. Therefore, it is necessary to develop a new relationship between society and environment, in which the limits of natural ecosystems in support human activities are taken into account.

This new approach requires deep mentality and behaviour changes to develop the environmental consciousness of societies. In fact, in a world with limited resources and many environmental impacts, it is obvious that environmental damage requires the integration of environmental factors in company budgets, promoting sustainable management strategies (Alves, 2006). In recent years, due to the growing global awareness and societal concern related to environmental issues, people perceive industrial products besides their look and performance. There are increasing concerns on the whole life-cycle of products, e.g. how they are made, used and disposed off. Therefore, in many countries, the environmental approach is becoming a relevant strategy to increase the market share of companies through sustainable products and processes.

In this context, during the last decades several companies have been introducing sustainable concepts in their management and processes through the development of tools and methods to introduce fully environmental concerns in their activities related to product design. Since the design phase receives and manages information from other phases, it influences the whole company, determining activities such as production, marketing, sales, styling etc (Jeswiet & Hauschild, 2005).

Many authors have pointed out the importance of design for company's success, linking it to the global product development (Fig. 1). It makes decisions during planning, which influence the efficiency of products and processes in their whole life-cycle, also leading the definition of the best way to meet customer needs (Goedkoop et al., 1999). In the 90s, researchers stressed the design phase as essential to define the success of products and companies, pointing out that only among 5% - 15% of the launched products were profitable for companies (Baxter, 1999). Moreover, according to Bettina (Stamm, 2003) a British survey found that in 2002, about 75% of the SMEs (Small and Medium Enterprises) recognized the design phase as a significant strategic phase. Compared to other product life-cycle phases, different studies indicate that design can anticipate new business opportunities and that about 70% of the product costs are decided in design phase (Deng, & Edwards, 2007).

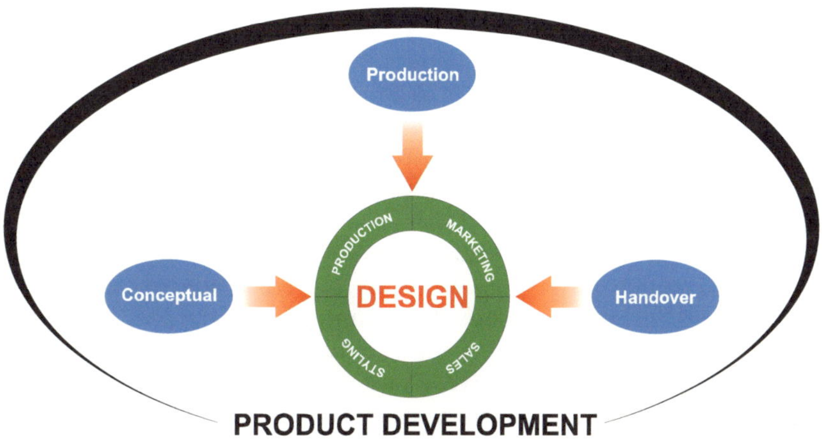

Fig. 1. Phases of the product development process

Therefore, the earlier the improvements are taken into account, the higher the potential for cost savings. However, the large set of different functions and activities of the design process highlighted that a single method or even more a single tool is not enough to address all issues. Many design methods and tools have been developed and mixed to overcome those differences. For example, Chan and Tong (2007) have developed a specific method for a materials selection stage within a design phase based on a life-cycle approach.

In this context, through the multidisciplinary nature of product design, companies can catalyze innovation and prevent pollution, since environmental issues need to be considered in the design phase. Moreover, design also represents an interface between people and companies that pressure and respond to it. Due to this, a collaborative design approach can better fulfill at the same time economic, social and environmental requirements. On the other hand, as discussed by Papaneck (1992), among others, design has been used as an instrument to motivate consumption. It has indirectly focused on profitable outcomes, driven by economic performance of products, in which such approach illustrates the predominance of economic rationality at the expense of environmental issues. Then, since the design phase is presented as the core of the product development, in the last years environmental design has been widely studied to ensure that efforts toward increased environmental rationality and the sustainability of industrial systems are made.

2. Ecodesign into the life-cycle thinking approach

Over the last two decades the interest in environmental issues has grown significantly. Ever increasing attention has been paid to the environmental impacts emerged from design activities, hence from manufacture, usage, and disposal of industrial products. Since the 70s one of the authors, Victor Papanek (1971), also has been discussing the relationship between projectual professions and the environmental impacts raised by their activities.

Twenty years later, Fiksel (1998) published one of the most famous works on the relationship between environmental and projectual issues. The book, illustrated with successful examples, states that projects can meet their environmental requirements without a decrease in other performances. He considers that designers must take into account environmental aspects in a broad life-cycle approach. In recent years, Manzini and Vezzoli (2002) also discussed about life-cycle thinking and the environmental effects from design.

Environmental issues are vital for modern societies, besides the obvious survival problem, because natural resources depletion also implies the drastic decline of the current industrial and economic systems. However, only in recent years people have perceived industrial products besides their look and performance, mainly due to the growth of global awareness and societal concern about environmental impacts. People are increasingly concerned with the whole product life-cycle, for instance how products are made and how they are disposed of. In a world with limited resources and many environmental impacts, sustainable societal and industrial styles have become essential, requiring the integration of the environmental issues in public and private budgets. Consequently, nowadays several companies have been considering the environment as a powerful factor and a business opportunity.

Recent surveys show that environmental issues now seem to be taken seriously by most business agendas. In the past, the majority of companies perceived environment as external or peripheral to their business activities, solving environmental issues by a "curative approach", pollutant emissions were found and cleaned up (end of pipe). Afterwards, a preventive approach was developed to prevent pollution, it showed better results saving costs and resources, and it also motivated life-cycle thinking, hence the ecodesign concepts. Ecodesign also lead designers to take into account environmental aspects, besides traditional aspects currently considered. It implies a holistic outlook (life-cycle thinking) on all phases of a product or system, well known as cradle-to-grave approach, by measuring environmental outcomes that form the basis of ecodesign. The basic idea of ecodesign is to find the source of impacts, increasing the products eco-efficiency still in the design phase. Due to its preventive life-cycle nature, ecodesign represents the best set of strategies to improve companies' market performance whereas it can be applied to develop "greener" projects for any industrial sector. Ecodesign has been recognized as an enabler of companies' transformation, it can lead to behaviour changes that are urgently needed to improve our future, addressing environmental issues and problems behind problems.

Although the design phase itself is a "clean phase", most environmental impacts are defined there, and once a project moves from design to production, its environmental performance is largely defined/fixed. Thus, when potential environmental loads are made explicit even in that phase, strategies may be developed to prevent them. In the Netherlands, a research suggested that the design process can reduce the current levels of pollution and resource consumption by about 25% to 50% per person (Hilton, 2008).

Many terminologies of ecodesign have emerged during the last years, such as Design for Environment (DfE), Sustainable Design (SD) and Design for Sustainability (DfS) etc. They represent just an evolution of the perspectives on how the subject is perceived and also to some extent an increasingly critical perspective on ecology and design in theory and practice. (Baumann et al., 2002). Despite many definitions, ecodesign emerges as a conciliatory proposal between sustainability and design concepts, also influencing potential life and behaviours of communities when facing the environment in which they live.

Unlike current design, used to stimulate consumption by creating desire for ownership and possession, ecodesign aims at a balance between tools and beliefs to establish a new ethic. Its concepts have potential to change consumer behaviours and ideas about satisfaction and quality of life. Moreover, ecodesign must be thought as a synergy towards a new aesthetical, technical and environmental reasoning, hence, toward new society beliefs.

The main advantage of ecodesign is taking into consideration the environment at the earlier phases, in which constrains are more flexible, once in latter stages despite the project knowledge is large, few changes are feasible since most critical decisions have been taken, decreasing the creativeness, hence reducing the innovation level.

3. Materials and ecodesign

The control, extraction and use of materials have always been closely linked to the human history. For instance, the evolutionary ages are represented by materials such as Stone Age, Iron Age etc, and currently it is possible to state that modern societies have been living in the Plastic and Silicon age, despite the significant introduction of new materials in market.

Nowadays, there are probably over 100.000 commercial materials on the market with different types of impacts on environment (Ljungberg, 2007). This long list demonstrates the problem related to the depletion of natural resources worldwide. Despite the increasing awareness of environmental problems in industrialized societies, society still depends on conventional materials to produce goods. The inability of the environment to support the constant abuse of human activities is becoming clear through various forms of pollution and natural resources depletion that threaten the survival of the planet. Due to this, some studies have demonstrated the depletion and scarcity of resources, even considering that the per capita consumption of natural resources in developing countries is about 1/6 of the consumption in developed countries (Tukker et al., 2000). For instance, the European waste generation is expected to increase about 43% by 2020, while its energy and resources intensity will decline (OECD, 2001).

In this sense, the discussion about the preservation of natural resources has led to the renewed interest concerning materials, motivating changes in product design to address this environmental issue (Papanek, 1995). Collaborative R&D efforts among material scientists and engineers, research institutions and government have been stimulated to find commercial application for renewable natural raw materials, employing them in technical products and/or structural parts. Thus, ecodesign is not just a newly coined buzzword, but it has guided the development of a new generation of natural materials. It shows the importante role played by ecodesign related to the natural resource depletion, since it dictates the type and quantity of raw materials and energy that is required to produce and dispose products. Hence, it also shows how important materials are also in product design

(Deng & Edwards, 2007). Despite being generally regarded for their functional attributes, they have become fundamental in a sustainability context. It explains why most eco-methods are implicitly related to material and/or energy management, directly linked with resource scarcity. Then, designers must find materials that can address environmental issues related to the natural resources depletion, besides other requirements, since earth has a limited capacity to cope with the current wide level of impacts raised by human activities.

3.1 Natural resource consumption (depletion)

Halog (2004) defined sustainability as the optimization of the human activity level within both the limit of renewable resources supply and the capacity of the environment to absorb waste. Although this is not the unique dimension of sustainability, it is clear that in the last decades the unsustainable human activities have exceeded the threshold of recovery of the planet, which threatens the welfare of future generations. Even religions, which are traditionally pro-natal, perceive the human being as conqueror of living things on earth, neglecting the environment and its limits (Ljungberg, 2007)

According to WWF (Hails et al.,2006), 40% of natural resources have been consumed since 80s and under the current consumption rate two planets like Earth would be needed to sustain the current patterns (Fig. 2). This was also quantified by the global Ecological Footprint (EF) which in 2003 was already about 2.2 global hectares per person. In other words, each person needs at least 2.2 hectares of productive land to support its activities.

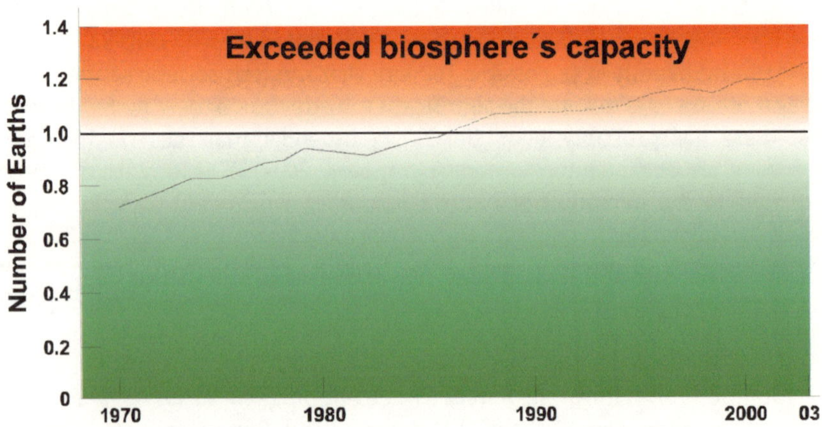

Fig. 2. Global Ecological Footprint. Adapted from (Hails et al., 2006, 2008)

Baudrillard (2005) states that nowadays the social and satisfaction values are fake, they are based on consumption and due to the eternal creation of new products and social needs they become an eternal dissatisfaction motivating more consumption ad infinitum. Easterlin (2006) explained the "eternal dissatisfaction" stating that there is no relationship between the increase of individual income (purchasing power) and the happiness increase. He presented surveys carried out in US and Japan that pointed out the limitation of income growth to produce satisfaction, it means that money and happiness are not directly related. Nevertheless, consumption has been growing worldwide, motivating the development of the global environmental policy related to sustainable consumption that also considers the existence of a maximum consumption to provide individual well-being, beyond a minimum threshold.

Facing this grim context, in which societies can neither give up development nor maintain the current environmental depletion, the last decades have been thought to be hard to replicate. The achievement of a sustainable state depends at least on three key factors: the population, the demand for human welfare and eco-efficiency of applied technologies. The eco-efficiency factor stands out to reach the condition toward sustainability, with a constant growth of population worldwide.

It is a fact that the world population is growing, according to the UN (2004) and the OECD report (2008), it is expected that the world will have about 8.2 billion and 10 billion of inhabitants by 2030 and 2050, respectively (Fig. 3). In Western Europe (WEU) and in South-Eastern Europe (SEE), the population will grow about 1.1% and 16%, respectively, from 2005 to 2030, while in North American countries it will increase about 21%. Moreover, Asian countries will continue increasing their already remarkable population. India with a population of about 1.3 billion inhabitants (growth of about 31%) tends to achieve China´s population by 2030 (OECD, 2008). This constant increase has become a constraint for global economy and social development. It will put more pressure on environment through increased consumption, and hence more search for natural resources to provide greater welfare for developing countries.

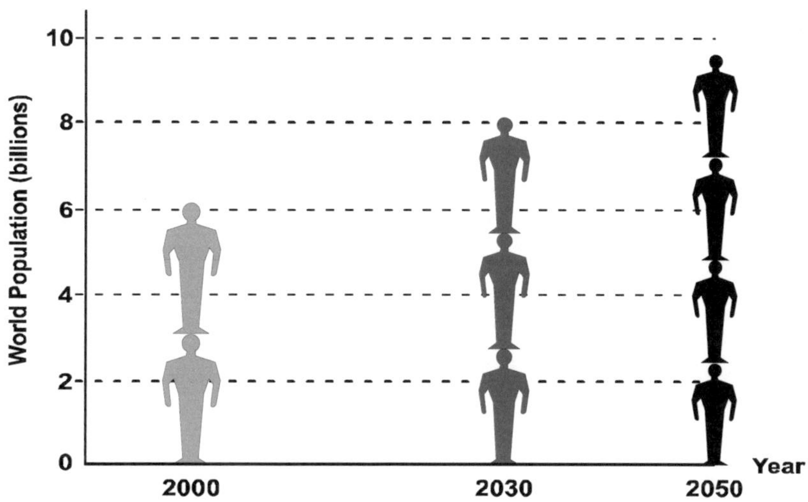

Fig. 3. Forecast of the global population growth. Adapted from (United Nations, 2004)

The situation is dramatized by the forecast of the global increase of the Gross Domestic Product (GDP) by 2030 (European Environment Agency, 2007). According to Field III et al (2001), at a GDP global growth rate of 3% per year, societies will produce and dispose of more objects in the next 25 years than in the whole history of the human being. However, nature will not be able to sustain such growth rates of up to 3% per year expected for the next years by 2030. Indeed, the GDP per capita of Central Europe (CEU) will grow about 141%, while India´s will grow about 169%. China will present the highest increase of about 200% by 2030, even though the USA and Canada will still have the highest GDP per capita (European Environment Agency, 2007).

The expected growth of the incomes is heavily dependent on natural resource use, so the demand for resources will increase, mainly for fossil fuels that are used to produce energy and products in many industrialized countries. Although they are essentially non-renewable

and limited resources, their usage has increased at least 20-fold per capita since the end of the 19th century, highlighting the necessity to reduce their consumption (Ljungberg, 2007).
Indeed, about 40% of the global energy is derived from petroleum and recently, European studies (OECD, 2008; EEA, 2007) indentified that food and beverage, as well as transport and housing account for 70% to 80% of the environmental impacts and 60% of the natural resource consumption. Also, the global resource extraction is expected to increase about 50% from 2002 to 2020 and almost 50% of it is related to non-metallic minerals. However, sustainability can be achieved e.g. by reducing the over-population and/or the consumption, even by eco-efficiency of industrial systems, using less energy or material per unit product. Whereas the expected population growth and therefore the larger consumption of natural resources, for ecodesign the eco-efficiency stands out as the best condition toward sustainability.

In this sense, to minimize environmental loads the ecodesign strategies focus on eco-efficiency, maximizing material productivity. Eco-efficiency measures have shown significant savings in most circumstances, using natural resources in a consistent manner combining environmental and economic performance. To embrace eco-efficiency some publications envisage a reduction in the material flows by Factor 4 or Factor 10 concepts (Schmidt-Bleek, 2008; Weizsäcker et al., 2009). They are based on the dematerialization idea of production and consumption in which all needed products and services as well as energy demands are to be met with less resources.

Figure 4 illustrates levels of eco-efficiency for a given innovation applied on a product against a sustainability degree. Accordingly, environmental improvements are classified into four categories in regards to different levels of eco-efficiency. A potential sustainable level can be achieved by systemic innovations, where new scenarios are proposed to meet sustainable lifestyles (Factor X). It focuses on cultural activities and new standards for quality of life, altering the industrial and social structure of behaviours.

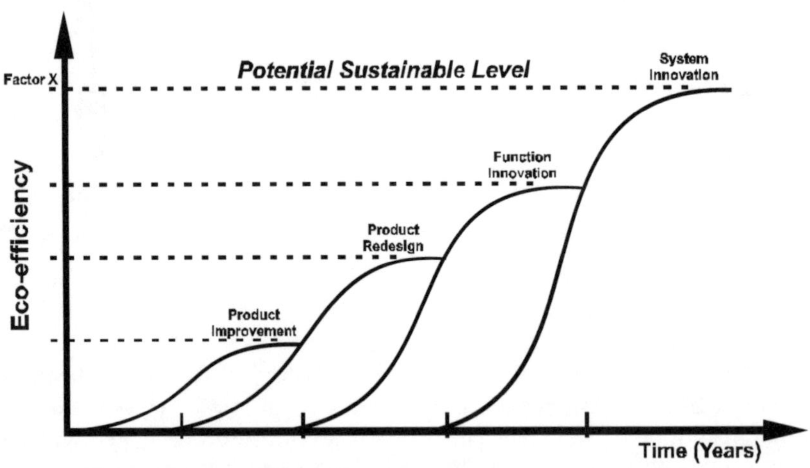

Fig. 4. The Brezet Model for environmental innovations.

Due to the over-consumption and the irreversibility of resource exhaustion, the scarcity of natural resources tends to reach permanent degradation. Renewable resources such as plants, air, water, have a short natural growth of few months, maybe years and can be replenished in a useful time. In contrast, non-renewable resources such as fossil fuel and ore have a long natural formation cycle. It is important to note that a renewable resource can become non-renewable if its replenishment rate is smaller than its consumption rate, exceeding its biological growth. Many empirical projects that use renewable sources have reduced environmental loads from resource depletion, keeping the quality of life (Figure 5).

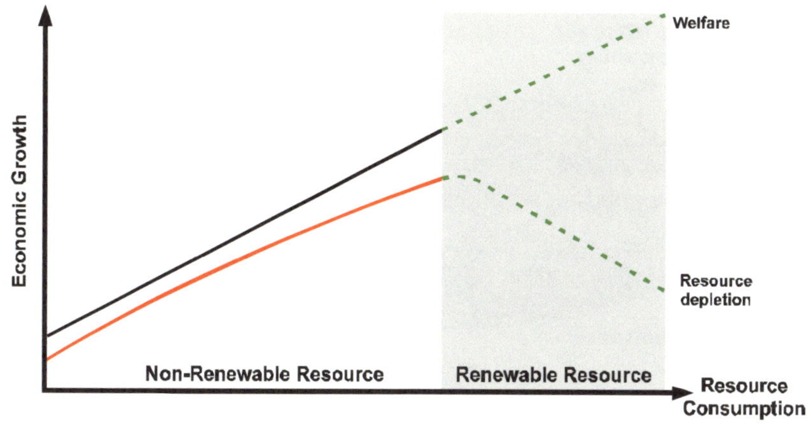

Fig. 5. Potential of renewable resources to reduce depletion. Adapted from (Community and Information Society Technology Programme, 2003)

Several researches (Datschefski, 2001; EEA, 2007) have pointed out Europe as the leading continent in investments related to the use of renewable sources in two important sectors: energy and transport. It has helped to reach targets established in Kyoto's protocol related to the reduction of the greenhouse gases (GHG).

3.2 Sustainable mobility through ecodesign of renewable materials

Nowadays, substantial environmental impacts are induced mainly by four economic sectors: energy, transport, agriculture and tourism. Among them, transport is considered to be the largest pollutant sector with pervasive negative impacts. It is the largest source of greenhouse gases (GHG) due to the rapid growth rate of the global fleet, compared with its progress in energy efficiency. The transport sector requires the introduction of new technologies, which can promote more eco-efficiency in the whole transport system, even changing consumer behavior to achieve a satisfactory reduction of environmental impacts (EEA, 2007).

In recent years, despite the adoption of national programs to reduce emissions, the global indices have shown that GHG concentration grew about 30% from 1980 to 2000 and it is projected to increase about 37% and 52% until 2030 and 2050 respectively compared to levels in 2005 (United Nations, 2009). It will engrave the impact load on environment (e.g. increasing the global warming), since transport sector is the second largest (about 24%) and the second fastest growing source of GHG and CO2 emissions after the energy sector (about 45%) (Fig. 6). Global emissions of CO2 from the transport sector are expected to double between 2005 and 2050, from 6.1 GtCO2 to 12.2 GtCO2. In OECD countries the transport's share represents 30% of CO2 emissions, in which road transport accounts for almost 80% of them (EEA, 2007; OECD, 2007).

This is why the transport sector is a greatest concern to the EU member states, specifically regarding to climate changes. Then, it is important the development of cleaner vehicles combining lower environmental impacts with affordable cost, or even to develop new concepts of transport systems based on eco-efficiency levels that a given innovation project aims to achieve.

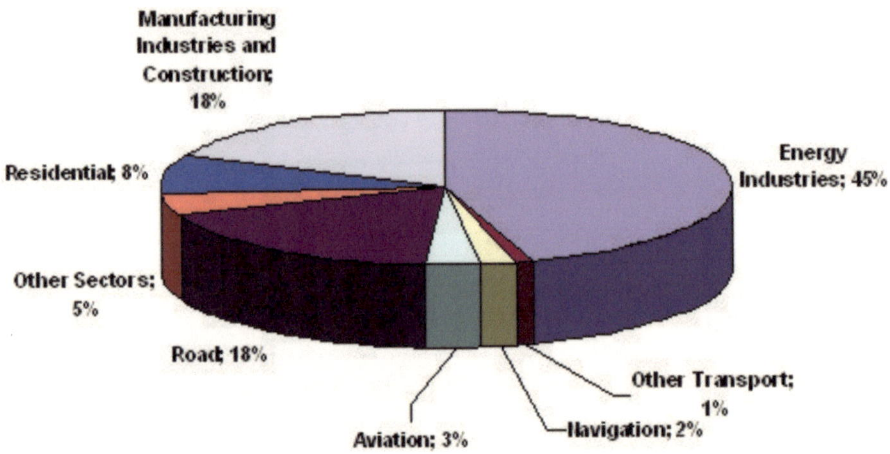

Fig. 6. CO2 emissions related to combustions. (OECD, 2007)

In this specific context, in which automotive companies are undergoing several environmental pressures, ecodesign and renewable materials converge to develop eco-friendly vehicles and/or components, achieving tighter environmental targets towards sustainable mobility. Among a wide range of materials made of renewable sources, there are the composites materials reinforced with vegetable fibers, which show a great potential for automotive usage.

They offer many advantages compared to synthetic fibers, reducing the dependence of non-renewable resources and the greenhouse emissions, they also present light weight, high specific strength and are biodegradable, which is crucial in their disposal phase. They corroborate with the questions and hypothesis of this research since they are a great candidate to replace synthetic fibers as reinforcement of polymeric matrices for industrial/structural automotive components.

3.3 Greener composite materials: vegetable reinforcements

The wide usage of glass fibers with thermoset matrices are considered critically. They have caused several environmental impacts mainly in their disposal phase related to the landfill and incineration control, besides the natural resources depletion, since glass fibers are made of non-renewable sources (e.g. silica and lime).

In this context, many researchers (Puglia et al, 2005; Suddell et al., 2002) have been studying natural fibers as replacement of glass fibers, improving the environmental performance of composites and products made of them. Products are classified based on their source: vegetable, animal or mineral, and they are commonly used as reinforcement of thermoplastic and thermosetting polymeric matrices.

Among them, vegetable fibers have emerged as realistic alternative and they are the most commercially used natural fibers, pointing out a great possibility for industrial application in composite materials. Furthermore, they are an important source of income for agricultural societies implying positive social impacts. For example, in Brazil there are many types of fibers such as jute, sisal, coir and curauá, all of them already with commercial applications. Brazil has a large potential to produce vegetable fibers that can be found natively or cultivated, becoming a source of income for several local communities (Alves, 2006).

Currently, automotive industry is the most avid and significant in terms of vegetable fibers usage in composite materials. They demand a shift of vehicle design from oil-derived polymers and synthetic reinforcements to natural materials, focusing on the environmental requirements of vehicles. Joshi et al (2004), shows that vegetable composites present better environmental performance than glass composites, e.g. their incineration consume lesser energy (about 45%), resulting in lower GHG emissions.

In Europe, automotive companies have developed components made of vegetable composites motivated mainly by regulations that have played an important role toward sustainable mobility. A good regulation example is the European Directive 2000/53/EC, in which vehicles have to be partially decomposable or recyclable (95%) by 2015 (Suddell et al., 2002). In USA, about 1.5 million vehicles are already using vegetable fibers as reinforcement of thermoplastic and thermosetting polymers (Faruk, 2009).

In Brazil some automotive initiatives are concerned with the selection of "greener" materials from renewable sources. For instance, in 1992 Mercedes-Benz of Brazil agreed to make an initial investment of US$1.4 million to research the vegetable fibers usage in its vehicles. This initiative translates into new jobs in the coconut fiber production including agricultural producers, and processing plant workers (Alves_a et al., 2009). It is expected that the use of natural fibers in automotive components will grow about 54 % per year, since European and American car makers have been already using them [194]. Projections point out an increase of the European demand for vegetable fibers in automotive industry from 70,000 tones in 2005 to 100,000 tones by 2010, which means a potential European market at about €100 millions (Faruk, 2009).

In Europe the use of natural fibers in the automotive sector increased from virtually "0" in 1994 to more than 28,000 tones in 2000, and in 2002 the global market share of vegetable composites reached 685,000 tones (US$775 million) mainly motivated by their low cost and density (Margets, 2002). In 2003 the automotive industry consumed about 45,000 tones of natural fibers, in which about 18,000 tones (40%) was accounted for German automotive industry with a linear consumption growth of about 30% (Karus et al., 2003). Indeed, Germany is the most significant customer of vegetable fibers (about 3.5kg per vehicle, excluding wood and cotton) it means about 2/3 of the total European consumption. Predominantly, using vegetable fibers results in lighter internal components (about 25% eu) with comparable properties to glass fibers composites. It reduces the fossil fuel consumption and then the GHG emissions of vehicles (Miller, et al, 2000), which comply with Environmental Directives and Kyoto's Protocol (Rouison, D. et al., 2006; Dietrich, A.B, 2005).

Undoubtedly, vegetable or 'green' composites have played a major role in the product design process, presenting many environmental advantages compared to synthetic composites, and offering a possibility for developing countries to use their own natural resources in a wide range of industries. Among various vegetable fibers available to the automotive sector one of the most used is jute fiber, which grows in sub-tropical countries such as Bangladesh, India and Brazil. Table 1 shows properties of some natural and conventional synthetic fibers.

Fiber	Density (g/cm^3)	Elastic Modulus (GPa)	Specific Elastic Modulus (E/ρ)	Tensile Strength (GPa)	Specific Tensile Strength (σ/ρ)	Elongation at break (%)	Cost($/kg) (1€=2.5R$)
E-Glass	2.6	73	28.07	1.8 – 2.7	0.69 – 1.04	2.5	6
Carbon (PAN)	1.8	260	144.44	3.5 – 5.0	1.94 – 2.78	1.4 – 1.8	60
Aramid	1.45	130	89.66	2.7 – 4.5	1.86 – 3.10	3.3 – 3.7	30
Jute	1.45	10 – 32	6.89 – 22.07	0.45 – 0.55	0.31 – 0.38	1.1 – 1.5	0.8
Sisal	1.45	26 – 32	17.93 – 22.07	0.58 – 0.61	0.40 – 0.42	3 – 7	0.8
Coir	1.33	4 – 6	3.01 – 4.51	0.14 – 0.15	0.11	15 – 40	0.8

Table 1. Mechanical properties of some fibers. (Alves_b et al., 2009)

Extracted from the stem of the jute plant (Corchorus capsularis and Corchorus olitorius), jute fibers grow mainly in India and Bangladesh, the largest global producers. They also grow natively in China, Nepal, Thailand and Brazil, and together all of these countries represent about 95% of the global jute fiber production (Alves_b et al., 2009). The global production of jute fibers have increased since last century reaching 1 million tones per year in 1900 and about 3,292 millions tones in 2004. In the last years, the jute production growth occurred mainly due to their production and consumption in India. India has a set of policies that have motivated the production and export of jute fibers and their derived products (Anderson et al., 2004).

Jute fibers have irregular cross-sections and their micro-cellular structure is composed of micro fibrils. They have an average length of about 0.5 - 6.0mm with fiber diameter varying from 0.01 to 0.04mm. Their chemical composition varies according to the plant age and the maceration, presenting as main constituents: lignin 12.5% - 13.5%, cellulose from 59% to 61%, fats and waxes 0.9% - 1.4%, minerals 0.5% - 0.79%, nitrogenous matter 1.56% - 1.87% (Alves, 2006).

4. Ecodesign parameters of the case study: The buggy

This work presents the Buggy vehicle as the case study (Fig. 7), defining its frontal bonnet, made of composite material, as the Functional Unit (FU) for technical and environmental analysis. Through the Buggy case study (CS-Buggy) this research intends to evaluate the feasibility of the jute fibers to replace glass fibers as the reinforcement of composite materials, also evaluating the social and environmental performance of the Buggy's enclosures made of jute composites. This work also presents the Sustainable Design Procedure (SDP), a systematic method to introduce environmental concerns in small and medium size companies (SMEs) through their materials knowledge expertise.

Since the design phase dictates most of inputs and environmental loads of a product or a process, composite materials are the innovation focus of the CS-Buggy, also introducing environmental concerns into SMEs planning, this work developed a Sustainable Design Procedure (SDP), for more details see Alves_a et al (2009). SDP is a systematic procedure that aims an "integration" of environmental concepts into the materials selection stage within the design phase. Since materials and their processes are the core business of SMEs, SDP can act as a strategic ecodesign procedure extending environmental awareness for the whole company from design to company policies.

Fig. 7. CS-Buggy vehicle.

SDP intends to influence different decision levels of companies beyond product development, providing a comprehensive and long term approach to achieve the potential sustainable level of eco-efficiency as discussed before. In this sense, a significant attention must be paid to the educational aspects of designers since SDP is based on the philosophy in which to do "sustainable design". One first needs to breed "sustainable designers". Subsequently, the environmental knowledge is expected, otherwise it becomes difficult to do any environmental improvement and/or innovation.

SDP aims to optimize the CS-Buggy regarding to the following factors: user needs, design requirements, production process, cost and environmental factors. The SDP structure is composed by qualitative and quantitative stages and it is presented as a sequential procedure in Figure 8. Even though it is a concurrent design approach, in which all stages are defined by traditional and environmental inputs, they can be combined in a simultaneous and interactive way.

Through a filter step, SDP can have as multiple feedback loops as required to re-evaluate previous decisions that have been made, ensuring a collaborative system in which all goals were reached. It is important to note that, to increase the innovation, environmental inputs must be taken into account from the beginning of the design process, and not as a final appendix. According to Manzini and Vezzoli (2002), environmental factors, besides their technical and economic advantages, change the professional perspective, creating an innovative environment.

In fact, environmental inputs improve the innovation as a new variable combined with traditional inputs, generating new ideas (environmental proposals) from a new environmental point of view. The qualitative phases Design Goals and Design Requirements are detailed in Alves_a et al (2009), in which the following total performances were obtained based on the five parameters (Table 2):

	Parameters					
	Environmental	Aesthetical	Technical	Economic	Process	Total
FC	12	8	25	2	17	**64.32**
FK	13	6	23	5	17	**64.31**
FJ	23	17	14	15	13	**80.39**
FC	23	17	13	17	13	**81.93**
FS	23	17	14	15	13	**80.50**

Table 2. Total performance (Γ) of the fibers reinforcement.

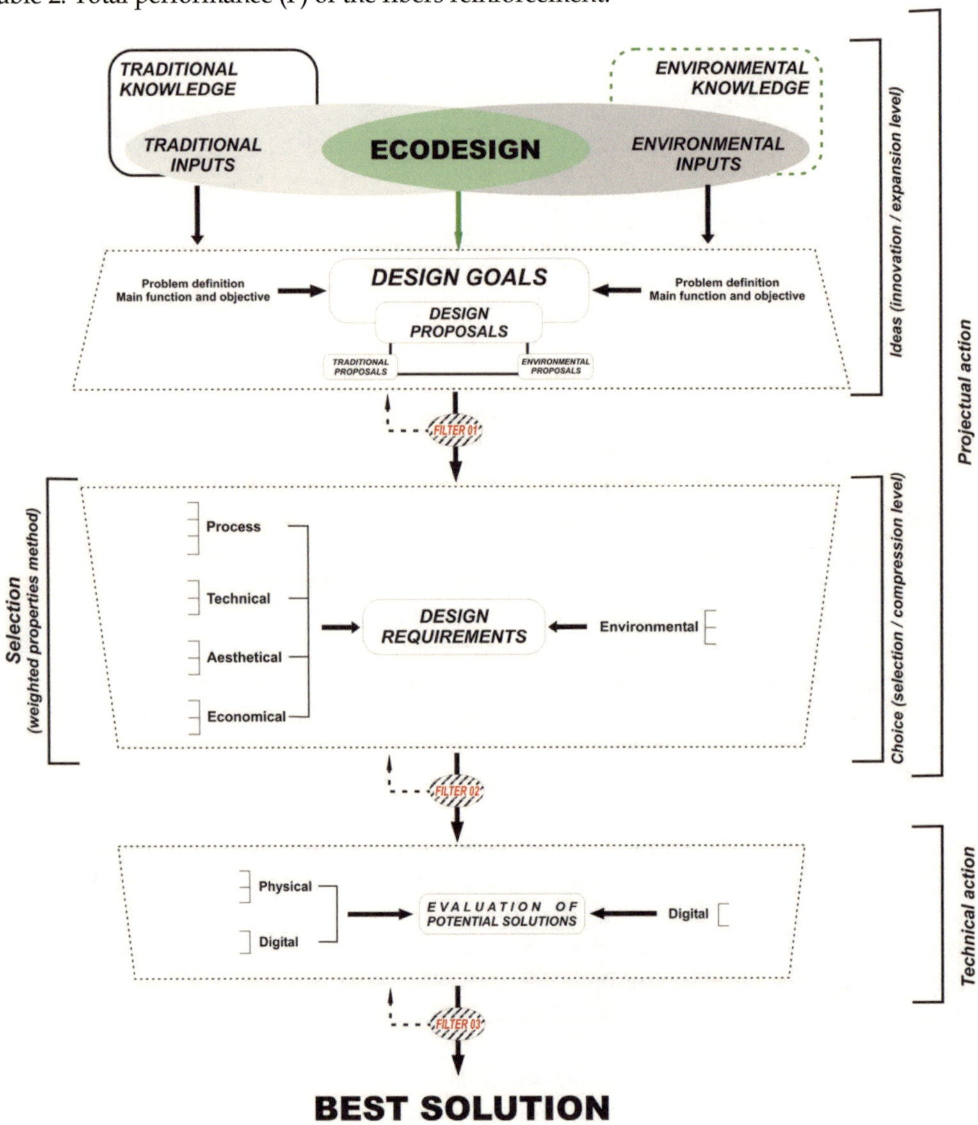

Fig. 8. Structure of the Sustainable Design Procedure.

Finally, it is important to note that the design requirements point out a possible solution. Therefore, after this stage it is necessary to carry out a quantitative analysis to evaluate the feasibility of the best choice and to ensure the success of the whole project, mainly when the best choice is a new and unknown material like in this case study (vegetable fibers). Thus, the remainder discussions are exclusively dedicated to the final SDP stage: evaluation and validation of the choice, due to its crucial influence on the final decision making.

5. Enclosures of the CS-Buggy: from sustainability to the use of vegetable fibers in vehicles

In the previous analysis, the total performance has shown vegetable fibers (sisal, jute and coir) as a potential replacement of glass fiber reinforcements usually used to produce the enclosures of concurrent buggies. Among selected vegetable fibers, jute fiber presents the lowest total performance (see Table 2), even tough it was defined as the best potential choice to be evaluated due to the following aspects:

- No significant difference among all vegetable fibers performance;
- Among selected vegetable fibers, only jute fiber allows an useful production of bi-axial and multi-axial fabrics.

5.1 Materials

The fiber reinforcements (Jute and Glass-E) used in this research to manufacture the reinforced polyester composites have two different fabric arrangements (bi-axial and multi-axial) (Fig. 9). The jute fibers were supplied by Castanhal Têxtil Inc from Amazonas State, Brazil. The glass fibers, used as the control material, were supplied by Matexplas Ltda. (Lisbon, Portugal). The standard thermosetting liquid resin used as matrix was the orthophthalic Unsaturated Polyester (UP) Quires 406 PA, and the peroxide methyl ethyl ketone (PMEK) used as the curing agent, was also obtained from Matexplas Ltda. (Lisbon, Portugal). Acetone (technical grade) was used as bleaching solvent to the superface treatment of the jute fibers.

Fig. 9. Fiber's fabrics. (a) Bi-axial glass fibers; (b) Multi-axial glass fibers; (c) Bi-axial jute fibers; (d) Multi-axial jute fibers.

5.2 Characterization and treatments of the jute fibers

Despite the good properties of the vegetable fibers, they are often considered only for applications that require low mechanical performance, due to their hydrophilic nature related to the presence of hydroxy groups in their cellulose structure, besides their natural oleines on the surface, raising their inadequate interface adhesion with polymeric matrices that present a hydrophobic character (Westerlind, & Berg, 1998; Belgacem & Gandini, 2005). These opposite features obstruct the contact between the vegetable fiber and polymeric matrix, resulting in a

poor efficiency to transfer loads across the composite. It implies the failure of the interface between matrix and fibers and accelerates the degradation of the composite. To obtain the percentage of the moisture content and other volatile compounds (mostly oleines) of the jute fibers as well as their thermal stability, a thermogravimetry analysis was performed (TG - weight loss versus temperature). The TG analysis was carried out under He flow (2.0 NL/h) from room temperature to 500ºC with a heating rate of 10ºC/min. All the tests used 50-60 mg of jute fibers placed in an alumina crucible (100µL), using a TG-DTA-DSC LabSys equipment. For the analysis three replicas were obtained. The thermogram for the jute fibers (Fig. 10) shows a small weight loss (about 8.7%) in the range 30ºC-125ºC. This weight loss can be ascribable to the loss of fiber moisture, and for temperatures higher than 240ºC the drastic weight loss can be ascribable to the jute fiber thermal degradation (Joseph et al., 2003).
In this context, in order to increase the wetting behavior of the jute fibers with apolar polyester, and thus improving the interface bonding fibers/matrix, jute fibers were subjected to two treatments to remove their moisture content and the oleines. In the first drying treatment, focused on moisture content in jute fibers, some bi-axial and multi-axial samples of jute fabrics were dried overnight (12h) at 140ºC (temperature based on TG analysis), using an universal oven. In the second bleaching/drying treatment, focused on oleines and waxes on the jute fiber surfaces, other samples were previously soaked in acetone (technical grade) during 24h, and were then dried according to the first treatment. The treated jute fabrics were designated as Jute Fibers Dried (JFD) and Jute Fibers Bleached/Dried (JFB/D), while untreated jute fibers were assigned as Jute Fibers Control (JFC) and glass fiber was assigned as Glass Fibers Control (GFC).

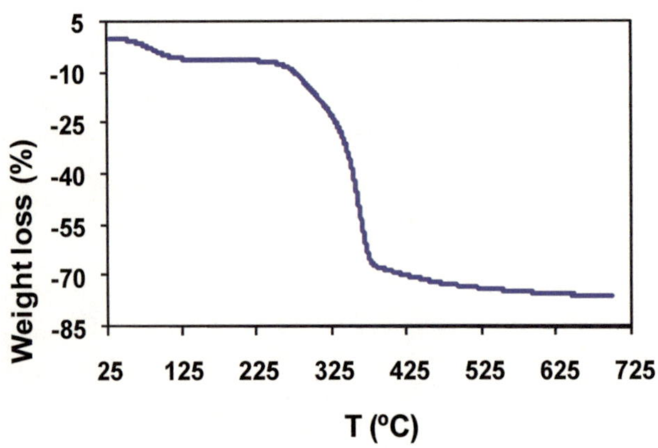

Fig. 10. Thermogram of the untreated jute fiber.

To understand the effects of the treatments on the surface of the jute fibers, an infrared spectra was carried out with a resolution of 16 cm^{-1}. It was performed using a Horizontal Attenuated Total Reflectance Infrared Spectroscopy (FTIR-HATR). Sixty-four scans were accumulated for each spectrum to obtain an acceptable signal-to-noise ratio. During spectra acquisition samples were pressed with 408 PSI. The absorbance of each spectrum was corrected with the Kubelka-Munk transform (Kruse & Yang, 2004).
Figure 11 presents the collected spectra from untreated and treated jute fibers. Several bands were obtained, in which the vibration modes were assigned according to the previously published researches (Ray & Sarkar, 2001).

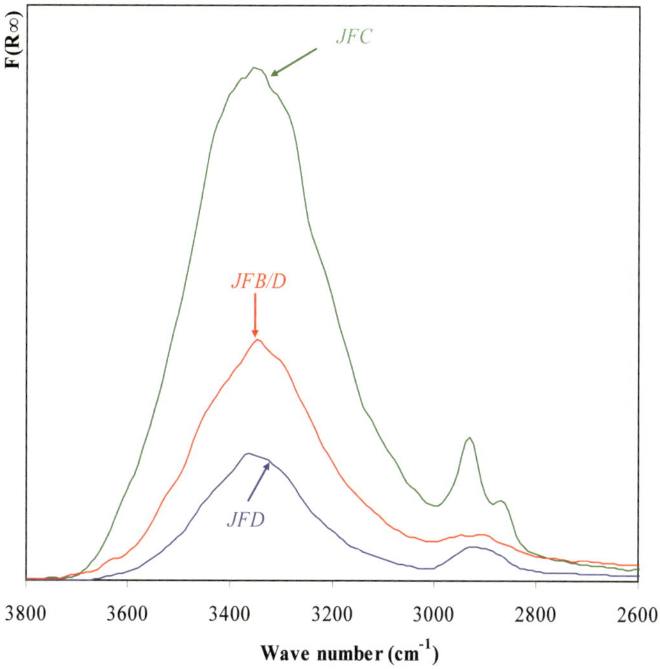

a)

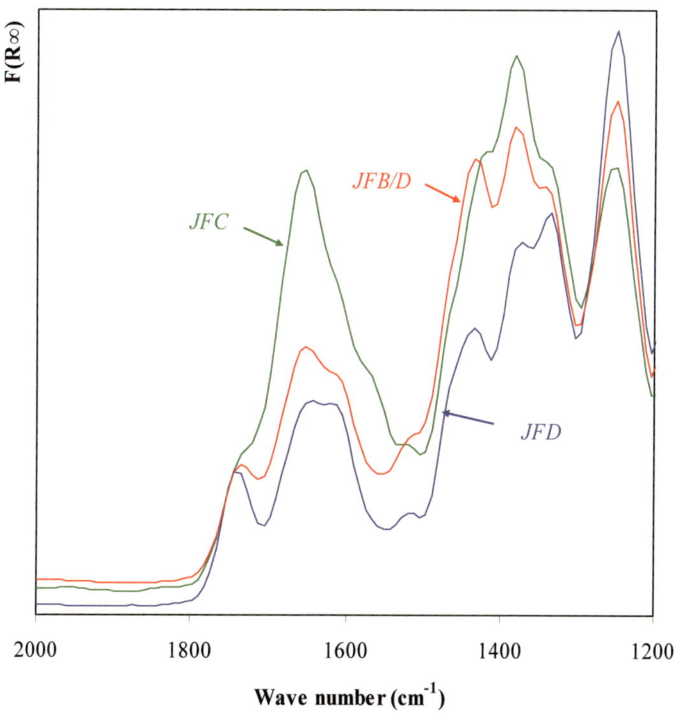

b)

Fig. 11. FTIR-HATR of untreated and treated jute fibers: (a) 3800 - 2600; (b) 2000 - 1200.

For the analyzed sample the major spectral differences were observed for the regions related to the -OH vibrations. Figure 11 (a and b) shows that for the JFD the O-H stretching band (3720-3000 cm-1) and the vibration of the adsorbed water (1640 cm-1) are significantly less intense than the respective bands for JFC and JFB/D. It can be concluded that the drying treatment was effective to decrease surface moisture content, contributing to improve the compatibility between jute fibers and unsaturated polyester matrix. On the other hand, it is possible to note that the bleaching/drying treatment reduced the efficacy of the drying treatment, since acetone removes waxes and oils from the jute fibers surface, which provide a protective layer for vegetable fibers. Thus, the removal of this natural protection exposes fibers surfaces, which increases their hydrophilic behavior.

Another thermogravimetry analysis was performed to investigate the effects of the treatments on the jute fibers, using the same set up of the first thermogravimetry, in which three replicas were obtained for each sample (JFC, JFD and JFB/D).

Figure 12 shows the main results from thermal analysis of JFC, JFD and JFB/D. The differentiated curves of weight loss are presented (DTG). The thermal decomposition profile was similar for all the analyzed samples. A small weight was observed in the range 30-200°C corresponding to dehydration of fibers. The JFC presents a moisture content of about 8.7% while JFD presented about 6.8%. It also points out the efficacy of the drying treatment, since it removed more than 20% of the fibers moisture content. On the other hand, JFB/D treatment as explained before, removed the protective layer made of waxes and oils from the jute fibers surface. In this sense, it presents fiber moisture content of about 7.6%, which means 11.7% higher than the moisture content found for JFD, pointing out its effect to decrease the efficacy of the drying treatment.

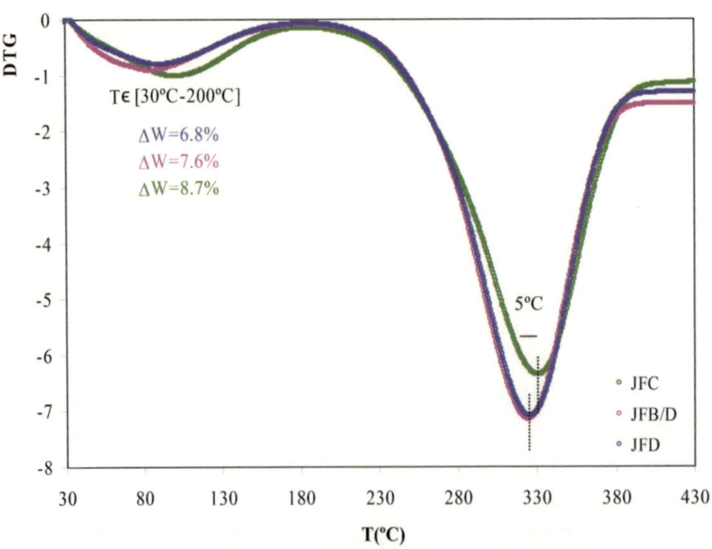

Fig. 12. Thermogravimetry analysis of untreated and treated jute fibers.

Thermogravimetry results are in accordance with FTIR data. In fact, the FTIR bands related to the -OH species are more intense for JFC and JFB/D samples. The thermal stability of the jute fibers was slightly affected by both treatments. For treated jute samples, the maximum temperature of the thermal decomposition process is 5°C lower than the maximum temperature observed for the untreated jute samples.

After the chemical/physical characterization of the jute fibers and the effects of their respective treatments, composites were manufactured with untreated and treated jute fibers (JFC, JFD and JFB/D) and glass fibers (GFC), and then specimens were obtained from composites and tested under tensile and bending tests, according to ASTM standard (D-3039 and D-790), and Dynamic Mechanical Analysis (DMA). The specimens were cut from composite plates, produced with both bi-axial and multi-axial fiber arrangements. They were produced by Resin Transfer Molding (RTM) process using a RTM UNIT obtained from ISOJET Equipments (France). Composites were prepared varying the fiber content (Vf) from 20% to 30% to reach the maximum volume fraction (Vf) of reinforcement that was used to balance RTM processability and the mechanical properties of the composites. Each Vf was obtained based on jute fibers as volume control, due to their larger filament's diameter (40 μm) compared with the glass fibers (14 μm).

Multi-axial plates were manufactured with one layer of fabrics, while bi-axial plates were manufactured with six layers according to the following stacking sequence [(0/90), (45/-45), (0/90)]S. Polyester matrix was then mixed with PMEK (0.25 % in volume) and the resin mixture was degassed under a vacuum of 10 mm Hg for 10 min before the impregnation of the fabrics. After that, it was allowed to pass through the mold under different pressures, which were optimized for each fabric arrangement. After the complete filling of the mold, the plates kept 1h curing inside the mold, and were then extracted from the mold and allowed to post cure at room temperature (about 300 h).

5.3 Mechanical behavior of the composites

Figure 13 and Table 3 present the results of the mechanical behavior of the composites, in which the data given for each property are the average of five specimens. For all specimens, the composite materials displayed nearly linear elastic behavior up to the fracture. In the bi-axial samples, GFC presents higher tensile strength (about 100%) than the JFC. It is not associated with the fiber content of the composites (Vf), since the GFC has a lower volume fraction (about 33%) compared to the maximum Vf reached for JFC, produced by RTM process. In fact, it is related to the nature of the fibers used to reinforce the polyester matrix. For multi-axial composites, the specimens have roughly equivalent strengths around 26 MPa. Like in the bi-axial composites, for multi-axial arrangement the tensile strength is not associated with the fiber content, since for GFC the Vf of the glass fiber is much lower (about 50%) than the maximum Vf achieved for JFC, produced by RTM process. Moreover, the Vf of the multi-axial GFC was of about 50% of the maximum volume fraction in which would be possible to produce it, implying a significant decrease of the mechanical properties of the multi-axial GFC composite.

Results also revealed that both treatments brought a significant increase on the stiffness of the jute composites, moving their elastic modulus from about 1.83 GPa for JFC to 5.29 GPa (about 189%) and 4.91 GPa (about 168%) for JFD and JFB/D, respectively. Both treatments provided a significant improvement on the interface bonding of bi-axial jute composites, decreasing significantly their strain (average 55%), in fact their strain became lower even than the strain of glass fiber composites (about 16%). Moreover, the coefficient of variation (CV) for bi-axial jute composites presents a very significant decrease, from 14.70% for JFC to 4.10% and 3.59% for JFD and JFB/D, respectively.

Despite treated composites still presenting lower elastic modulus (about 26%) than that obtained from Classical Theory of Laminated - CTL (6.89 GPa), the results make clear that both treatments provided really great effects related to the interface bonding of bi-axial jute composites. Nevertheless, results also point out an unsuitability of the CTL to predict the mechanical properties of the bi-axial vegetable composites. Unlike for the stiffness, the treatments did not bring a significant increase for the strength of treated composites (average 18%). Indeed it increased from 27.76 MPa (JFC) to 30.38 MPa and 35.33 MPa for JFD and JFB/D, respectively (Table 3). Thus, based on the fact that the elastic modulus is determined from the slope of the stress versus strain curves, its large increase after the treatments can be explained by the improvement of the interface jute/polyester, due to the significant decrease in the maximum strain of the composites.

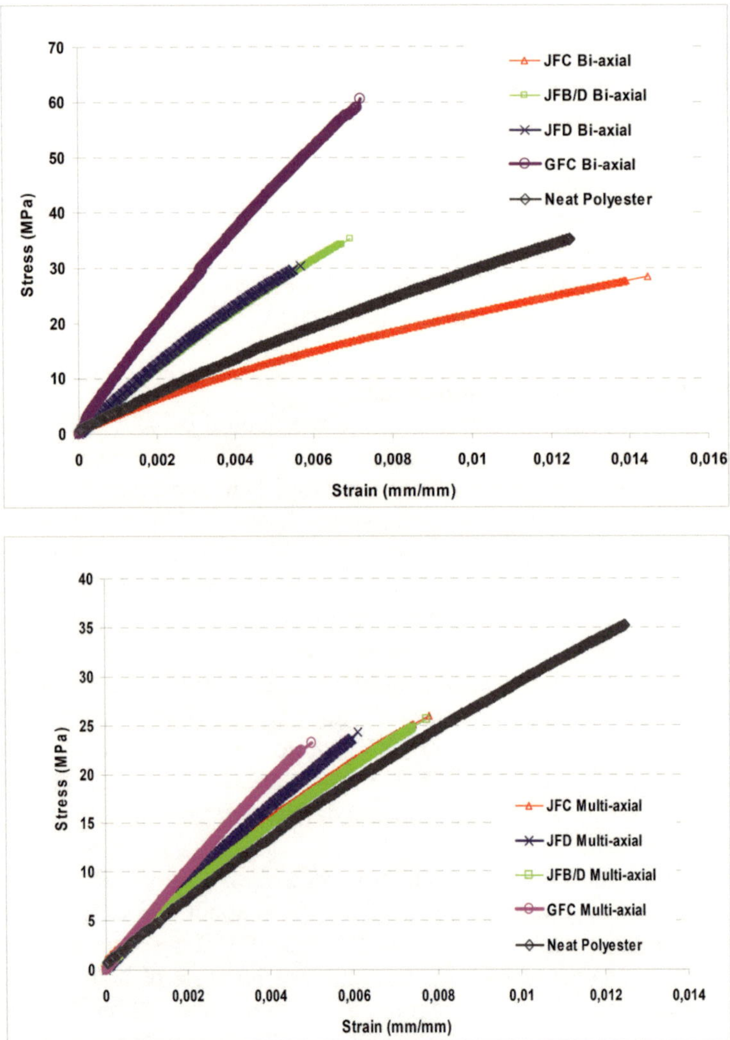

Fig. 13. Evolution on tension of the composites (Bi-axial and Multi-axial, each curve is a plot of a particular specimen whose behavior is representative of its group).

Composites	Fiber Arrangement	V_f (%)	Maximum Stress (MPa)	Maximum Strain (%)	Elastic modulus (GPa)	Coefficient of Variation for modulus (%)
GFC	Bi-axial	21	60.52	0.69	8.81	6.02
	Multi-axial	9	23.21	0.51	4.69	4.81
JFC	Bi-axial	31	27.76	1.49	1.83	14.70
	Multi-axial	14	26.41	0.83	3.19	5.34
JFD	Bi-axial	28	30.38	0.58	5.29	4.10
	Multi-axial	11	24.39	0.60	4.23	5.14
JFBD	Bi-axial	23	35.33	0.72	4.91	3.59
	Multi-axial	9	25.58	0.80	3.55	3.91

Table 3. Mechanical properties of the composites.

On the contrary, for multi-axial fiber composites, both treatments did not imply significant improvements on their mechanical properties. Unlike the bi-axial jute composites, treatments implied no significant change in the elastic modulus of the multi-axial composites (average 22%), moving it from about 3.19 GPa (JFC) to 4.23 GPa and 3.55 GPa for JFD and JFB/D, respectively. Since this fabric's arrangement does not require fibers in tow form, their wettability is much more efficient than the wettability found for bi-axial arrangement, confirming that the arrangement of jute fabrics has large influence on the fiber impregnation. Related to the maximum stress, again treatments did not imply significant changes on it, decreasing from 26.41 MPa (JFC) to 24.39 MPa (JFD) and 25.58 MPa (JFB/D) (about 6%).

Figure 14 emphases the fracture cross section of the JFC specimens using a Scanning Electron Microscope (SEM). The rupture was accompanied by a clear withdrawal of the fibers from matrix (pull-out effect), leaving holes that indicate the very poor interface bond (Fig. 14 b). Besides the weak interface, Figure 14 (a and b) also shows that the fibers in the bi-axial JFC composite are not completely involved by matrix, indeed it makes clear the poor wettability in the center of the jute tow.

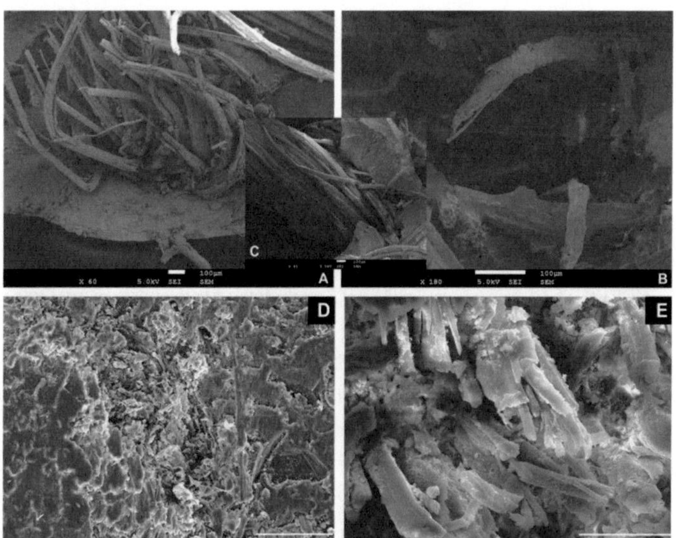

Fig. 14. SEM of the bi-axial jute composites. (a, b and c) untreated; (d and e) treated.

Table 3 also shows that the treatments brought an increase of the matrix volume fraction (Vm) of the jute composites. It is important to remark that JFD and JFB/D present higher elastic moduli than JFC, even with a decrease in their fiber content (Vf). This effect is associated with the better impregnation of the jute fibers by matrix, emphasized by Figure 14 (d and e) that shows the cross section surfaces of the treated bi-axial JFD and JFB/D specimens. After both treatments and on the absence of the moisture content, the tows of the jute fibers are completely impregnated by polyester matrix even into their center, unlike the bi-axial JFC composites. Sydenstricker et al (2003) analyzed sisal fibers after treatments and also found an effective improvement in interfacial adhesion, decreasing the pull-out effect.

Since the results of the mechanical properties of both treated jute composites showed no significant difference between the effects raised by both treatments, drying treatment was assigned as the best choice due to its lower costs and environmental impacts. Thus, DMA tests were performed on JFD composites to refine the effects of the drying treatment. The DMA shows that for both fiber arrangements the activation energies present an increase for both JFD composites compared with their respective JFC composites (44% and 21% for multi-axial and bi-axial), which confirms the better interaction between jute/polyester, requiring more activation energy to flow the matrix (Table 4).

The activation energy observed for both treated jute composites is higher than for untreated jute composites, by about 22% and 45% for bi-axial and multi-axial, respectively (Table 4). Compared to the neat polyester matrix, the activation energies of the treated jute composites are higher by about 57% and 22% for multi-axial and bi-axial respectively. In this sense, it is clear that the drying treatment improved the interface bonding, and increased the interaction jute/polyester. All of these results corroborate the previous results, as discussed before. Finally, all results show that both treatments were responsible for a significant improvement on the mechanical behaviors of the jute composites by extraction of moisture and other compounds from jute fiber. In fact, the treatments improved the wetting behavior of the twisted tow of the bi-axial jute fibers, improving the interface bonding jute/polyester.

Composite		T_g em E" (°C)			E_a (kJ.mol^{-1})	E´ (MPa)
		1 Hz	5 Hz	10 Hz		
Polyester Matrix		51.75	57.24	60.00	252.72	2,456 (23°C) 2,538 (10°C)
Multi - axial	JFC	34.73	39.16	41.54	274.75	1,971 (23°C)
	JFD	65.06	69.50	70.32	398.04	2,987 (23°C)
Bi - axial	JFC	41.5	46.8	49.2	252.52	1,322 (10°C)
	JFD	60.65	66.11	67.45	307.64	2,754 (10°C)

Table 4. Activation energy of the neat polyester and the composite materials.

5.4 Numerical analysis of the jute composites: design optimizations

In the experimental evaluation of the composites (quantitative analysis), results have shown vegetable fibers as the potential solution, corroborating with qualitative analysis of SDP. Given the bi and multi-axial JFD composite as the best choice, they were carried out through numerical evaluation using ABAQUS 6.7 software. The frontal bonnet of the CS-Buggy with thickness at 4 mm was assigned as the Functional Unit (FU) to predict the behavior of the glass and jute composites during their usage, investigating the suitability of jute fibers to manufacture technical parts. The control bonnet was defined based on the current glass

composite used to produce a concurrent buggy. It is made of multi-axial glass fiber composite with about 23% of fiber volume fraction (Vf) and about 4 mm of thickness, and was assigned as Glass Bonnet Control (GBC). The candidate bonnet made of JFD composites was assigned as Jute Bonnet Composite (JBC).
The boundary conditions of the model can be seen on Figure 15, in which the pressure load was about 800 N (80 kg). The pressure area was assumed as circular with the diameter of about 200 mm placed at the center of the bonnet.

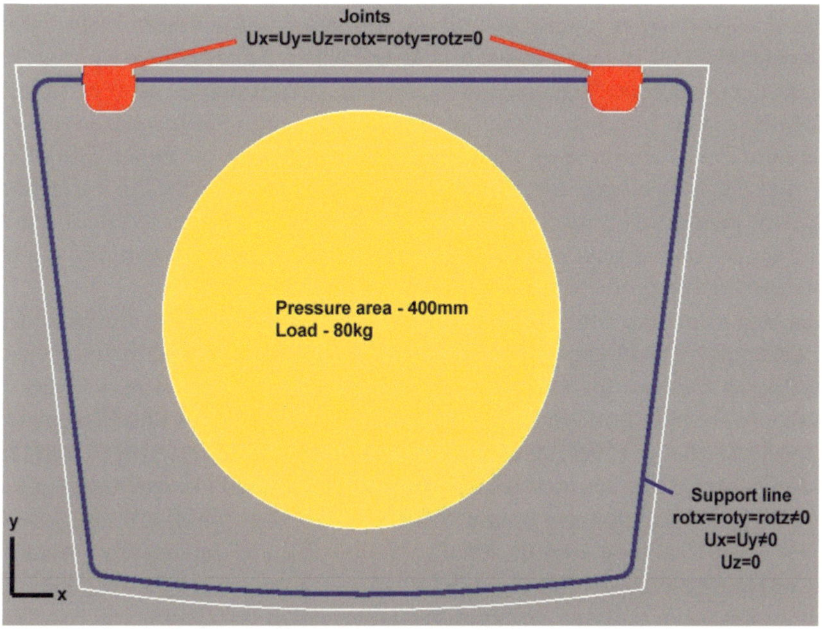

Fig. 15. Boundary conditions of the FEA of the frontal bonnet of the CS-Buggy.

Although the lower mechanical properties of the jute fiber composites comparing with glass fiber composites, and despite their current applications being somewhat limited to non-structural components, the experimental and numerical results pointed out jute fibers as a useful possibility to replace glass fibers in automotive components, satisfying the needs of the end customer. The results of the optimization of the bi-axial JBC show that the bi-axial arrangement of the jute fibers supports the load pressure of the project without implies any change in the design (dimensions and styling) of the bi-axial JBC, besides the change in its layers stacking sequence from $[(0^o/90^o), (45^o/-45^o), (0^o/90^o),]_s$ to $[(0^o/90^o), (0^o/90^o), (45^o/-45^o)]_s$.

6. Environmental performance of the jute composites

The main goal of this work, based on the Triple Bottom Line concepts (Alves, 2006), is to obtain the equilibrium among social, environmental and economic performance of the jute fiber composites to produce technical automotive components. In the previous paragraphs it was possible to evaluate and confirm, through numerical and experimental analysis of the composites, the technical and economic feasibility of the jute fibers in replacing of the traditional glass fibers as reinforcement of composite materials. Thus, to ensure the sustainability and ecodesign concepts based on the Triple Bottom Line, a Life Cycle

Assessment (LCA) was performed to assess the environmental impact of using jute fiber composites and their required treatments for automotive design applications to manufacture the enclosures of the CS-Buggy. The results were compared with the impacts raised by current enclosures made of glass fiber composites over the entire life cycle of the CS-Buggy, assessing the consequences of replacing glass fibers for untreated and treated jute fibers on the overall sustainability of this specific and important automobile sector in Brazil (leisure and tourism).

Like the previous numerical analysis, in the LCA evaluation the frontal bonnet of the CS-Buggy was also assigned as functional unit of the analysis, or in other words, the functional unit can be stated as "the engine cover of 0.35 m2 which achieves the required mechanical and structural performance". Since the LCA was performed to achieve environmental impacts related to the composite materials used to produce the frontal bonnet of the CS-Buggy, its boundary conditions is the entire life cycle of the bonnets made of composite materials and their influence for whole CS-Buggy vehicle, from the extraction of raw materials, over production processes and the use phase to the end-of-life of the vehicle. It includes all the needed transportations as well as the infrastructure to apply the treatments to the jute fibers and to produce the bonnets and to dispose of them.

The inputs regarding the jute fibers cultivation and production were provided by the supplier Castanhal Têxtil Inc, nevertheless they can also be estimated based on the literature. Inputs related to the polyester matrix, glass fibers and vehicles used for transportation were based on SimaPro 7.0 database in its IDEMAT and Ecoinvent libraries. Inputs related to the production of all bonnets were based on the production of the composites (Table 5). The journey logistic inputs were based on the supplier's database, while electric energy inputs were obtained from Coltro et al (2003) and they are related to the Brazilian electric energy system. Finally, the landfill and incineration scenarios of the end of life of the bonnets were based on Brazilian government reports (Alves_b et al., 2009), the recycling scenario was based on experimental results of the mechanical recycling. Figure 16 shows the schematic diagram of the assumed life-cycle to the functional unit, in which green colored inputs were obtained by the authors and black colored inputs were obtained in the SimaPro database.

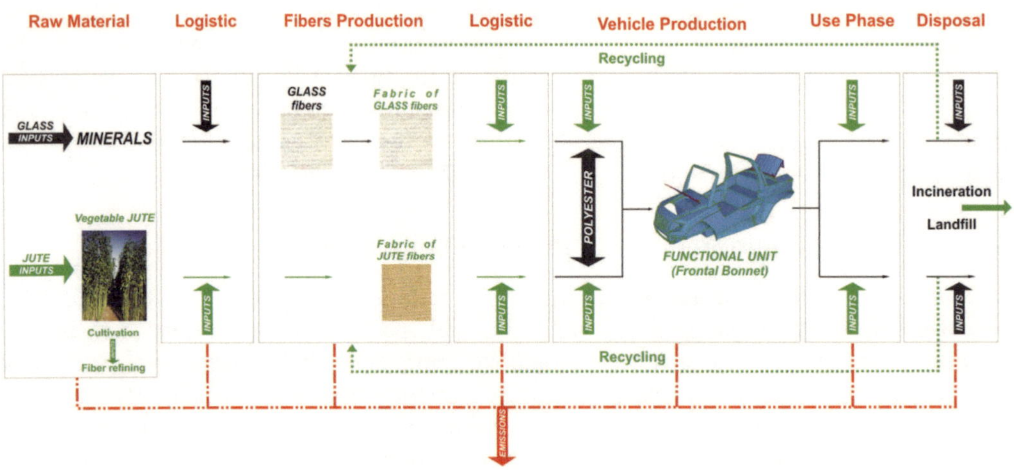

Fig. 16. Boundaries assumed in the LCA.

Bonnet	Injection flow (cc/min)	Volume of the fiber (%)	Mass of the bonnet (kg)	Mass of the fiber (kg)	Injection time (seg)	Total energy consumption (kW.h)
Jute Fibers (untreated and treated)	45	31	1.77	0.65	353	18.5
Glass Fibers	50	21	2.02	0.74	364	17.9

Table 5. Inputs of the bonnet´s production.

For the use phase the fuel consumption was taken into account to identify how influential is the replacement of the glass composites for the lighter jute fiber composites. Through the lower density of the jute fibers in comparison to glass fiber, it was possible to calculate the percentage of reduced weight of the bonnet made of jute fibers (about 15 %) and of whole vehicle (0.048%). In this sense, based on literature (Ljungberg, L.Y, 2007; Miller, et al., 2000), the decreasing fuel consumption of the CS-Buggy due to the use of the jute bonnet was estimated at about 0.029 %, which means about 7.71 L (5.55 kg) for an expected life of 265,500 km. This expected use phase life is based on Sindipeças reports in which is established the average life of a Brazilian vehicle at about 20 years and its average annual use of about 13,275 km/year (Alves_b et al., 2009). It was estimated a current fuel consumption of about 10 km/L for a total weight of the CS-Buggy of about 600 kg. In this sense, the fuel consumption assigned to the bonnets made of glass and jute fibers was respectively about 64.36 kg and 58.81 kg taking into account the density of the petrol at 0.72 kg/L. Regarding the scenario of the final disposal of the enclosures, it will be explained later.

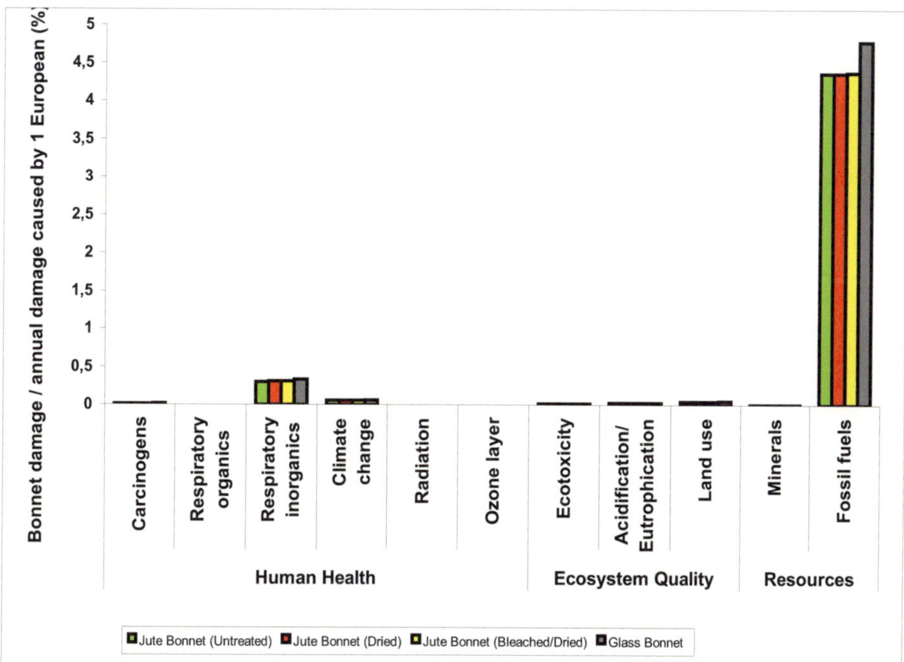

Fig. 17. Impact categories of the bonnets (Total Life-Cycle).

Regarding to the total life-cycle of the bonnets, Figure 17 and Table 6 show the total damage caused by the environmental impacts in their total life-cycles. Overall, it is clear that the use phase is significantly more pollutant than production and disposal phases (about 1,000%), in fact disposal phase represents just about 3% of the total damage, being raised by energy consumption of the recycling scenario. The significant impacts are raised by the use phase (about 97%), since its values are very close of total life cycle and most of impacts are related to the resources damage category due to the consumption of fossil fuel, while 3% are related to the production phase and its energy consumption, which raises respiratory inorganics impacts. In the whole life cycle, glass bonnet presents larger environmental damage (average 9%) comparing with damage raised by all jute bonnets, due to its higher weight and fuel consumption. About the treatments, Table 6 shows that comparing to the untreated jute bonnets, both drying and bleaching/drying treatments decrease the environmental performance of bonnets at about 1% and 2% respectively. In other words, both treatments are high pollutant until the production phase, in which dried and bleached/dried jute bonnets have 18% and 42% more environmental impacts than untreated jute bonnets. After the use phase, the consumption of the fossil fuel (more pollutant) becomes the treatments no significant to the total damage. Finally, results show that in spite the high importance of the production and disposal phases for the life cycle of vehicles, in this CS-Buggy the use phase is more pollutant and more important to focus the design improvements. It confirms researches (Ashby & Johnson 2002) in which the use phase is the most pollutant phase of a vehicle.

Damage category		Production Phase - Use Phase	Disposal Phase
Human Health	UJB	0.36529	0.01012
	DJB	0.37383	0.01012
	B/DJB	0.37572	0.01015
	GB	0.39440	0.01797
Ecosystem Quality	UJB	0.08450	0.00161
	DJB	0.08589	0.00161
	B/DJB	0.08604	0.00161
	GB	0.09179	0.00281
Resources	UJB	4.38789	-0.01428
	DJB	4.38789	-0.01428
	B/DJB	4.39931	-0.01428
	GB	4.80267	-0.01483

Table 6. Damage categories of the bonnets (Total life-cycle).

Related to the total enclosures of the CS-Buggy, results show that the replacement of all glass fibers for jute fibers improves the environmental performance of the vehicle at about 15%, while the frontal bonnet means an improvement of about 9%. Thus, a much more significant effect could be reached by switching to light-weight design of vehicles by design of composite materials. About treatments, unlike the treated jute bonnets in which treatments decreased in the environmental performance of them (about 1% and 2%), for total enclosures, the treatments implied lower differences among their environmental performance. It proves that treatments of jute fibers are a great choice, improving the mechanical performance of the jute composites without imply environmental impacts.

6.1 Social and economic analysis

In regards to the social requirements, jute fiber plays an important role from fiber cultivation of the plant to the production of the bonnet. In its cultivation phase jute is an important income source to the local farmer communities contributing to the sustainability of the region, avoiding the rural exodus hence its social problem in industrial cities. In the production phase, jute fiber causes fewer health risks and skin irritation than glass fibers for the employees that are directly involved in the production of the components. In the use phase, the social advantage of the jute fibers is related to the human health since jute fibers imply lower fuel consumption than glass fibers, and then raising lower GHG emissions and their environmental impacts. The social advantages of the disposal of jute fibers are also related to the human health, since they are biodegradable for landfill scenarios, while for the recycling scenario they require less energy compared with glass composites (about 50%).

Related to the economic advantages, in Brazil, jute fibers cost about seven times less than glass fibers, while production costs are almost the same, since it is possible to produce either jute or glass composites with almost the same setup and production processes. Using jute fibers also implies lower fuel consumption, so it means an economic advantage for owners of the vehicle. Still, the potential global market for natural fibers in the automobile industry is expected to increase. Nowadays in the USA more than 1.5 million vehicles are the substrate of choice of bio-fibers such as kenaf, jute, flax, hemp and sisal and thermoplastic polymers such as polypropylene and polyester (Faruk, 2009; Margets, 2002).

Finally, this LCA analysis presents the consequences of the replacement of the glass fibers by the jute fibers as reinforcement of composite materials to produce automotive structural components. In regards to the composite materials, CS-Buggy demonstrated that jute fiber composite presents the best solution enhancing the environmental performance of the CS-Buggy's enclosures, hence improving the environmental performance of the whole vehicle. However, it is important to remark that, despite jute fibers being well known as natural, and hence expected to present lower environmental impacts than glass fibers, the LCA showed that until the production phase of the composites, jute fibers imply higher environmental impacts, since they require more energy for manufacturing the composites. Indeed, only from the use phase of the CS-Buggy jute fibers present lower impacts than glass fibers, in which the fuel consumption becomes lower due to the weight reduction of the vehicle.

LCA also pointed out some unknown impacts in production and disposal phases of the bonnets, specifically related to the logistic transports of the jute fibers and the recycling scenario of the composites. It provides to designers an overview scenario of the whole issue that help to make decisions, besides those traditional inputs usually used in the product design, working in partnership with suppliers to improve the logistic of the jute fibers and focusing on the most pollutant phases to prevent potential environmental effects.

7. Conclusions

This work presented a comprehensive and integrated approach of the ecodesign and sustainability concepts through using friendly eco-composite materials, reinforced with jute fibers. As explained at the beginning, the life-cycle approach used here provided a larger point of view of ecodesign. Through the Sustainable Design Procedure, as a strategic ecodesign method, it was possible to show how the integration of the environmental inputs really improve the level of innovation of the current product design, by interconnecting them with traditional inputs such as the properties of materials and economic factors. In fact, the environmental inputs denoted a new approach of the problem, motivating the inclusion of vegetable fibers and hence jute fibers as candidate to replace glass fibers as reinforcement of composite materials.

The results show that jute fibers need some treatment to improve the mechanical behaviour of the composites, since they present significant moisture content. On the other hand, unlike several chemical treatments of fibers obtained in the literature, in this research two treatments were performed and showed that a simple and inexpensive drying of the fibers is enough to improve the composite properties. In fact, the treatments improved the wetting behaviour of the twisted tow of the bi-axial jute fibers, and then, they improve the interface bonding jute/polyester. After the treatments the volume fraction of matrix into the composite shows an increase due to the completely impregnation of jute fiber tows by matrix, also pointing out the improvement of the interface bonding due to the increase of the interface area.

Related to the environmental performance of the jute composites, the case study confirmed them as the best solution enhancing the environmental performance of the buggy's enclosures and hence improving the environmental performance of the whole vehicle, inspite of their respective treatments. Despite the higher energy consumption to dry the jute fibers, their lighter weight characteristic ensures their better environmental performance compared to the glass fibers. Since the use phase of vehicles was shown to be the most pollutant phase, the lighter weight of jute fibers implied a decrease of the fuel consumption of the vehicle used as case study. Also, LCA pointed out some unknown impacts in production and disposal phases of the bonnets, specifically related to the logistic transports of the jute fibers and the recycling scenario of the bonnets. It is important to remark that results show that automotive components made of vegetable composites need to be lighter than glass composites to present better environmental performance. Otherwise, they do not present environmental advantages, raising more impacts than glass composites.

Finally, this work can be considered a first step towards the sustainability of the Brazilian industry of buggies, since it can be a motivation for other companies to produce more sustainable vehicles, toward the sustainability of this mobility market. It can even drive users awareness for more environmentally friendly consumption behaviour.

8. References

Alves, C. (2006). *Design sustentável: a importância das fibras de juta, sisal e coco, no planejamento de produtos e éticas sustentáveis*. Universidade Estadual Paulista, Master Degree Thesis, Bauru.

Alves, C. (a); Ferrão, P.; Freitas, M.; Silva, A. J.; Luz, S. M.;Alves, D. E. (2009). Sustainable design procedure: The role of composite materials to combine mechanical and environmental features for agricultural machines. *Materials & Design*, 30, 10, p. 4060-4068. ISSN 0261-3069.

Alves, C.(b); Ferrão, P.; Silva, A. J.; Reis, L. G.; Freitas, M.; Rodrigues, L. B.; Alves, D. E. (2009). Ecodesign of automotive components making use of natural jute fiber composites. *Journal of Cleaner Production*. 18, 4, p. 313-327. ISSN 0959-6526.

Anderson, J., Bishop, G., Jansz, A. (2004). Green guide to composites: an environmental profiling system for composite materials and products. Watford BRE and NetComposites. ISBN 1 86081 733 5

Ashby, M.; Johnson, C. (2002). *Materials and design: the art and science of material selection in product design*. 1 ed., Butterworth-Heinemann, ISBN 0750655542, Oxford.

Baudrillard, J. (2005). *The System of Objects*. Verso Books, ISBN 978 1 84467 053 6, New York.

Baumann, H.; F. Boons; A. Bragd (2002). Mapping the green product development field: engineering, policy and business perspectives. *Journal of Cleaner Production*, 10, 5, p. 409-425, ISSN 0959-6526.

Baxter, M. (1999). *Product Design: Practical Methods for the Systematic Development of New Products (Design toolkits)*. Stanley Thornes, ISBN 0748741976, London.

Belgacem, M.N.; Gandini, A. (2005). Surface modification of cellulose fibres. *Polímeros*, 15, p. 114-121, ISSN 0104-1428.

Bertalanffy, L.v. (1969). *General system theory: foundations, development, applications*, George Braziller, ISBN 0-7139-0192-6, New York.

Chan, J.; Tong T. (2007). Multi-criteria material selections and end-of-life product strategy: Grey relational analysis approach. *Materials & Design*, 28, 5, p. 1539-1546, ISSN 0261-3069.

Coltro, L.; Garcia, E.; Qeiroz, G. (2003). Life Cycle Inventory for Electric Energy System in Brazil. *Int J LCA*, 8, 5, p. 290-296.

Datschefski, E. (2001). *The Total Beauty of Sustainable Products*. Rotovision. ISBN 2880465451, East Sussex.

Deng, Y.; Edwards, K. (2007). The role of materials identification and selection in engineering design. *Materials & Design*, 28, 1, p. 131-139, ISSN 0261-3069.

Dietrich, A.B. (2005). Materiales compuestos: Procesos de fabricación de embarcaciones. UPC, ISBN 84-8301-820-9, Barcelona.

Community and Information Society Technology Programme (2003). *Virtual dematerialisation: ebusiness and factor X*. Digital Europe, Available from: <http://www.forumforthefuture.org/files/digitaleuropeVirtualdematerialisation.pdf>, Accessed in 12/03/2008.

Easterlin, R. (2006). Life cycle happiness and its sources: Intersections of psychology, economics, and demography. *Journal of Economic Psychology*, 27, 4. p. 463-482, ISSN 0167-4870.

EUROPEAN ENVIRONMENT AGENCY. Europe´s environment: the fourth assessment. 2007, EEA, ISBN 978-92-9167-932-4, Copenhagen.

Faruk, O. *Cars from Jute and Other Bio-Fibers*. Available from: <http://biggani.com/files_of_biggani/mashiur/interview/omar_faruk.pdf>, Accessed in 03/05/2009.

Field III, F.; Clark, J.; Ashby, M (2001). *Market Drivers for Materials and Process Development in the 21st Century*. Available from: <http://www.mrs.org/s_mrs/bin.asp?CID=9284&DID=197217&DOC=FILE.PDF> Accessed in 23/07/2009.

Fiksel, J. (1998). *Design for Environment: Creating Eco-Efficient Products and Processes.* McGraw-Hill, ISBN 0070209723, New York.

Goedkoop, M.; van Halen, C.; te Riele, H.; Rommens, P. (1999). *Product Service Systems, Ecological and Economic Basics.* Available from: <http://www.pre.nl/pss/download_PSSreport.htm>, Accessed in 23/07/2008.

Hilton., M. (2008). *Sustainable Consumption Facts and Trends.* World Business Council for Sustainable Development, ISBN 978-3-940388-30-8, Brussels.

Hails, C.; Loh, J.; Humphrey, S. (2006). *WWF: Living planet report 2006.* Aconda Verd WWF. ISBN 2-88085-272-2.

Hails, C.; Loh, J.; Humphrey, S. (2008). *WWF: Living planet report 2008.* Aconda Verd WWF. ISBN 978-2-88085-292-4

Halog, A. (2004). An approach to selection of sustainable product improvement alternatives with data uncertainty. *Journal of Sustainable Product Design,* 4, 1, p. 3-19, ISSN 1573-1588.

Jeswiet, J.; M. Hauschild (2005). EcoDesign and future environmental impacts. *Materials & Design,* 26, 7, p. 629-634. ISSN 0261-3069.

Joshi, S. V.; Drzal, L. T.; Mohanty, A. K.; Arora, S. (2004). Are natural fiber composites environmentally superior to glass fiber reinforced composites? *Composites Part A: Applied Science and Manufacturing,* 35, 3, p. 371-376. ISSN 1359-835X.

Joseph, P.; Josepha, K.; Thomas, S.; Pillaic, C.; Prasadc, V.; Groeninckxd, G.; Sarkissovad, M. (2003). The thermal and crystallization studies of short sisal fibre reinforced polypropylene composites. *Composites: Part A - Applied Science and Manufacturing,* 34: 253-266.

Karus, M.; Kaup, M.; Ortmann, S. (2003). Use of natural fibres in composites in the German and Austrian automotive industry, Nova-Institut GmbH: Hürth.

Kruse, B; Yang, L. (2004). Revised Kubelka–Munk theory. I. Theory and application. *J. Opt. Soc. Am. A,* 21, 10, p. 1933-1941.

Ljungberg, L.Y. (2007). Materials selection and design for development of sustainable products. *Materials & Design,* 28, 2, p. 466-479, ISSN 0261-3069.

Manzini, E.; Vezzoli, C. (2002). *O desenvolvimento de produtos sustentáveis.* EDUSP, ISBN 85-314-0731-1, São Paulo.

Margets, R (2002). DEFRA, Annual report of the Government-Industry Forum on Non-Food Uses of Crops. London: Department for Environment, Food and Rural Affairs.

Meadows, D.H.; Meadows, D. L.; Randers, J.; Behrens III, W. (1972). *The Limits to Growth: A Report to the Club of Rome.* Available from: <http://www.clubofrome.org/docs/limits.rtf>, Accessed in 06/10/2007.

Miller, W. S.; Zhuang, L.; Bottema, J.; Wittebrood, A. J.; De Smet, P.; Haszler, A.; Vieregge, A. (2000). Recent development in aluminium alloys for the automotive industry. *Materials Science and Engineering A,* 280, 1, p. 37-49. ISSN 0921-5093.

ORGANISATION FOR ECONOMIC CO-OPERATION AND DEVELOPMENT (2001). *Environmental Outlook,* OECD, ISBN 92-64-18615-8, Paris.

ORGANISATION FOR ECONOMIC CO-OPERATION AND DEVELOPMENT (2007). *Cutting Transport CO2 Emissions: What Progress?.* European Conference of Ministers of Transport, ISBN 92-821-0382-X, Paris.

ORGANISATION FOR ECONOMIC CO-OPERATION AND DEVELOPMENT (2008). *Environmental Outlook to 2030.* OECD, ISBN 978-92-64-04048-9, Paris.

Papanek, V. (1971). *Design for the Real World: Human Ecology and Social Change.* Pantheon Books, ISBN 0-394-47036-2, New York.

Papanek, V. (1995). *The Green Imperative: Natural Design for the Real World.* Thames and Hudson, ISBN 0-500-27846-6, London.

Puglia, D.; Biagiotti, J.; Kenny, J.M. (2005). A Review on Natural Fibre-Based Composites: Application of Natural Reinforcements in Composite Materials for Automotive Industry. *Journal of Natural Fibers,* 1, 3, p. 23-65. ISSN 1544-0478.

Ray, D.; Sarkar, B. K. (2001). Characterization of alkali-treated jute fibers for physical and mechanical properties. *Journal of Applied Polymer Science,* 80, 7, p. 1013-1020.

Rouison, D.; Sain, M.; Couturier, M. (2006). Resin transfer molding of hemp fiber composites: optimization of the process and mechanical properties of the materials. *Composites Science and Technology,* 66, 7-8, p. 895-906, ISSN 0266-3538.

Schumacher, E.F. (1989). *Small is beautiful: economics as if people mattered,* Harper & Row Publishers. ISBN 0060916303, New York.

Schmidt-Bleek, F. (2008). Factor 10: The future of stuff. *Sustainability: Science, Practice, & Policy.* Available from: <http://ejournal.nbii.org/archives/vol4iss1/editorial.schmidt-bleek.html.> Accessed in 12/03/2008.

Stamm, B.v. (2003). *Managing innovation, design & creativity.* John Wiley & Sons Ltd, ISBN 0-470-84708-5, West Sussex.

Suddell, B.C; Evans, W. J. (2002). A Survey into the Application of Natural Fiber Composites in the Automotive Industry. In: Fourth international symposium on natural polymers and composites (ISNaPol), p. 455-1, Brazil, 09/2002, Sao Pedro.

Sydenstricker, T. H. D; Mochnaz, S; Amico, S. C. (2003). Pull-out and other evaluations in sisal-reinforced polyester biocomposites. *Polymer Testing,* 22, p. 375-380.

Tukker, A.; Eder, P.; Charter, M. (2000). *Eco-Design: European State of the Art - Part I: Comparative analysis and conclusions.* Available from: <http://ftp.jrc.es/EURdoc/eur19583en.pdf>, Accessed in 04/07/2007.

Uexküll, J.v. (1982). Jakob von Uexküll´s The theory of meaning. *Semiotica,* 42, 1, 25-82, ISSN 0037-1998.

United Nations (1987). *Report of the World Commission on Environment and Development: Our Common Future.* Available from: <http://www.un-documents.net/wced-ocf.htm>, Accessed in 12/06/2008.

United Nations (1992). *Agenda 21: report of the United Nations conference on environment and development.* Available from: <http://www.un.org/esa/dsd/agenda21/index.shtml>, Accessed in 10/23/2008.

United Nations (1996). *Indicators of Sustainable Development: Framework and Methodologies.* Available from, <http://www.un.org/esa/sustdev/csd/csd9_indi_bp3.pdf>, Accessed in 15/06/2008.

United Nations (2002). *Report of the World Summit on Sustainable Development.* United Nations. ISBN 92-1-104521-5, New York.

United Nations (2006). *World Population Prospects The 2004 Revision: Analytical Report.* Department of Economic and Social Affairs. ISBN 92-1-151409-6, New York.

United Nations (2009). *WATER IN A CHANGING WORLD*. UNESCO, ISBN 978-9-23104-095-5, United Kingdom.

Westerlind, B.S.; Berg, J.C. (1988). Surface energy of untreated and surface-modified cellulose fibers. *Journal of Applied Polymer Science*, 36, 3, p. 523-534, ISSN 1097-4628.

Weizsäcker, E.v.; Hargroves, K.; Smith, M. H. (2009). *Factor four: doubling wealth, halving resource use.* Earthscan, ISBN 978-1-84407-591-1, London.

15

Are Skill Design Structure Matrices New Tools for Automotive Design Managers?

Jean-Pierre Micaëlli[1] and Éric Bonjour[2]
[1]*Université de Lyon, INSA Lyon, ITUS Research Team,*
1, rue des Humanités, F-69621 Villeurbanne Cedex
[2]*FEMTO-ST – AS2M Institute*
24, rue Alain Savary, F-25000 Besançon
France

1. Introduction

The 2000s have been marked by significant change both in the nature of the vehicle and in its design process. The car satisfies an ever-present need. It longitudinally, autonomously and safely carries a reduced number of passengers and goods. In the future, acceptable vehicles must achieve requirements like reliability, safety, drivability, low gas consumption, minimal environmental footprint, low cost... The satisfactory solutions the designers are expected to offer can not be considered as pure mechanical systems. They integrate coupled functional modules that are embodied in multi-physical components (mechanical components, electronic or electrical devices, embebbed software...). Their design requires skills that are "new" from the automotive design managers' viewpoint. The issue concerning the identification, evaluation, building and modelling of skill networks opens promising ways for researchers and practitioners. Thus the purpose of this chapter will be to define the concept of skill network and to explain how it can be mapped by using "Design Structure Matrices" (DSMs) (Browning, 2001).

This chapter will be illustrated with an example concerning a French automaker's design office (Bonjour & Micaëlli, 2010). It aims at developing vehicle organs. These ones concern the powertrain system and the chassis. Their life cycle exceeds two decades. Thus designing them consists in developing a product family compliant with different platforms, models or generations of vehicles. These organs are mass produced. Since 1997, the mentioned design office of over 5,500 designers has been structured according to the systems engineering principles and processes (ISO 15288). Depending on the project, its technical activity partially or totally covers the Vee cycle. In this chapter, we shall not address the issue of skill network identification, mapping and building from a global viewpoint, but from a local one focused on the intermediate layer of the Vee cycle, namely the design of functional architectures. We therefore describe how the skill networks related to this task are restructured.

The remainder of this chapter is structured as follows. Section 2 defines the concept of skill network. It also proposes a conceptual framework including close concepts (job position, profession, core competence...). Section 3 outlines examples of skill networks and the structuring principles. Section 4 presents the principles that help to identify and structure skill networks. Section 5 describes the proposed method and its application in the case of a powertrain design office and finally, section 6 discusses perspectives concerning the use of this approach for developing specialized knowledge and related skill networks.

2. What is a skill network?

Since Wheelwright and Clark's work (1992), design managers have considered that matrix management complies with the organization of design activities. A design office is seen as a structure combining a portfolio of design projects and a portfolio of skills. Project managers bundle several teams in a given project and different skills in a given team. It has also become usual for researchers to develop sophisticated methods optimizing design project scheduling or team building. Little work has been done to explicit the concept of skill. Authors have most often an impoverished vision of this notion. A skill would be a stock of "commodities" corresponding to the knowledge workers store in their brain (Gherardi, 2007). The project manager would pick up required skills in this stock, as does the consumer to a supermarket shelf. The project manager would have in mind a good deterministic model. For such a project, for such a list of requirements, he/she perfectly knows the kind of tasks, teams, skills, internal designers or providers required to achieve it. He/she would behave as an "arbitrageur" (Lachmann, 1986). He/she would balance the value and the cost of each skill and assess the ability level of his design office. If a given skill has a poor value, if the design office ability is low, and if the "transaction costs" (Williamson, 1985) are low, then he/she will outsource it. Otherwise, he/she will behave in miser. He/she will consider the skill as a core competence. He/she will protect it the best he/she can. Table 1 represents the project manager's alternatives that extend the 'make or buy' choice. It shows an option in which he faces with a dilemma. Another option is also very difficult to manage. It concerns the situation in which co-design and long-term partnership with a supplier is required. This situation tends to become dominant in the automotive industry.

Design Office ability	Value of the skill	
	Low	*high*
low	If the transaction costs are low, then outsource the skill (buy)	If the transaction costs are low, then co-develop the skill
high	Dilemma	Develop the skill (make)

Table 1. Project manager's make or buy choice.

The vision of the project manager as arbitrageur is based on three implicit assumptions: 1- the designed product is modular, 2-the content of projects, tasks and teams' work can be completely defined, 3- skill is a static entity. According to the pattern described above, a bundle of skills develop a single module, and this module satisfies an isolated set of requirements. Therefore, the project manager can easily define the required skills and coordinate them. If the product architecture is "integrative" (Sosa et al., 2003), then the organizational problem he/she copes with is more complex. Skills integration in design projects does not consist in buying and using separated skills. The project manager must create and implement workplaces facilitating intra-team and inter-teams learning. His main question is not: what skills to buy? But: how to facilitate shared "exploration" (March, 1995) of the design problem and cross-learning between teams or skills (Lester & Piore, 2004)? What is important then for him/her, is not the static attributes of the skill (what is made for? Who possesses it?? What is its level of expertise?), but its evolutionary ones (what is its potential of learning? How to change it during the project? By linking with what other skills?). This list of questions leads the manager to rethink the notion of skill. It can no longer remain ill-defined. The way we suggest is to define a set of separate entities closed to the notion of skill:

- a skill is defined as a functional concept. It consists in specialized knowledge "owned" by a design actor, who can be a designer or a design manager. Its main attributes is its function,
- a profession is understood as an evolutionary concept. It can be seen as an evolving set of specialized knowledge possessed by individuals, shared by a working community (or skill network) - socially recognized -, and re-built due to long-term processes. The profession gives them a common and perennial identity. They see themselves as a set of peers,
- a job position defines the workplace in which the design actor performs his/her "working activity" (Engeström, 1987) in a given organization called design office, project team or design department,
- a skill network is both an evolutionary and a functional concept. It consists in a community within which expertises are developed. The boundaries of a community may be contained in those of the firm (bounded community) or be more extensive (boundless community),
- a professional path describes the potential transfer from one job position to another one.

Every organizational entity of a design office can be seen as a skill network. Thus a design actor may participate to three skill networks: the project team, the department, and the "community of practice" (Wenger et al., 2002). The design manager can formalize a skill network as a functional department and as a project team. The project team is an organizational entity that has a limited life. Its goals are operative. They are focused on short or medium term. The department is a perennial organizational entity. It can support a "design core competence" (Bonjour & Micaëlli, 2010). The life of a community of practice may exceed that of a team or a functional department. This structure is also less hierarchical and fuzzier than a formalized organization (Wenger et al., 2002), e.g. a project team or a design department.

The design manager can bundle designers within these different skill networks by using various criteria (see later). DSMs promoters underline the fact that this tool can be used to structure the design office (Sosa et al., 2004). They pay their attention to the operative level. Each designer has a job position corresponding to a role he plays into a project team. Therefore an Organization_DSM can be used to map the expected flow of data between job positions. It is helpful to bundle job positions within a same team and reduce coordination costs within the project. A job position is also responsible for performing expected tasks. A Process_DSM can model the precedence between tasks within a design process. This last variety of DSM can minimize the feedback loops in the project. A design task contributes to

the design of a given component. A Product_DSM can represent the interfaces between components. One clusters them into modules in order to minimize the interfaces within the system. These three types of DSM have been commonly studied in the research community working on the DSMs (Browning, 2001).
Fig. 1 depicts the conceptual framework we propose to better understand the concept of skill network. This framework is presented by means of the Unified Modeling Language (UML) class diagram. The left part of the figure shows the cognitive, communitarian and organizational entities previously mentioned (profession, job position, skill, skill network, community of practice, design office, project team, department…). Its right part contains the design management domain (design manager, DSM…).
Until now, DSMs have not been used to cluster specialized knowledge according to their cognitive proximity, in order to identify relevant skill networks. The design manager can use the results provided by these tools in at least three ways. Firstly, he/she may decide to perpetuate the identified skill network as a functional department. Secondly, he/she may encourage experts to build a community of practice to explore a particular problem. Thirdly, he/she may suggest his/her company and its suppliers to consider the institutional conditions for recognition of a new profession. In all cases, he/she does not act directly on the knowledge 'owned' by individuals. He/she structures their future working activities. Table 2 can help him/her to identify the skill networks within the design office he manages. This table reuses the entities of the "model of the activity" proposed by Engeström (1987).

Question	Entity	Current Case Study
Who possessed the skill?	Design actor and his community	System architect
Who recognizes the skill?	Alter	Design manager
What is its content?	Object	To develop a powertrain system satisfying key requirements
What does the professional?	Tasks	Requirements analysis and architectural design
What are his tools?	Tools	Systems engineering standard, models, software
What is the content of his specialized knowledge?	Knowledge, expertise	Architecture principles, technological knowledge that enables to link requirements and architectural elements
Who are his peers?	Community	Systems engineering community
With what other communities are the links?		Communities of specialists of embodiment design, software engineers…
In which organizational entity does his job position fit?	Division of labor	Design office, department, team…

Table 2. Activity-based definition of a skill network.

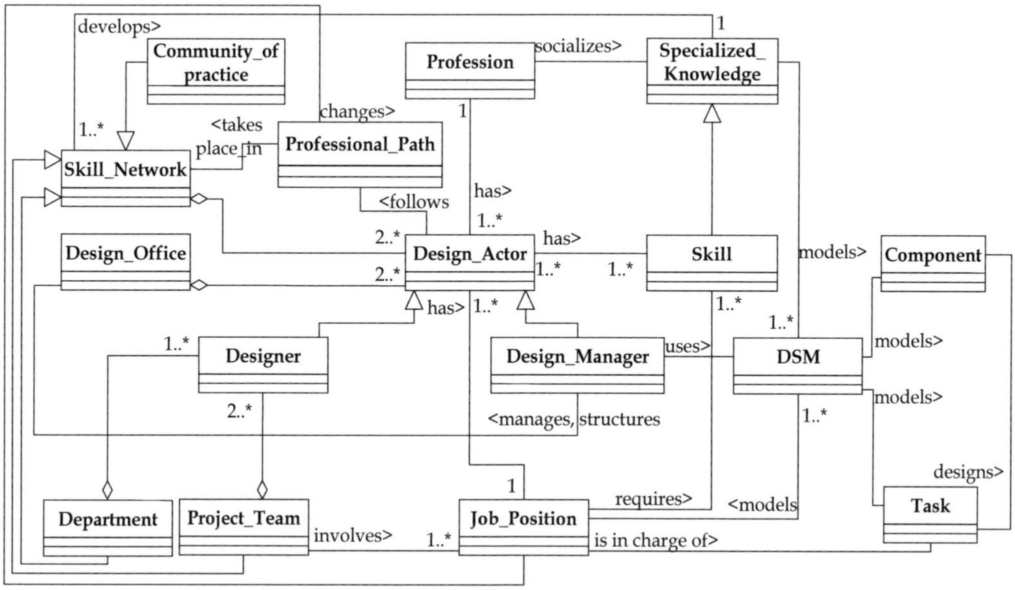

Fig. 1. Skill network conceptual framework.

3. Examples of skill networks

In 2003, the design office that was in charge of the chassis design was decomposed into functional departments. The design manager used two criteria to structure this organization:
1. the nature of the design task,
2. the nature of the designed object.

For instance, a first functional department was responsible for the architectural design (design task) of chassis (designed object). Another one was responsible for the integration and validation of chassis. Chassis were seen as a system composed of axles, suspensions, steering and brakes.

Therefore, functional departments were responsible for the architectural design of axles, of suspensions, power steering and brakes. Other departments were in charge of the integration and validation of these organs. The department related to the architectural design of axles was composed of the following job positions: axle system architect, modelling and simulation of mechanical systems, CAD (Computer Aided Design) definition (profession known as projector), specification of simulations and physical tests for risk mitigation, team leader. A project team was composed with designers who played these different roles. These job positions were found in other departments such as the architectural design of the suspension. Of course, some differences exist between functional departments, due to the difference between designed objects.

Only one skill network related to CATIA V5 was recognized through these two functional departments. Its purpose was to develop specialized knowledge related to the recognized profession (projector) who was in charge of the CAD definition. Several reasons may explain this fact. Firstly, design managers underestimated the need for boundless skill networks.

Secondly, they did not recognize the other job positions as professions. Thirdly, their knowledge was not judged important enough to justify that designers spent time to create and develop communitarian knowledge. The harmful consequence was that the experience feedback throughout the chassis projects was rather poor. Moreover, we diagnosed another delicate situation. Acoustics was specialized knowledge to help designers meet a key requirement related to the customers' comfort (reduction in noise and vibrations). An expertise in this domain requires at least a decade of experience. But the turnover that was imposed to engineers led to a dissemination of the skilled individuals. An effective community of practice, with leaders, experts, junior engineers, apprentices, should have been built and reinforced.

A second example of skill networks identification can be given. Within the design office that is in charge of the design of powertrain and chassis, we can cite the following job positions: requirements analysis leader, system architect (responsible for system architectural design), design project manager. These job positions are linked to systems engineering processes (ISO 15288). Within the functional department that is responsible for the powertrain system design, the design actors form a recognized skill network. Its purpose is to develop world class knowledge in powertrain engineering: specification, architecture, modelling and technical synthesis (acoustics, chemistry...), integration and validation of powertrain. Career paths (syn. professional paths) within this skill network are possible across these job positions.

Within this design office, the profession of project manager has been also officially recognized. A specific department, called engineering management, has been formed in order to use and to develop specialized knowledge related to project management at the system level. Different names have been attributed to these job positions, e.g. product-process pilot. He/she coordinates the design of sub-systems. According to the system decomposition level, different job positions have been identified. Project leaders intervene at the sub-system level. Team leaders operate at the level of the components. Project leaders and team leaders are assigned to functional departments. Together they form a community of practice. Last but not least, professional paths exist between those job positions.

4. Skill network mapping

How to put Skill_DSM into practice? A skill network is supported by something which is shared by several design actors. It may be a designed object (engine, gearbox, chassis…), a design task (requirements analysis, architecture, validation…), a disciplinary field (chemistry, acoustics, reliability, project management…), a shared-cost tool (CAD, test benches…)… Expressed differently, all design activity entities (designed object, design task, disciplinary field, tool…) can be used as skill network identification criteria. The design manager can use one or the other. A single well-defined criterion does not exist. If he/she adopts a bottom-up approach, then he/she will consider first the profession. If he/she adopts a top-down approach, then he/she will consider first the designed objects, the design tasks or the tools.

If one returns to the example of the design office responsible for chassis development, one can see that its design manager has followed the following steps to structure it:

- the chassis was divided into several functional modules (product breakdown structure). Thus the design manager adopted an object-based approach (top-down approach),
- the design tasks were defined following Systems engineering standard (ISO 15288),
- the job positions were both defined following Systems engineering and automotive professional standards,
- each functional department was defined by mapping a module to a set of tasks, so a set of job positions.

This organizational design facilitated "dialogue" (Lester & Piore, 2004) between different designers sharing a same object, i.e. a given functional module. However, this design world (skill network) was separate from the validation world that was responsible for physical tests and chassis design evaluation. The main criterion that explained this separation was related to cost-shared tools. It has a major drawback. Designers were acting in a virtual world. They make little connections with the physical world. A community of practice was created (but it was not a boundless community) and professional paths were facilitated between these two worlds to mitigate this drawback. "Engineering liaisons" (Bonjour & Micaëlli, 2010) roles or job positions were clearly defined in some design departments, for instance, specification of simulations and physical tests for risk mitigation (see the example 1 above).

5. Skill network reengineering

Once skill network identification criteria are adopted, it is then possible to create what we call a Skill_DSM. We propose a method for identifying knowledge clusters which are relevant to build new departments, teams or communities of practice.
This method is structured into the following steps:

- list the design tasks,
- estimate the cognitive proximity between tasks by estimating the knowledge or the methods shared by designers. The proximity is estimated on a scale [0, 10],
- build the corresponding numerical DSM matrix,
- apply a clustering algorithm to highlight clustered tasks,
- interpret and check the consistency of each cluster as an interesting skill network.

Data are obtained through interviews with design managers, project managers and experts. The managers are more oriented towards the identification of departments. The experts are more interested in identifying communities of practices.
We applied the previous method to depict the skill networks related to the functional architecture of hybrid powertrains. Fig.2 shows a real size Skill_DSM. For privacy reasons, the picture of this DSM was blurred (empty cells are equal to 0).
Several interpretations of this DSM can be made.
Firstly, one can be focused on its static aspects. Each module depicts a closed skill network the design manager can recognize as a functional department or a team. For example, the fifth cluster represents the functional department responsible for a key requirement of powertrains: reductions in polluting emissions in compliance with Euro VI regulation. Expertises, routines and specialized knowledge belonging to this skill network contribute to a current automaker's design core competence (Bonjour & Micaëlli, 2010). This skill network is based on specialized knowledge related to design (functional design, fuel specification…),

to chemistry (fuel chemistry, combustion, catalysis...), to purchases and outsourcing (partnerships with exhaust pipe suppliers...)... The presented DSM also points out potential job positions related to engineering liaisons between this cluster and the cluster 2 (another potential skill network).

Secondly, one can extract some evolutionary phenomena from this matrix. It shows professional paths within a given skill network or between skill networks. These paths lead to three different types of knowledge:

- a narrow and deep expertise belonging to a specific cluster (syn. skill network),
- an expertise in engineering liaison,
- an expertise in integrative knowledge.

Integrative knowledge is a knowledge that is common to almost all the other knowledge in a given cluster. A novice can manage few specialized knowledge whereas an expert can navigate between different knowledge related to the same cluster.

Those different interpretations of the Skill_DSM show how this model proposes a very rich semantics.

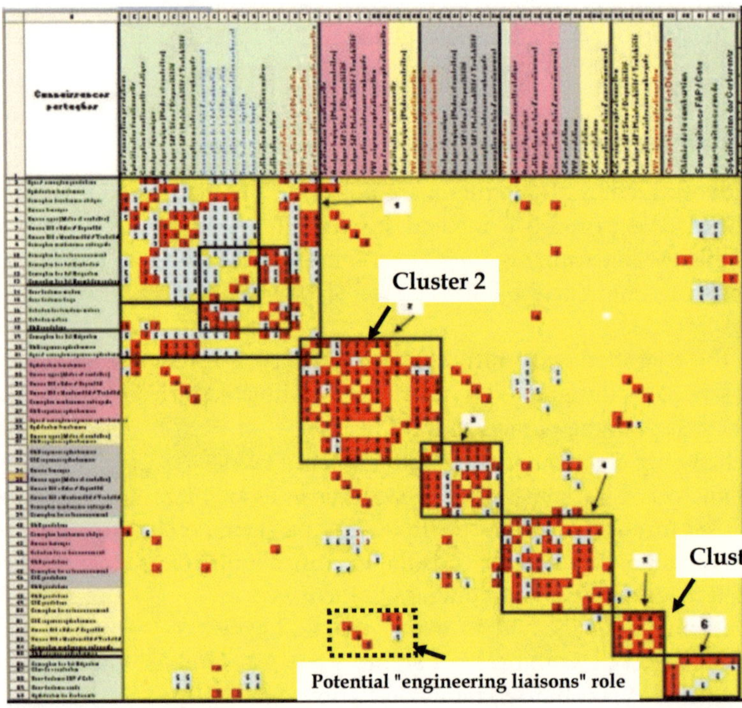

Fig. 2. Example of a Skill_DSM.

6. Perspectives

We have proposed a bottom-up approach to help design managers to identify potential key skill networks by using Skill_DSM. However, a top-down approach could be envisaged. It consists in analysing firms' design core competence and determining which skill networks could enhance skills, abilities or routines that largely contribute to core competence. This

approach should be developed to provide design managers a global skill network management approach. It is based on identification, structuring and evaluation tools.
This chapter has outlined the way of identifying potential skill networks. Its aim has not been to evaluate their contribution to core competencies. This lack is paradoxical because DSMs are primarily managerial tools and not only optimization-based representations. The main question is not: how to optimize such clustering algorithms to cluster such DSMs? But rather: What services do these tools offer to the concrete design managers' "activity" (Engeström, 1987)? Managerial issues that are related to this key question concern design dialogies (two characteristics which are contradictory and must be considered at the same time): Can they use Skill_DSM to balance the division of labour and the coordination between skill networks, the operative performance of the design project and the skills or competences development, the "exploitation" of existing skill networks and the "exploration" to create new boundless communities (March, 2008)? Can design managers use DSMs to integrate benchmarking and best practices? Can they use them to stabilize professional paths or to facilitate the evolution of professions?
Thus numerous extensions of skill DSMs are necessary to improve their integration in concrete design offices.

7. Acknowledgments

The authors would like to thank the design managers of the automaker's design office for their fruitful collaboration.

8. References

Bonjour, É., Micaëlli, J-P., (2010). Design Core Competence Diagnosis: A Case from the Automotive Industry. *IEEE Transactions on Engineering Management*, Vol. 57, N° 2, 323–337.

Browning, T-R., (2001). Applying the design structure matrix to system decomposition and integration problems: a review and new directions. *IEEE Transactions on Engineering Management*, vol. 48, 292–306.

Engeström, Y., (1987). *Learning by Expanding: An Activity-Theoretical Approach to Developmental Research*. Helsinki, FIN, Orienta Konsultit.

Gherardi, S., (2007). *Organizational Knowledge: The Texture of Workplace Learning*. Malden, MA: Blackwell Publishing, 2007.

Hamel, G., & Prahalad, C.K. (1994). *Competing for the Future*. Boston, MA: Harvard Business School Press.

International Standard Organization (ISO), (2000). *15288 Standard*. Geneva, CH.

Lachmann, L-M., (1986). *The Market as an economic Process*. Oxford, UK: Basil Blackwell.

Lester, R., & Piore, M., (2004). *Innovation: The Missing Dimension*. Cambridge. MA: Harvard University Press.

March, J-G., (2008). *Explorations in Organizations*. Stanford, CA: Stanford Business Book.

Sosa, M.E., Eppinger, S.D., & C. Rowles, (2003). Identifying modular and integrative systems and their impact on design team interactions. *Transactions of the ASME Journal of Mechanical Design*, N°125, 240-252.

Sosa, M.E., Eppinger, S.D., & C. Rowles, (2004). The misalignment of product architecture and organizational structure in complex product development. *Management Science*, Vol.50, N°12, 1674–1689.

Wenger, E., McDermott, R., & Snyder, W-M., (2002). *Cultivating Communities of Practice: A Guide to managing Knowledge*. Boston, MA: Harvard Business School Press.

Wheelwright, C., & Clark, (1992). *Revolutionizing Product Development: Quantum Leaps in Speed, Efficiency, and Quality*. New York, NY: The Free Press.

Williamson, O-E., (1985). The Economic Institutions of Capitalism: Firms, Markets, Relational Contracting. New York, NY: The Free Press.

Part 5

Materials: Analysis and Improvements

16

Effects of Environmental Conditions on Degradation of Automotive Coatings

Mohsen Mohseni, Bahram Ramezanzadeh and Hossain Yari
Department of Polymer Eng. and Color Tech.,
Amirkabir University of Technology
P.O.Box 15875-4413, Tehran,
Iran

1. Introduction

Two main goals are expected when coatings are applied to substrates. The main one is protection of substrate from various aggressive environments such as sunlight and humidity. The second is to impart color and aesthetic to the substrate to be coated. In some applications such as automotive coatings, these two are highly important. Exposure for a long time to different permanent (sunlight, rain & humidity) and occasional (acid rains and various biological substances) parameters during the service life of these coatings results in loss of performance. Such phenomena not only render the coating to degrade also lead to depreciation of appearance attributes of the finished car. Automotive coatings are usually multi-layered systems in which each layer has its predefined function. These make the whole system resist to various environmental factors. Figure 1 shows a typical automotive coating system.

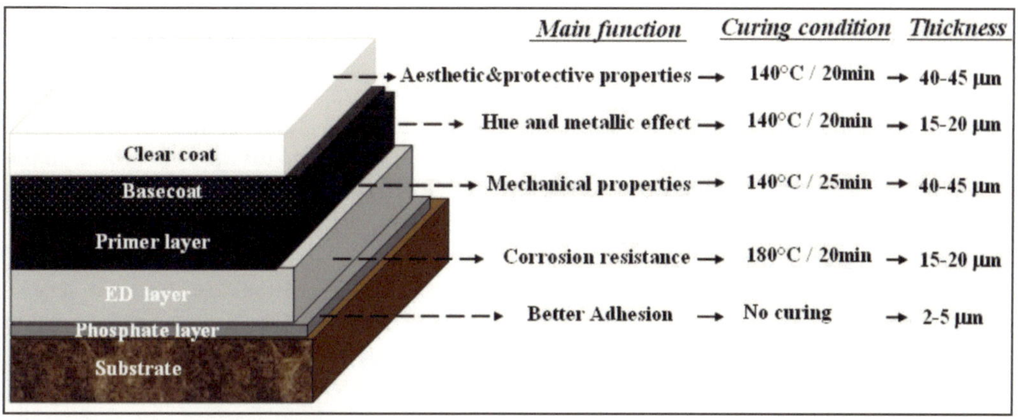

Fig. 1. Specifications of a multilayer automotive system

As figure 1 describes, the substrate is initially coated by a conversion layer such as phosphate or chromate to enhance the adhesion and corrosion protection of the metallic substrate. Then, an electro deposition (ED) coating, usually based on epoxy-amine containing anticorrosive pigments and zinc powders, is applied to protect the coating from corrosion. The primer surfacer which is a polyester melamine coating is then applied. The main function of this layer is to make the coating system resist against mechanical deformations such as stone chipping. The color and special effects, such as metallic luster are obtained using a basecoat layer which is typically an acrylic melamine resin pigmented with metallic and pearlescent pigments. To protect the basecoat, a non-pigmented acrylic melamine clear coat is applied over this layer. This latter layer is responsible for the gloss and smoothness of the coating system. On the other hand, the clear coat, apart from creating a highly glossy surface, is intended to protect the underneath layers, even the substrate, against various aggressive weathering (i.e. humidity and sunlight) and mechanical (i.e. mar and scratch) factors during service life.

It should be noted that all layers are applied when the previous layer has dried, except for the clear coat that it is applied through a wet-on-wet method in which it is applied on the wet basecoat layer after a short time for flashing off the solvents. The curing processes of all layers are presented in figure1.

In order to fulfill the required properties, automotive coating systems are required to remain intact during their service life, because they are extremely vulnerable to deteriorate (Nguyen et al., 2002 a; b; 2003; Yari et al., 2009a). There are various environmental factors which can potentially be fatal for these coatings and may cause loss of appearance and protective aspects of the system. The consequences of these factors are discoloration, gloss loss, delamination, crack propagation, corrosion, and gradually building up coating degradation. Acid rain, hot-cold shocks, UV radiation, stone chips, car washing, fingernail and aggressive chemical materials are among those parameters rendering the coatings to fail in short and/or long exposure times to environment. These would lead to dissatisfaction of customers. Therefore, it is vital to enhance the resistance of the coating against environmental factors.

In the following part of this chapter, different environmental conditions and their effects on various aspects of coating have been presented. Preventive methods will be given where necessary. Among the environmental factors, the influence of biological materials will be explained with more details because their effects have not been discussed elsewhere.

2. Environmental factors

Environmental factors are those substances or conditions imposed by the environment to which the automotive coatings are exposed. As such, different chemical and/or mechanical alterations (degradation) may result. Here, they have been divided to three main subcategories, i.e.; mechanical, weathering, and biological factors.

2.1 Mechanical damages

Automotive coatings can be encountered different outdoor conditions during their service life. Mechanical objects can put severe effects on these coatings. Depending on the type of imposed stress to these coatings various kinds of degradation can be observed (Shen et al., 2004). The most important of these can be seen in Figure 2.

2.1.1 Chipping resistance

The ability of multi-layer automotive coatings to withstand against foreign particles without being damaged is named stone-chip resistance. It is found that, when stone particles attack a coating they have velocity near to 40-140 km/h. This can cause coating delimitation from the

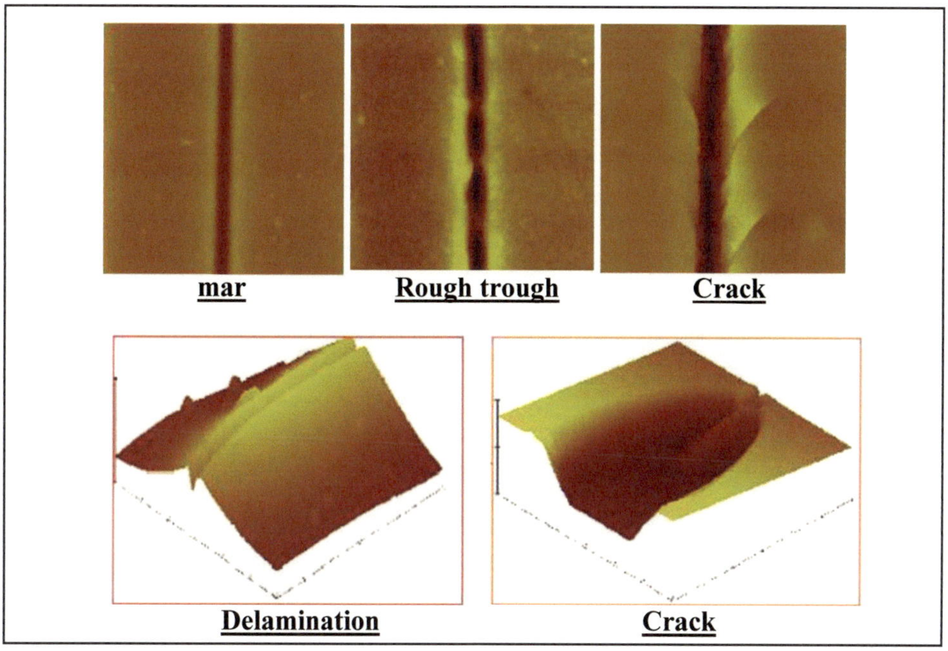

Fig. 2. Different type of mechanical damage occurring on automotive coatings (Shen et al., 2004).

paint-substrate interface (Lonyuk et al., 2007; Buter & Wemmenhove, 1993). For multi-layer system, coating layers interadhesion, coatings mechanical properties and coating interaction to substrate are the most important factors affecting chip-stone resistance. These can make the chipping resistance of these systems very complicated. It has been demonstrated that, the mechanical properties of each layer can affect their chip resistance. In this regard, it has been found that glass transition temperature of the primer layer is the main factor controlling coating chipping resistance. The greater glass transition temperature may cause adverse performance. The temperature at which this measurement is conducted is also very influential. The failure appeared during chipping in a multi-layer coating system can be both adhesive and/or cohesive failure. It was found that when the strength between two layers exceeded, the defect was mainly adhesive failure. As a result of this, delaminating, flaking or peeling will occur. On the other hand, crack initiation and propagation within a coating layer across the other layers can cause cohesive failure (Lonyuk et al., 2008) (Figure 3).

2.1.2 Abrasion resistance

Basecoat/clear coat systems create an outstandingly high glossy appearance in comparison to other automotive paint systems. However, such a high gloss makes mechanical damages more visible when they appear. Scratch and mar are the most important of these failures. They are micrometer deep surface damages that may ruin the initial appearance of automotive finishes. The difference between mar and scratch is mainly in their different sizes and morphologies. Scratch is a consequence of tribological events encountered by automotive clear coats. The size for this type of damages is 1-5 µm (Courter, 1997; Tahmassebi et al., 2010). To show how these types of damages influence coating appearance, the visual performance of coating before and after scratching are shown in Figure 4.

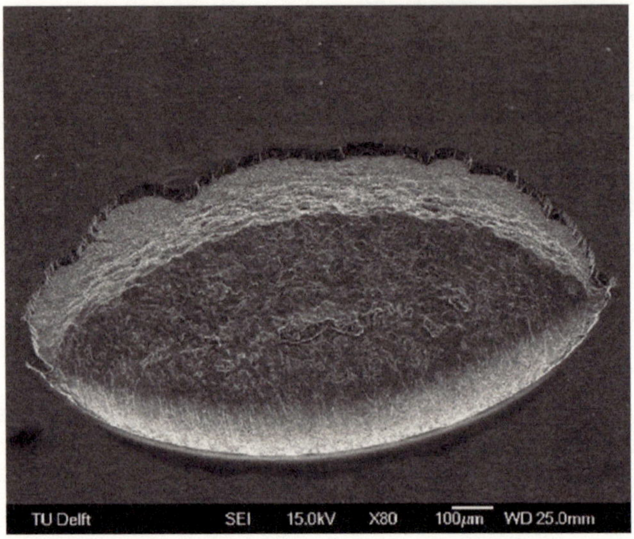

Fig. 3. The SEM micrograph of the chipped surface of coating (Lonyuk et al., 2008).

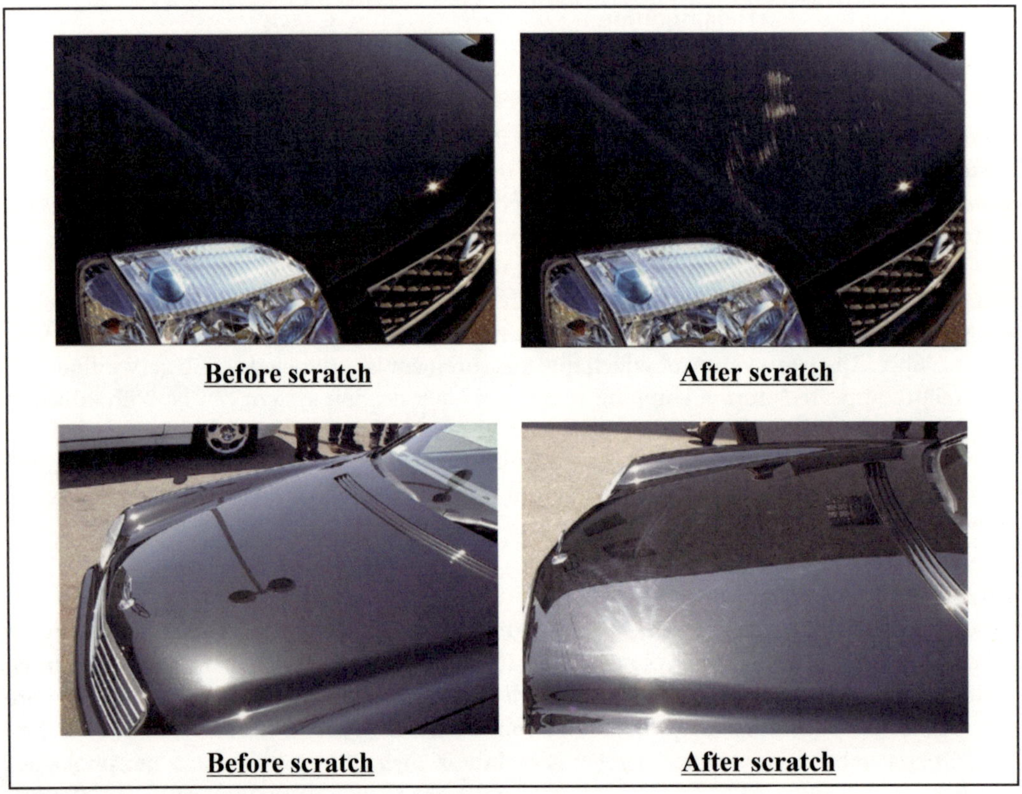

Fig. 4. Visual differences of automotive coating before and after scratching.

Mechanical damages of these types may be caused by polishing equipments, carwash bristles, tree branches and sharp objects such as keys (Tahmassebi et al., 2010).

2.1.3 Scratch type

The performance of automotive coatings is further complicated by nature of the created scratches, which in turn is influenced by the viscoelastic properties of the clear coat itself, and the conditions under which they are created. In this regard, when an external stress is applied to coating, there would be three different kinds of coating responses: elastic deformation, plastic deformation and fracture deformation (Tahmassebi et al., 2010; Lin et al., 2000; Hara et al., 2000). Elastic deformation has limited effect on the appearance of a coating, therefore determination of plastic and fracture deformation seem more important. Some scratches are irregular and of a fractured nature (Figure 5-a) and may involve material loss, while others are smooth (Figure 5-b), regular and involve plastic deformation of clear coats (Lin et al., 2000; Ramezanzadeh et al., 2010; Jardret & Morel, 2003; Jardret & Ryntz, 2005; Jardret et al., 1998).

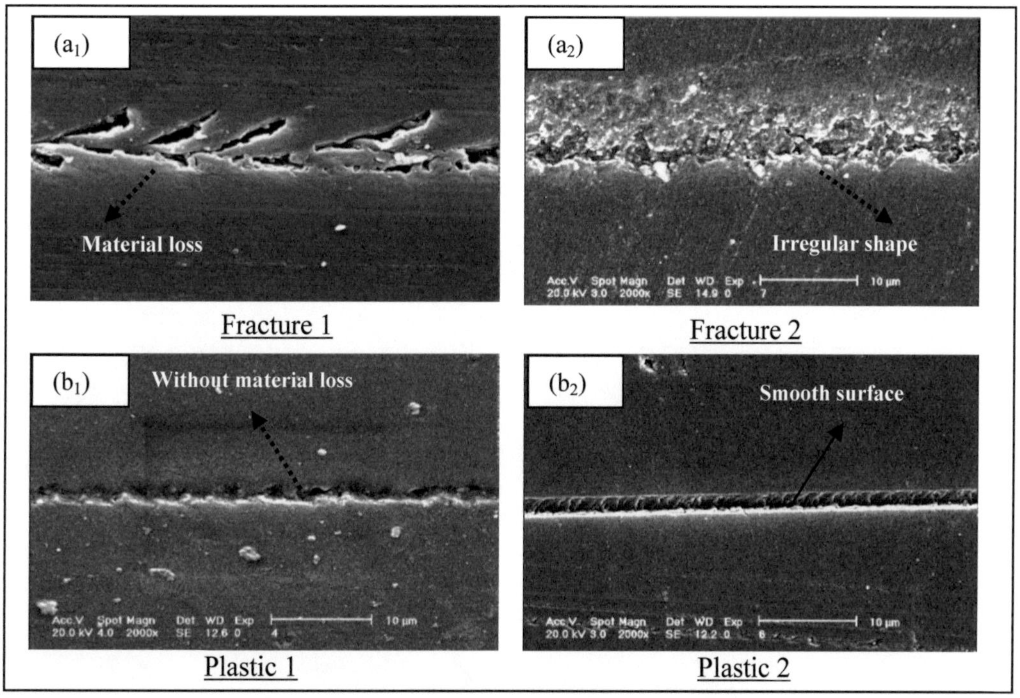

Fig. 5. SEM micrographs of two types of (a) fracture and (b) plastic scratches (Tahmassebi et al., 2010; Ramezanzadeh et al., 2010).

Various parameters such as scratch force, scratch velocity and environmental temperature would influence the type and form of scratch produced.

There are many differences between these two types of scratches. First, fracture types are irregular and may involve material loss (Figure 5-a), while others are smooth, regular with no material loss (Figure 5-b). The visibility of fracture-type scratches is independent on the direction of incident light and illumination. Conversely, plastic-type scratches are not visible if the longitudinal direction of the scratch coincides with the direction of the lighting. These differences are schematically shown in Figure 6-a and b (Lin et al., 2000).

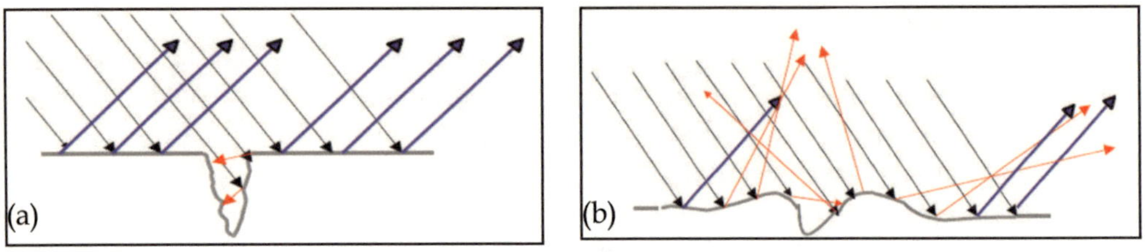

Fig. 6. Schematic illustration of (a) fracture and (b) plastic type's scratches

Elastic or plastic behaviors of a clear coat result in spontaneous or retarded recovery of the created scratches, respectively. This is usually named as healing ability of clear coat. Fracture behavior, on the other hand, arises from tearing apart of polymer chains contained within the clear coat, therefore recovery or healing of the created scratches would not be possible. The mechanism by which scratch can be formed by a scratch indenter are shown in Figure 7 (Hara et al., 2000).

According to figure 6, different parameters like indenter tip morphology (tip radiance and stiffness), tip velocity and coating viscoelastic properties affect the coating response against applied stress. As shown in this figure, applied force can be divided into tangential and vertical vectors. Tangential forces cause compression and stretching in the clear coat in front and behind of such particles, respectively. Tensile stresses produced behind such particles can cause cracks in the clear coat and/or aid in scratch formation. Consequently, the tensile stress/strain behavior of clear coats can be used to predict scratch behavior. This phenomenon has been shown by Jardret and Morel in detail (Jardret et al., 2000; Jardret & Morel, 2003).

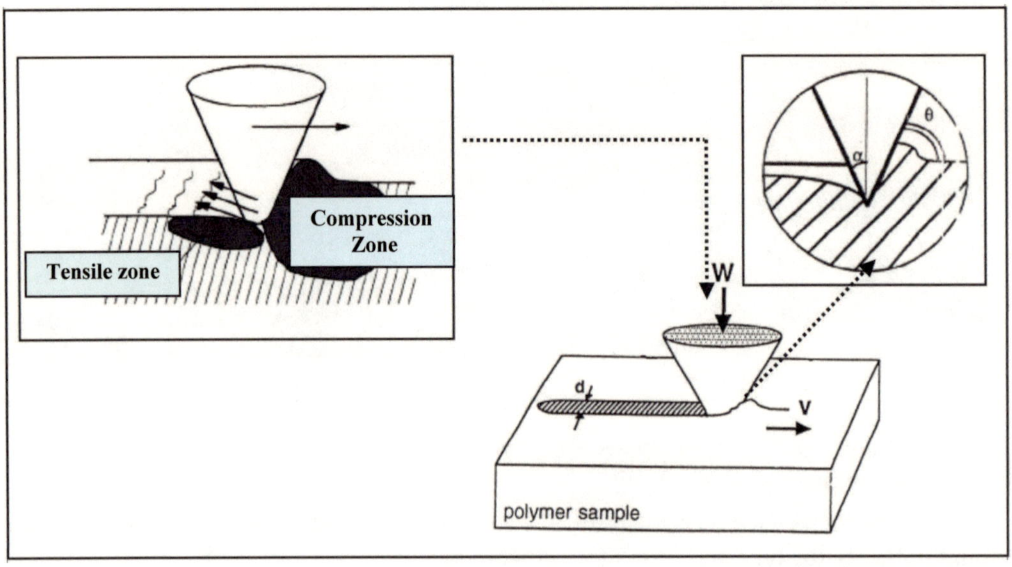

Fig. 7. Schematic illustration of how scratch indenters affect coating deformation type (Hara et al., 2000).

2.1.4 Methods to improve coating scratch resistance

Based on the above explanations, improving scratch resistance and variations in scratch morphology are of utmost importance in the research and development departments of the automotive finishing industry. Accordingly, researchers have proposed various methods for improving the scratch resistance of automotive clear coats. The proposed methods include procedures to increase surface slippage and hardness, as well as enhancing cohesive forces within clear coats that modify the viscoelastic properties of clear coats as a whole. Increasing surface slippage and hardness inhibit the penetration of scratching objects into clear coats, thereby increase the force necessary to create scratches. If forces generated by scratching objects exceed that of the cohesive forces within a clear coat, then polymer chains of the clear coat tear apart and show a fracture-type (Hara et al., 2000). There are many methods to improve coating viscoelastic properties including changing clear coat chemistry and using different pigments (in both nano and micro size) and additives (like polysiloxane additives). However, changing the chemical structure of a clear coat would not guarantee modification of its viscoelastic properties. Furthermore, changing the chemical structure of a clear coat may incur unwanted adverse effects on other properties of the resultant clear coat and will in most cases, increase its price. Consequently, attempts have been made in many research programs to modify viscoelastic properties by physical incorporation of various additives into a clear coat of known chemical structure. Controlled use of these additives could ensure minimization of unwanted variations in other properties of the resultant clear coat as well as being an attractive and economically viable alternative (Tahmassebi et al., 2010; Ramezanzadeh et al., 2010; Zhou et al., 2002; Ramezanzadeh et al., 2007; Ramezanzadeh et al., 2007; Jalili et al., 2007).

2.1.5 Methods to evaluate coating scratch resistance

Several methods have been used to evaluate the scratch and mar resistance of clear coats. Scratch-tabber is one of the most traditional used methods for analyzing coating scratch resistance. This method can predict coating scratch resistance based on the weight loss of coating during scratch test (Lin et al., 2000). Laboratory car wash simulator is another method which has been used in recent years. This is a useful method based on an appropriate simulation from a real scratching process in an outdoor condition (Tahmassebi et al., 2010). Nano and micro-indentation are powerful methods to evaluate both scratch resistance and morphology of coating. In addition, use of these methods could be favorable for analyzing clear coat scratch resistance, deformation type of the clear coat (plastic or fracture) and viscoelastic properties (Tahmassebi et al., 2010). Gloss-meter and goniospectrophotometer have been used to evaluate the effects of scratches produced on the appearance of clear coat (Tahmassebi et al., 2010). Microscopic techniques including optical, electron and atomic microscopes have been used to investigate scratch morphology.

2.2 Weathering factors

Weathering factors are those that are applied to the coating by weathering (or climate), and cause alteration in chemical structure (Nguyen et al., 2002 a; b; 2003, Bauer, 1982), affecting various aspects of the coating properties such as physical (Osterhold & Patrick, 2001), mechanical (Tahmassebi & Moradian,2004; Nichols et al., 1999; Gregorovich et al., 2001; Nichols & Darr, 1998; Nichols,2002; Skaja, 2006) and electromechanical (Tahmassebi et al., 2005) properties. The severity of degradation caused by weathering factors depends strongly on climatic condition. Sunlight and humidity are the most important weathering factors. It is almost impossible to prevent automotive coatings being exposed to sunlight.

2.2.1 Sunlight

Sunlight reaching the earth contains a wide range of wavelengths from 280 to 1400nm (Valet, 1997). The most harmful part is the uv range (less than 380 nm). Most polymers are sensitive to this part of the sunlight. For example polyesters and alkyds have absorption peaks around 315 and 280-310 nm, respectively (Valet, 1997). The absorbed energy can cause a kind of degradation called "photodegradation", the mechanism of which is known and has been extensively discussed in litreatures (Pospısil & Nespurek, 2000; Valet, 1997). A brief description of photodegradation is given here. The absorbed energy by some chromophoric groups (ch) of the polymer turns it to an excited state (ch*). This excited state is able to induce formation of various free radicals. The following equations present different free radicals produced during photodegradation.

A) Initiation

$$\text{Polymer (p)} \xrightarrow{\text{Sunlight}} \text{Free radicals } (P\bullet, PO\bullet, HO\bullet, HOO\bullet, \ldots)$$

B) Propagation

$$P\bullet + O2 \longrightarrow POO\bullet$$

$$POO\bullet + PH \longrightarrow POOH + P\bullet$$

$$PO\bullet + HO\bullet \longrightarrow POOH$$

$$2\,POOH \longrightarrow PO\bullet + POO\bullet + H_2O$$

$$PO\bullet + PH \longrightarrow POH + P\bullet$$

$$PH + HO\bullet \longrightarrow P\bullet + H_2O$$

C) Termination

$$P\bullet + P\bullet \longrightarrow P\text{-}P$$

$$POO\bullet + P\bullet \longrightarrow POOP$$

$$P\bullet + PO\bullet \longrightarrow POP$$

$$PO\bullet + PO\bullet \longrightarrow POOP$$

As a consequence, chain scission and formation of various stable and unstable spices such as peroxide, hydroperoxide, hydroxyl and carbonyl groups are the most important reactions involved in photodegradation. Formation of different polar species leads to an increase in surface energy of the coating (Tahmassebi & Moradian, 2004). These produce hydrophilic groups in the coating and increase the susceptibility for water diffusion. Finally, this leads to greater potential of underneath layer to be corroded.

2.2.2 The effect of basecoat pigmentation

Due to significant role of the clear coat on weathering and mechanical properties of automotive coatings, most of the previous studies have focused on an isolated clear coat layer. But there are reasons to believe that the basecoat greatly affects the weathering performance of its attached clear coat. In order to illustrate how a basecoat could vary the weathering performance of a clear coat, it is necessary to clarify how a basecoat reacts to incident light. As stated before, common basecoat contains colored pigments and/or metallic flakes. Colored pigments absorb and/or scatter incident visible light reaching the bulk of a basecoat, according to their color, size and refractive index. Metallic flakes, based on their level of orientation, reflect and/or scatter incident light only at the surface of the clear coat. In this manner, fractions of returned incident light passing through the clear coat are decisive in causing chemical changes in the clear coat structure, leading to alterations in the clear coat properties.

In order to elucidate the influence of basecoat pigmentation on degradation of a typical automotive clear coat during accelerated weathering tests, using two different basecoats (i.e. silver and black) can be useful. Amongst common commercial basecoats, silver and black seem to be two extreme basecoats. In other words, a silver basecoat is characterized by the presence of high loads of aluminum flakes (acting as a reflective source of visible light), and a lack of colored pigments, in which the chance of reflecting incident light is high and the chance of absorbing incident light is minimal. While the black basecoat, is characterized by the presence of high loads of a black pigment (acting as an absorbent of visible light), and a lower load of aluminum flakes; this means that the reflection or scattering chances of incident light are low and its absorption is high. Figure 8 schematically shows how two different basecoat pigmentations react to incident light.

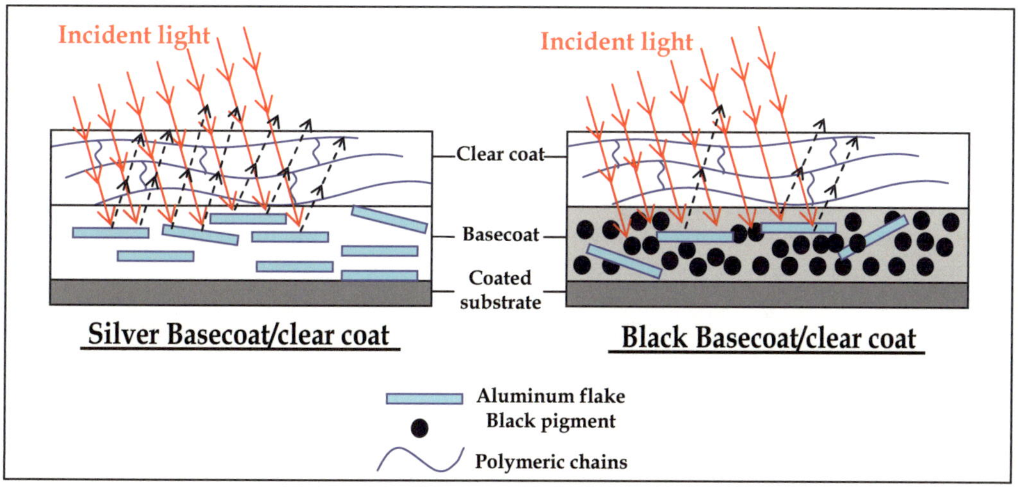

Fig. 8. The reaction of two different basecoat pigmentations to incident light.

Therefore, these two basecoats seems to be two extreme examples in their reaction to incident light. Other basecoats, depending on their ability to reflect or absorb light could be ranked to be somewhere between the black and silver.

The rate of variations in carbonyl groups of a coating during weathering can in fact be considered as the photodegradation rate of that coating (Mielewski et al., 1991). Figure 9 shows normalized absorbances of carbonyl bands of clear coats attached to silver or black basecoats.

It is clearly obvious that the photodegradation rate of the clear coat having a silver basecoat is greater than that of the black one during weathering. Such results indicate the higher ability of silver basecoat to induce photodegradation reactions in the clear coat during weathering exposure (Yari et al., 2009a).

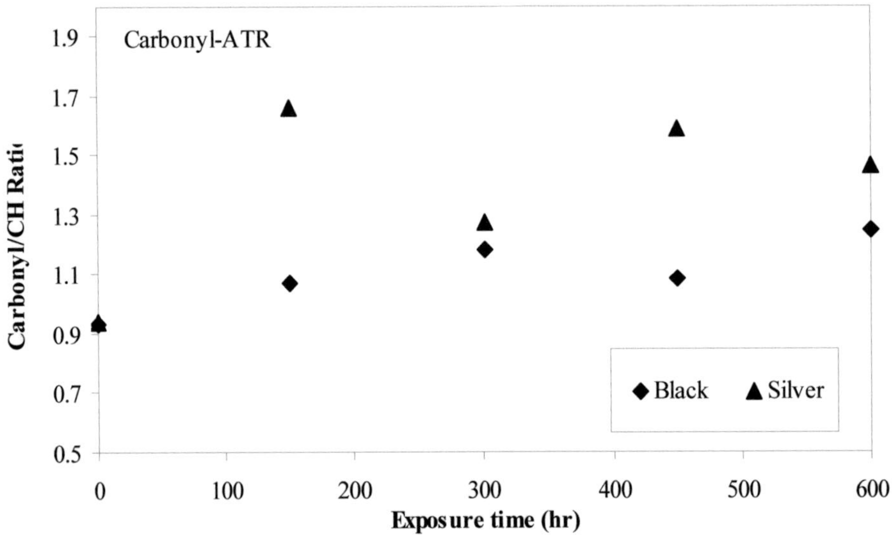

Fig. 9. Normalized absorbances of carbonyl bands of clear coats attached to silver or black basecoats.

Various approaches are available for lower photodegradation mechanisms given above. The first method is to prevent the UV rays from being reached the coating chromophores by adding substances which are able to strongly absorb and filter the UV wavelengths (Valet, 1997; Bauer, 1994). These materials are called Ultra Violet Absorber (UVA). The conventional UVAs are benzotriazoles, triazines and bezophenones. Nowadays, by advances obtained in nanotechnology, new generation of materials have been achieved that not only are capable to absorb UV rays, but also can improve the mechanical, thermal and electrochemical performance of the coating (Peng et al, 2008; Dhoke et al., 2009; Xu & Xie, 2003) . The best choices for this purpose are titanium dioxide, zinc oxide, cerium oxide, iron oxide or even silica nanoparticles. Because of the high surface area of these nanoparticles the absorption efficiency of these materials has been promoted considerably. Figure10 shows AFM topographic images of two acrylic melamine clear coats containing 0 and 3.75% nanosilica after 1000 hours exposure times (Yari, 2008).

Figure 10 also clearly reveals that the most variations is assigned to neat polymer while nanocomposite tolerates less variation in surface topology, meaning less weathering degradation. This indicates that incorporation of nano silica into acrylic melamine not only has not any effect on weathering durability, it enhances its resistance during weathering. The better weathering performance of clear coats containing nanosilica is assigned to the ability of nano silica particles to absorb the ultra violet and visible light, resulting in less degradation in nano silica-containing clear coats (Jalili, 2007; Zhou, 2002).

Another preventive strategy for improving the resistance of coatings against photodegradation is the use of quenchers and radical scavengers. Quenchers are materials that can transfer the excited state of ch* to themselves. They then become excited. Their excited state is not able to produce free radicals. Radical scavengers convert the active free radicals to inactive ones and are unable to participate in photodegradation reactions. Hindered amine light stabilizers (HALS) are the most typical kinds of additives for this purpose(Bauer et al., 1992; Seubert, 2003; Mielewski et al., 1993). Synergestic effect of HALS and UVA have made a significant improvement in photostability of the coatings.

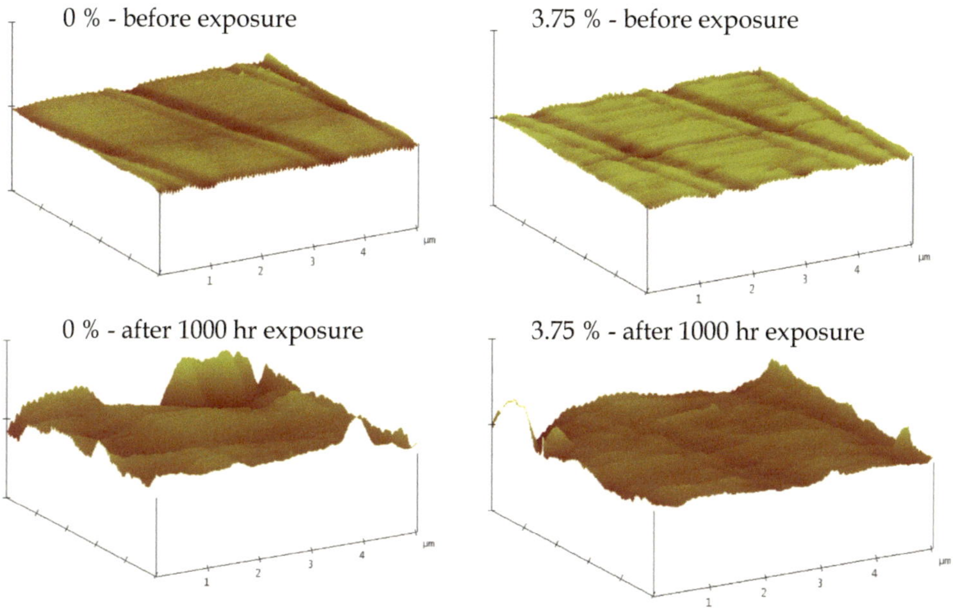

Fig. 10. AFM topographic images of different clear coats after various exposure times.

2.2.3 Water and humidity

Raining, car-washing, and dew formation are conditions by which water is in contact with automotive coatings during its service life. While, most polymers are hydrophobic and are not affected by water and humidity, some polymers that have water-sensitive linkages in their structure can be hydrolyzed by water or humidity. Acrylic/melamine as the most typical structure used in automotive clear coats, is vulnerable to water and well susceptible to hydrolytically degrade.

Figure 11 depicts different reactions happening in hydrolytic degradation of a typical acrylic melamine.

In these hydrolytic degradations, various etheric, esteric and methylene bridges are broken, creating various OH&NH-containing products, i.e. methylol melamine and primary or secondary amines (Nguyen et al., 2002 a; b; 2003). Meanwhile, other reactions called self-condensation reactions occur between methylol melamine groups present either in initial structure of clear coats or formed during early times of reactions. As a result of self-condensation reactions, different melamine-melamine linkages i.e. new methylene or etheric bridges (reactions c and d in figure 11) are formed. These new formed linkages have less flexibility than the initial linkages. This results in a higher glass transition temperature.

It has been demonstrated that chemical structure (like the ratio of acrylic/melamine or polyol/isocyanate) and cross-linking density of the clear coat have a significant impact on the intensity of the hydrolytic degradation (Yari et al., 2009b). The lower the cross-linking density, the greater is water permeation and blister formation. The assessment of the resistance of the coating against humidity is carried out by saturated humidity test. The results of blister formation and the visual appearance of two different coatings (with high and low cross-linking densities) are shown in Figure 12.

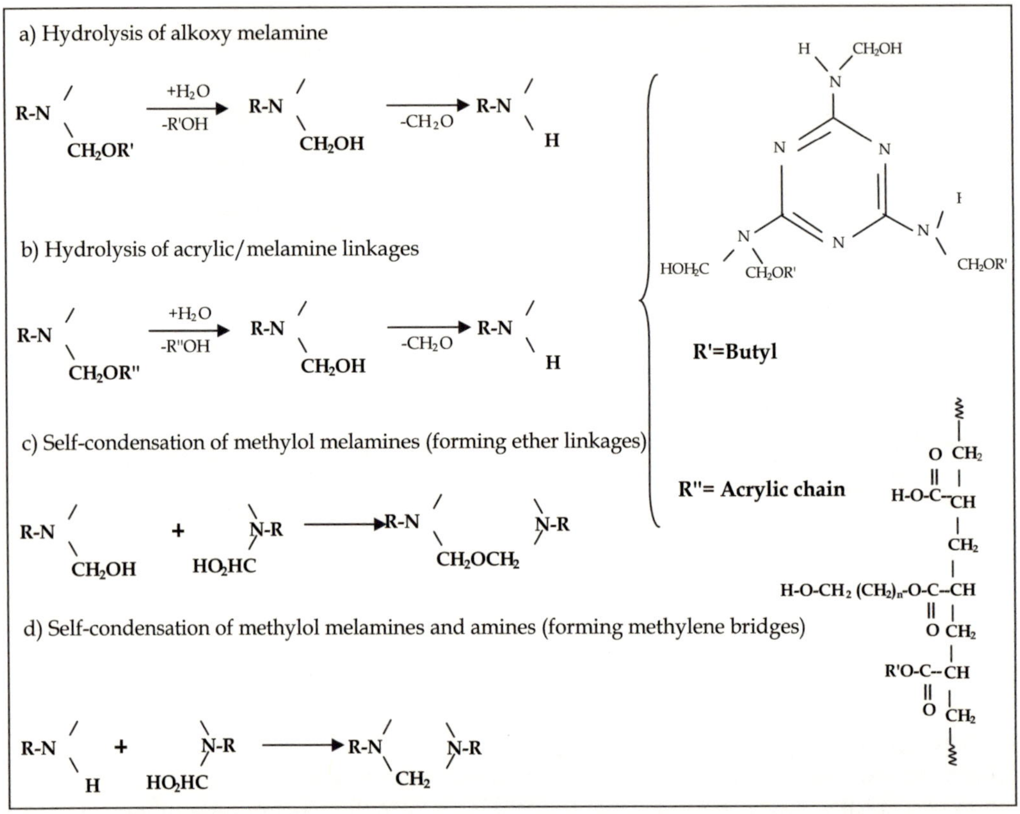

Fig. 11. Degradation and self-condensation reactions for a typical acrylic melamine.

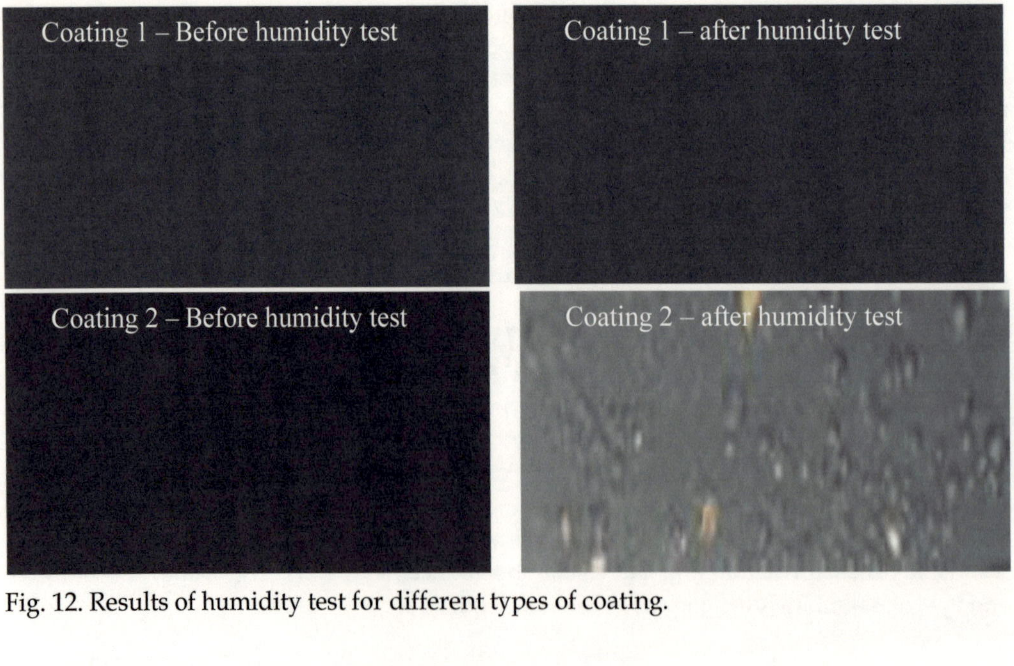

Fig. 12. Results of humidity test for different types of coating.

Coating 1 is an automotive type with high cross-linking density (v_e= 0.002673 mol/cm^3) and coating 2 is the same one with lower cross-linking density(v_e= 0.000486 mol/cm^3). In contrary to coating1, which shows no blistering, severe blisters are seen on the surface of coating2. Blistering is a result of diffusion of water and other soluble materials into coating.

2.2.4 Acid rain

Acid rain which is a very common phenomenon in urban and industrial areas is a catalyzed type of hydrolytic degradation. Acidic environment catalyzes the hydrolysis reactions. Various gases like SO_2 produced in the polluted areas are converted into sulfuric acids which makes the precipitates acidic. These acidic rains when fall on the coatings catalyze the hydrolysis reaction of acrylic melamine clear coat. The acid catalyzed hydrolysis has been investigated in several works (Mori et al., 1999; Schulz, et al, 2000; Palm& Carlsson, 2002). It has been found that the acid rain and the acid catalyzed hydrolysis are most likely to occur at moderate to strong acidic environments. For example, the results reported by Schulz and co-workers (Schulz, et al, 2000) showed that, the pHs of a real acid rain even at the aggressive environments (Jacksonville, Florida) lied in the range of 3.5-4.5. Acid rain etches the acrylic melamine and strongly decreases the coating surface.

Different strategies can be adopted to increase the hydrolytic resistance of an acrylic melamine coating; decreasing the ratio of melamine, use of hydrophobic chains, decreasing melamine solubility, decreasing the basic strength of melamine and partially replacing of melamine with other amino resins.

2.3 Biological materials

Biological materials are those substances produced from bio sources. These are the most important environmental factors which affect the chemical, mechanical and visual performance of automotive coatings. These mainly include insect bodies, tree gums and bird droppings. Whilst, the influence of sunlight, humidity and acid rain on automotive coatings, especially on clear coat has been studied thoroughly, the effect of biological materials has not been dealt with in more details. In this regard, an automotive coating is repeatedly exposed to different biological materials such as bird-droppings, tree gums and insect bodies. Therefore, the investigation of the influence of such materials and the coating degradation mechanism seems inevitable. Stevani and co workers (Stevani et al., 2000) studied the influence of dragonfly eggs, a native insect of north and south America, on an acrylic melamine automotive clear coat. They found that hydrogen peroxide released during hardening of eggs, oxidizes the cysteine and cystine residues present in the egg protein, leading to the formation of sulfinic and sulfonic acids. The acids produced catalyze the hydrolytic degradation.

2.3.1 Bird droppings

In different papers, the effects of bird droppings on appearance and thermal-mechanical properties of coating have been investigated (Ramezanzadeh et al., 2009; Ramezanzadeh et al., 2010 a). Typical defects observed on the clear coats influenced by bird droppings were investigated by different techniques as shown in figure13 (Ramezanzadeh et al., 2010 a; Yari et al., 2010).

The optical microscope images of clear coats show that even at a relatively short exposure time to bird droppings and pancreatin, the clear coat surfaces have been etched severely.

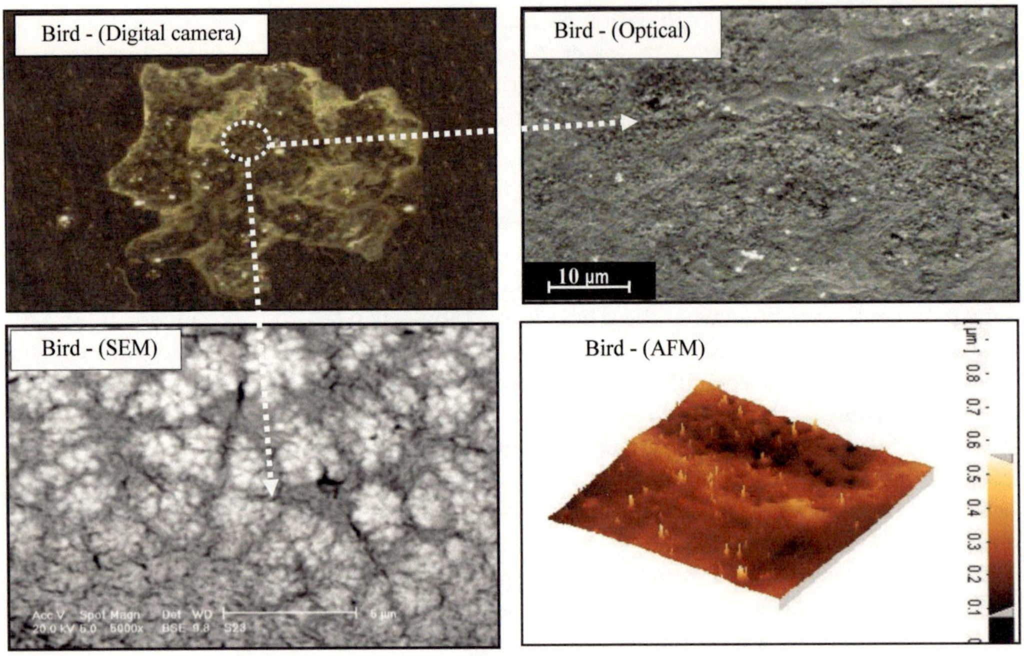

Fig. 13. Appearance of defects created after being exposed to bird droppings.

These images may confirm that chemical reactions have occurred at the surface, leading to dissolved and etched areas. It was found that bird droppings decreased the appearance parameters of clear coat, i.e. gloss, distinctness of image (DOI) and color values, therefore affecting the aesthetic properties of the coating system (Ramezanzadeh et al., 2009). Thermal-mechanical studies also showed that hardness, glass transition temperature and cross-linking density of degraded clear coats decreased in the presence of bird droppings (Ramezanzadeh et al., 2010 a). Also, the influence of aging method (pre-aging or post-aging) and chemical structure of clear coats against such bio attacks, were reported (Ramezanzadeh et al., 2009; Yari et al., 2009 c) [11,12]. It was observed that post-aging process, which simultaneously exposes bird droppings and UV radiation to coatings, degraded the clear coat much more intensively than the pre-aging one, in which only bird droppings on pre-weathered clear coats was exposed (Ramezanzadeh et al., 2009). The investigation of clear coat chemistry revealed, that incorporating higher ratios of melamine cross-linker, in spite of resulting a higher cross-linking density, led to an inferior biological resistance (Yari et al., 2009 c).

Although the main process was a hydrolytic cleavage, it was also a catalyzed hydrolytic degradation. The mechanism of this bio-attack is shown in figure14.

It has been reported that bird droppings consists of amylase, lipase and protease which are all hydrolase enzymes and are responsible for cleavage of C-O-C (for example in starches), COO esteric linkage (for example in glycerin) and CO-NH peptide amide linkages (for example in proteins), respectively. Enzymes are amino-acid molecules that their function is to catalyze various chemical reactions in biological environments, e.g. in the human body or animals. The rate of most enzyme-catalyzed reactions is millions of times faster than those of comparable un-catalyzed reactions.

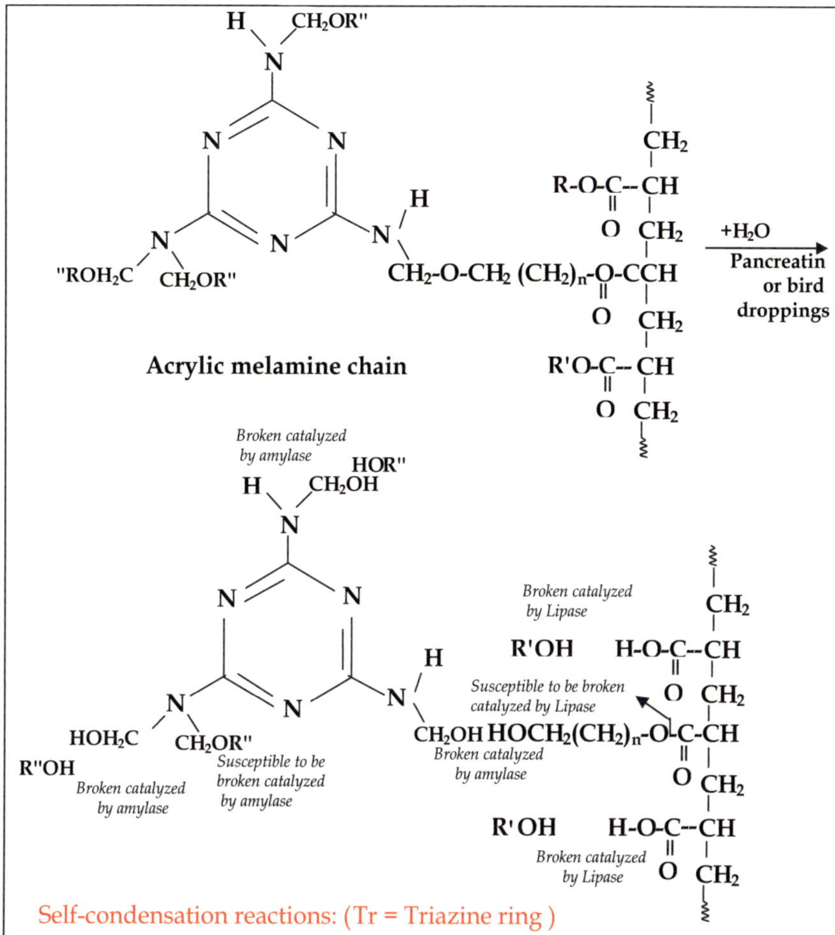

Self-condensation reactions: (Tr = Triazine ring)

A) $2\,Tr\text{-}NH\text{-}CH_2\text{-}OH \longrightarrow Tr\text{-}NH\text{-}CH_2\text{-}O\text{-}CH_2\text{-}NH\text{-}Tr$ (formation new etheric linkages)

Or

B)

B-1) $Tr\text{-}NH\text{-}CH_2\text{-}OH \xrightarrow{Fast} Tr\text{-}NH_2 + CH_2O$

B-2) $Tr\text{-}NH\text{-}CH_2\text{-}OH + Tr\text{-}NH_2 \rightarrow Tr\text{-}NH\text{-}CH_2\text{-}NH\text{-}Tr$ (formation new methylene bridges)

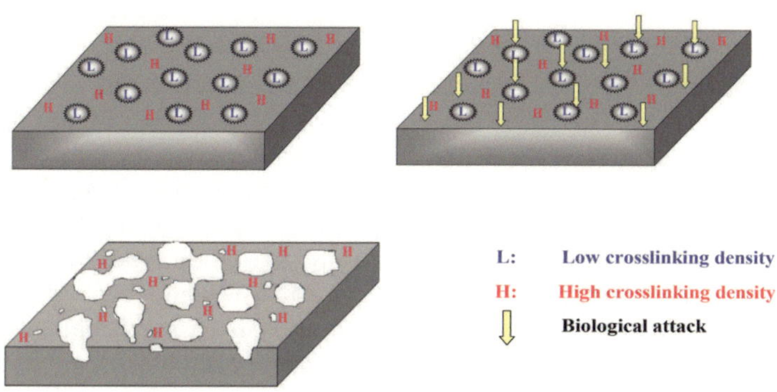

Fig. 14. Degradation Mechanism of a typical acrylic melamine caused by bird droppings.

After pancreatin or bird-droppings deposition on clear coat surface, the hydrolysis reaction can take place. The enzymes present in these materials catalyze the hydrolysis reaction. Among these enzymes, protease due to the absence of amide linkages (-CONH-) is relatively inactive on acrylic melamine. However, amylase and lipase enzymes act on etheric and esteric linkages respectively, accelerating the cleavage of these bonds. Due to the presence of high active sites (etheric and esteric linkages) in acrylic melamine, the cross-linked network is cleaved. This leads to formation of soluble products and releasing from the coating, leaving etched area on the surface. The clear coat consists of high cross-linking and low cross-linking regions. The latter are more vulnerable against hydrolytic degradations (Sangaj & Malshe, 2004) and are more affected.

As seen in SEM images of figure13, there are some micro cracks at degraded areas. This may be attributed to an ion-induced oxidation due to the presence of metal ions (Ratner et al., 1997).

Moreover, extensive studies on the similarity of bird droppings and pancreatin using X-ray fluorescence and Fourier Transform Infrared Analyses (Yari et al., 2010) showed that the chemical structures are generally similar. So same effects are created on coating after being exposed to bird droppings and pancreatin. Therefore, the use of pancreatin instead of natural bird dropping seems an alternative.

2.3.1.1 The effect of clear coat chemistry

The monomer types of acrylic resin, the functional groups of melamine cross-linker and the acrylic/melamine ratio, are the main factors which affect the curing (and inevitably its performance) in the resultant coating. However, due to the presence of esteric and etheric linkages in the structure of these resins, the occurrence of hydrolytic reaction seems probable, leading to inferior chemical and weathering resistance. It has been found that the chemistry of clear coat affects the coating performance against bird-dropping. It was shown that two acrylic melamine clear coats differing in melamine ratio had different resistance against bird dropping. Figure 15 shows the optical Images of two different partially methylated acrylic/melamine clear coat (Cl-1 and Cl-2) which only differ in acrylic:melamine ratios. Cl-1 has more melamine portion in its formulation.

The comparison of optical images of both clear coats shows that the Cl-1 undergoes more catastrophic etching compared to Cl-2. It may be attributed to higher portion of melamine component of Cl-1 which is more susceptible to hydrolysis reaction and therefore, a higher etching. whereas Cl-2 sample, with less amount of melamine, experiences lower etching (Yari et al., 2009 c).

2.3.1.2 The effect of basecoat pigmentation.

It has been demonstrated that basecoat pigmentation via affecting the efficiency of curing process of its attached clear coat influences the biological resistance of automotive coating system. In seeking the reason why the degrees of cure are different, the effect of pigmentation on heat transfer should be considered. In Figure 16 various mechanisms of heat transfer during the curing process are schematically shown (Ramezanzadeh et al., 2010 b).

The typical ovens used for curing of the coatings utilize hot air conditioning as well as IR lamps. It may also be expected that convection and radiation heat transfer are more important during such curing processes. The difference in curing behavior of clear coats attached to black and silver basecoats (two extreme basecoats) can be explained by emissivity factor of these basecoats. Emissivity factor of a material is the relative ability of its surface to emit energy by radiation. It is defined by the ratio of energy radiated by a

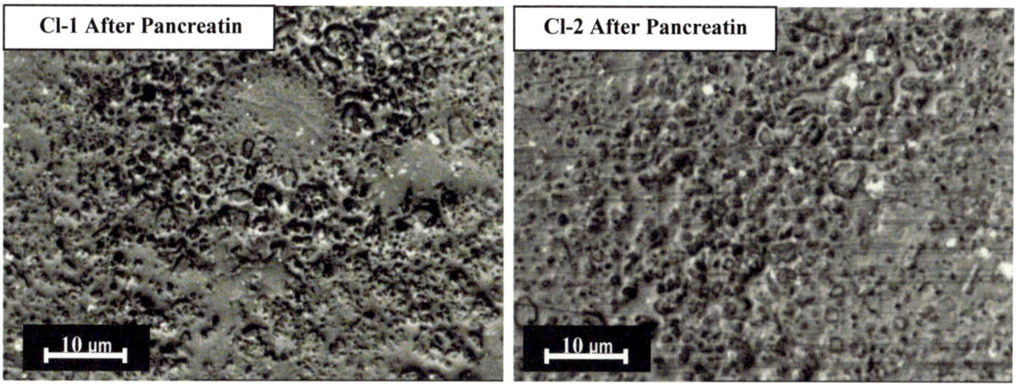

Fig. 15. Optical microscope micrographs of different samples differing in melamine ratio (Cl-1 has more melamine portion) degraded by pancreatin (or bird droppings) .

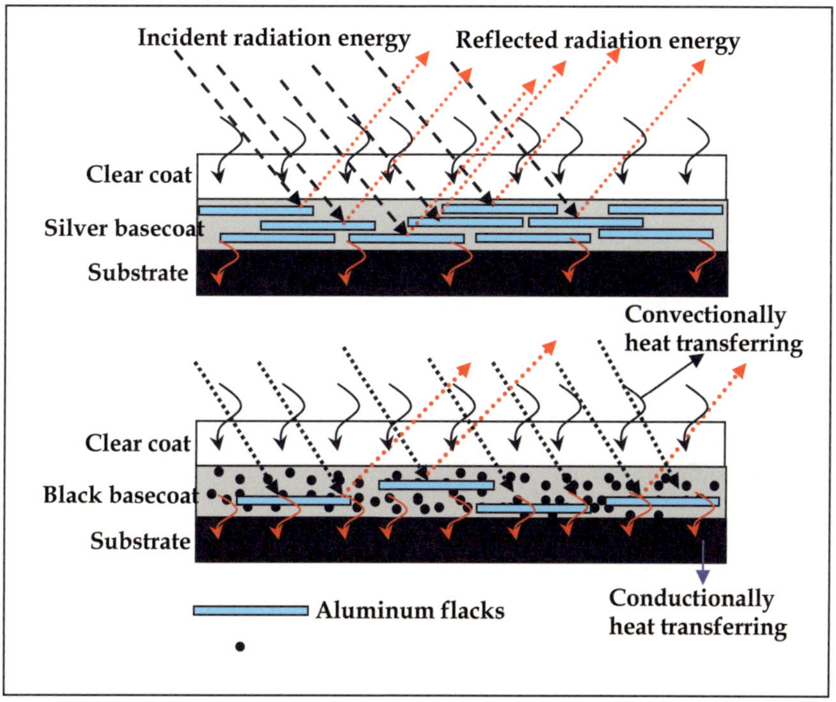

Fig. 16. Schematic representation of heat transfer during the curing process

particular material to energy radiated by a black body. According to Thomas (Thomas, 2005) the infrared emissivity factor of basecoats containing a typical carbon black pigment or a typical non-leafing aluminum pigment are 0.88-0.9 and 0.29-0.33, respectively. The greater the emissivity factor of a coating the lower is its infrared reflection. Additionally, it is highly likely that a silver basecoat would contain larger loads of an aluminum pigment compared to a black basecoat. Therefore, it is probable that the clear coat attached to a silver basecoat would be exposed to extra infrared radiation than that attached to a black basecoat. This extra energy may in turn induce a more complete degree of cure in the clear coat attached to a silver basecoat than that attached to a black basecoat.

Effects of basecoat pigmentation on visual performance of clear coats experiencing bird dropping attack can clearly be observed in Figure17.

Fig. 17. Optical micrographs of the clear coat samples on silver and black basecoats exposed to biological materials

It is seen that the more efficient curing on clear coat having silver basecoat results in better performance compared to that of having a black basecoat.

2.3.1.3 The effect of aging process

As the biological materials affect the coating both in aged samples (exposed to environment) and its freshly prepared form, the aging conditions used to study the effect of these materials included a pre-aging and post-aging. Pre-aging means that before the exposure of clear coat to biological attack a four-stage aging process is performed. This multi-stage aging is conducted according to PSA D27 5415 standard. The details of stages have been explained elsewhere (Ramezanzadeh et al., 2009). In summary, these stages are schematically presented in figure 18. In post-aging method, the coating is subjected to both aging and biological attacks simultaneously.

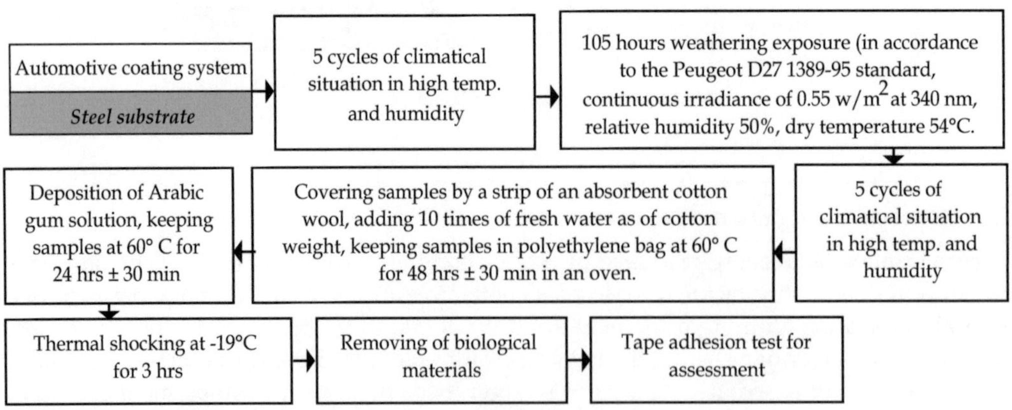

Fig. 18. A brief description of test method PSA Peugeot - Citroen D27 5415.

It was found that clear coats which have experienced simultaneous weathering and biological materials(post-aging) was more degraded than those being initially experienced weathering condition followed by exposure to biological materials. It is due to the intensifying effect of UV radiation as well as sunlight.

Different methods can be pursued to prevent from degradation caused by bird droppings. Making the surface more hydrophobic, using clear coat with fewer esteric or etheric reactions can also be useful.

2.3.2 Natural tree gum

It is a general belief, that cars should be better kept under the shadow of trees in order to prevent them from a direct sunlight exposure. However, in this case the effects of gums extracted from the tree may be simply neglected. This can be seen from Figure 19.

Fig. 19. Effect of Arabic gum on car body.

The visual effect which is produced under tree gum attack can be shown in Figure 20.

According to Figure 20, the clear coats exposed to natural tree gums show considerable surface cracks indicating a severe physical effect. In addition, etching behavior of both materials, shown itself as numerous holes on the surface, can be also observed in the case of this kind of degradation. According to the above explanation, the general effects of gums can be appeared in both physical and chemical on the surface of coating. In addition, SEM micrographs of this kind of degradation can reveal sub-cracks inside of macro cracks shown in optical images. It can be also found that, the affected area (inside the crack) is lighter than the unaffected parts of coating. This can illustrate different elemental composition inside and outside the cracks. This can, similar to bird droppings, reveal the presence of metal compounds in gums structure. This phenomenon will be discussed later. The increased roughness in nano-scale of degraded parts of coatings (exposed to Arabic gum) can be obtained from the AFM micrograph (Ramezanzadeh et al., 2010b).

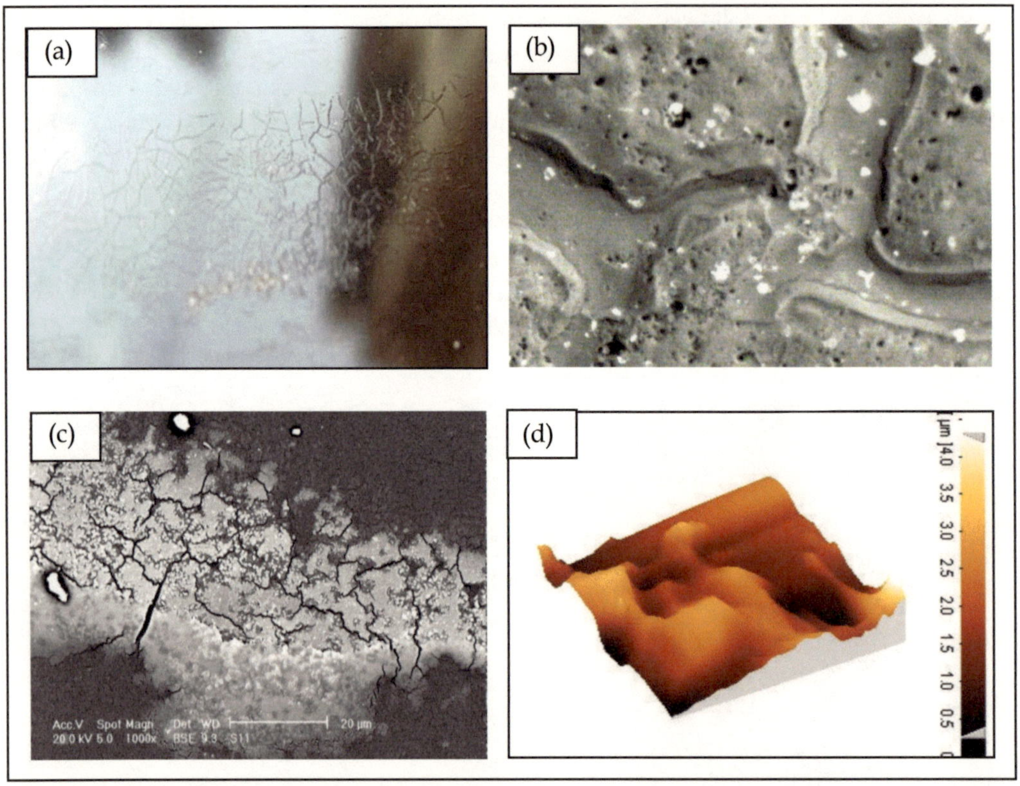

Fig. 20. Visual performance of coating after the tree gum attack, (a) visual performance, (b) optical image, (c) SEM micrograph and (d) AFM micrograph (Ramezanzadeh et al., 2010c).

2.3.2.1 Tree gum characteristics

Tree gums are completely soluble in water and have a sticky behavior in this state. The pH of this material is about 4.5 in a slurry state. Arabic gum has been used as a synthetic equivalent for tree gums. This was due to the similar acidic nature, physical state in water and their similar chemical structures (shown in Figure 21). However, many parameters i.e. soil nature, climatic condition (which trees grow there) and the type of tree can influence these characteristics (Ramezanzadeh et al., 2010b; Ramezanzadeh et al., 2010c).

The solubility of these gums can be explained by the presence of high amount of -OH groups (as shown in Figure 21) making them soluble in water. Due to these OH groups, Arabic gum has a sticky behavior in the slurry state. Therefore, when gum is exposed to the clear coat surface, based on the surface chemistry (hydrophobicity or hydrophilicity) a good adhesion can be obtained between the polar groups of coating and gum. When this system is exposed to higher temperatures, water is gradually vaporized. During the gum drying process, a significant stress can be imposed by the gum to the surface. To have a more understanding on how these applied stresses act, the visual performance of gum exposed clear coats are given in Figure 22 (Ramezanzadeh et al., 2010c).

According to the observations made in Figure 22, two different phenomena can be observed on the free films and the full system on metal plates. In the latter, a severe crack formation can be seen for the dried gum exposed samples, causing similar cracks on the clear coat layer. On the other hand, gum applied on free films has made the films to shrink. For the

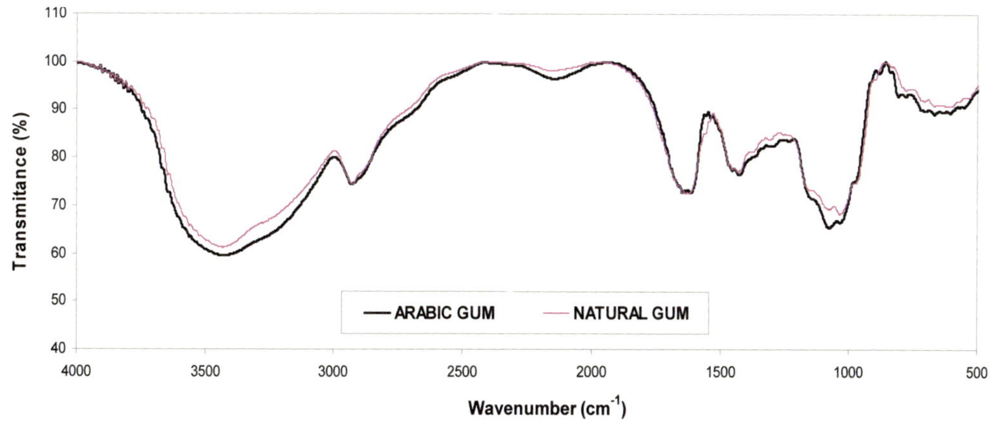

Fig. 21. FTIR spectra for natural and synthetic (Arabic) tree gum (Ramezanzadeh et al., 2010c).

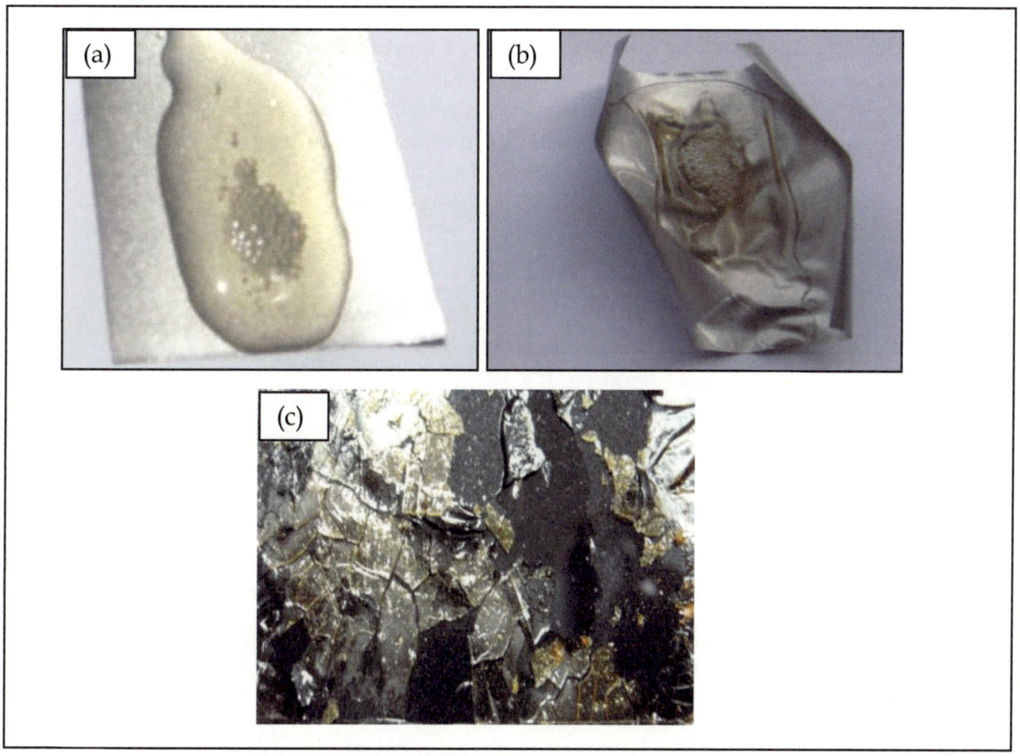

Fig. 22. The visual effect of gum attacked to (a) fully coated system on metal substrate and (b) free film of basecoat/clear coat (Ramezanzadeh et al., 2010c).

clear coat applied on the full automotive system, due to the great adhesion of clear coat to basecoat layer, and basecoat layer to other layers which are in contact with metal substrate, the stress cannot overcome the adhesion force and therefore, it causes surface cracks to propagate. However, for the free films, due to the lack of adhesion to the substrate, the greater cohesion force, in comparison to its adhesion, would turn the film to shrink. Different factors including aging condition, clear coat surface chemistry and basecoat pigmentation can influence this kind of degradation which will be briefly discussed later. Regarding the above explanations, the main source of producing this kind of degradation is the stress formation during gum drying. The stress can overcome adhesion force (between coating and substrate or clear coat and the other coating layers in a multi layer system) and/or cohesion of the clear coat and/or basecoat layers. The ability to store such stress and dissipate it can be depended on many factors, mainly coating viscoelastic properties and temperature. These will be explained later.

2.3.2.2 Physical attack by tree gum

2.3.2.2.1 Effect of aging condition on gum attack

In a real outdoor condition, coatings properties are continually affected by aging conditions. These effects could irretrievably change the chemical and mechanical properties of coatings. Aging process has been shown as an important factor which significantly influences the clear coat properties before and after the biological attack by bird droppings (Yari et al., 2010c; Ramezanzadeh et al., 2010). As previously discussed, two different ideas of the effect of aging on gums attack can be available. Aging condition can influence the degradation occurring during gum attack by affecting coating properties before the test. The second idea is the effect of weathering condition imposed to clear coat in contact with gum. The mechanisms by which these two aging conditions affect gum attack severity are completely different. To show how aging, before or after gum attack, can influence coatings properties, the visual performances of samples experienced these are given in Figure 23.

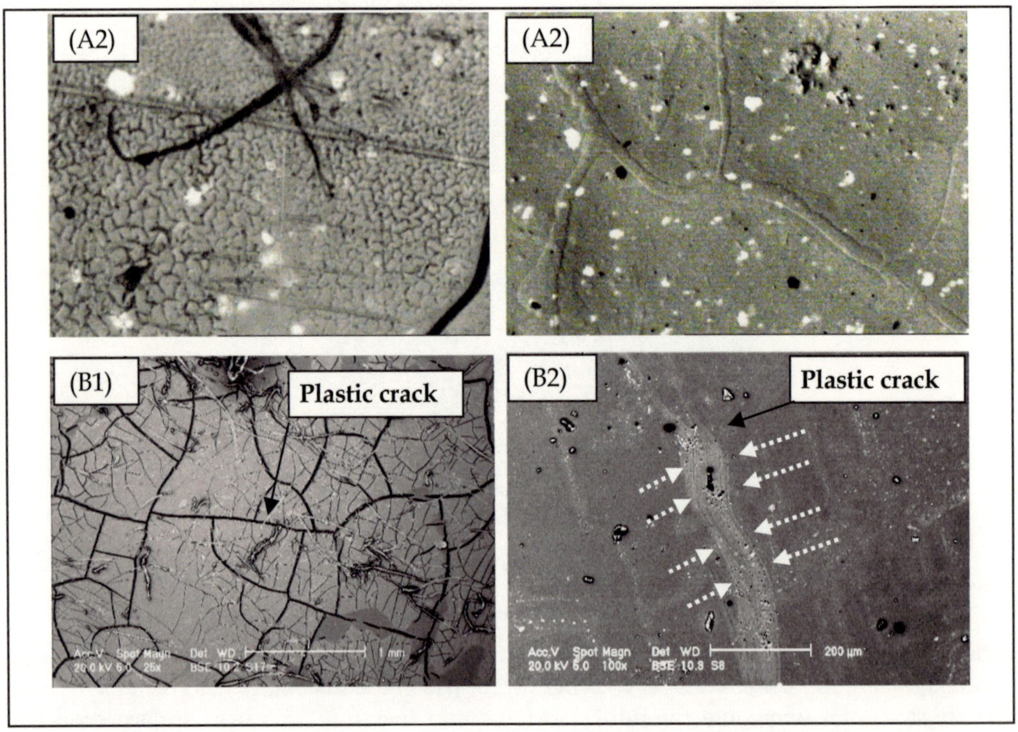

Fig. 23. Samples attacked by gum in A1 (optical micrograph) and B1 (SEM micrograph) post-aging and A2 (optical micrograph) and B2 (SEM micrograph) pre-aging processes (Ramezanzadeh et al., 2010c).

According to Figure 23, a greater surface crack has been produced on the samples exposed to gum at the post-aging process, in comparison to the pre-aging one. These results can reveal that, aging is an important parameter influencing the clear coat biological behavior. This explains that the effects of gum to give rise in surface attack may not only show the crack density or size differences but also reveals that the cracks produced on the samples experienced pre-aging condition have a plastic morphology, whilst cracks created under post-aging show a fracture nature. Highly fractured cracks observed on the clear coats exposed to gum under post-aging process, in comparison to the plastically deformed ones shown in the samples exposed to pre-aging, clearly reveal the importance of aging process on the crack morphology evolution (Ramezanzadeh et al., 2010c).

A significant increase of Δ[NH/NH_2 and OH]/ [CH] after pre-aging process (before biological test) has occurred. This indicates that aging may significantly affect the clear coat by a chemical degradation mechanism. The increase in surface OH groups in the pre-aging condition, leads to significant increase of clear coat hydrophilicity (Ramezanzadeh et al., 2010c), and therefore, stronger interaction with gum. In addition, decreased Tg and crosslinking density of clear coat were obtained at this condition. These changes can negatively influence coating properties against the stress performed by gum. On the other hand, during the biological test in the post-aging, using water sprayed to the clear coat surface (during 300 h of xenon test), a greater interaction of gum and clear coat surface can be created. This can cause a more severe crack on the samples experienced post-aging. A decrease in drying process of gum can result in a greater interaction to clear coat, and therefore enhanced surface attack (Ramezanzadeh et al., 2010c).

2.3.2.2.2 Effect of coating chemistry on gum attack

As it was previously shown (Ramezanzadeh et al., 2010b; Ramezanzadeh et al., 2010c), due to the sticky behavior of gums in the slurry state, it makes a good adhesion to clear coat surface. Many researchers have tried to distinguish the main source of this adhesion. In fact, the tendency of Arabic or natural tree gum to adhere to the surface, results from the polar groups existed in this material. Therefore, the adhesion of Arabic gum to coating can directly depend on the clear coat surface energy (balance of hydrophilicity and hydrophobicity). According to the above explanations, the effect of gums on the clear coat can be directly corresponded to the strong adhesion before the experiment, as well as to the weak attachment after the drying. This behavior causes a great stress to the clear coat, which in turn is responsible for the physical degradation of coating, as shown in Figures 22 and 23. The failure which this stress can perform to clear coats can depend on both clear coats compositions and the undercoat layers mechanical and viscoelastic properties. In addition, the effects of this stress on coating performance can be discussed by two different phenomena as (i) stress restoring and (ii) crack propagation. Based on coating viscoelastic properties, different behaviors of the coating against the inserted stress is predictable. The greater toughness and elastic properties of a coating, the higher the ability is for stress restoring, leading to relaxation during a period of time. In this case, stress can not affect the coating properties. However, most coatings have a viscoelastic properties rather than elastic. The viscose part of the coating does not have restoring and relaxing behavior against applied stress. So, the stress causes a failure on the coating. When the applied stress is not able to overcome the adhesion forces between coating layers, it can affect the cohesion. Different factors may affect the coating cohesion properties, especially the cross-linking density. A lower cross-linking density can cause a lower cohesion. In a real condition, coating surface contains different areas having different cross-linking densities. Theses parts of coatings have a lower elastic behavior and, therefore are able to restore the stress. In addition, the lower cross-linking density of some parts of clear coat may be attributed to a lower curing degree. Therefore, it may be expected that, these parts of coating have a more polarity than other parts due to the presence of unreacted functional groups. This may cause a stronger interaction of gums polar groups to clear coat surface at these areas. Based upon the above explanations, the stress inserted to the clear coat can affect some parts of the coating more intensively than the other parts. Therefore, stress can be propagated from the weak points. In this way, the applied stress may be dissipated by the crack formation (Ramezanzadeh et al., 2010c).

2.3.2.2.3 Effect of basecoat pigmentation on gum attack

It was previously demonstrated that (Ramezanzadeh et al., 2010c; Yari et al., 2009a), basecoat pigmentation can considerably influence the mechanical properties of an automotive clear coat. This behavior was attributed to the curing degree and post reactions occurred. Therefore, according to these results, outdoor weathering conditions may well affect the clear coat properties based on the type of basecoat. Biological resistance of an automotive clear coat can also be expected to show different behavior depending on the basecoat pigmentation type. To show how basecoat pigmentation affect coating behavior against gum attack, the effect of gum on fully coated and free films of the clear coats applied over silver and black basecoat are shown in Figure 24 (Ramezanzadeh et al., 2010c).

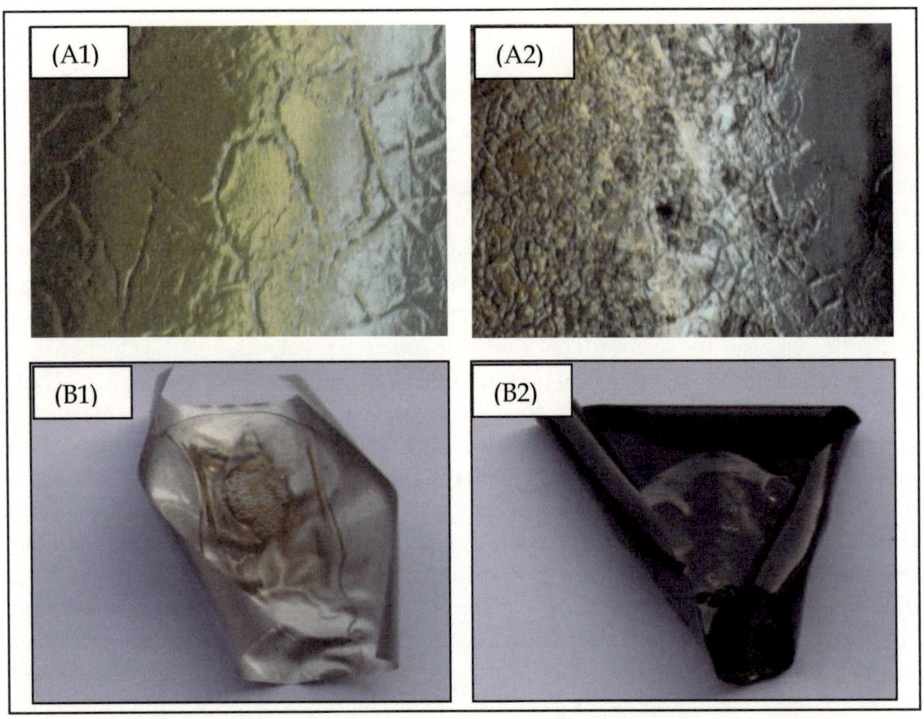

Fig. 24. Effect of basecoat pigmentation on their biological performance in the case of gum attack on (A1) and (A2) full coated and (B1) and (B2) free films (Ramezanzadeh et al., 2010c).

It can be seen that the surface cracks produced by gums over the clear/black system are smaller in size. However, fewer cracks being greater in size for the clear coat on the silver basecoat can be observed. Smaller cracks appeared in the black coating system revealed the greater ability to restore and relax the stress. To show how gum can differently affect clear coats applied over different basecoats, the drying process of gum on these two different samples are shown in Figure 25 (Ramezanzadeh et al., 2010c).

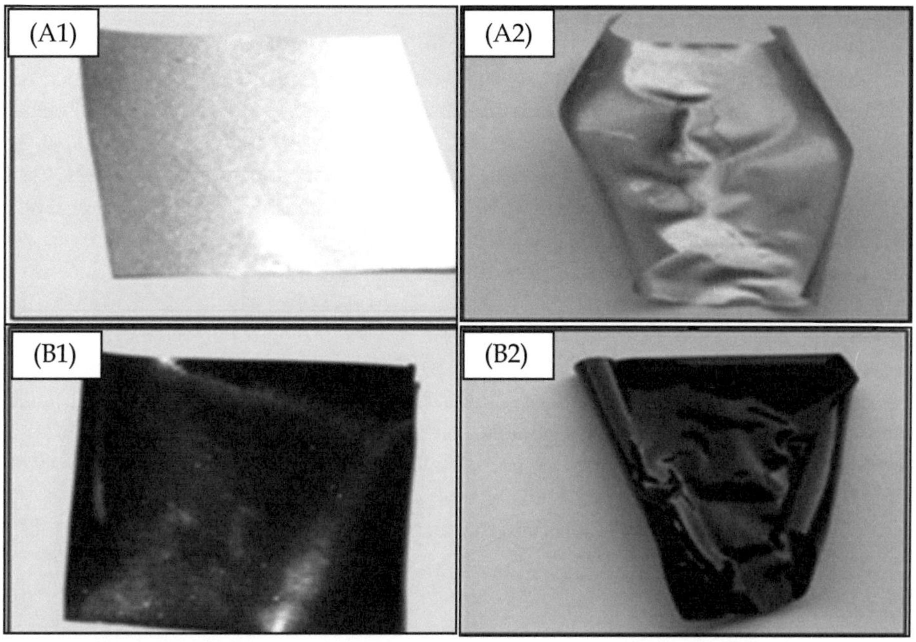

Fig. 25. Effect of gum on (A1) silver and (B1) black (before exposure) samples, and (A2) silver and (B2) black samples after exposure (Ramezanzadeh et al., 2010c).

According to Figure 25, greater shrinkage of black sample can be obtained. These differences can be resulted from the many different factors mainly the difference between the chemical structures of the clear coats applied over these basecoats and the mechanical properties of basecoat layer. In a silver system, due to the lower emissivity factor of basecoat, a greater curing can be obtained resulting in a higher cross-linking density and toughness. In addition, as a result of this better curing, less hydrophilicity and therefore adhesion of gum to clear coat surface can be obtained. The higher Tg of the clear coat on the silver system, as well as its greater cross-linking density, would result in different mechanical properties (Ramezanzadeh et al., 2010c). Moreover, a greater clear coat storage modulus of the silver system (at the temperature of biological test), can reveal different mechanical properties of this clear coat, which can effectively influence the biological attack to tackle the stress. The result is a higher capability of this coating to restore and further dissipate the stress performed by Arabic gum. This means that the greater cross-linking density of this sample (silver one) causes a higher cohesion. Therefore, the ability of the clear coat to distribute the stress on the entire film and other layers is prevention of the stress concentration and formation of cracks, the consequence of which is that the mechanical properties of basecoat layer are affected. Hence, as the mechanical properties of basecoat layers are different, the higher vulnerability of the black system in biological attack is probable. The greater aluminum flakes presented in the silver basecoat, due to the formation of a stronger physical network, causes a higher toughness, in comparison to the black basecoat, causing greater resistance of coating against the applied stress for this system. In addition, the presence of aluminum flakes in the basecoat can cause a greater damping behavior of this layer by preventing the stress concentration on coating (Ramezanzadeh et al., 2010c).

2.3.2.3 Chemical attack by tree gum

As shown in Figure 20, several etched areas can be observed on the clear coats which only experienced 300 h exposure to simultaneous weathering and gums. The presence of these etched areas in a relatively short exposure time indicates that, gums can accelerate the hydrolytic degradation of acrylic melamine, leading to extensive formation of soluble products which are easily released from the coating, leaving spotted etched areas. As stated before, the pHs of Arabic and natural gums are acidic (4.7 and 4.28, respectively). The acidic environment created by gums may also account for the occurrence of such accelerated etching phenomenon. It has been found that, acidic solutions can affect and catalyze the hydrolysis reactions in the same way (Zhou et al., 2002; Schulz et al., 2000). Several researchers have studied this condition for acrylic melamine, in terms of degradation caused by ''acid rain'', which is also a very common phenomenon. The pH of acid rain is around 3.5–4.5. A stronger acidic environment causes more catastrophic degradations. Therefore, the grater variations in FTIR spectra of clear coats exposed to natural gum, compared to that of samples in contact with gum, can be explained by the more acidic nature of this material. These observations can illustrate that, gums can influence coating properties both in chemical and physical ways but mainly in physical direction (Ramezanzadeh et al., 2010c).

3. Concluding remarks

The properties and characteristics of automotive coatings have been discussed. The complicated conditions imposed to these systems need to be well understood in order to enhance their resistance against environment. Photo and hydrolytic degradations are the two common phenomena occurring under external conditions. In addition, the viscoelastic behavior of coatings is also detrimental for a proper mechanical performance. Above all, biological degradation is as important as the other types of failures.

To highlight this kind of degradation the effects of bird droppings and tree gums on an automotive clear coat have been studied. Results showed an irretrievable effect of these biologicals on the visual performance, mechanical and chemical properties of clear coat. Effect of clear coat chemistry on its biological performance, exposed to natural and synthetic biological materials, has been studied. The effects of basecoat pigmentation and aging condition on the biological performance of the clear coat have been also investigated. The general conclusions obtained are shown below:

1. It has been found that the catalytic hydrolysis of etheric and esteric bonds are the reasons for coating degradation when exposed to bird droppings. It was found that natural bird droppings, due to containing some digestive hydrolyse enzymes such as amylase and lipase, are able to catalyze the hydrolytic cleavage of etheric and esteric linkages of acrylic melamine clear coat. The consequence of these cleavages is the release of water soluble products from the coating, leaving etched areas and local

defects as well as decreased appearance on clear coat surface. Results clearly revealed that bird droppings considerably affect the clear coat mechanical properties. According to these results, Tg and elastic modulus were negatively decreased. In addition, the decreased micro hardnesses of clear coats exposed to these biological materials was a further observation indicating the severe effects of biological materials on the mechanical properties of clear coats.

2. The pronounced effect of natural tree gum was a severe crack formation and shrinkage on fully coated systems and free film samples, respectively. It was also shown that, gum could strongly attach to clear coat surface before a drying process commenced. During gum drying, significant stress can be applied on the coating layers, especially the clear coat. Based on the coating properties, i.e. viscoelastic and toughness, different behaviors of coatings against applied stress, such as stress relaxation and/or coating failure were observed.
3. It has been demonstrated that many parameters mainly surface chemistry and viscoelastic properties of clear coat (the balance of surface hydrophobicity/ hydrophilicity), aging condition (post or pre aging) and basecoat pigmentation (metallic or non-metallic) can influence coatings biological performance.

4. Future trends

It would be interesting to further study the effects of surface chemistry (hydrophilicity/hydrophobicity balance) on the biological resistance of automotive coatings. Also investigating the influences of viscoelastic properties of coating systems need more attention. Use of nano-based materials such as additives and pigments seem to be effective.

5. References

Bauer, DR. (1982). Degradation of organic coatings I. hydrolysis of melamine formaldehyde/acrylic copolymer films. *Journal Applied Polymer Science,* 27., 3651-3662

Bauer, DR. (1994). Chemical criteria for durable automotive topcoats. *Journal of coatings and technology,* 66., 835., 57-65

Buter, R.; Wemmenhove, A. (1993). Automotive waterborne surfacer with improved stone-chip resistance. *Progress in Organic Coatings,* 22., 83-105

Courter, JL. (1997). Mar Resistance of Automotive Clear coats: I. Relationship to Coating Mechanical Properties. *Journal of Coatings Technology,* 69, 866., 57-63.

Dhoke, SK.; Khanna, AS.; Sinha, TJM. (2009). Effect of nano-ZnO particles on the corrosion behavior of alkyd-based waterborne coatings. *Progress in Organic Coatings,* 64., 371-382

Gregorovich, BV.; Adamsons, K.; Lin, L. (2001). Scratch and mar and other mechanical properties as a function of chemical structure for automotive refinish coatings. *Progress in Organic Coatings,* 43., 175–187

Hara, Y. Mori, T. Fujitani, T. (2000). Relationship Between Viscoelasticity and Scratch Morphology of Coating Films. *Progress in Organic Coating,* 40., 39-47.

Jalili, MM.; Moradian, S.; Dastmalchian, H.; Karbasi, A. (2007). Investigating the variations in properties of 2-pack polyurethane clear coat through separate incorporation of hydrophilic and hydrophobic nano-silica. *Progress in Organic* Coating, 59., 81-87.

Jardret, V.; Zahouani, H.; Loubet, JL.; Mathina, T. (1998). Understanding and Quantification of Elastic and Plastic Deformation During a Scratch Test. *Wear*, 218., 8–14.

Jardret, V.; Lucas, BN.; Oliver, W. (2000). Scratch Durability of Automotive Clear Coatings: A Quantitative, Reliable and Robust Methodology. *Journal of Coatings Technology*, 72, 907., 79-88.

Jardret, V.; Morel, P. (2003). Viscoelastic Effects on the Scratch Resistance of Polymers: Relationship Between Mechanical Properties and Scratch Properties at Various Temperatures. *Progress in Organic Coatings*, 48., 322–331.

Jardret, V.; Ryntz, R. (2005). Visco-Elastic Visco-Plastic Analysis of Scratch Resistance of Organic Coatings. *Journal of coatings and technology Research*, 2, 8., 591-598.

Lin, L.; Blackman, GS.; Matheson, RR. (2000). A new approach to characterize scratch and mar resistance of automotive coatings. *Progress in Organic Coatings*, 40., 85-91.

Lonyuk, M.; Bosma, M.; Riemslag, AC.; Zuidema, J.; Bakker, A.; Janssen, M. (2007) Stone-impact damage of automotive coatings. single-impact tester. *Progress in Organic Coatings*, 58., 241–247.

Lonyuk, M.; Bosma, M.; Vijverberg, CAM.; Bakker, A.; Janssen, M. (2008). Relation between chip resistance and mechanical properties of automotive coatings. *Progress in Organic Coatings*, 61., 308–315.

Mielewski, DF.; Bauer, DR.; Gerlock, JL. (1991). The role of hydroperoxides in the photo-oxidation of crosslinked polymer coatings. *Polymer Degradation Stability*, 33., 93-104

Mori, K.; Tachi, K.; Muramatsu, M.; Torita, K. (1999). Mechanism of acid rain etching of acrylic/melamine coatings. *Progress in Organic Coatings*, 36., 34-38

Nguyen, T.; Martin, J.; Byrd, E.; Embree, N. (2002a). Relating laboratory and outdoor exposure of coatings: II. Effects of relative humidity on photodegradation and the apparent quantum yield of acrylic melamine coatings. *Journal of coatings and technology*, 74., 65-80

Nguyen, T.; Martin, JW.; Byrd, E.; Embree, N. (2002b). Relating laboratory and outdoor exposure of coatings III. Effect of relative humidity on moisture-enhanced photolysis of acrylic-melamine coatings. *Polymer Degradation Stability*, 77., 1–16

Nguyen, T.; Martin, J.; Byrd, E. (2003). Relating laboratory and outdoor exposures of acrylic melamine Coatings, IV. Mode and mechanism for hydrolytic degradation of acrylic- melamine coatings exposed to water vapor in the absence of light. *Journal of coatings and technology*, 75., 37-50

Nichols, ME.; Darr, CA. (1998). Effect of weathering on the stress distribution and mechanical performance of automotive paint systems. *Journal of coatings and technology*, 70., 885., 141–149.

Nichols, ME.; Gerlock, JL.; Smith, CA.; Darr, CA. (1999). The effects of weathering on the mechanical performance of automotive paint systems. *Progress in Organic Coatings*, 35., 153–159.

Nichols, ME. (2002). Anticipating paint cracking: the application of fracture mechanics to the study of paint weathering. *Journal of coatings and technology*,74., 924., 39–46.

Osterhold, M.; Patrick, G. (2001). Influence of weathering on physical properties of clear coats. *Progress in Organic Coatings*, 41., 177–182.

Palm, M.; Carlsson, B. (2002). New accelerated weathering tests including acid rains. *Journal of coatings and technology*, 74., 69-74

Peng, G.; Li, Q.; Yang, Y.; Wang, H.; Li, W. (2008). Effects of nano ZnO on strength and stability of unsaturated polyester composites. *Polymer Advanced Technology*, 19., 1629–1634.

Pospısil, J.; Nespurek, S. (2000). Photostabilization of coatings. Mechanisms and performance. *Progress Polymer Science*, 25., 1261–1335.

Ratner, B.; Haffman, AS.; Schoen, FJ.; Lemons, JE. (1997) Biomaterials science. An introduction to materials in medicine. Academic press; 1st edition, May.

Ramezanzadeh, B.; Moradian, S.; Tahmassebi, N.; Khosravi, A. (2007a). The Effect of Different Additives on Scratches and Mar Resistance of Automotive Clear Coats. In: Proceeding ICE, 2007, Toronto, Canada.

Ramezanzadeh, B.; Moradian, S.; Tahmassebi, N.; Khosravi, A. (2007b). Modifying the scratch and mar resistances of automotive clear coats: In: Proceeding International Seminar on Polymer Science & Technology, 2007, Tehran, Iran.

Ramezanzadeh, B.; Mohseni, M.; Yari, H.; Sabbaghian, S. (2009). An evaluation of an automotive clear coat performance exposed to bird droppings under different testing approaches. *Progress in Organic Coatings*, 66., 149-160

Ramezanzadeh, B.; Mohseni, M.; Yari, H.; Sabbaghian, S. (2010a). A study of thermal-mechanical properties of an automotive coating exposed to natural and simulated bird droppings. *Journal thermal analysis calorimetry*, in press, DOI 10.1007/s10973-009-0442-4

Ramezanzadeh, B.; Mohseni, M.; Yari H. (2010b)The role of basecoat pigmentation on the biological resistance of an automotive clear coat. *Journal Coating Technology Research*, DOI 10.1007/s11998-010-9254-5

Ramezanzadeh, B.; Mohseni, M.; Yari, H. (2010c). The Effect of Natural Tree Gum and Environmental Condition on the Degradation of a Typical Automotive Clear Coat. *Journal Polymer Environment*, DOI 10.1007/s10924-010-0201-4

Ramezanzadeh, B.; Moradian, S.; Khosravi, A.; Tahmassebi, N. (2010d). Effect of polysiloxane additives on the scratch resistance of an acrylic melamine automotive clear coat. *Journal of coatings and technology Research*, DOI 10.1007/s11998-010-9239-4

Sangaj, NS.; Malshe, VC. (2004). Permeability of polymers in protective organic coatings. *Progress in Organic Coatings*, 50., 28–39.

Schulz, U.; Trubiroha, P.; Schernau, U.; Baumgart, H. (2000). The effects of acid rain on the appearance of automotive paint systems studied outdoors and in a new artificial weathering test. *Progress in Organic Coatings*, 40., 151–165

Shen,W.; Jiang, B.; Scholten, A.; Schwenke, R.; Mi, L.; Seal, C.; and Wang, P. (2004) quantitative index for mar and scratch resistance of materials for automotive glazing applications and quantitative evaluation of damages by different scratching modes. *Tribology Letters*, 17, 3., 637-644.

Skaja, A.; Fernando, D.; Croll, S. (2006). Mechanical property changes and degradation during accelerated weathering weathering of polyester urethane coating. *Journal of coatings and technology Research*, 3.,1., 41-51.

Stevani, CV.; Faria, DLA.; Porto, JS.; Trindade, DJ.; Bechara, JH. (2000). Mechanism of automotive clear coat damage by dragonfly eggs investigated by surface enhanced Raman scattering. *Polymer Degradation Stability*, 68., 61-66

Tahmassebi, N.; Moradian, S. (2004). Predicting the performances of basecoat/clear coat automotive paint systems by the use of adhesion, scratch and mar resistance measurements. *Polymer Degradation Stability*, 83., 405–410.

Tahmassebi, N.; Moradian, S.; Mirabedini, SM. (2005). Evaluation of the weathering performance of basecoat/clear coat automotive paint systems by electrochemical properties measurements. *Progress in Organic Coatings,* 54., 384-389.

Tahmassebi N.; Moradian, S.; Ramezanzadeh, B.; Khosravi, A.; Behdad, S. (2010). Effect of addition of hydrophobic nano silica on viscoelastic properties and scratch resistance of an acrylic/melamine automotive clear coat. *Tribology International,* 43., 85–693.

Thomas, RK. (2005). High Emissivity Coatings for Improved Performance of Electric Arc Furnaces. Presented at AISTech 2005 Iron and Steel Technology Conference May.; Charlotte.; North Carolina.

Valet, A. (1997). Light stabilizers for paints. poppdruck publication,.

Xu, T.; Xie, CS. (2003). Tetrapod-like nano-particle ZnO/acrylic resin composite and its multi-function property. *Progress in Organic Coatings,* 46., 297–301.

Yari, H. (2008). Ms thesis. Amir Kabir University of Technology.

Yari, H.; Moradian, S.; Ramezanzadeh, B.; Kashani, A.; Tahmasebi, N. (2009a). The effect of basecoat pigmentation on mechanical properties of an automotive basecoat/clear coat system during weathering. *Polymer Degradation Stability,* 94., 1281-1289

Yari, H.; Mohseni, M.; Ramezanzadeh, B. (2009b). Comparisons of Weathering Performance of Two Automotive Refinish Coatings: A Case Study. *Journal Applied Polymer Science,* Vol 111, Issue 6., 2946-2956

Yari, H.; Mohseni, M.; Ramazanzade, B.; Naderi, N. (2009c). Use of analytical techniques to reveal the influence of chemical structure of clear coat on its biological degradation caused by bird-droppings. *Progress in Organic Coatings,* 66., 281-290

Yari, H.; Mohseni, M.; Ramazanzade, B. (2010). A mechanistic study of degradation of a typical automotive clear coat caused by bird droppings. *Journal of coatings and technology Research,* DOI 10.1007/s11998-010-9273-2

Zhou, S.; Wu, L.; Sun, J.; Shen, W. (2002). The change of the properties of acrylic-based polyurethane via addition of nano-silica. *Progress in Organic Coating,* 45., 33-42.

17

Modern Automotive Gear Oils - Classification, Characteristics, Market Analysis, and Some Aspects of Lubrication

Waldemar Tuszynski, Remigiusz Michalczewski,
Witold Piekoszewski and Marian Szczerek
Institute for Sustainable Technologies - National Research Institute (ITeE-PIB)
Poland

1. Introduction

1.1 Classification of automotive gear oils

Modern gear oils can be divided into three main groups (Chwaja, Marko, 2010):

- manual transmission fluids (MTF) intended for lubrication of automotive manual gear boxes, transaxles, axles, and differentials,
- automatic transmission fluids (ATF) which lubricate: step-type automatic transmissions (AT), double clutch transmissions (DCT), and continuously variable transmissions (CVT),
- oils for lubrication of gears in off-road vehicles (e.g. tractors).

In European automotive market manual transmission fluids (MTF) are predominant with over 75% of the market compared to only 25% of ATFs. However, in USA and Japan the situation is reversed: ATFs take respectively 92% and 75% of the market, leaving 8% and 25% to MTFs.

This chapter concerns solely the first group of oils, i.e. manual transmission fluids (MTF); for simplification, the equivalent name "automotive gear oils" will be used throughout the text.

There are two different classifications of automotive gear oils.

The first one specifies lubricant service designations or the so-called performance levels of automotive gear oils. It has been provided by the American Petroleum Institute (API) in the API Publication 1560, 7th Edition, published in July, 1995. The API classification divides automotive gear oils into 7 performance levels. Four performance levels are in current use, three are not. The reason for the performance level not to be in current use results from the unavailability of test equipment and does not mean that such products have been withdrawn from the market.

The API classification has been described in Tab. 1 (API designations in current use) and Tab. 2 (designations not in current use).

Apart from the designations from Tabs. 1 and 2, there is also a class denoted as API GL-5(LS) or GL-5+. Gear oils that meet the requirements of this class contain special friction modifiers (FM) preventing from the *stick-slip* occurrence under conditions of limited slip (LS). As such, GL-5(LS) oils are intended for lubrication of limited slip differentials.

To reduce the number of various gear oils in the market and in turn simplify oil selection, many lubricant manufacturers implement more universal (multi grade) gear oils. In this group gear oils denoted as API GL-4/GL-5 or GL-4+ predominate.

API service designation (performance level)	Application and short characterisation
GL-1	Manual transmissions operating under such mild conditions that straight petroleum or refined petroleum oil may be used satisfactorily. Not satisfactory for many passenger car manual transmissions. GL-1 oils may contain oxidation and rust inhibitors, defoamers, and pour depressants. Friction modifiers (FM) and extreme pressure (EP) additives shall not be used.
GL-4	Axles with spiral bevel gears operating under moderate to severe conditions of speed and load or axles with hypoid gears operating under moderate speeds and loads. GL-4 oils may be used in selected manual transmission and transaxle applications where MT-1 lubricants are unsuitable. GL-4 oils contain up to 4% of extreme pressure (EP) additives.
GL-5	Gears, particularly hypoid gears, in axles operating under various combinations of high-speed/shock load and low-speed/high-torque conditions. GL-5 oils contain up to 6.5% of extreme pressure (EP) additives.
MT-1	Nonsynchronised manual transmissions used in buses and heavy-duty trucks. API MT-1 does not address the performance requirements of synchronized transmissions and transaxles in passenger cars and heavy-duty applications. API MT-1 oils provide protection against the combination of thermal degradation, component wear, and oil-seal deterioration, which is not provided by lubricants in current use meeting only the requirements of API GL-1, 4, or 5.

Table 1. API service designations of automotive gear oils in current use, according to API Publication 1560

The second classification is the SAE viscosity classification according to SAE J306:2005 standard. It divides automotive gear oils into 11 grades basing on their rheological properties - tab. 3.

For lubrication of automotive gears the most often multiviscosity-grade oils are employed, e.g. SAE 80W-90. This designation means that such an oil meets the requirements of both a low-temperature (SAE 80W) and a high-temperature grade (SAE 90).

1.2 Characteristics of automotive gear oils on the base of the „market analysis"

The authors have performed a comprehensive „market analysis" of the automotive gear oils. The analysis was related to a search for commercial automotive gear oils offered (mainly in the European market) by the most recognised lubricant manufacturers, including Polish companies. On the base of the up-to-date information presented on the webpages a database has been collected including the manufacturer name, trade name of the oil, API GL performance level, SAE viscosity grade, kind of the base oil, application, fulfilled

API service designation (performance level)	Application and short characterisation
GL-2	Automotive worm-gear axles operating under such conditions of load, temperature, and sliding velocities that lubricants satisfactory for API GL-1 service will not suffice. GL-2 oils contain anti-wear or film-strength improvers specifically designed to protect worm gears.
GL-3	Manual transmissions operating under moderate to severe conditions and spiral-bevel axles operating under mild to moderate conditions of speed and load. GL-3 oils provide load-carrying capacities exceeding those satisfying API GL-1 but below the requirements of API GL-4 oils. GL-3 oils are not intended for axles with hypoid gears. Contain up to about 3% of antiwear (AW) additives.
GL-6	Hypoid gears designed with a very high pinion offset. EP properties typically better than of GL-5 oils. GL-6 oils contain up to 10% of extreme pressure (EP) additives.

Table 2. API service designations of automotive gear oils not in current use, according to API Publication 1560

SAE viscosity grade	Maximum temperature for viscosity of 150 000 cP, °C	Kinematic viscosity at 100°C, cSt (mm^2/s) Minimum	Kinematic viscosity at 100°C, cSt (mm^2/s) Maximum
70W	-55	4.1	-
75W	-40	4.1	-
80W	-26	7.0	-
85W	-12	11.0	-
80	-	7.0	<11.0
85	-	11.0	<13.5
90	-	13.5	<18.5
110	-	18.5	<24.0
140	-	24.0	<32.5
190	-	32.5	<41.0
250	-	41.0	-

Table 3. Automotive gear oil viscosity classification, according to SAE J306:2005

requirements, certificates achieved, recommendations, etc. The collected database includes information on over 330 automotive gear oils. The offer of the following 35 lubricant manufacturers (given in an alphabetical order) has been analysed: *Agip (Eni Group), Amsoil, Aral, BP Lubricants, Carlube, Castrol, ChevronTexaco, Comma, Exxon Mobil Corporation, FL Selenia Poland, Fuchs Oil Corporation, Gulf Oil International, Kager Products, Kroon-Oil, Liqui Moly, Lotos Oil S.A., Lukoil-Permnefteorgsintez, Magna Industrial Co. Ltd, Motul, Naftochem,*

Neste Oil, Oel Brack A.G., Orlen Oil, PPHTU Adwa, Quaker State, Red Line Synthetic Oil Corp., Royal Dutch Shell Group, Silesia Oil, Specol Lubricants, Statoil, Sunoco, Tedex Production, Total SA, Valvoline (Ashland Inc.), Zakl. Chem. Organika S.A. It is worth to add that among the analysed offers the widest offer of automotive gear oils proposes Exxon Mobil, followed by the offer of ChevronTexaco, Orlen Oil (Poland), Fuchs, Aral, Gulf, Neste Oil, Kroon-Oil, Agip, and Castrol.

The database enabled the authors to identify the most often used and popular API GL performance levels of the automotive gear oils, and their percentage in the general offer - Fig. 1.

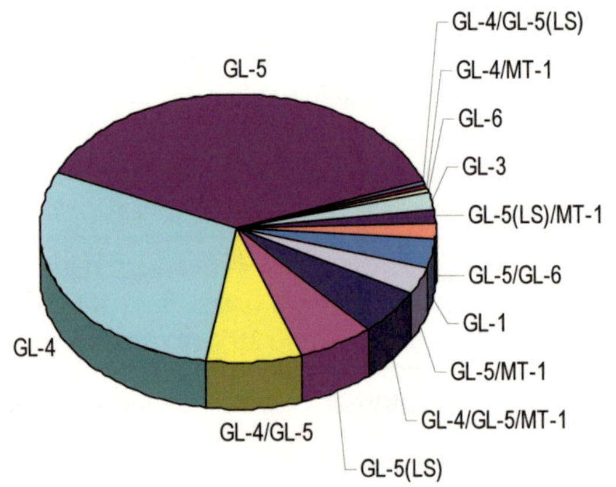

Fig. 1. The percentage of gear oils of particular API GL performance levels (including universal - multi grade oils) in the general offer of automotive gear oils in the market

From Fig. 1 it is apparent that automotive gear oils denoted as API GL-5 and GL-4 definitely predominate. There is also a considerable offer of API GL-5(LS) gear oils, known also as GL-5+ oils.

As concerns universal (multi grade) gear oils, for some time they have been constituting a substantial part of the market. In this group oils denoted as API GL-4/GL-5 or GL-4+, as well as API GL-4/GL-5/MT-1 predominate.

Regarding API MT-1, the authors have not found any products labelled with only this designation. It is used complementarily with other designations of the universal oils, e.g. API GL-4/GL-5/MT-1.

The designations API GL-2 and GL-6 may be treated as obsolete. There are very few manufacturers of GL-6 gear oils. The GL-2 designation is practically no longer used at present; the lubricant manufacturers tend to use the ISO VG classification instead to identify oils for worm gears.

The next step of the analysis of the database was to identify the most often used and popular SAE viscosity grades of the automotive gear oils, and their percentage in the general offer - Fig. 2.

From Fig. 2 it can be seen that multiviscosity-grade oils definitely predominate. In this group SAE 80W-90, 75W-90 and 85W-140 oils constitute the most substantial part.

As concerns single-grade oils, the biggest segment of the market belongs to SAE 80W oils.

The last step of the analysis of the database was to identify the percentage of the types of base oils used for formulation of modern automotive gear oils - Fig. 3.

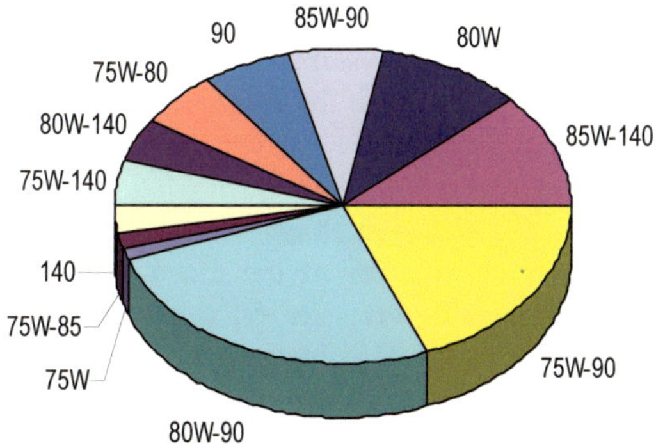

Fig. 2. The percentage of gear oils of particular SAE viscosity grades (including multiviscosity-grade oils) in the general offer of automotive gear oils in the market; only those grades of which percentage was at least 1% are considered

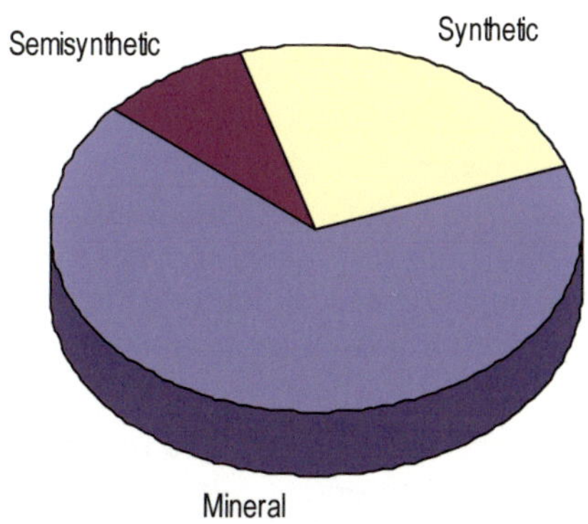

Fig. 3. The percentage of the types of base oils used for formulation of automotive gear oils

From Fig. 3 it is apparent that mineral, automotive gear oils still definitely predominate in the market, followed by synthetic oils. The smallest segment of the market is occupied by gear oils with semisynthetic bases. It is interesting that some lubricant manufacturers, e.g. American firms like Amsoil and Red Line Synthetic Oil Corp. use exclusively synthetic base oils to formulate automotive gear oils.

1.3 Some aspects of degradation of automotive gear oils

During exploitation of lubricating oils in machines significant changes of oil physico-chemical properties take part due to ageing. The ageing rate depends on the working temperature, load, speed and environment. They lead to oil degradation which causes deterioration of oil performance and may even result in a necessity for the oil change.

The following processes in the bulk of the aged oil can be identified: oxidation and polymerisation of hydrocarbons, production of acids and resins, decrease in the content of additives due to precipitation of their oxidised products, mechanical destruction of viscosity improvers leading to a drop in the viscosity index, rise in the concentration of solid contaminants coming from mating components (particles of steel, bronze, metal oxides, rubber), increase in the concentration of contaminants coming from the environment (dust, soil, water), and contamination with products of bacteria activity (Baczewski & Hebda, 1991/92), (Luksa, 1990). The technical systems like transmissions of road vehicles are especially exposed to an adverse influence of the environment.

1.4 Example of gear oil degradation in a car

The authors investigated physico-chemical changes in the commercial API GL-3 gear oil lubricating the transmission of a small car at the mileage of 50 000 km. The following measures were determined: kinematic viscosity (ν) measured at 40 and 100°C, viscosity index (VI), total acid number (TAN), contamination with water (H_2O), and contamination with solid particles (SP). The comparison of the fresh and used oil is presented in Tab. 4.

	State of gear oil	ν_{40} mm s^{-2}	ν_{100} mm s^{-2}	VI	TAN mg KOH g^{-1}	H_2O ppm	SP vol. %
	Fresh	164	15	90	0.8	23	≈0.0
	Used	126	12	85	1.1	129	0.2

Table 4. Changes of measures of physico-chemical properties of the API GL-3 gear oil due to deterioration during lubrication of the car transmission

As could be expected, in the used oil at the mileage of 50 000 km a significant increase in the content of solid particles and water was identified; TAN rose as well. Both kinematic viscosity and viscosity index significantly decreased due to mechanical degradation of polymer chains in the viscosity improvers and/or hydrocarbon chains in the oil at the contact zone of mating components (gears, rolling bearings, synchronizers). These adverse processes have an influence on wear of components of automotive transmissions.

1.5 Wear of transmissions components - gears and rolling bearings

One of the dominating forms of wear of toothed gears is scuffing. Scuffing is a form of wear typical of highly-loaded surfaces working at high relative speeds. Another form of wear is rolling contact fatigue (pitting). Pitting is a form of wear typical of highly-loaded surfaces working at a sliding-rolling and rolling contact, e.g. such components in transmissions like toothed gears and rolling bearings. It is caused by the cyclic contact stress, which leads to cracks initiation. The lubricant is pressed into the cracks at a very high pressure (elastohydrodynamic lubrication), making them propagate. Finally, cyclic stress results in breaking a piece of material off the surface (Pytko & Szczerek, 1993), (Lawrowski, 2008). The mentioned forms of wear are shown in Fig. 4.

The resistance to scuffing and pitting depends on many various factors. They are: material properties, surface machining, geometry of the tribosystem, working conditions, as well as physico-chemical properties of the lubricant which significantly change due to ageing. Although an effect of the aged oil degradation on its performance has been described in the literature (Yamada et al., 1993), (Hohn et al., 2001), few data concern differentiation between behaviour of automotive gear oils having different API GL performance levels, particularly in the aspect of the oil-surface interactions.

In the further part of this chapter the authors compare two different automotive gear oils of different chemistry and performance levels - API GL-3 and GL-5. The aim is to find an effect of their ageing on deterioration of their properties, and in turn on scuffing and pitting prevention. To simulate oil deterioration taking part during exploitation of machines (e.g. cars), laboratory ageing of the oils was performed.

a)

b) c)

Fig. 4. Most dangerous forms of wear of transmission components: a) scuffing of a gear, b) pitting of a gear, c) pitting of a bearing ball

2. Research equipment

2.1 Scuffing tests

The scuffing tests were performed using a four-ball testing machine, denoted as T-02, designed and manufactured by ITeE-PIB - Fig. 5.

Test specimens were bearing balls made of chrome alloy bearing steel, with diameter of 12.7 mm. Surface roughness was R_a = 0.032 µm and hardness between 60 and 65 HRC. The sliding four-ball tribosystem and some its important dimensions are shown in Fig. 6.

The tribosystem consists of the three stationary balls (2) fixed in the ball pot (4) and pressed at the required load P against the top ball (1). The top ball is fixed in the ball chuck (3) and rotates at the defined speed n. So, between the balls pure sliding appeared, which created conditions for investigation of scuffing. The contact zone of the balls was immersed in the tested oil.

After tribological experiments the worn surface of the bottom balls was analysed using a scanning electron microscope (SEM) and energy dispersive spectrometer (EDS). EDS analyses were performed at the accelerating voltage of 15 kV. Prior to analyses test balls were washed for 5 mins in n-hexane using an ultrasonic washer.

Fig. 5. T-02 Four-ball testing machine

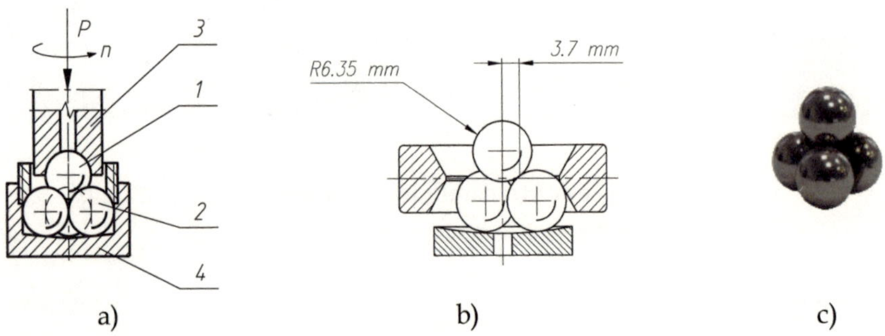

Fig. 6. Sliding four-ball tribosystem: a) drawing (1 - top ball, 2- stationary balls, 3 - ball chuck, 4 - ball pot), b) some important dimensions (wear track radius and ball radius), c) photograph

2.2 Pitting tests

The surface fatigue (pitting) tests were performed using a four-ball rolling tester, also designed and manufactured by ITeE-PIB, denoted as T-03 - Fig. 7.

The rolling four-ball tribosystem and some its important dimensions are shown in Fig. 8.

The tribosystem consists of the three bottom balls (2), free to rotate in the special race (3), and pressed at the required load *P* against the top ball (1). The bottom balls are driven by the top ball which is fixed in the ball chuck and rotates at the defined speed *n*. The contact zone of the balls was immersed in the tested oil.

The test balls were the same as used in the scuffing tests.

The T-03 tester was equipped with a vibration monitoring system. At the onset of the fatigue failure, when vibrations increased sharply, the system turned the rig down.

Fig. 7. T-03 Four-ball pitting tester

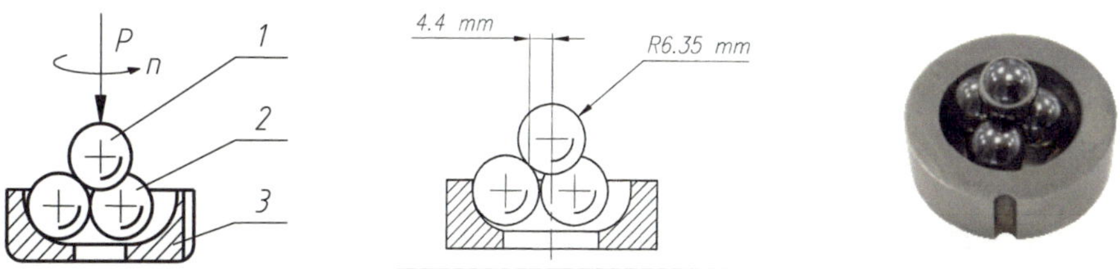

a) b) c)

Fig. 8. Rolling four-ball tribosystem: a) drawing (1- top ball, 2 - bottom balls, 3 - race), b) some important dimensions (wear track radius and ball radius), c) photograph

The worn surface on the upper ball was analyzed using a scanning electron microscope (SEM), energy dispersive spectrometer (EDS) and atomic force microscope (AFM). EDS analyses were performed at the accelerating voltage of 15 kV. Prior to analyses the test balls were washed for 5 mins in n-hexane using an ultrasonic washer.

3. Test methods

3.1 Scuffing tests

The properties of the tested lubricants related to prevention of scuffing are called the extreme pressure (EP) properties. In this work the extreme pressure properties of the tested oils are characterised by the so-called limiting pressure of seizure, denoted as p_{oz}. This measure is determined according to a test method developed in the Tribology Dept. of ITeE-PIB, having been presented in the literature (Piekoszewski et al., 2001), (Szczerek & Tuszynski, 2002), (Burakowski et al., 2004). A unique feature of the test method is related to continuously increasing load until scuffing and then seizure occurs, and analysis of scuffing propagation.

Test conditions are: load increase 409 N s^{-1}, initial load 0, maximum load about 7400 N, load increase time approximately 18 s (until the highest load is reached), rotational speed 500 rpm (sliding speed 0.19 m s^{-1}).
It is assumed that the test finishes when seizure takes place, i.e. at the time of exceeding 10 N m friction torque (this quantity is calculated on the base of measurements from a force transducer located at the distance 0.15 m from the test shaft axis). When seizure is not detected, the attaining of maximum load (about 7400 N) finishes the test.
For the tested lubricant the limiting pressure of seizure (p_{oz}) is calculated from the equation (1):

$$p_{oz} = 0.52 \frac{P_{oz}}{d^2} \quad \left(N\,mm^{-2}\right) \tag{1}$$

where:
P_{oz} - load that causes seizure (or maximum load when seizure does not appear), the so-called seizure load, N,
d - average wear scar diameter, from the measurements on the three bottom balls in the direction parallel and perpendicular to the "striations", mm.
The rounded value 0.52 results from the four-ball geometry.
So, the limiting pressure of seizure (p_{oz}) is a nominal pressure at the time of seizure (or at the end of a run) exerted on the wear scar area between two contacting balls. The bigger p_{oz} value, the better extreme pressure properties of the tested lubricant.
For each tested oil at least 3 runs were performed and the results averaged. The outliers were rejected on the base of Dixon test, for the significance level $\alpha = 5\%$.

3.2 Pitting tests

The resistance to pitting was characterised by the so-called 10% fatigue life, denoted as L_{10}. The procedure of its determination is presented in IP 300 standard. The value of L_{10} represents the life at which 10% of a large number of test balls, lubricated with the tested oil, would be expected to have failed.
Test conditions, adopted from IP 300, were as follows: rotational speed 1450 rpm, applied load 5886 N (600 kgf), run duration until pitting occurs, number of runs 24. Only those runs were accepted for which pitting occurred on the top ball (requirement of IP 300 standard). In each run the time to pitting failure occurrence was measured.
After test completion the 24 values (failure times) were plotted in the Weibull co-ordinates, i.e. the estimated cumulative percentage failed against the failure time. Then, a straight line was fitted to the points. From the line the 10% life L_{10} was read off.

4. Gear oils tested and their ageing

Two mineral, automotive gear oils of API GL-3 and GL-5 performance levels were used. The oils were formulated and delivered by the Central Petroleum Laboratory (CLN) in Warsaw, Poland.
In the GL-3 oil the commercial package of lubricating additives was based on zinc dialkyldithiophosphate (ZDDP), classified as antiwear (AW) and partly extreme pressure (EP) additives. GL-5 oil contained a package of EP additives based on organic sulfur-phosphorus (S-P) compounds.

The gear oils were contaminated with a special test dust (3 samples with various dust concentrations), distilled water (3 samples with various water concentrations) and were laboratory oxidised at 150˚C (3 samples oxidised at various times) - Fig. 9.

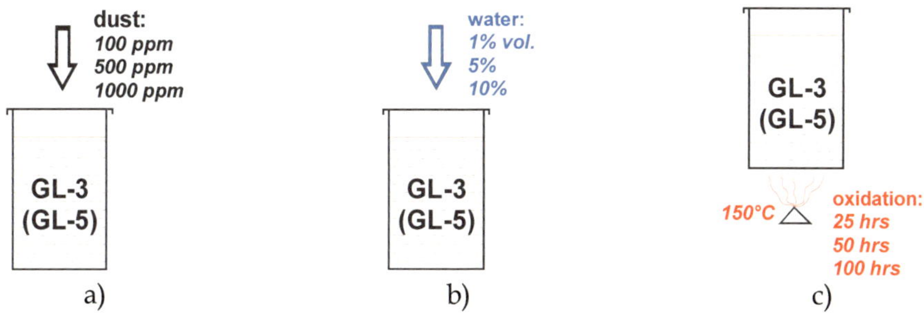

Fig. 9. Laboratory ageing of the API GL-3 and GL-5 gear oils: a) contamination with dust, b) contamination with water, c) oxidation

The main components of the test dust were SiO_2 grains (72.4% wt.) and Al_2O_3 (14.2% wt.). Maximum grain size did not exceed 0.08 μm. The granulometric composition of the test dust is given in Tab. 5.

Grain size, μm	Grain share, wt. %
0.08 - 0.04	9.1
0.04 - 0.02	19.5
0.02 - 0.01	14.7
0.01 - 0.005	19.7
0.005 - 0	37.0

Table 5. The granulometric composition of the test dust

Prior to pouring in the oils, the dust had been dried at 100ºC for 6 hrs.

Oxidation of the oils was performed using a special oil bath at 150ºC, without air flow, nor a catalyst. After oxidation for a given time, basic physico-chemical properties of the oil sample were determined, for example total acid number (TAN) and changes in infrared (IR) spectra, i.e. changes of areas under peaks characteristic for interesting chemical bonds in the lubricating additives. IR spectra were obtained using Fourier transform infrared microspectrophotometry (FTIRM). It is worth mentioning that TAN is the quantity (expressed in mg) of potassium hydroxide (KOH) needed to neutralize the acid in 1 g of oil. So, TAN indicates the amount of oxidation that the oil has undergone.

Before tribological tests each oil sample was stirred for 30 mins to equalise their bulk composition. In case of water contamination, oil-water emulsions were obtained.

5. Results and discussion - scuffing tests

5.1 Testing of dust-contaminated gear oils

Fig. 10 presents the values of the limiting pressure of seizure (p_{oz}) obtained for the gear oils of API GL-3 and GL-5 performance levels - pure and contaminated with the test dust at increasing concentrations. Interval bars reflecting the repeatability of the used test method have been added to the graphs.

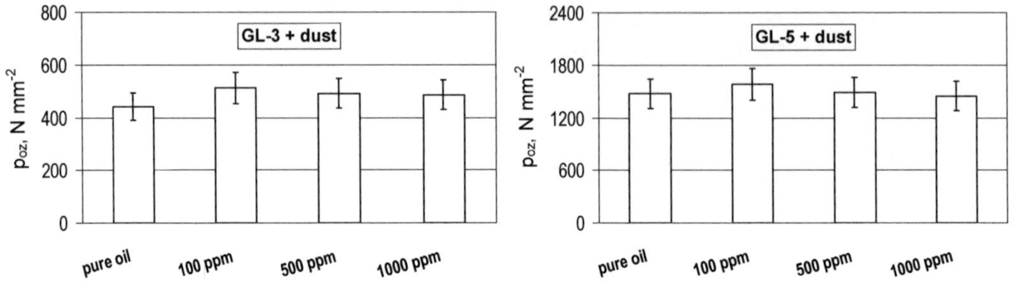

a) b)

Fig. 10. Limiting pressure of seizure (p_{oz}) obtained for the gear oils - pure and contaminated with the test dust: a) GL-3 oil, b) GL-5 oil

From Fig. 10 it can be observed that the contamination of the oil with the dust practically does not affect the oil extreme pressure properties. The reason is that under severe friction conditions wear is so intensive that abrasive action of the dust does not matter.

It should also be noted that the GL-3 gear oil gives about threefold lower values of p_{oz} than GL-5. This much less efficiency of the GL-3 oil under severe friction conditions can be attributed to action of AW type lubricating additives (ZDDP) which are used in such oils. It is known that AW additives shows much poorer performance under severe conditions than EP ones (S-P compounds) which are used in GL-5 gear oils.

5.2 Testing of water-contaminated gear oils

Fig. 11 presents the values of the limiting pressure of seizure (p_{oz}) obtained for the gear oils of API GL-3 and GL-5 performance levels - pure and contaminated with water at increasing concentrations.

a) b)

Fig. 11. Limiting pressure of seizure (p_{oz}) obtained for the gear oils - pure and contaminated with the water: a) GL-3 oil, b) GL-5 oil

The contamination of the GL-3 gear oil by water at the concentration of 1% has a significant, deleterious effect on the oil extreme pressure properties. Further increasing the water contamination has no effect on p_{oz} values. In comparison, GL-5 gear oil shows less "sensitivity" to water contamination - lower concentrations of water do not exert any effect and a drop in the extreme pressure properties is visible only when 10% of water is added to the oil.

For interpretation of the obtained results the wear scars on the bottom balls were analysed using SEM/EDS. SEM images of the worn surface and EDS maps for sulfur and phosphorus

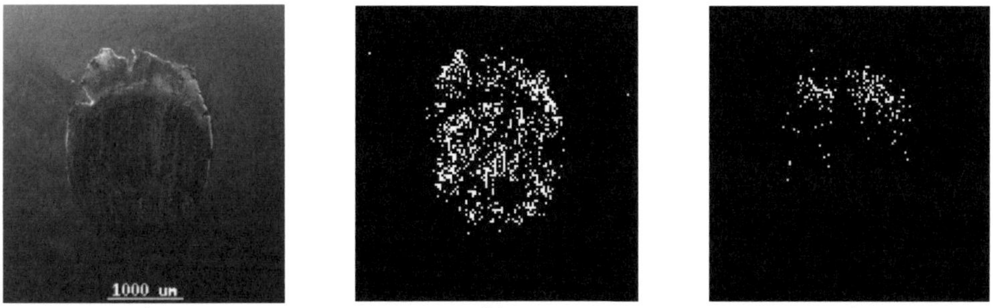

a) b) c)

Fig. 12. Pure GL-5 oil - SEM image of the wear scar (a) and EDS maps for: b) sulfur, c) phosphorus

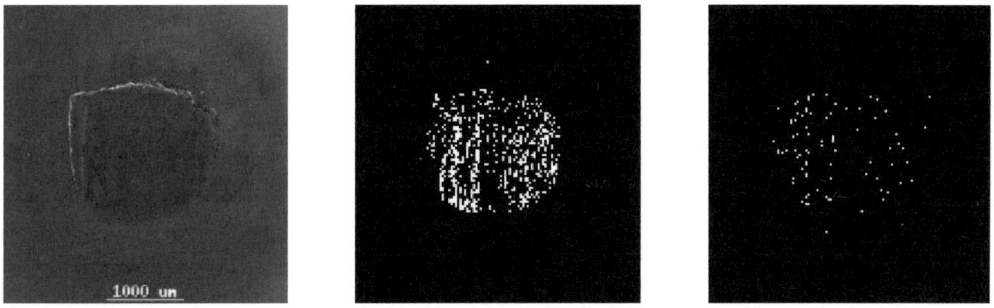

a) b) c)

Fig. 13. GL-5 oil contaminated with water at 10% concentration - SEM image of the wear scar (a) and EDS maps for: b) sulfur, c) phosphorus

in the surface layer are shown in Figs. 12 and 13 for pure GL-5 oil and this oil contaminated with 10% water.

From Figs. 12 and 13 it is evident that water contamination affects the oil-surface interactions - one can observe a decrease in phosphorus content in the tribochemically modified surface layer of the wear scar.

The next step of analysis was to quantitatively examine the wear scar surface layer using EDS. Fig. 14 shows the weight concentration of sulfur and phosphorus in the surface layer for both the gear oils contaminated with water. The analyses were performed at three different points of the wear scar. The graphs present the average values of elemental concentration.

From Fig. 14 it is apparent that for GL-3 gear oil contaminated with 1% or more water a significant decrease in the concentration of sulfur and phosphorus takes place. For the contaminated GL-5 oil the concentration of sulfur remains practically constant but a drop in phosphorus concentration occurs in case of the highest rates of water contamination.

It is well known that prevention of scuffing is realised by sulfur and phosphorus compounds (Godfrey, 1968), (Forbes, 1970), (Stachowiak & Batchelor, 2001). These compounds are formed owing to physical and chemical adsorption, followed by chemical reactions of active lubricating additives with the steel surface. The sulfur and phosphorus compounds prevent creation of adhesive bonds or enable their shearing. A great role is played here particularly by inorganic compounds like FeS.

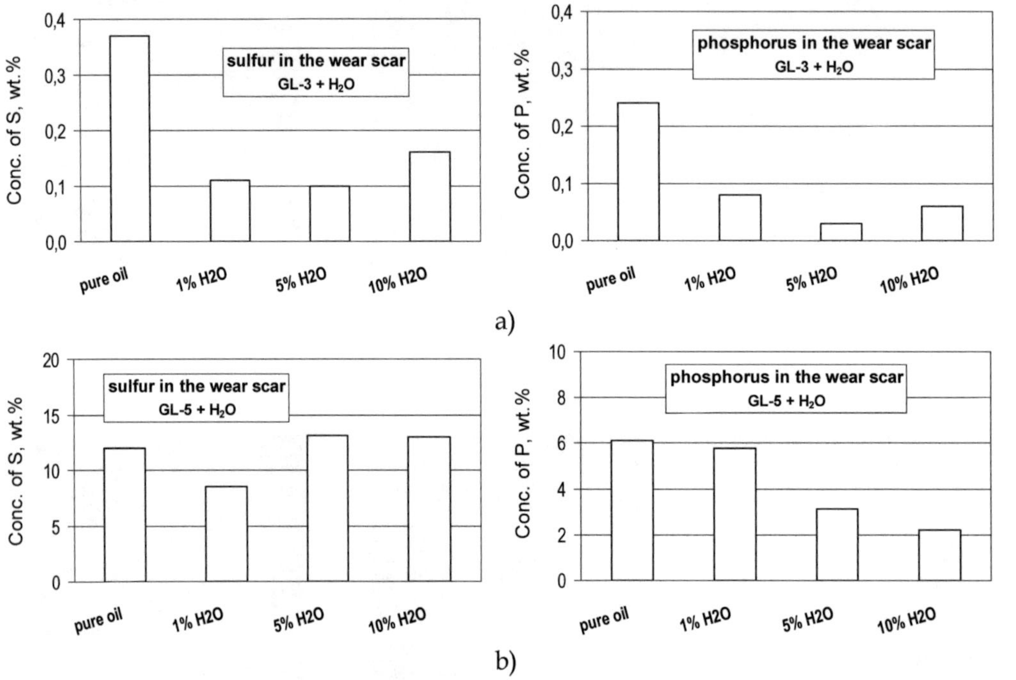

Fig. 14. Average concentration of sulfur and phosphorus in the surface layer of the wear scar for the gear oils contaminated with water: a) GL-3 oil, b) GL-5 oil

So, a significant decrease in the concentration of sulfur and phosphorus in the surface layer of the wear scar for GL-3 gear oil contaminated with 1% or more water is responsible for a dramatic deterioration of its extreme pressure properties (Fig. 11 a). For GL-5 gear oil poorer scuffing performance observed not sooner than for 10% water contamination (Fig. 11 b) can be attributed to a drop of phosphorous visible in case of the highest water content.

It should also be noted that for all samples of GL-5 gear oil incomparably higher concentration of sulfur and phosphorus can be found in the wear scar surface layer than for GL-3 oil. This is a result of more effective action of EP additives in GL-5 oils than AW additives in GL-3 oil, hence much better extreme pressure properties of the sooner.

5.3 Testing of oxidative degradation of gear oils

Fig. 15 presents the values of the limiting pressure of seizure (p_{oz}) obtained for the gear oils of API GL-3 and GL-5 performance levels - pure ("fresh") and oxidised for longer and longer time.

Fig. 15 shows that the oil oxidation exerts in general a positive effect on extreme pressure properties of both the tested gear oils. For GL-3 oil the values of p_{oz} increase with extending oxidation time. Only after the longest oxidation time a sudden drop in the oil performance occurs. For GL-5 oil its oxidation also exerts a rather positive effect on extreme pressure properties - a slow but sustained rise in the values of p_{oz} is observed with extending oxidation time. The only exception is GL-5 oil oxidised for 50 hrs, giving an unexpected, noticeable drop in its performance.

For interpretation of the obtained results the wear scars on the bottom balls were analysed using SEM/EDS. SEM images of the worn surface and EDS maps for sulfur and phosphorus

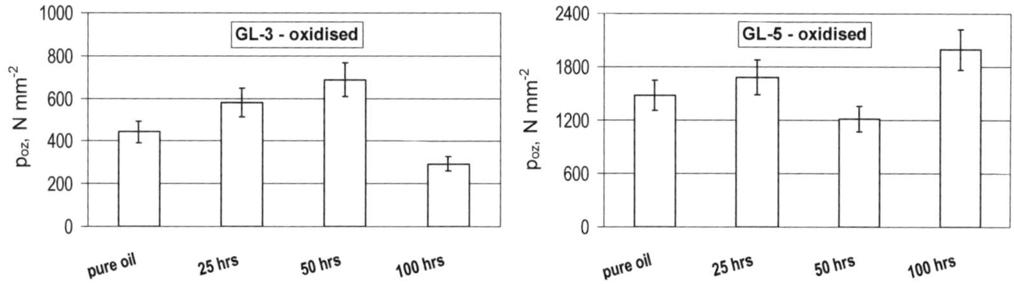

Fig. 15. Limiting pressure of seizure (p_{oz}) obtained for the pure and oxidised gear oils: a) GL-3 oil, b) GL-5 oil

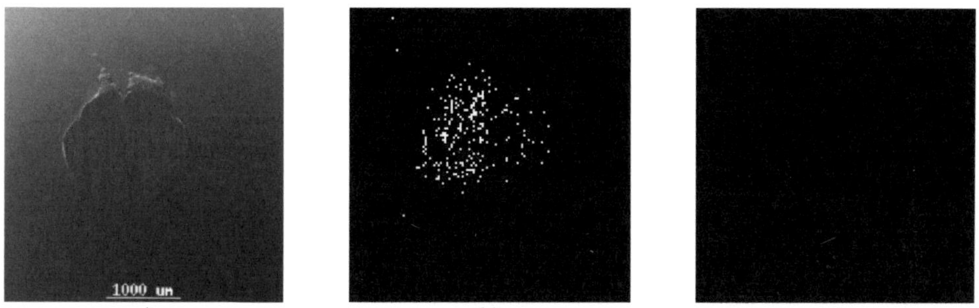

a) b) c)

Fig. 16. GL-5 oil oxidised for 100 hrs - SEM image of the wear scar (a) and EDS maps for: b) sulfur, c) phosphorus

in the surface layer are shown in Fig. 16 for GL-5 oil oxidised for 100 hrs. Respective images obtained for the pure GL-5 oil have been shown earlier in Fig. 12.

From Figs. 12 and 16 it is evident that oil 100 hrs-long oxidation affects the oil-surface interactions - one can observe a noticeable decrease in phosphorus content in the tribochemically modified surface layer of the wear scar. The map of phosphorus is 'empty' for the reason of its very little concentration in the surface layer, less than 1% wt. (a sensitivity threshold of EDS mapping is in practice about 1% wt.).

The next step of analysis was to examine the wear scar surface layer quantitatively using EDS. Fig. 17 shows the weight concentration of sulfur and phosphorus in the surface layer for the both oxidised gear oils.

From Fig. 17 it can be seen that for GL-3 gear oil oxidised for 25 and 50 hrs the concentration of sulfur and phosphorus in the surface layer of the wear scar is much higher than for the pure oil. A dramatic drop in their concentration, down to unidentifiable values is noticed not sooner than for the longest time of oxidation (100 hrs). So, the concentration of these elements in the surface layer in some way correlates with the tribological results (Fig. 15 a). One can thus infer that their concentration increase is beneficial to the extreme pressure properties of the oxidised oil and the respective mechanisms of such an action have been described earlier.

In case of GL-5 gear oil irrespective of the oxidation time the concentration of sulfur in the surface layer of the wear scar is high and does not change. A small drop in sulfur concentration is noticed only for the middle time of oxidation (50 hrs). The concentration of phosphorus significantly decreases for the longest oxidation times. It is the decrease in

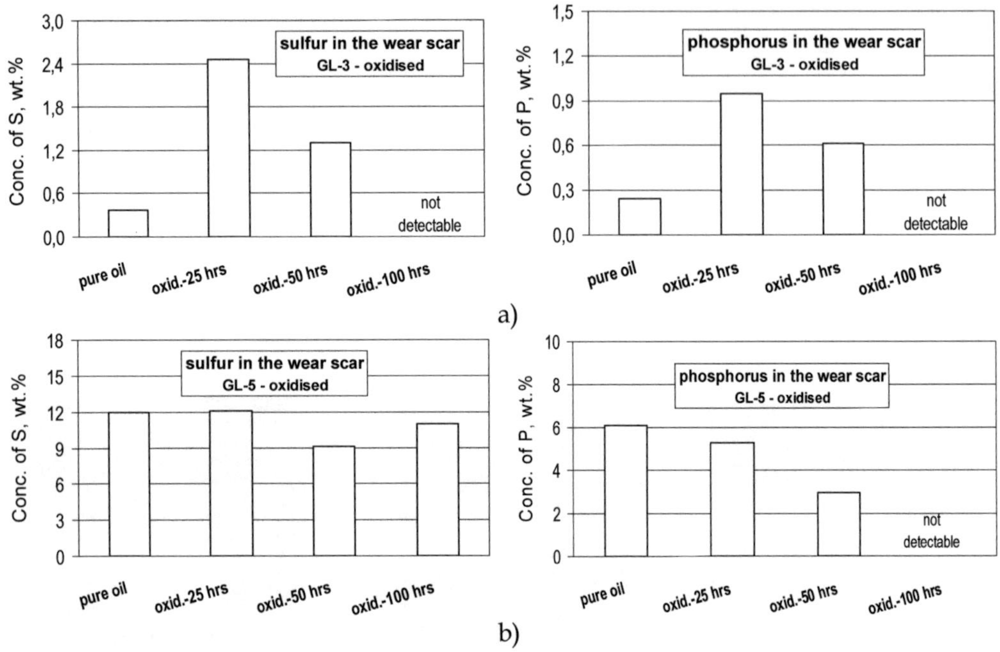

Fig. 17. Average concentration of sulfur and phosphorus in the surface layer of the wear scar for the oxidised gear oils: a) GL-3 oil, b) GL-5 oil

sulfur that may be a reason for an unexpected drop in the extreme pressure properties observed for GL-5 oils oxidised for 50 hrs (Fig. 15 b).

A dramatic drop in the concentration of sulfur and phosphorus in the wear scar surface layer in case of GL-3 oil oxidised for 100 hrs, accompanied by deterioration of its extreme pressure properties (Fig. 15 a) comes from a decrease in the lubricating additives in the oil due to precipitation of their oxidised products in the form of sludge, which has been postulated in the literature (Yamada et al., 1993), (Makowska & Gradkowski, 1999).

The changes in the physico-chemical properties due to oxidation were investigated by determination of TAN and FTIRM analysis of the tested oils. The values of TAN for the pure and oxidised oils are shown in Fig. 18, and the IR spectra - in Figs. 19 and 20.

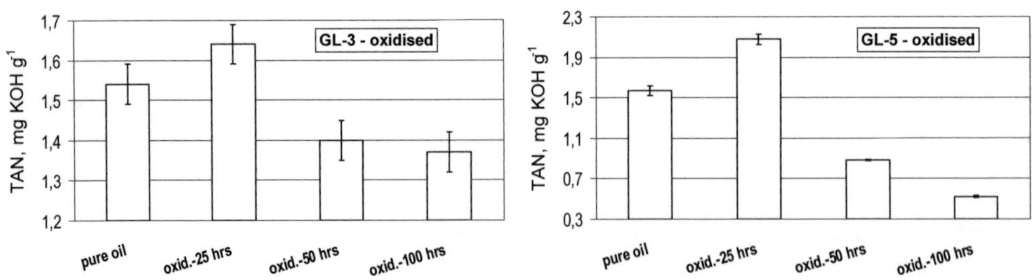

Fig. 18. TAN for the pure and oxidised gear oils: a) GL-3 oil, b) GL-5 oil

From Figs. 18 to 20 it is apparent that the symptoms of additives decrease in the oxidised GL-3 oil are: 10% drop in TAN and a very big decrease in the area under the peak at 965 cm^{-1}

in the IR spectrum; such a peak is typical of P-O-C bonds in the lubricating additives (ZDDP) used in GL-3 oils.

A decrease in the content of lubricating additives due to precipitation was also noticed for the oxidised GL-5 oil, which was identified by threefold drop in TAN of the oil oxidised for the longest time in comparison with the pure oil (Fig. 18 b). This much reduced the content of phosphorus in the worn surface, but because the concentration of sulfur (which is the most important element in the EP additives) practically did not change (Fig. 17 b) the extreme pressure properties of the oil oxidised for 100 hrs did not deteriorate.

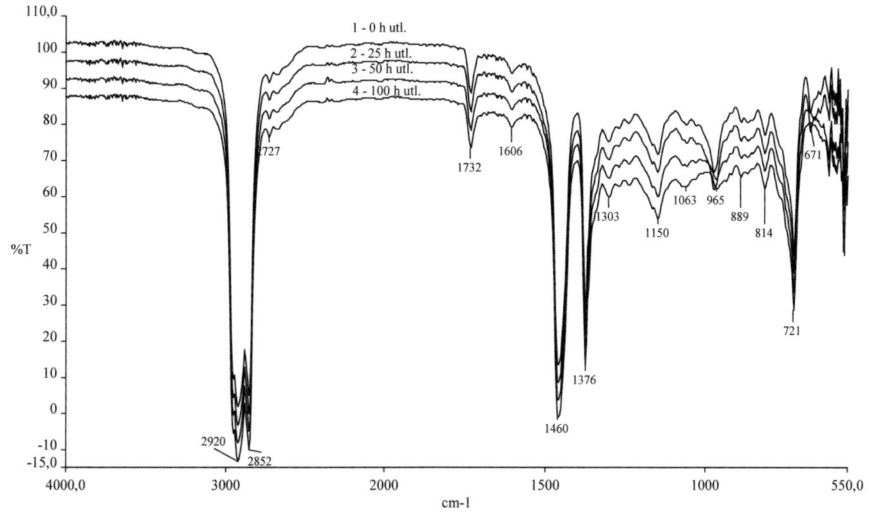

Fig. 19. IR spectrum for the pure and oxidised GL-3 oil; 1 - pure oil, 2 - oxidation for 25 hrs, 3 - 50 hrs, 4 - 100 hrs

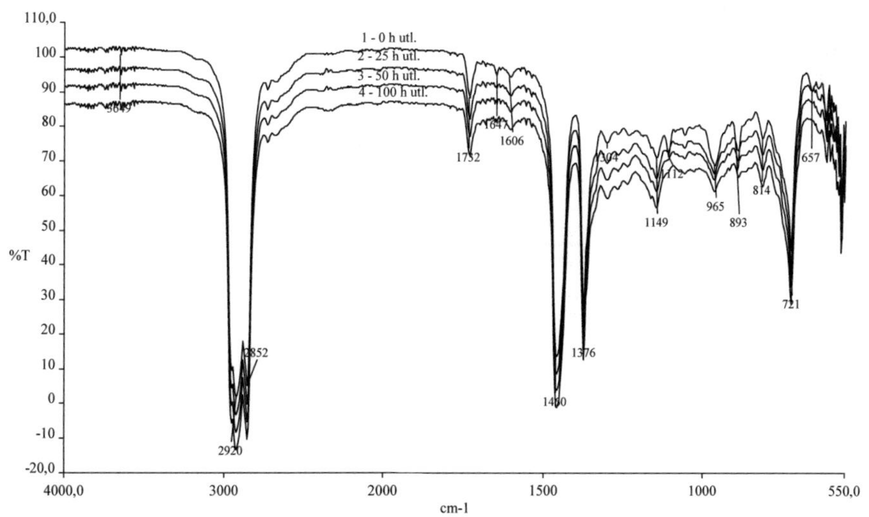

Fig. 20. IR spectrum for the pure and oxidised GL-5 oil; 1 - pure oil, 2 - oxidation for 25 hrs, 3 - 50 hrs, 4 - 100 hrs

6. Results and discussion - pitting tests

6.1 EHD oil film thickness during pitting tests - calculations

Because knowledge of the conditions in rolling contact will be helpful for further analyses, the authors have calculated the oil film thickness during pitting tests.

In the first approach the authors adopted a purely elastic model of the point contact for calculation. The calculated minimum film thickness was about 0.02 μm. However, the load between the balls gave unrealistic maximum Hertzian pressure 8.5 GPa, which would much exceed the yield strength of the material of the bearing balls (roughly assumed to be about 3 GPa, i.e. about one third of the average hardness expressed in GPa).

Because inspection of the wear track surface on the upper ball using profilometry revealed that the material was plastically deformed, the assumption of the point contact was no longer justified. So, an elastic model of the line contact of rolling elements was adopted for calculations with the well-known Dowson and Higginson's formulae compiled in the book (Winer & Cheng, 1980). It should be emphasized here that the contact of the four balls creates a circular wear track on the upper ball (plastically deformed), while the three bottom balls contact with the upper one randomly - over their entire surfaces.

The input data used for calculation of the minimum oil film thickness are given in Tab. 6 and some important dimensions of the four-ball rolling tribosystem are shown in Fig. 8 b.

In Tab. 6 the length L denotes the width of the plastically deformed zone between two mating balls and was averaged from measurements of the wear track profile on the upper ball made by a profilometer. As concerns rheological properties of the oils, they were determined at the temperature of 80°C, typical of relatively long (a few hours) tests in rolling movement. Pressure-viscosity coefficient was adopted from (Wang et al., 1996) for a mineral oil.

Quantity, unit	GL-3 oil	GL-5 oil
Radius R_1, mm	6.35	
Radius R_2, mm	6.35	
Length L, mm	1.3	
Load w, N	2649	
Velocity u_1, m s^{-1}	0.67	
Velocity u_2, m s^{-1}	0.67	
Modulus of elasticity E_1, GPa	210	
Modulus of elasticity E_2, GPa	210	
Poisson's ratio ν_1	0.3	
Poisson's ratio ν_2	0.3	
Oil viscosity (at 80°C) μ_0, Pa·s	0.0203	0.0199
Pressure viscosity coefficient (at 80°C) a, Pa^{-1}	$1.1 \cdot 10^{-8}$	

Table 6. Input data for calculation of the oil film thickness during pitting tests; symbols taken from (Winer & Cheng, 1980).

The calculated minimum lubricating film thickness h_{min} formed during the pitting tests for the pure gear oils is about 0.04 μm and is similar to values obtained by other authors for this kind of the tribosystem, e.g. (Libera et al., 2005). It should be noticed that the calculated film thickness is much thinner than occurring in service of machines. It is a result of relatively low velocity as well as disregarding an effect of viscosity improvers in the oil on the pressure-viscosity coefficient.

6.2 Testing of dust-contaminated gear oils

Fig. 21 presents the values of the 10% fatigue life (L_{10}) obtained for the gear oils of API GL-3 and GL-5 performance levels - pure and contaminated with the test dust at increasing concentrations. Confidence intervals calculated for the probability 90% have been added to the graphs.

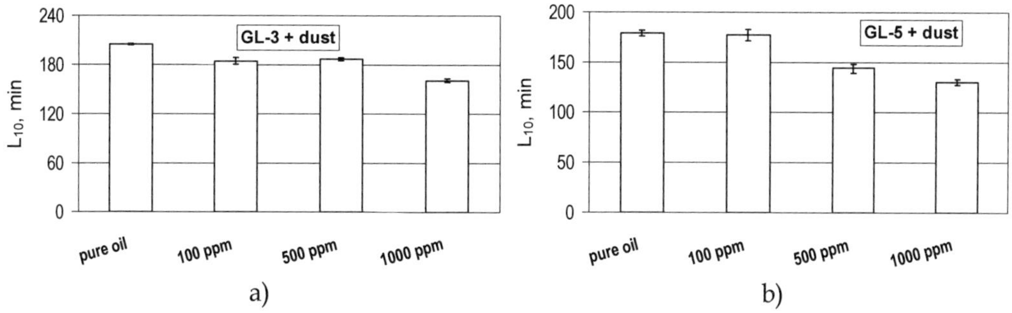

Fig. 21. Values of 10% fatigue lives (L_{10}) obtained for the gear oils - pure and contaminated with the test dust: a) GL-3 oil, b) GL-5 oil

From Fig. 21 it is apparent that the both contaminated gear oils give shorter fatigue lives with increasing concentration of the test dust.
The micro/nanotopography of the wear track surface on the top ball for the pure and dust contaminated GL-5 oil was inspected using AFM - Fig. 22.

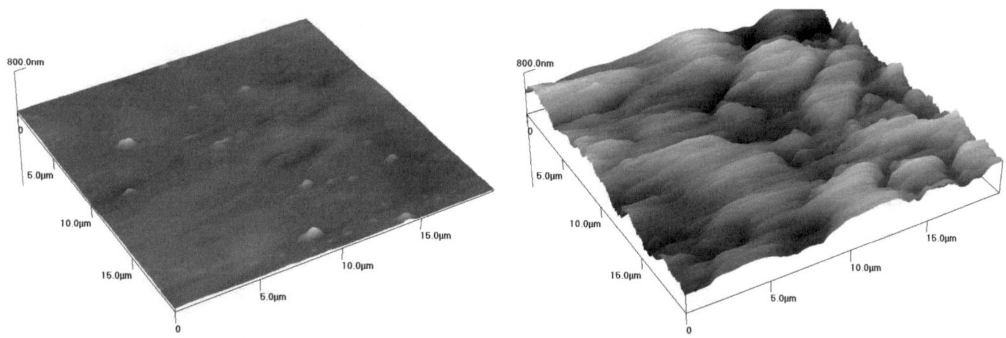

Fig. 22. AFM images of the wear track: a) pure GL-5 oil, b) GL-5 oil contaminated with the test dust at a concentration of 1000 ppm

It can be seen that the dust in the oil due to its abrasive action makes the worn surface rough and produces numerous surface defects. These defects act like stress raisers and accelerate initiation of surface fatigue cracks in this way. The abrasive action of dust particles resulted from their maximum size of 0.08 µm, which was much bigger than the minimum oil film thickness (0.04 µm).

6.3 Testing of water-contaminated gear oils

Fig. 23 presents the values of the 10% fatigue life (L_{10}) obtained for the gear oils of API GL-3 and GL-5 performance levels - pure and contaminated with water at increasing concentrations.

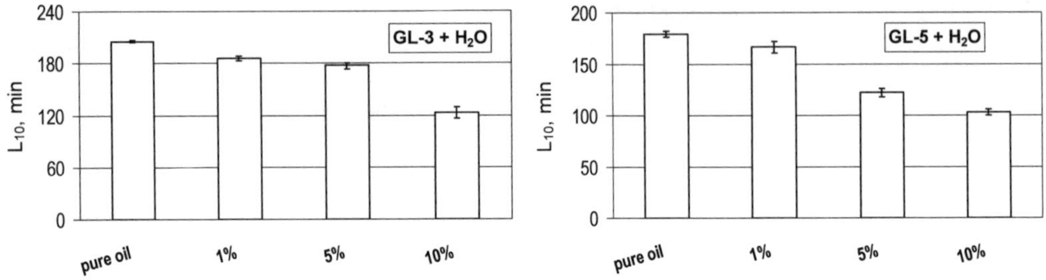

Fig. 23. Values of 10% fatigue lives (L_{10}) obtained for the gear oils - pure and contaminated with water: a) GL-3 oil, b) GL-5 oil

From Fig. 23 it is apparent that the both contaminated gear oils give shorter fatigue lives with increasing concentration of water. This is particularly noticeable for 10% water contamination in GL-3 oil as well as 5% and higher water content in GL-5 oil.

For interpretation of the obtained results the wear tracks on the top balls were analysed using SEM/EDS. SEM images of the worn surface and EDS maps for sulfur, phosphorus and zinc in the surface layer are shown in Figs. 24 and 25 for pure GL-3 oil and this oil contaminated with 10% water.

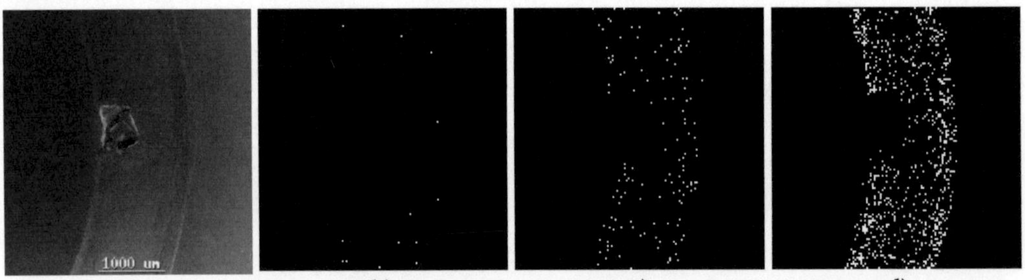

a) b) c) d)

Fig. 24. Pure GL-3 oil - SEM image of the wear track (a) and EDS maps for: b) sulfur, c) phosphorus, d) zinc

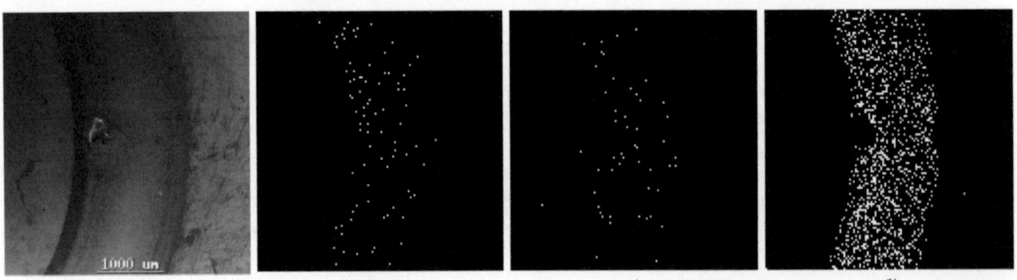

a) b) c) d)

Fig. 25. GL-3 oil contaminated with water at 10% concentration - SEM image of the wear track (a) and EDS maps for: b) sulfur, c) phosphorus, d) zinc

From Figs. 24 and 25 it is evident that water contamination affects the oil-surface interactions - one can observe a rise in sulfur and zinc content in the tribochemically modified surface layer of the wear track.

The next step of analysis was to quantitatively examine the wear track surface layer using EDS. Fig. 26 shows the weight concentration of sulfur and zinc (GL-3 oil) as well as sulfur and phosphorus (GL-5 oil) in the surface layer for both the gear oils contaminated with water. The analyses were performed at three different points of the wear track. The graphs present the average values of elemental concentration.

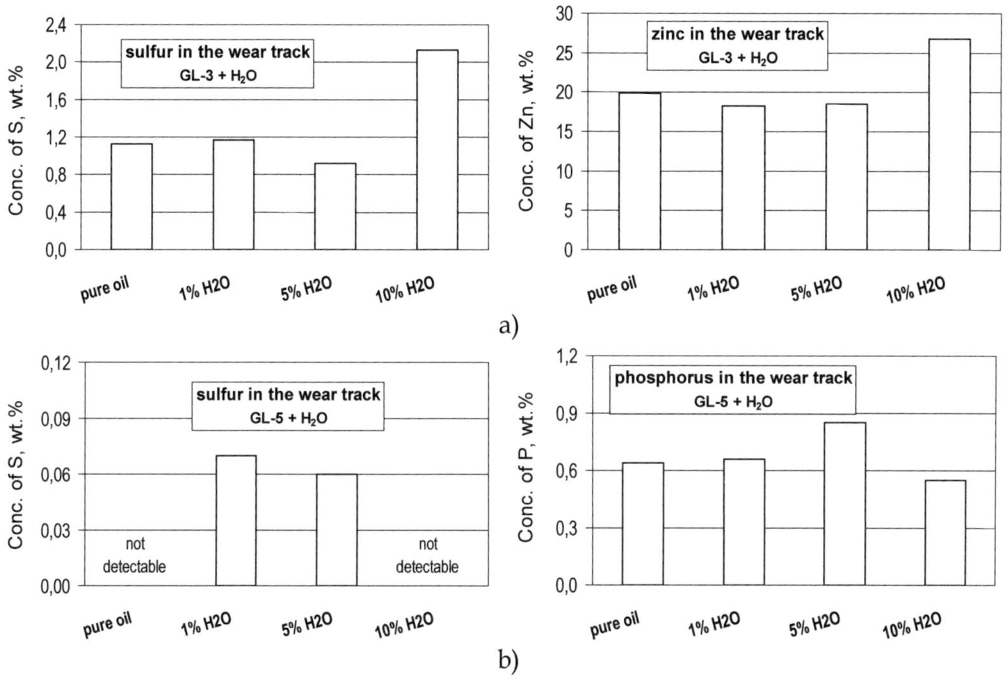

Fig. 26. Average concentration of sulfur, zinc, and phosphorus in the wear track surface layer for the gear oils contaminated with water: a) GL-3 oil, b) GL-5 oil

From Fig. 26 it is apparent that AW additives in GL-3 oil, having relatively low temperature of thermal decomposition, i.e. 200-300°C (Kawamura, 1982) under mild test conditions of rolling movement incomparably better tribochemically modify the wear track surface layer then EP ones. EP additives, present in GL-5 oil, with their much higher temperature of thermal decomposition, i.e. 400-500°C (Wachal & Kulczycki, 1988), have an incomparably lower chemical impact on the surface.

As can also be seen from Fig. 26, only for GL-3 gear oil contaminated with 10% water a significant change in the concentration of sulfur and zinc takes place in the wear track surface layer. For the contaminated GL-5 oil the concentration of sulfur is very low, within the limit of the sensitivity of the EDS technique. The content of phosphorus is also small and changes insignificantly. So, there is no evident correlation between the fatigue lives given by the water contaminated gear oils and elemental concentration of the tribochemically modified surface of the wear track.

Thus, for the oils contaminated with water a mechanism responsible for the drop in the fatigue life must be related to a decrease in the oil viscosity. This is followed by a drop in the thickness of EHL film leading to more frequent action of surface asperities; almost all of the load is carried in the plastically deformed tracks by asperity contact. More frequent cyclic stress results in a shorter fatigue life. Hypothetically, hydrogen embrittlement may also be at stake in case of oils contaminated with water, which is postulated elsewhere (Rowe & Armstrong, 1982), (Magalhaes et al., 1999).

6.4 Testing of oxidative degradation of gear oils

Fig. 27 presents the values of the 10% fatigue life (L_{10}) obtained for the gear oils of API GL-3 and GL-5 performance levels - pure ("fresh") and oxidised for longer and longer time.

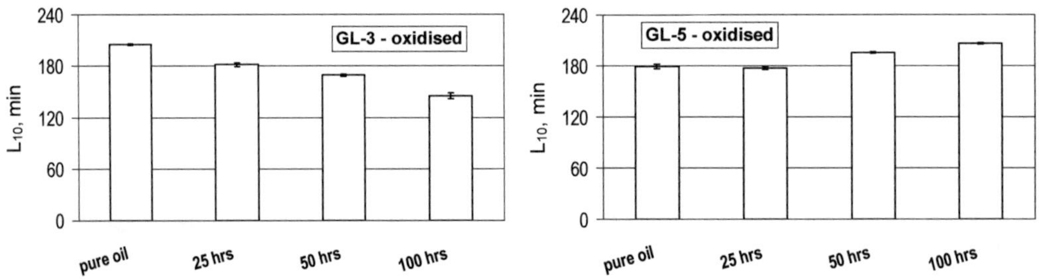

Fig. 27. Values of 10% fatigue lives (L_{10}) obtained for the pure and oxidised gear oils: a) GL-3 oil, b) GL-5 oil

From Fig. 27 it can be seen that the oil oxidation of GL-3 oil has an adverse effect on the fatigue life - its values steadily drop with increasing oxidation time. An opposite trend is shown by the oxidised GL-5 oil - the values of L_{10} increase with extending oxidation time, which is especially noticeable for the longest times.

For interpretation of the obtained results the wear tracks on the top balls were analysed quantitatively using EDS. Fig. 28 shows the weight concentration of sulfur and oxygen in the surface layer for the both oxidised gear oils.

From Fig. 28 it is apparent that for the oxidised GL-3 gear oil the concentration of sulfur in the surface layer of the wear track is much lower than for the pure oil. It comes from a decrease in the lubricating additives in the oil due to precipitation of their oxidised products in the form of sludge, which has been postulated in the literature (Yamada et al., 1993).

The changes in the physico-chemical properties due to oxidation were investigated by determination of TAN and FTIRM analysis of the tested oils. The values of TAN for the pure and oxidised oils are shown earlier in Fig. 18, and the IR spectra - in Figs. 19 and 20.

It has been already mentioned that the symptom of additives decrease in the oxidised GL-3 oil is a dramatic, several-fold drop in the area under the peak at 965 cm^{-1} in the IR spectrum; such a peak is typical of P-O-C bonds in the lubricating additives (ZDDP) in GL-3 oils.

In the literature a mechanism of the surface asperity softening due to a significant tribochemical modification is often attributed to fatigue life improvement achieved for lubricating additives. In this way surface asperities may be flattened, which reduces contact stress and in turn improves the fatigue life (Wang et al., 1996). So, worsening fatigue lives observed for the oxidised GL-3 oil (Fig. 27 a) may be attributed to the decrease in the concentration of sulfur in the worn surface (Fig. 28 a).

Another reason for reduction in the fatigue life for the oxidised GL-3 oil is related to the very high content of oxygen in the wear track surface layer (Fig. 28 a). Presumably, this comes from iron oxides. The role of such compounds seems rather deleterious as they can contribute to creation on the lubricated surface numerous corrosive micropits, being potential nuclei of fatigue cracks.

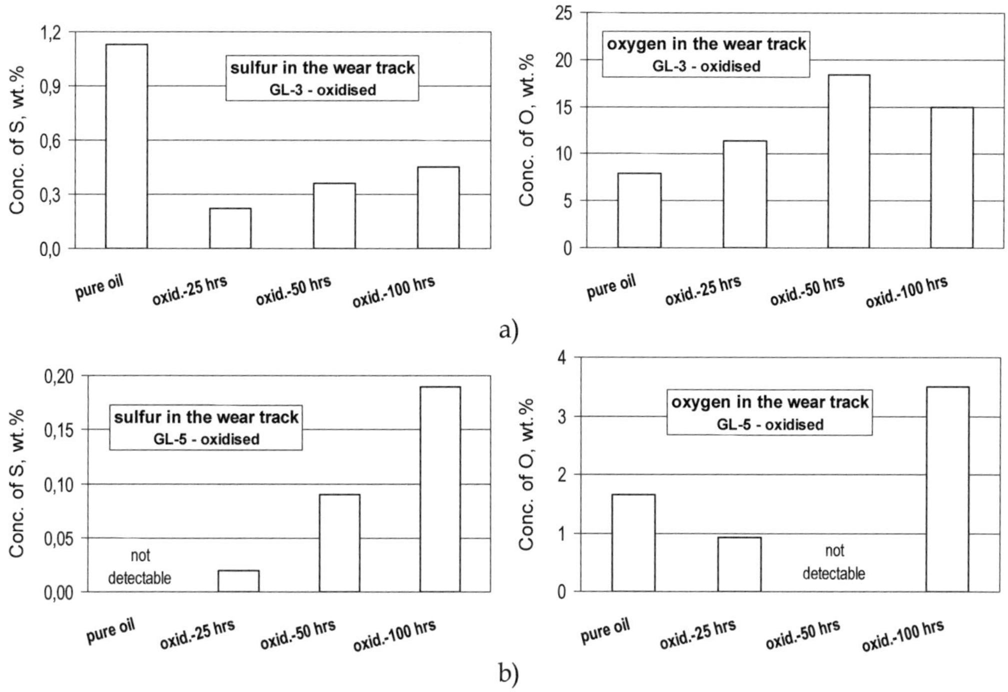

Fig. 28. Average concentration of sulfur and oxygen in the surface layer of the wear track for the oxidised gear oils: a) GL-3 oil, b) GL-5 oil

In case of the oxidised GL-5 oil, in the surface layer of the wear track a steady rise in the sulfur concentration takes place, although it is rather small (Fig. 28 b). A beneficial role of sulfur compounds has been mentioned earlier, so it may be a reason for fatigue life improvement observed for the oxidised GL-5 oil (Fig. 27 b).

The rise in fatigue lives given by the oxidised GL-5 oil can also relate to a decrease in the lubricating additives in the oil due to precipitation of their oxidised products. The symptoms of additives decrease in the oxidised GL-5 oil are: threefold drop in TAN for the longest oxidation time (Fig. 18 b) as well as nearly threefold drop in the area under the peak at 965 cm^{-1} in the IR spectrum (Fig. 20). The beneficial action of EP additives decrease is explained below.

EP type lubricating additives used in GL-5 gear oils are known for their high corrosion aggressiveness. It leads to creation on the lubricated surface numerous depressions and micropits due to corrosive wear, being potential nuclei for bigger "macropits". In this way the chance of failure increases, hence the fatigue life lubricated by EP additives tends to be reduced (Torrance et al., 1996). So, unlike in case of the oxidised GL-3 oil, the EP additives decrease in GL-5 oil due to oxidation exerts a beneficial influence on the surface fatigue life.

Like in case of the water contaminated oils, an adverse role of hydrogen embrittlement should not be neglected in case of oxidised gear oils.

7. Summary and conclusions

7.1 Scuffing tests

The contamination of the automotive gear oils of API GL-3 and GL-5 performance levels with the test dust practically does not affect their extreme pressure properties.

The contamination of the gear oils by water has a deleterious effect on their extreme pressure properties, however GL-3 oil is much more vulnerable to water contamination.
Oxidation exerts in general a positive effect on the both oils, however GL-3 oil shows a significant decrease in its extreme pressure properties after oxidation for the longest time.
SEM and EDS surface analyses show that there is a relationship between the extreme pressure properties of the aged gear oils and elemental concentration (sulfur and phosphorus) of the tribochemically modified surface of the wear scars.
So, from the point of view of the resistance to scuffing the most dangerous contaminant in automotive gear oils is water. However, ageing of such oils may even have a positive effect, like in case of the oxidised GL-5 oil.

7.2 Pitting tests

The ageing of the automotive gear oils generally exerts an adverse effect on the surface fatigue life (resistance to pitting). The only exception is for the oxidised API GL-5 oil - the fatigue life significantly improves for the longest periods of oil oxidation.
SEM, EDS and AFM analyses of the worn surface made it possible to identify factors having a deleterious (or beneficial) effect on the surface fatigue life due to action of the aged oils. So, dust in the oil produces numerous surface defects acting like stress raisers and accelerating initiation of surface fatigue cracks in this way. Water causes a drop in the oil viscosity, followed by a decrease in the EHL film thickness, leading to more frequent action of surface asperities, hence shorter fatigue life. For the oxidised GL-3 oil the fatigue life reduction results from a drop in the sulfur concentration in the worn surface; sulfur compounds formed by oil-surface interactions play a positive role in fatigue life improvement. A beneficial effect of oxidation of GL-5 oil on the fatigue life is related to a decreasing content of highly corrosive EP type lubricating additives due to precipitation of their oxidised products.
Although not investigated here, an adverse role of hydrogen embrittlement and iron oxides produced on the worn surface may also be at stake in case of oils contaminated with water and oxidised.
So, from the point of view of the resistance to rolling contact fatigue the most dangerous contaminants in automotive gear oils are dust and water.

7.3 Conclusions

Like in case of scuffing, also from the point of view of the resistance to pitting the GL-5 oil is generally more resistant to deterioration due to ageing than GL-3 oil.

8. References

Baczewski, K. & Hebda, M. (1991/92). *Filtration of working fluids*, Vol. 1, MCNEMT, ISBN 83-85064-17-6, Radom (in Polish)

Burakowski, T.; Szczerek, M. & Tuszynski, W. (2004). Scuffing and seizure - characterization and investigation, In: *Mechanical tribology. Materials, characterization, and applications*, Totten, G.E. & Liang, H., (Ed.), pp. 185-234, Marcel Dekker, Inc., ISBN 0-8247-4873-5, New York-Basel

Chwaja, W. & Marko, E. (2010). Driveline - What's happening, what's new, *Proc. III International Conference „Lubricants 2010" (proc. on flash memory)*, Rytro, Poland, 2010

Forbes, S. (1970). The load carrying action of organo-sulfur compounds - a review. *Wear*, Vol. 15, pp. 87-96, ISSN 0043-1648

Godfrey, D. (1968). Boundary lubrication, In: *Interdisciplinary approach to friction and wear*, Ku, P.M., (Ed.), pp. 335-384, Southwest Research Institute, Washington D.C.

Hohn, B.R.; Michaelis, K. & Weiss, R. (2001). Influence of lubricant ageing on gear performance. *Proc. 2nd World Tribology Congress*, p. 363, ISBN 3-901657-08-8, Vienna, 2001, the Austrian Tribology Society

Kawamura, M. (1982). The correlation of antiwear properties with the chemical reactivity of zinc dialkyldithiophosphates. *Wear*, Vol. 77, pp. 287-294, ISSN 0043-1648

Lawrowski, Z. (2008). *Tribology. Friction, wear and lubrication*, Oficyna Wydawnicza Politechniki Wroclawskiej, ISBN 978-83-7493-383-4, Wroclaw (in Polish)

Libera, M.; Piekoszewski, W. & Waligora, W. (2005). The influence of operational conditions of rolling bearings elements on surface fatigue scatter. *Tribologia*, Vol. 201, No. 3, pp. 205-215, ISSN 0208-7774 (in Polish)

Luksa, A. (1990). *Ecology of working fluids*, MCNEMT, ISBN 83-85064-13-3, Radom (in Polish)

Magalhaes, J.F.; Ventsel, L. & MacDonald, D.D. (1999). Environmental effects on pitting corrosion of AISI 440C ball bearing steels - experimental results. *Lubrication Engineering*, Vol. 55, pp. 36-41, ISSN-0024-7154

Makowska, M. & Gradkowski, M. (1999). Changes of zinc dialkyldithiophosphate content in lube oils during oxidation. *Problemy Eksploatacji*, Vol. 35, No. 4, pp. 127-133, ISSN 1232-9312 (in Polish)

Piekoszewski, W.; Szczerek, M. & Tuszynski, W. (2001). The action of lubricants under extreme pressure conditions in a modified four-ball tester. *Wear*, Vol. 249, pp. 188-193, ISSN 0043-1648

Pytko, S. & Szczerek, M. (1993). Pitting - a form of destruction of rolling elements. *Tribologia*, Vol. 130/131, No. 4/5, pp. 317-334, ISSN 0208-7774 (in Polish)

Rowe, N.C. & Armstrong, E.L. (1982). Lubricant effects in rolling-contact fatigue. *Lubrication Engineering*, Vol. 38, No. 1, pp. 23-30, 39-40, ISSN-0024-7154

Stachowiak, G.W. & Batchelor, A.W. (2001). *Engineering tribology*, Butterworth-Heinemann, ISBN 0-7506-7304-4, Boston-Oxford-Auckland-Johannesburg-Melbourne-New Delhi

Szczerek, M. & Tuszynski, W. (2002). A method for testing lubricants under conditions of scuffing. Part I. Presentation of the method. *Tribotest*, Vol. 8, No. 4, pp. 273-284, ISSN 1354-4063

Torrance, A.A.; Morgan, J.E. & Wan, G.T.Y. (1996). An additive's influence on the pitting and wear of ball bearing steel. *Wear*, Vol. 192, pp. 66-73, ISSN 0043-1648

Wachal, A. & Kulczycki, A. (1988). Thermogravimetric assessment of sorption of sulfur additives on the surface of iron. *Trybologia*, Vol. 97, No. 1, pp. 15-18, ISSN 0208-7774 (in Polish)

Wang, Y.; Fernandez, J.E. & Cuervo, D.G. (1996). Rolling-contact fatigue lives of steel AISI 52100 balls with eight mineral and synthetic lubricants. *Wear*, Vol. 196, pp. 110-119, ISSN 0043-1648

Winer, W.O. & Cheng H.S. (1980). Film thickness, contact stress and surface temperatures, In: *Wear Control Handbook*, Peterson, M.B. & Winer, W.O. (Ed.), pp. 81-141, ASME, New York

Yamada, H.; Nakamura, H.; Takesue, M. & Oshima, M. (1993). The influence of contamination and degradation of lubricants on gear tooth failure, *Proc. 6th International Tribology Congress EUROTRIB'93*, Vol. 2., pp. 241-246, Budapest

18

Development of a New 3D Nonwoven for Automotive Trim Applications

Nicole Njeugna[1], Laurence Schacher[1], Dominique C. Adolphe[1],
Jean-Baptiste Schaffhauser[2] and Patrick Strehle[2]
[1]Laboratoire de Physique et Mécanique Textiles EAC 7189 CNRS,
University of Haute Alsace
[2]N. Schlumberger
France

1. Introduction

Nowadays, the automotive manufacturers have to take into account the legislation on End Life Vehicle (ELV), especially the European Directive 2000/53/CE which constraints all automotive products to be at 85% recyclable and at 95% reuseable by January 2015 (EU Directive, 2000). The automotive multilayer structure used for automotive trim applications, fabric (PET) / foam (PU) / backing fabric (PA), does not offer ability for recycling or reusing and the question that has to be asked is "Could the PU foam used in the automotive trim applications be replaced by a mono component spacer material?" One answer is to propose an eco-friendly solution presenting a mono material product. Moreover, this new product has to answer to the automotive specifications in terms of lightness, formability and cost. Some solutions for PU foam replacement have been proposed, such as spacer fabrics presenting a vertical orientation of the yarns (weaving and knitting technologies) or a vertical orientation of the fibers (nonwoven technology). The vertical orientation of the fibers will improve the mechanical properties of the fabric especially for the compressional ones. Critical analyses between the different 3D textiles technologies show that the nonwoven technology provides the best industrial solution in terms of cost and productivity. Regarding the 3D nonwoven products, the "on the market" ones present drawbacks that do not allow them to answer positively to the initial question concerning the replacement of the PU foam. Indeed, the structure of these 3D nonwovens does not present a perfect vertical orientation of the fibres (Njeugna, 2009). Consequently, these products do not offer a maximal resilience in terms of compression properties.
In this context, a French consortium composed of research laboratory (LPMT as project leader), textile industrialists (N. Schlumberger, AMDES, Protechnic, Landolt, Dollfus & Müller, Rhenoflex Dreyer), textile technical centre (IFTH[1]) has been formed to develop an eco-friendly 3D nonwoven which would not present the previous drawbacks. This new 3D nonwoven could be used to replace polyurethane foam classically used in automotive trim applications. This consortium has been supported by the Alsace Textile Cluster, the Alsace

[1] IFTH : Institut Français du Textile Habillement, www.ifth.org

Region and the "Département du Haut-Rhin". This collaborative research project, named VERTILAP, has been labelled by the French competitiveness cluster "Vehicle of the Future" in 2006 and the French "Fibres Innovative cluster" in 2009.
This chapter will present the state of the art of the technical textiles classically used as automotive trim such as seat and door panel upholsteries. The manufacturing processes and the specifications of these automotive multilayer fabrics will be exposed. Their methods of characterization will be presented. The description of the PU foam and the problem it raises will be highlighted. The state of the art of the existing 3D textiles for PU foam substitution, processes and products will be detailed. This chapter will also present the principle of the VERTILAP® process and the experimental procedure which has been used to realise the VERTILAP® products. Methods and tools of characterization that have been developed in order to evaluate the physical and compression properties of this new material will be exposed. The comparative study that has been carried out between the VERTILAP® products and the classical automotive fabrics in the case of monolayer and multilayer structures will be detailed too.

2. Bibliographical study

2.1 Textiles used for automotive upholsteries

The textile fabric is an interesting material for automotive industry regarding its functionality (lightness, acoustic and thermal insulation, etc.) and its mechanical behaviour. It is used in three main components of the car: the interior, the engine compartment and the pneumatics (Némoz, 1999). The car interior has significantly evolved since the last decade and has become one of the key elements of the customer purchasing. Nowadays, the consumer pays special attention to the environment inside the car. Therefore, the factors of comfort, beauty (harmony of colours and designs) and security have become main factors in the sale of a vehicle. Since 90s, the car manufacturers have significantly increased the use of textiles in the interior trim. Actually, the weight of an European vehicle includes 11 kg of textiles on a surface of 16 m^2. Textile fabrics used for the seat are employed on a visible surface of 3.8 m^2 while those used for the door panel are employed on a visible surface of 1.7 m^2. (DGE, 2005), (Fung & Hardcastle, 2001)
This study aims to present the state of the art on the technical textiles classically used as seat and door panel upholstery in the car interior. Examples of automotive seat and door panel are illustrated on Fig. 1 and 2.

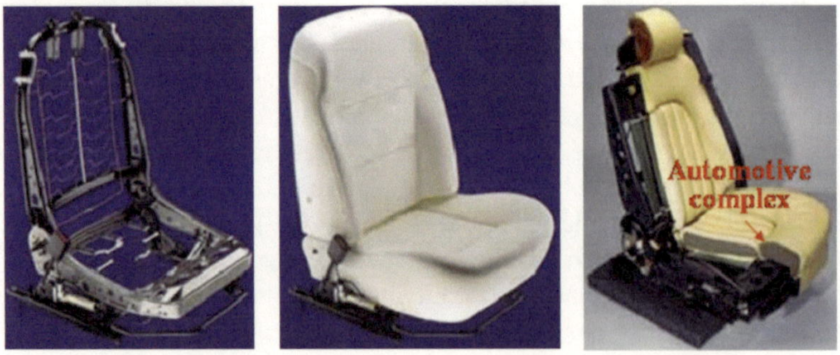

(a) (b) (c)

Fig. 1. Automotive seat: structure (a), foam cushion (b), automotive complex (c)

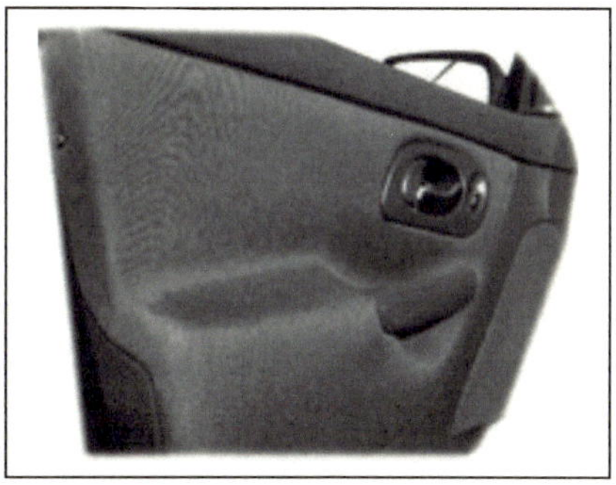

Fig. 2. Example of an integral door panel

Different methods of construction of seat and door panel are listed in the literature review (Fung & Hardcastle, 2001). The seat trimming can be realised thanks to the "foam in fabric" technique, the direct joining technique or the injection moulding technique. The "foam in fabric" technique consists on slipping the automotive complex on the seat cushion. The direct joining technique consists on spraying a solvent adhesive either on the automotive complex, either on the foam cushion or both in order to link them together. In the case of injection moulding technique, the foam is directly injected into the automotive complex previously placed in a mould. Textile-insert low pressure moulding, using polypropylene resin, is used to produce a covered door panel in a single operation.

The automotive complex (Fig. 3) is usually composed of a decorative fabric made of polyester, polyurethane foam and a backing fabric made of polyamide. The polyurethane foam is generally a thin layer with a thickness between 2 mm to 8 mm and a mass per unit area of about 200 g/m². The foam gives the flexibility and the soft touch while the backing fabric gives the dimensional stability to the multilayer structure. In case of "foam in fabric" technique, the backing fabric contributes to facilitate the slippage of the cover laminate on the foam cushion. The backing fabric is not necessary used in the case of door panel upholstery. (Caudron, 2003), (ITF, 1990)

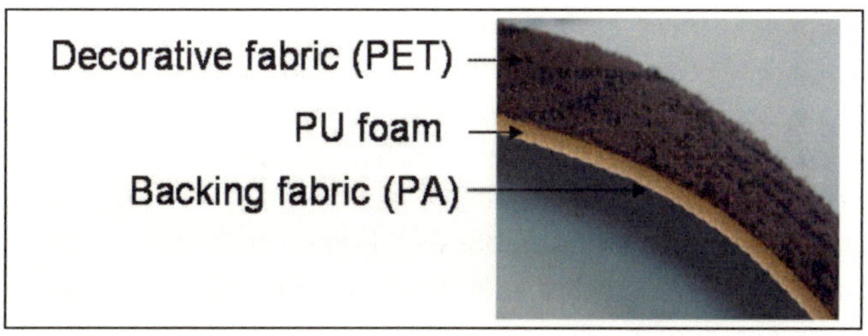

Fig. 3. The automotive complex

The automotive complex can be produced thanks to different techniques (Hopkins, 1995). Some of them are well known as the flame lamination and the dry lamination processes. In the flame lamination process (Fig. 4), the textile layers and the PU foam are linked together using the PU foam as an adhesive. This process has the disadvantage to generate toxic gases. The maximal speed can reached 25 m/min. In the dry lamination process (Fig. 5), hot melt adhesives (web, film, powder) are used to bind the textile layers and the PU foam. This process does not generate toxic gases as the flame lamination one but its main drawback is its cost. The maximal speed can reached 16 m/min.

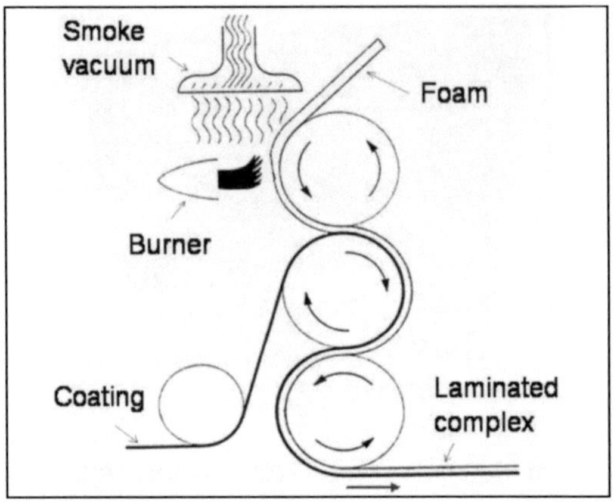

Fig. 4. The flame lamination process

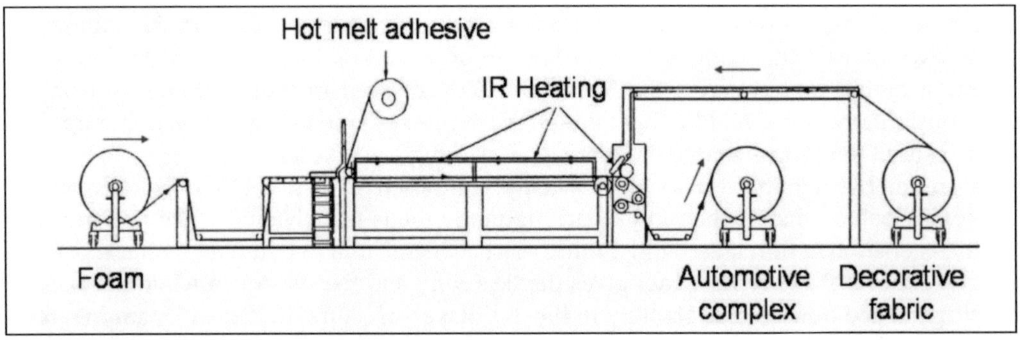

Fig. 5. The dry lamination process

It is important to note that the specifications and the characterisation tools of the automotive complex are specific to each car manufacturer. These specifications take into account the legislation of the markets, the security, the quality of the products and their cost (Faucon, 1995). For example, they have to be fire-proof, as light and as cheap as possible. Their quality is evaluated thanks to specific characterisation such as the mechanical behaviour (compression, tensile, flexibility, etc.), the physical behaviour (colour fastness, air permeability, etc.), the fogging, etc. International standard methods of characterisation of flexible cellular polymeric materials used in the automotive industry are well known such as:

- Determination of stress-strain characteristics in compression (ISO 3386/1, 1986)
- Determination of tensile strength and elongation at break (ISO 1798, 1983)
- Determination of compression set (ISO 1856, 2000)
- Determination of burning behaviour of interior materials. (ISO 3795, 1989)
- Etc.

2.2 The problem of the PU foam

The PU foam, thanks to its specific characteristics, is the key element of the multilayer fabric in terms of comfort and mechanical behaviour especially for the compression ones. It is obtained thanks to a chemical reaction between an isocyanate and a polyol (Fig. 6). The expansion of the foam is due to the reaction between the isocyanate and water. After this expansion, the foam will present a cellular structure which can be characterised by opened or closed cells (Fig. 7). (Recticel, 2009), (Berthier, 2009)

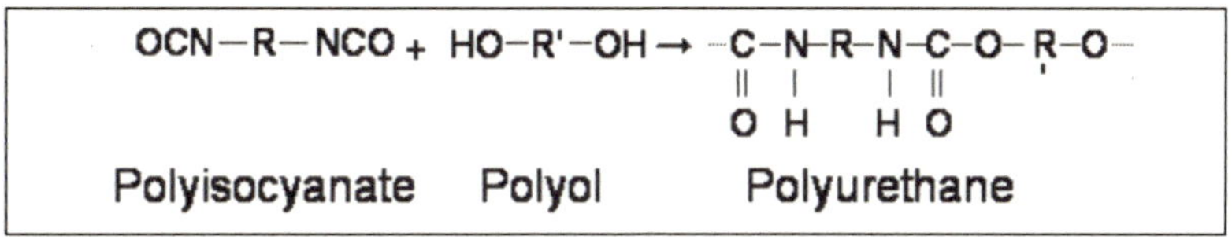

Fig. 6. Chemical polyaddition reaction of the formation of the PU foam

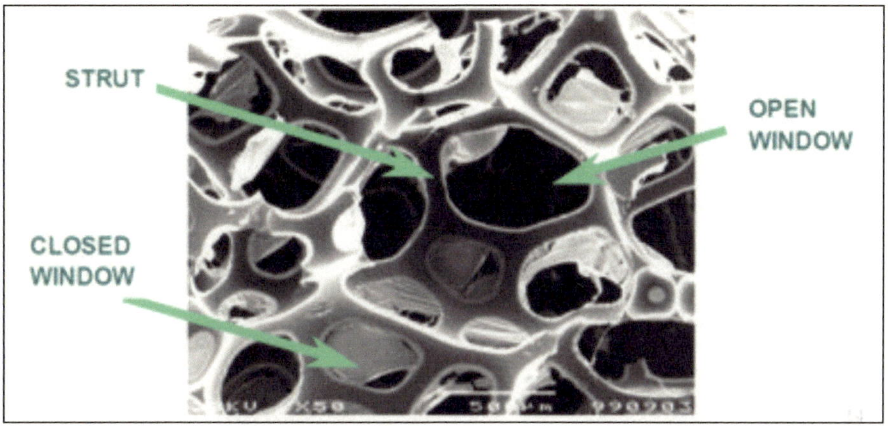

Fig. 7. Microscopic structure of the PU foam

The main problem of the PU foam is partly the toxic gases it generates during its manufacturing process as previously mentioned but also the recycling of the automotive complex at the end life vehicle. In fact, the recycling processes of such products require a delamination step of the different layers (PET, PU, PA). This operation is not optimal because some PU foam remains on the textile fabrics. It is also important to note that the machines used for the recycling are very expensive. On another hand, it is difficult to completely recycle the PU foam in spite of the developments which have been carried out on this way. Nowadays, some foam manufacturers like RECTICEL is developing new method to produce PU foam by using biochemical compounds (Persijn, 2008). It is already the case with their foam PURECELL® which contains at least 20% of natural compounds. Beyond this new development stay the ethical problem of the massive agricultural exploitation for the industry.

The PU foam has many serious drawbacks such as flammability, gases emissions due to the laminating processes. These problems lead to the question of its replacement by a new product. A key aspect of this new product is not to alter the product functionality. It means that the new product should present at least mechanical properties, especially compressional properties closed or equal to the actual automotive multilayer fabric. Another key aspect is to propose an environmentally friendly solution for complex fabric composed of a mono material product. This new product has to answer to the automotive specifications in terms of weight, formability and cost. In this context, industries and researchers all around the world are developing new products which could substitute the PU foam. (Kamprath, 2004), (Persijn, 2008)

2.3 Existing solutions to the PU foam replacement

The 3D textiles offer a good solution to the recycling issue of the multilayer products using PU foam because of their specific structure as spacer fabric. In fact, they present a vertical orientation of the yarns (weaving and knitting technologies) or a vertical orientation of the fibres (nonwoven technology). This vertical orientation will provide a good mechanical behaviour especially in term of compression. Analyses of the existing solutions have been carried out by textile industrialists and the obtained results show that the 3D textile technologies offer the best solution in terms of product quality and cost. It appears that the nonwoven technology provides the most interesting solution in terms of mechanical properties, cost and productivity. The nonwoven products issued from the 3D technology are known as (Struto, 2007), (Santex, 2007), (Karl Mayer, 2007), (Vasile et al., 2006). They can be divided in three categories: carding and vertical lapping processes, stitch-bonded processes and needle-punched processes.

- Carding and vertical lapping processes

STRUTO®, Santex WAVEMAKER® and V-Lap® technologies are vertical lapping system whereby a carded web is pleated in order to create 3D structure (Fig. 8 and 9). A thermal treatment is applied on the pleated structure in order to obtain the final product. The V-Lap® technology is closed to the STRUTO® one.

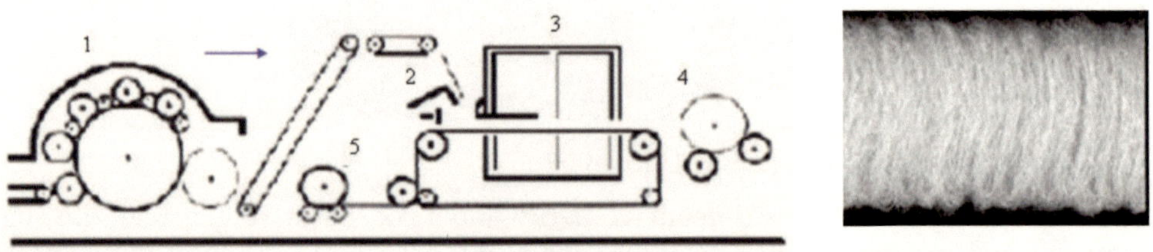

1 Card; 2 Vertical lapping system; 3 Oven; 4 3D nonwoven; 5 laminating layer.

Fig. 8. The STRUTO® process (left) and product (right)

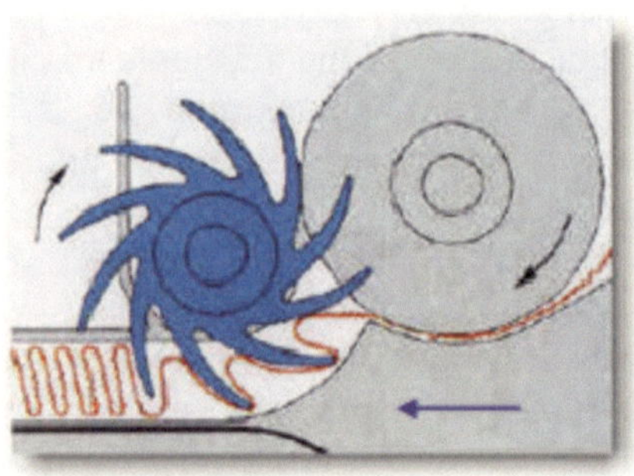

Fig. 9. The Santex WAVEMAKER® process

- Stitch-bonded processes

KUNIT and MULTIKUNIT are stitch-bonded technologies developed by Karl Mayer Textilmaschinenfabrik GmbH. The principle of these techniques is based on the principles of the stitching and the knitting technologies (Fig. 10). The KUNIT fabric presents a stitch side and a pile side. This fabric is used as base material for MULTIKUNIT production.

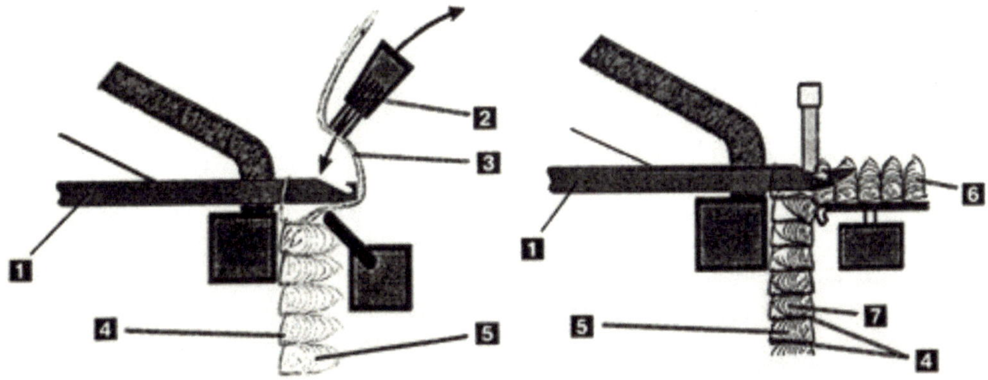

1 Compound needle; 2 Brush bar; 3 Fibre web; 4 Stitch side; 5 Pile side; 6 KUNIT nonwoven; 7 MULTIKUNIT nonwoven.

Fig. 10. The KUNIT (left) and MULTIKUNIT (right) processes

- Needle-punched process

NAPCO® is a needle-punched technology developed by the textile machinery manufacturer LAROCHE. The NAPCO® process consists to link two pre-needle nonwovens thanks to a fibrous bridge (Fig. 11). The obtained 3D structure is mainly used for composite application.

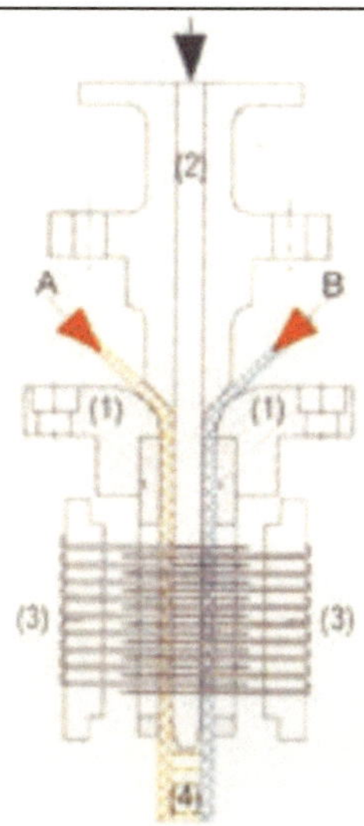	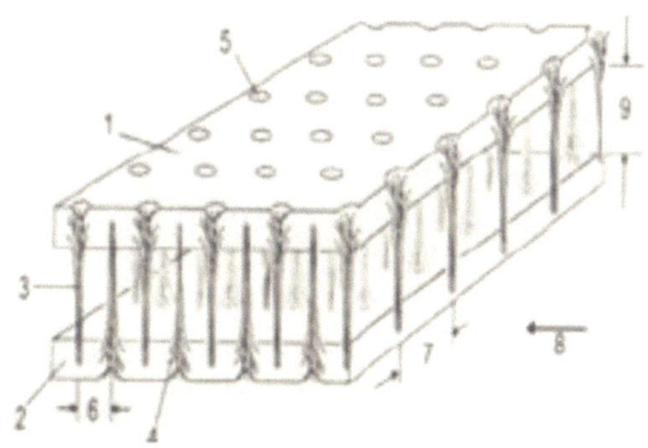
A and B are pre-needle nonwovens; 1 Stripper plate; 2 Spacer tables; 3 Needles' area; 4 Fibres' bridges.	1 Top layer; 2 Bottom layer; 3 Connecting layer (bridge fibres from 1); 4 Bridge fibres from 2; 5 Needle stitch; 6 Distance between bridge fibres depending on stitch depth; 7 Distance between bridge fibres depending on needle density; 8 Take-out direction; 9 Product thickness depending on the spacer's width.

Fig. 11. The NAPCO® process (left) and the obtained 3D structure (right)

The 3D nonwoven technologies allow producing bulky nonwoven presenting a low density with a maximal resilience. However, the "on the market" 3D nonwovens obtained through the existing vertical lapping processes present drawbacks (structure behaviour) that do not allow them to answer positively to our initial question concerning the replacement of the PU foam. Indeed, their vertical orientation is not optimum and the structure could be crushed when vertically compressed with a significant shear moment between top and bottom surfaces. Consequently, this work aims to answer positively to this question by developing a new 3D nonwoven obtained through a patented process VERTILAP® of the N. Schlumberger Company (Dumas et al., 2007). The VERTILAP project aims to develop a new pleated 3D nonwoven. This project has been conducted in order to involve the different partners when their skills and know-how were needed in the project. Automotive upholsteries for headrest and door panels have been also realised in order to demonstrate the taylorability and formability of the VERTILAP® 3D nonwoven.

3. Presentation of the VERTILAP® process

The VERTILAP® process (Fig. 12) is a vertical lapping system whereby a tow or a web is pleated thanks to folding elements. The process is composed of four main functions:

- The opening and defibering of the tow
- The verticalisation of the tow
- The extraction and condensation of the pleats
- The thermobonding and the lamination of the 3D pleated structure.

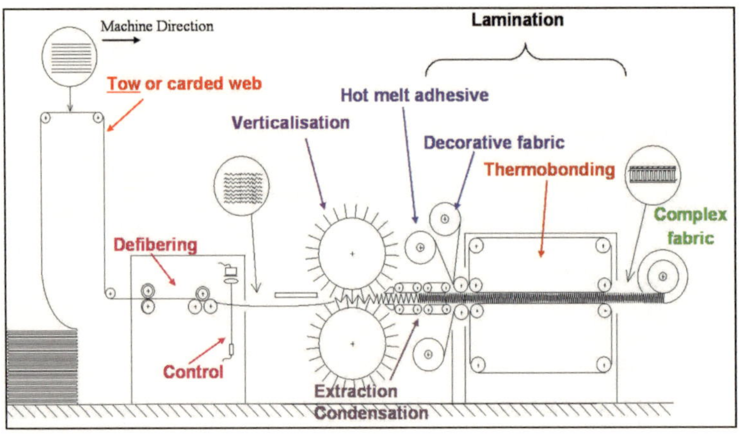

Fig. 12. The VERTILAP® process

The opening of the tow will allow spreading the filaments on the creel. To obtain a good product's homogeneity, the tow's section must be spread as evenly as possible.

The defibering function is a filament separating zone. It is necessary to individualise the filaments inside the tow. The defibering principle (Fig. 13) consists to separate the filaments by driving them into a tensioning separating zone. The filament separating cylinder set is composed of a cylinder with square threading, topped by a rubber coated pressure roller on which a pneumatic pressure is applied. Along the contacting generator, zones where the filaments are alternately nipped and released are successive. If we consider two neighbouring filaments, one will be tightened a little before the other and release a little before the other, so that their crimping are not any longer facing each other and will not reimbricate any more. The tension to carry out in this defibering zone must be lower than the filament elastic limit. This function of the machine has a considerable influence on the tow quality. (NSC, 2007)

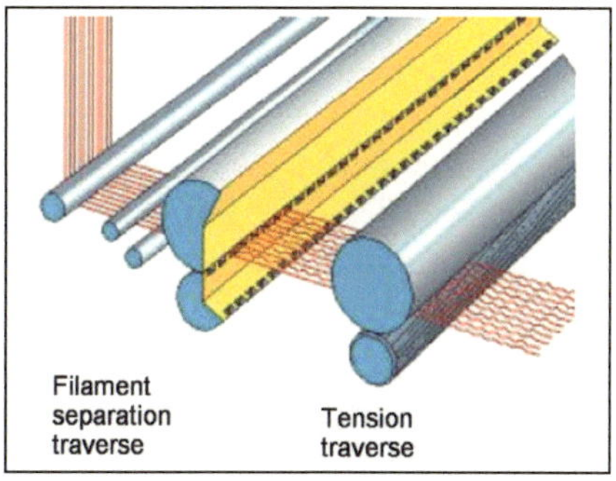

Fig. 13. The defibering principle

After being defibered, the tow is verticalised in order to create the pleats. These last ones are then extracted from the verticalisation zone and condensed to obtain the pleated structure. At this step, additional layers can be joined by thermo binding on the 3D pleated structure to fix it and to obtain the final multilayer fabric. In this process, the thermal treatment is essential in the formation and the fixation of the pleats and the 3D structure.

In this study, the VERTILAP® process is presented as an experimental prototype of 20 cm width. As input, tow was a bi-component co-polyester/polyester presenting a count of 90 ktex and a filament's count of 4.4 dtex. The experimental prototype is suitable for tows presenting a count lower than 30 ktex. A filament separation technique has been developed to divide the initial tow of 90 ktex into finer ones. The tow has been pleated under the glass transition temperature of the co-polyester sheet which is 73°C. The obtained 3D nonwovens present a thickness of 6 mm. The laminating function was done separately thanks to a flatbed laminating system (Meyer Company, 2007) provided by Protechnic Company (Fig. 14). The 3D nonwovens have been laminated with external layers made of polyester and co-polyester hot melt adhesives. The VERTILAP® experimental prototype has been controlled through the following parameters: tow's count, speeds before and after the verticalisation zone, temperature of the verticalisation zone. The laminating process has been regulated through the speed, the pressure and the temperature.

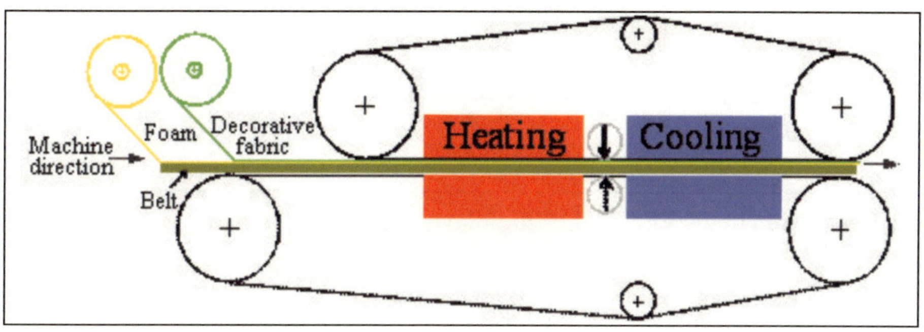

Fig. 14. The laminating process

Two kinds of VERTILAP® products have been manufactured: the monolayers and the multilayers. The obtained multilayer products have always been made of 100% polyester in order to facilitate their recycling.

4. The experimental study

Experimental study has been made in two campaigns of production, A and B, followed by complete characterisation test of the manufactured products. For each campaign, the VERTILAP® products have been compared to automotive PU foams. The tested materials, the methods and tools of characterisation and the obtained results of the comparative study between the VERTILAP® products and the PU foams are presented below. (Njeugna, 2009)

4.1 Tested materials

Five different monolayer 3D nonwovens (NT1, NT2, NT3, NT4 and NT5) have been manufactured in campaigns A and B. From each of them, multilayer products have been prepared using needle-punched, spun-bonded nonwovens and knitted fabric as external layers. Two different monolayer foams (m1, m2) classically used by car manufacturers and representing two kinds of comfort have been tested. The tested automotive multilayer product (Cm) is composed of three layers, decorative fabric (PET) / PU foam / backing fabric (PA). The tested samples are presented in Table 1. The description of the different types of the VERTILAP® multilayer samples is presented in Table 2.

	VERTILAP® products		**PU foams**
	Campaign A	**Campaign B**	
Monolayer	NT1, NT2, NT3, NT4	NT5	m1, m2
Multilayer	L1, L2	L3	Cm

Table 1. The tested samples

NT1, NT2, NT3 and NT4 are 3D nonwoven.

	Samples	Laminating components
Campaign A	L1	NT40 / 3D nonwoven / NT40
	L2	NT44 / 3D nonwoven / NT44
Campaign B	L3	NT40 / 3D nonwoven / T200

Table 2. Description of the VERTILAP® multilayer samples

With:

- NT40 is a needle-punched nonwoven presenting a mass per unit area of 40 g/m^2.
- NT44 is a spun-bonded nonwoven presenting a mass per unit area of 44 g/m^2.
- T200 is a knitted fabric presenting a mass per unit area of 200 g/m^2.

4.2 Methods and tools of characterization

4.2.1 Physical characterization

The pleated structure of the 3D nonwoven obtained thanks to VERTILAP® process has been geometrically described as a triangle after the verticalisation (Fig. 15) and as a loop after the laminating process (Fig. 16). In both cases, the shape has been characterised by the thickness (e_0), the pleat's angle (θ), the rate of condensation, the pitch (p) and the fibrous wall thickness (e_p) which has been neglected in order to simplify the model.

The pleat's angle will indicate the vertical orientation of the pleat. In the case of the triangular shape, the vertical orientation will be reach with a value of the pleat's angle closed to 0°. In the case of the loop shape, the vertical orientation will be reach with a value of the pleat's angle closed to 90°.

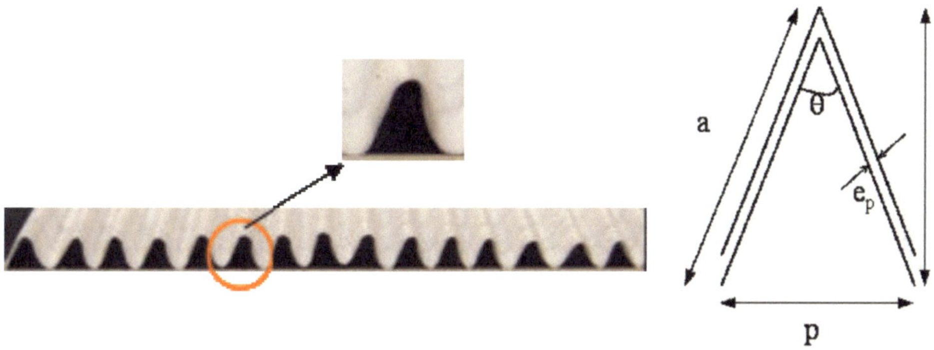

Fig. 15. Geometrical modelling of the pleat after verticalisation process

After the verticalisation process, the geometrical parameters of the pleat have been defined by the following equations:

$$p = \frac{l_r}{n_p} \tag{1}$$

$$a = \sqrt{(e_0)^2 + (\frac{p}{2})^2} \tag{2}$$

$$l_a = 2.a.n_p \tag{3}$$

$$\theta = 2.\arctan(\frac{p}{2.e_0}) \tag{4}$$

$$Tx_c = 100.\frac{l_a - l_r}{l_a} \tag{5}$$

Where:

a is the hypotenuse

l_a is the apparent length of the sample when its pleated structure is flattened

l_r is the real length of the sample in its pleated structure

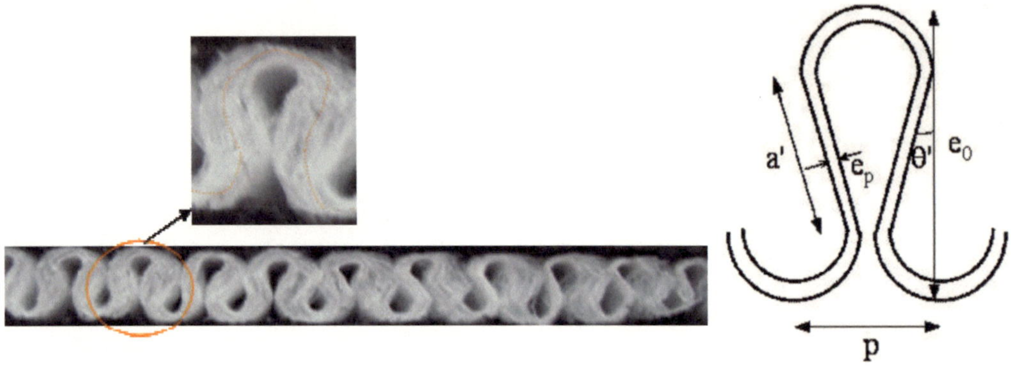

Fig. 16. Geometrical modelling of the pleat after the laminating process

After the laminating process, the geometrical parameters of the pleat have been defined by the following equations:

$$p = \frac{l'_r}{n_p} = 2.r \tag{6}$$

$$a' = \sqrt{(e_0 - p)^2 + (\frac{p}{2})^2} \tag{7}$$

$$l_{loop} = \pi.p + 2.a' \tag{8}$$

$$l'_a = n_p.l_{loop} \tag{9}$$

$$\theta' = \arctan\frac{p}{2(e_0 - p)} \tag{10}$$

$$T'x_c = 100.\frac{l'_a - l'_r}{l'_a} \tag{11}$$

Where:
a is the hypotenuse
l'_a is the apparent length of the sample when its pleated structure is flattened
l'_r is the real length of the sample in its pleated structure
l_{loop} is the length of the loop

In the case of the foam, the cellular structure has been geometrically modelled as a pentagon (Fig. 17) thanks to an adapted method developed from *VISIOCELL®* used by the OEM (Original Equipment Manufacturer) (Recticel, 1999), (Drean, 2006). The foam has been characterised thanks to the horizontal and vertical mean cell sizes. The sizes are measured on different area of the sample and on a group of five adjacent cells. The characteristics of the tested foams are presented in Table 3.

The physical characterisation of the 3D nonwoven has been extended to comfort evaluation which can be evaluated thanks to air permeability (BS 5'636, 1990) and thermal insulation

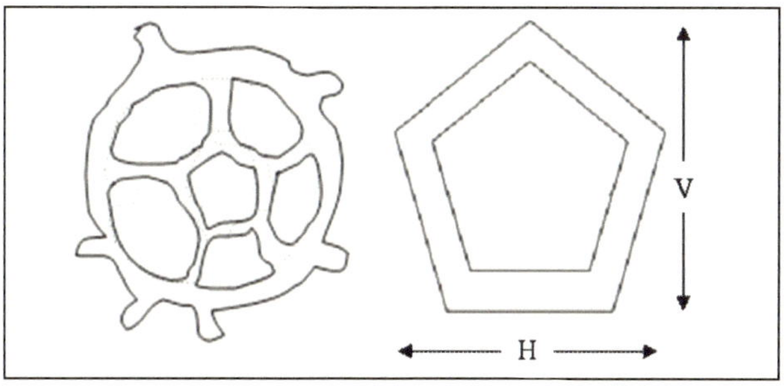

Fig. 17. Geometrical modelling of the PU foam

	m1	m2	Cm
Thickness (mm)	5	5	4
Weight (g/m²)	182	180	386
Density (kg/m³)	36	36	87
Vertical cell mean size, V(μm)	0.25	0.21	0.25
Horizontal cell mean size, H (μm)	0.25	0.31	0.30

Table 3. Characteristics of the tested foams

(Kawabata, 1980). The air permeability measurement has been performed by using the air permeability tester FX3300 under a pressure of 98 Pa on a surface of 5 cm². The coefficient of thermal conductivity (K, unit in W/m.K) has been measured thanks to the KES-FB7 thermolab II of Kawabata Evaluation System for Fabrics. The measurements were performed at 23°C during 60s on a sample surface of 25 cm². The apparatus have been customized in order to minimise the air leakage and the heat losses on the lateral edges (Fig. 18).

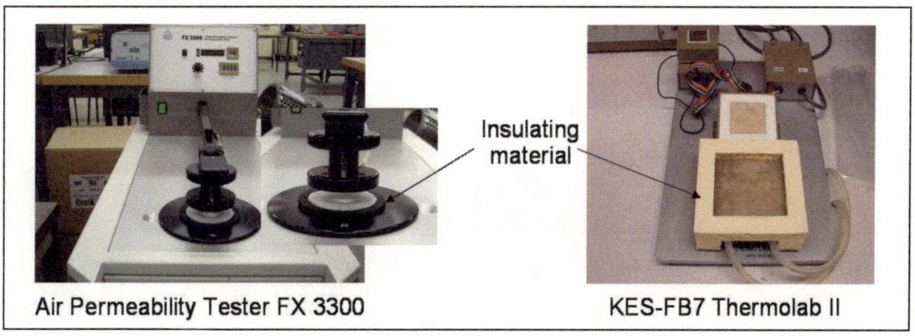

Fig. 18. Customization of testing apparatus for air permeability and thermal conductivity

4.2.2 Compression characterization

The mechanical characterisation has been focused on the compression behaviour because it is the most important property to analyse on the new 3D product if compared with the PU foam. The compression behaviour has been evaluated thanks to two different testing methods; a first method based on the Kawabata recommendations and a second method based on automotive standard ISO 3386/1: 1986.

The first testing method has been carried out on the KES-FB3 module. For that, two procedures have been successively defined, the first one using the standard conditions of Kawabata and the second one derived from these conditions. In fact, the standard configuration of the apparatus highlighted during the test an indentation phenomenon on the pleated material (Fig. 19) which was due to the small surface (2 cm^2) of the compression plate in regards with the testing sample structure (100 cm^2): only one or two pleats were under the pressure foot during the test. (Njeugna et al., 2008)

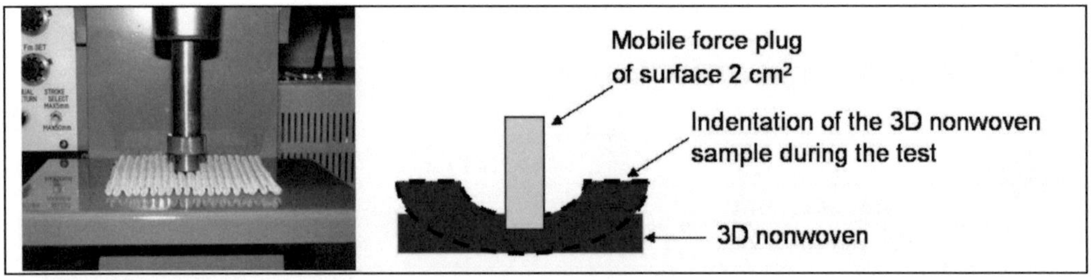

Fig. 19. Standard configuration of the KES-FB3 compression tester

The second procedure consisted on modifying the surface of the compression plates in order to compress the testing sample on its whole surface (Fig. 20). The obtained results avoid indentation phenomenon observed on initial tests. They have also shown the resilient property of the pleated 3D nonwoven. The second procedure has been validated for the compression characterisation. The samples have been compressed under a maximal load of 3 kPa at a speed of 12 mm/min during one cycle. The results give information on the thickness, the compressibility, the dissipated energy and the resilience of the material.

Fig. 20. Customization of the KES-FB3 compression tester

The second testing method has been carried out on a universal screw driven testing machine (Instron 33R4204) fitted with 5 kN load cell (Fig. 21). The solicitation speed was at 12 mm/min. The tests have been performed in static mode. A sanding paper has been fixed on the surface of the fixed compression plate of the Instron machine in order to eliminate any slippage of the sample during the test. The samples have been compressed up to 50% of their initial thickness then decompressed at the same speed until the plates come back to their initial locations. Five cycles of compression have been performed with a rest time of 10s between each cycle. The stress deformation curves have been plotted. The maximal stress at 50% deformation of the initial thickness and the dissipated energy have been determined.

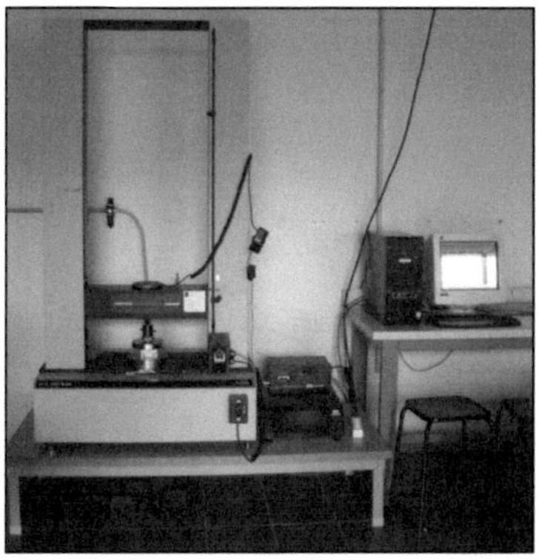

Fig. 21. INSTRON 33R4204 testing device

4.3 Comparative study: VERTILAP® products vs. PU foams

4.3.1 Campaign A

Production of the VERTILAP® products has been made in two steps respectively dedicated to the preparation of the feeding tow and to the manufacturing of the 3D nonwoven. The feeding tows have been prepared thanks to a simple manual technique whereby the initial tow of 90 ktex has been divided into finer tows presenting a count from 9 ktex to 18 ktex. The obtained tows have been defibered in a converting machine (NSC, 2007) in order to improve the quality of the filament opening. During the manufacturing process, the speeds before and after the verticalisation zone have been varied. A digital camera has been used to observe the products throughout the processing range. These observations have shown irregularities in the formation of the compacted 3D structure. It has also been observed that the outgoing product was still hot at the output of the machine. This observation has allowed showing that the condensation process was not fully controlled. The single 3D nonwovens have been laminated under a speed of 2 m/min at 150°C. The hot melt adhesive was a 25 g/m^2 co-polyester web with a melting temperature of 120/125°C.
The geometrical modelling of the pleated 3D nonwoven has shown that they present a pleat's angle of 41°, a rate of condensation of 65% and a number of pleats/cm of 2.2. The pleats in the laminated structure present an angle of 57° and a rate of condensation of 77%. The pleat's angle and the rate of condensation have respectively increased of 28% and 16% after the laminating process.
The results of the physical characterisation have shown that, in the case of monolayer products (Fig. 22), the 3D nonwovens are thicker and more comfortable in terms of air permeability than the PU foams. They are less comfortable in terms of thermal insulation compared to m1 sample. The PU foams are twice lighter than these new products.
In the case of the laminated products (Fig.), the VERTILAP® products are thicker and more comfortable in terms of air permeability and thermal insulation than the tested multilayer foam (Fig. 23). They are also heavier than the tested foam.
The results of the compression behaviour on one cycle test (KES-FB3) have shown that, in the case of monolayer products (Fig. 24), the 3D nonwovens and the PU foams globally present the same resilient behaviour. The PU foams are more compressible than the tested 3D nonwovens and they dissipated less energy. This last result shows that the 3D nonwovens will present better characteristics in term of soft touch compared with the PU foams.

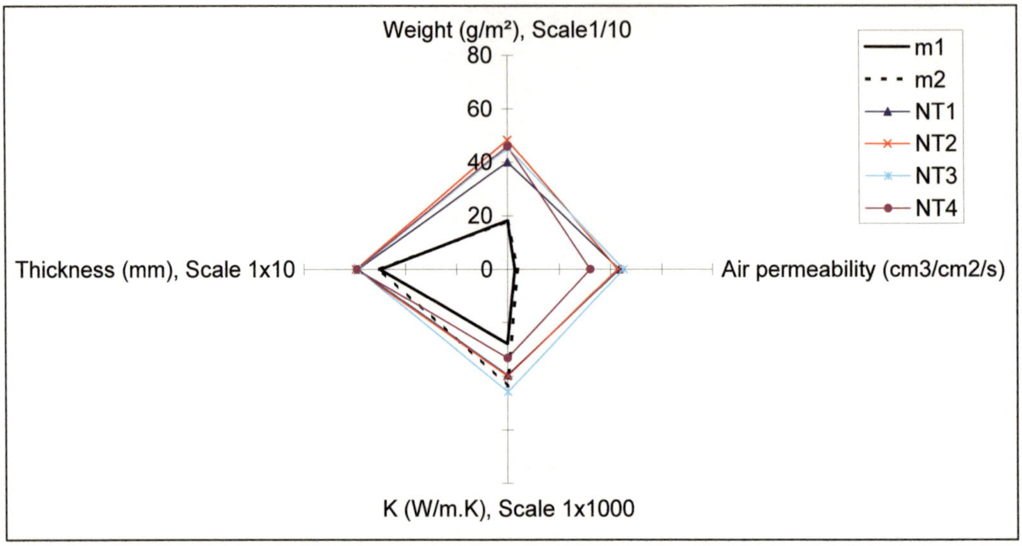

Fig. 22. Physical characteristics of the tested monolayer samples

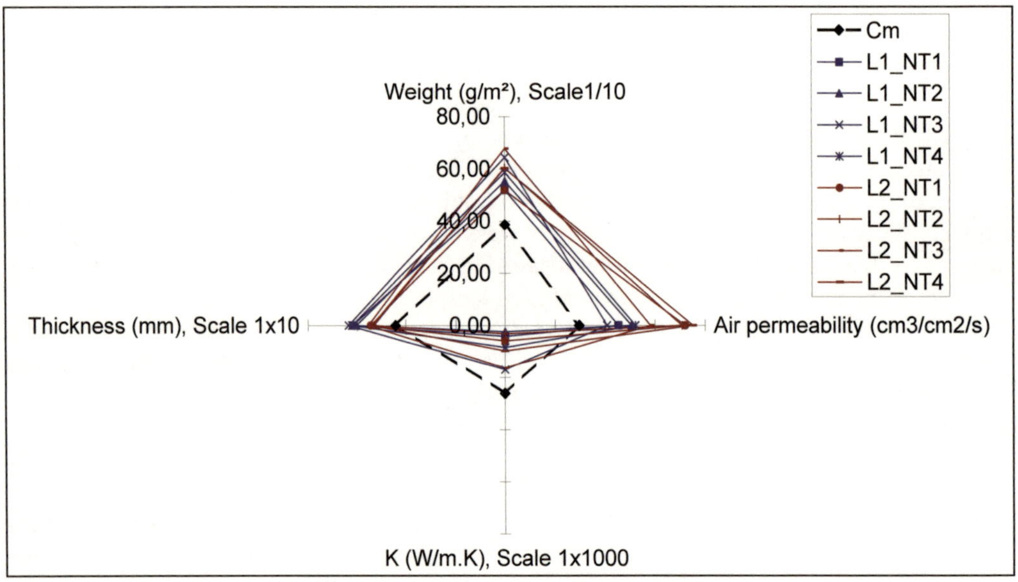

Fig. 23. Physical characteristics of the tested multilayer samples

In the case of the laminated products (Fig. 25), the VERTILAP® products laminated with the needle-punched nonwovens (L1 samples) and the tested foam globally present the same resilient property while the VERTILAP® products laminated with spun-bonded nonwovens (L2 samples) are the most resilient. The foam is more compressible than the VERTILAP® products. The L2 samples and the foam globally present the same characteristic in term of dissipated energy while the L1 samples dissipate the most energy. It can be said that the L1 samples present the same resilient property than the tested foam but they will be more comfortable in term of soft touch.

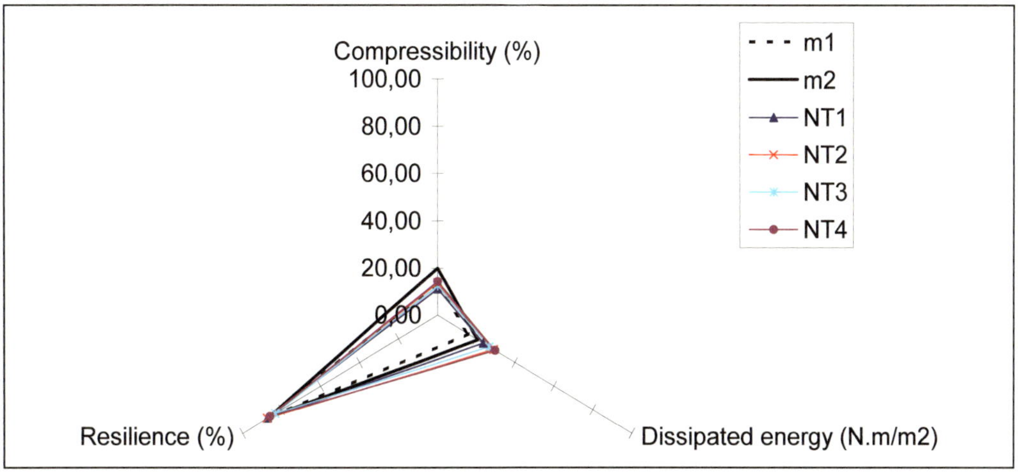

Fig. 24. Compressional characteristics of the tested monolayer samples (KES-FB3)

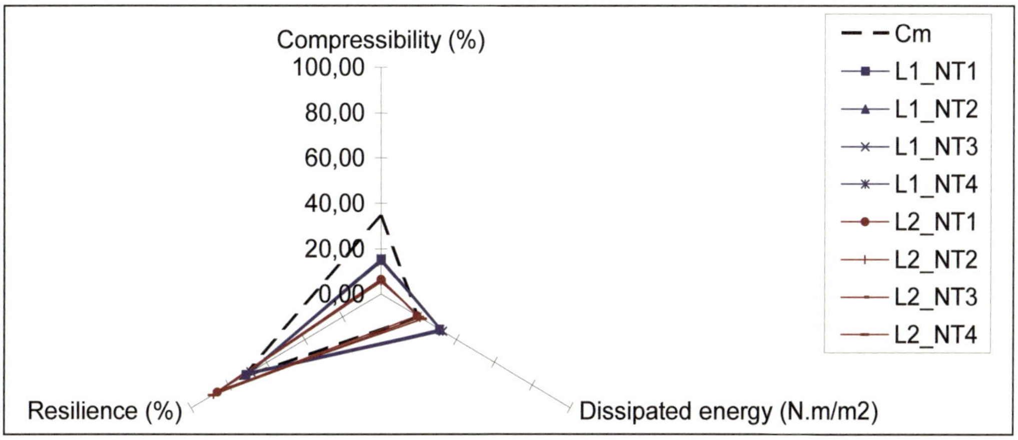

Fig. 25. Compressional characteristics of the tested multilayer samples (KES-FB3)

Regarding the compression test on five cycles, it has also been observed that the VERTILAP® products are more resilient and dissipate more energy than the tested PU foams. These observations have been done in both cases of the monolayer and laminated products (Fig. 26 - 29). The analysis of the raw results has shown differences between the behaviour of the 3D nonwoven and the PU foam. It has been observed an important reorganisation of the fibrous structure in the case of the 3D nonwoven while the cellular structure of the PU foam remained more constant. This reorganisation displays different individual behaviours of the filaments inside the pleated structure.

The results of this campaign have shown interesting properties of the VERTILAP® products in terms of comfort and mechanical behaviour compared with the tested PU foams. At this step, the main drawback of this new 3D nonwoven is its weight and its poor reproducibility. In fact, the obtained results have shown high dispersion values in the case of the VERTILAP® products. A second campaign has been carried out in order to reach the goal of the weight reduction of the VERTILAP® products.

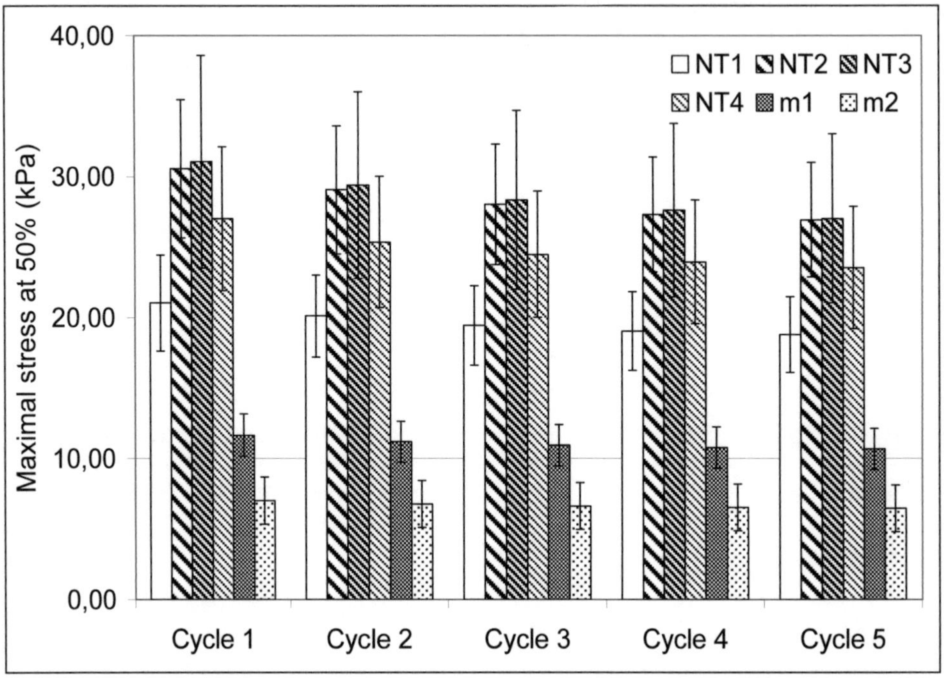

Fig. 26. Maximal stress at 50% deformation of initial thickness of the tested monolayer samples

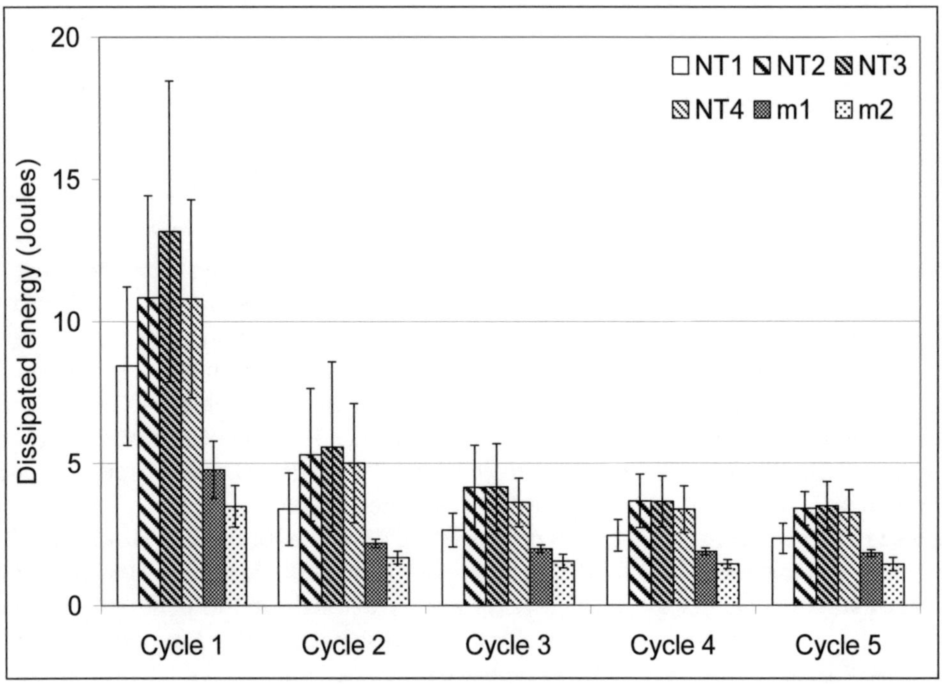

Fig. 27. Dissipated energy of the tested monolayer samples

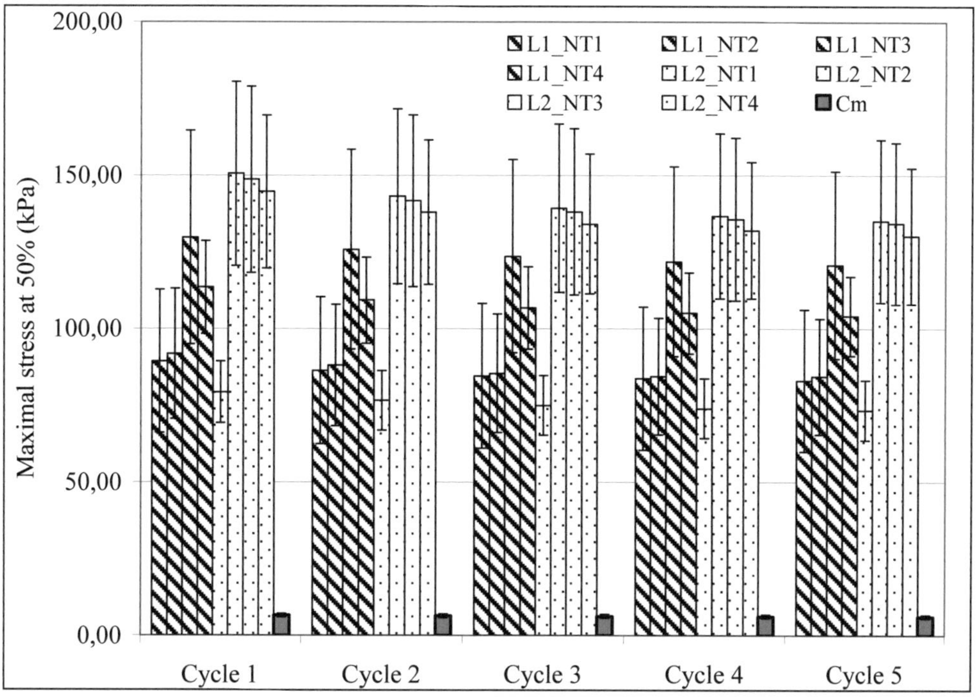

Fig. 28. Maximal stress at 50% deformation of initial thickness of the tested multilayer samples

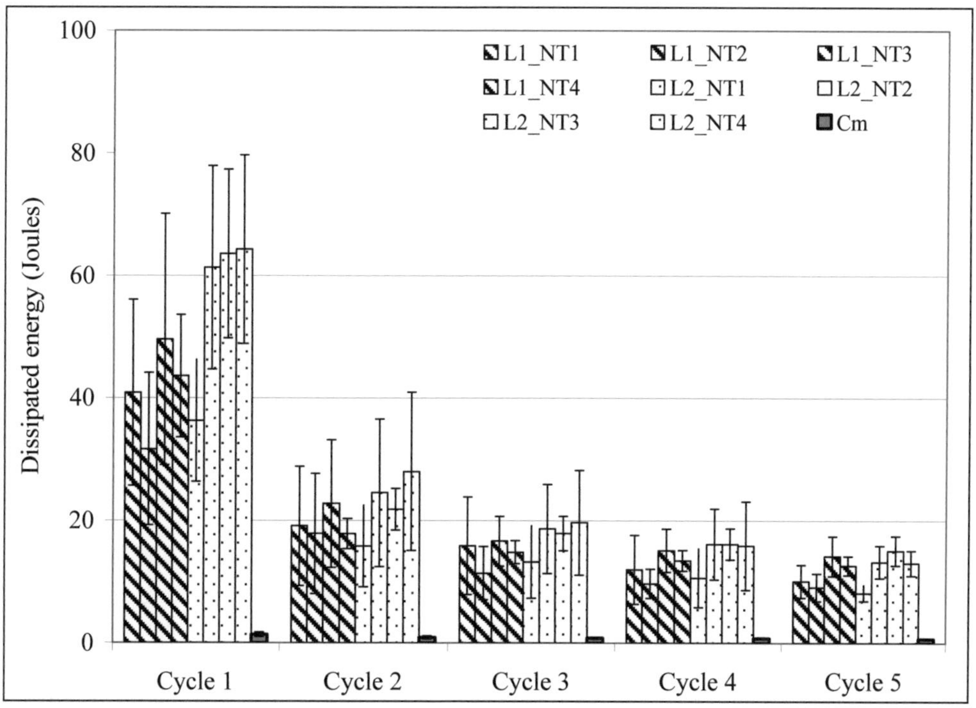

Fig. 29. Dissipated energy of the tested multilayer samples

4.3.2 Campaign B

In this experiment, the previous production procedure has been applied to manufacture the VERTILAP® products of this campaign but the technique to divide the initial tow of 90 ktex has been improved by spreading the tow between two beams in order to apply a minimal tension necessary for the filaments separation. Tows presenting a count from 7 ktex to 10 ktex have been pleated. During the manufacturing process, the speed before the verticalisation zone has been varied. The obtained single 3D nonwovens have been laminated at a speed of 5 m/min at 120°C. The hot melt adhesive was a 20 g/m^2 co-polyester web with a melting temperature of 60/75°C. It is also important to note an increase of 60% of the laminating speed compared to the previous samples (NT1, NT2, NT3 and NT4). This result enables to validate the products/process procedure.

The results of characterisation have shown a decrease of the weight of the 3D nonwovens compared to the previous samples. Indeed, the single 3D nonwovens present a mass per unit area of 164 g/m^2 while the mass per unit area of the laminated ones is 484 g/m^2. Structure's irregularity has been observed on the manufactured 3D nonwovens. This irregularity is mainly due to the irregularity in the tow. In fact, finer the tow, the more irregular the structure is as expressed in the Martindale's law (Martindale, 1945).

Regarding the physical characteristics (Fig. 30) in the case of the monolayer products, the objective of lightness has been reached and the 3D nonwoven, NT5, is also more comfortable in term of air permeability compared with the tested foams (m1, m2). NT5 also presents a better thermal insulation property compared with m1 sample. In the case of the multilayer products, the foam (Cm) present better physical characteristics compared with the laminated 3D nonwoven (L3 sample).

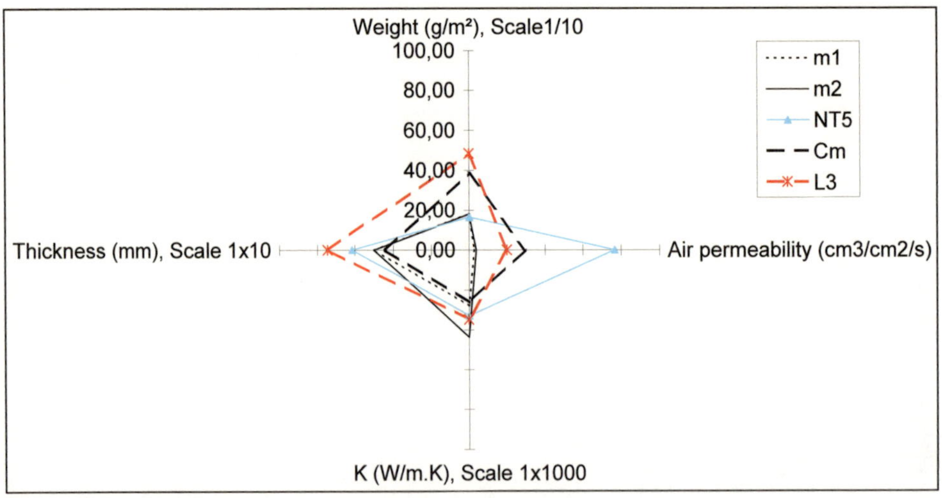

Fig. 30. Physical characteristics of the tested samples

Regarding the compression properties on one cycle (Fig. 31), a balance has been observed between the resilience and the dissipated energy in the case of single and laminated 3D nonwovens. This result shows that this new product presents, simultaneously, good resilient property and suitable comfort (soft touch). Except the problem of structure's irregularity, the characteristics of the obtained 3D nonwovens have been significantly improved. In both cases of monolayer and multilayer products, it has been observed that the

VERTILAP® products and the foam present globally the same resilient property but the foams dissipated less energy. It can be said that, the VERTILAP® products present better characteristic in term of comfort (soft touch).

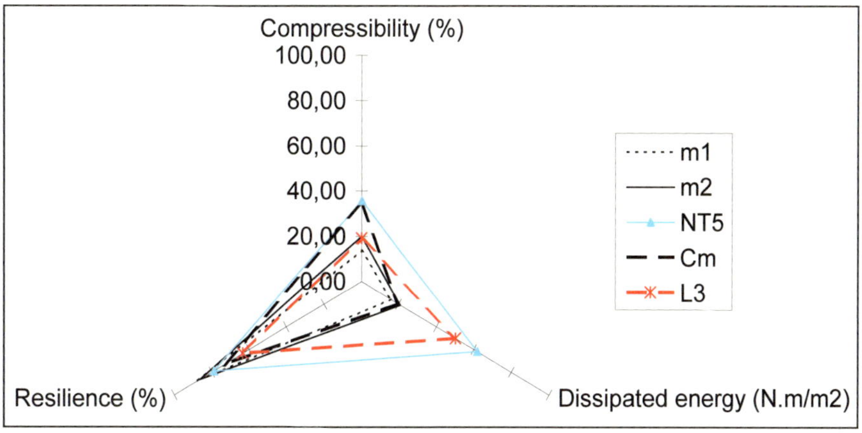

Fig. 31. Compressional characteristics of the tested samples

The compression curves of the tested samples are presented on Fig. 32.

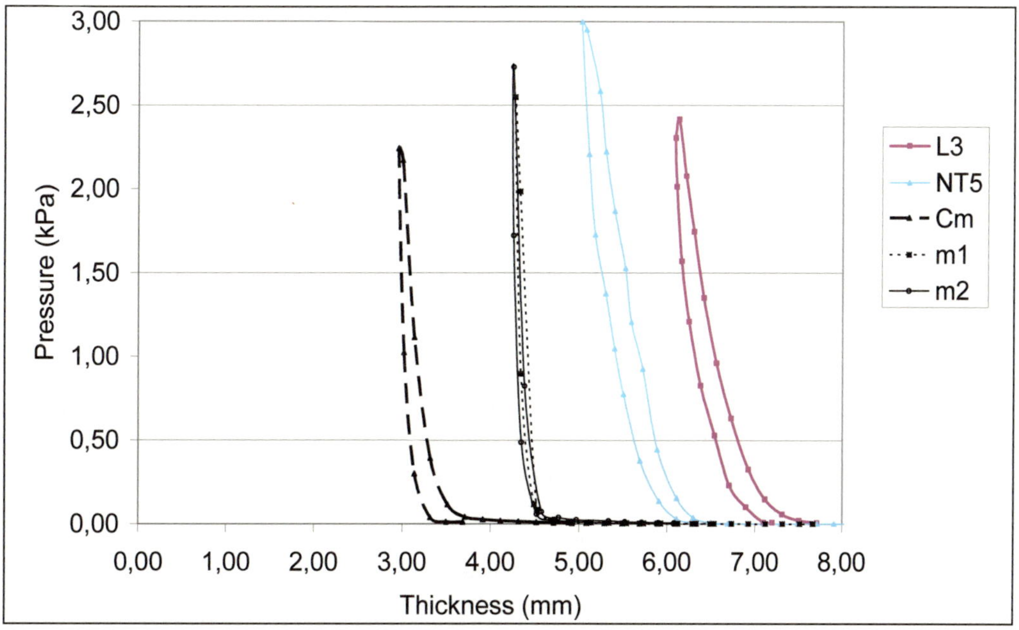

Fig. 32. Compression curves on one cycle (KES-FB3) of the tested samples

In addition to the previous characterization, the study of the tailorability of these new products has been carried out. The tailorability of the VERTILAP® 3D nonwoven has been positively validated through the execution of upholsteries for a headrest and door panels (Fig. 33). These automotive prototypes have been visually and tactically assessed thanks to sensory panelists (Philippe et al., 2004) and textile industrialists.

Headrest (Peugeot 205)

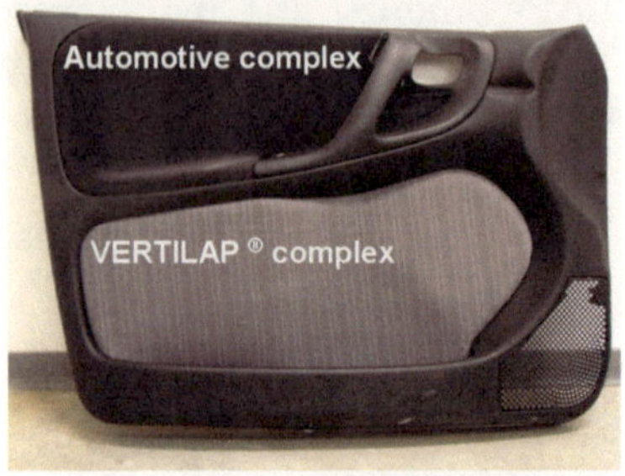

Door panel (Opel Astra)

Fig. 33. Automotive prototypes with VERTILAP® products

At the end of this campaign, the initial question of PU foam replacement has found a positive answer. Indeed, the development of the experimental prototype has allowed improving the quality of the final product especially in terms of weight and comfort in the case of the monolayer products. Nevertheless, the feeding material presents the problem of the structure's irregularity. The final results show that the developed products/process procedure has been successfully implemented and has permitted to improve the process and the expected products.

5. Conclusions and outlook

One original point of the VERTILAP project is the cluster that has been built for it (scientists, textile companies and competitiveness clusters). This cluster has made possible the development of an innovative 3D nonwoven. This work has contributed to increase know-how on the VERTILAP® process and knowledge on the obtained pleated 3D nonwoven in terms of methods and tools of characterisations. This study has shown that the new 3D nonwoven present good qualities, in terms of compressional behaviour and comfort (soft touch, air permeability and thermal insulation), compared to the current automotive PU foam. The realisation of the automotive parts (headrest and door panel) with this new product has shown that the VERTILAP® products present good suitable taylorability properties. At this step of the work, the question initially asked "can the PU foam be replaced by the VERTILAP® 3D nonwoven?" has found a positive answer and the recyclability problem has been solved. It can be said that the VERTILAP® 3D nonwoven could be a good candidate to replace certain PU foam in automotive trim applications.

Moreover, the obtained results during this work have generated data that will be used to develop a new VERTILAP® prototype of 1m width. This new prototype will be manufactured by the new subsidiary company NSC Environnement of the NSC Group. This new machine will allow conducting industrial testing campaign at high speeds of production. Different feeding materials such as nonwoven web or carded web will be used in order to obtain a good homogeneity of the product. The obtained 3D nonwovens thanks to this new prototype will be characterised through more investigations. In fact, characteristics such as the behaviour modelling, the acoustic insulation, the comfort through sensory analysis and the taylorability could be realised.

The forthcoming of the VERTILAP project has been initiated in order to extend the development of the new 3D nonwoven beyond automotive applications. This second phase has been labelled, in 2009, by the French competitiveness "Fibres Innovative Cluster". New industrial partners (Freudenberg Politex, Paul Hartmann, DIROY, Jacob Holm Industries, Albany International, Steelcase) have joined the project VERTILAP for this industrial phase.

6. Acknowledgment

This work has been done thanks to the financial support of Alsace Region, the Département du Haut-Rhin and OSEO.

7. References

Berthier, J-C. (2009), Polyuréthanes PUR, *Techniques de l'ingénieur*, (janvier 2009), pp 1-20, AM3425v2

BS 5'636 (1990), Determination of air permeability of textile fabric

Caudron, J.C. (2003), Etude du marché du polyuréthane et Etat de l'art de ses techniques de recyclage, *Rapport de l'ADEME (Agence de l'Environnement et de la Maîtrise de l'Energie)*, (27 juin 2003)

DGE (2005), Etude sur les Textiles Techniques, *Rapport de la Direction Générale des Entreprises* (DGE), France, (Juin 2005)

Drean, E. (2006), *Contribution to the development of piezoelectric sensors for the mechanical characterisation of textile fabrics*, PhD Thesis, University of Haute Alsace, Mulhouse, France

Dumas, J-L.; Schaffhauser, J-B. (2007). Patent N° WO2007125248, N.Schlumberger Company

EU Directive (2000), Directive 2000/53/CE of the European parliament and council of 18th September 2000 related to the End Life Vehicle, *Official journal of the European Communities*, 2000

Faucon, C. (1995), Les exigences fonctionnelles des matériaux de garnissage dans l'automobile, *Actes du 61ème congrès de l'ACIT*, pp 65-80, Lille, juin 1995, France

Fung, W., Hardcastle, M. (2001), Product engineering - Interior trim, Textiles in automotive engineering, In: *Textiles in automotive engineering*, The Textile Institute, pp 194-211, Woodhead Publishing Limited, ISBN 1 85573 493 1, Cambridge, England

Hopkins, J. (1995), A comparative analysis of laminating automotive textiles to foam, *Journal of coated fabrics*, (January 1995), pp 250-267

ISO 1798 : 1983, Flexible cellular polymeric materials - Determination of tensile strength and elongation at break, Ed.2

ISO 3386/1: 1986, Polymeric materials, cellular flexible - Determination of stress-strain characteristics in compression - Part 1: Low density materials

ISO 3795: 1989, Road vehicles, and tractors and machinery for agriculture and forestry - Determination of burning behaviour of interior materials

ISO 1856: 2000, Flexible cellular polymeric materials - Determination of compression set, Ed. 3

ITF (1990), Les matériaux textiles utilisés dans les habitacles des véhicules de transport, *Extraits du stage des 23 et 24 octobre 1990, 12*, Institut Textile de France, Lyon, France, (octobre 1990)

Kamprath, A. E. (2004), End-of-Life vehicles Recovery and Recycling polyurethanes Car Components Options Analysis, Recticel, http://www.idcpuresearch.com/downloads.htm

Karl Mayer Group (2007), Technical textiles, The Karl Mayer guide to Technical Textiles, http://www.karlmayer.com/internet/docs/MALIMO_EN.pdf, consulted in January 2007

Kawabata, S. (1980), The standardization and analysis of hand evaluation, (Ed. 2), The Textile Machinery Society of Japan, Osaka

Martindale, J. G. (1945), A new method of measuring the irregularity of yarns with some observations on the origin of irregularities in worsted slivers and yarns, *Journal of the Textile Institute*, Vol.36, (March 1945), T38-T47

Meyer Company, Flatbed laminating system, Maschinen Fabrik Herbert Meyer GmbH, www.meyer-machines.com, consulted in November 2007

Némoz, G. (1999), Les textiles (presque) partout dans l'automobile, *TUT*, N° 32., (2nd quarter 1999) pp16-18

Njeugna, N. (2009), *Contribution to the development and the industrialisation of a 3D nonwoven system*, PhD Thesis No. 2009/23, University of Haute Alsace, Mulhouse, France

Njeugna, N., Adolphe, D. C., Schacher, L., Schaffhauser, J-B., Strehle, P. (2008), Modification of compressional testing procedures for 3D nonwoven system for automotive interior applications, *Proceedings of the 4th International Textile Clothing & Design Conference*, pp 859-863, ISBN 978-953-7105-26-6, Dubrovnik, October 2008

NSC (2007), The TT12 crush cutting Converter, *Technical notice*, http://www.nsc-fibretoyarn.com, consulted in September 2007

Persijn, B. (2008), PU-foams in automotive, *Proceedings of Textile & plastics, 6th International Conference on Automotive and Transport Interior Decoration*, Dec'autex 2008, Mulhouse, France, november 2008

Philippe, F., Schacher, L., Adolphe, D., and Dacremont, D. (2004), Tactile Feeling: Sensory Analysis Applied to Textile Goods, *Textile Research Journal*, 74 (12), 1066-1072

Recticel (2009), What is PU?, Publication of the International Development Centre of Recticel company,
http://www.idcpuresearch.com/downloads.htm

Recticel (1999), A new method to measure the cell diameter of polyurethane foam, Visiocell, *Technical Foams*, Business Line Management Technical Foams, (Ed. 1), pp 4 - 8, Damstraat, Belgium

Struto International Inc. (2007), Struto® Nonwoven, http://www.struto.com/, consulted in January 2007

Santex Group (2007), Wavemaker® Nonwoven,
http://www.cavitec.ch/en/?menu=produkteprogramm, consulted in January 2007

Vasile, S., Langenhove, L. V., de Meulemeester, S. (2006), Effect of Production Process Parameters on Different Properties of a Nonwoven Spacer produced on a 3D Web Linker®, *Fibres & Textiles in Eastern Europe*, Vol. 14, N°4 (58), (October/November 2006), pp 68-74

19

Automotive Catalysts: Performance, Characterization and Development

Nelcy Della Santina Mohallem,
Marcelo Machado Viana and Ronald A. Silva
Universidade Federal de Minas Gerais
Brazil

1. Introduction

Nowadays, automotive catalysts have been used to reduce atmospheric emissions, due to significant parcels of the global emissions of pollutants agents provoked by vehicles. Automotive exhaust catalysts were introduced in the 70's decade, because some countries established restricting laws related to emissions of carbon monoxide (CO), nitrogen oxides (NOx) and hydrocarbons (HC) by the engines. These products generated by the combustion process are extremely harmful to health and the environment (Massad et al, 1985). For example, CO combined with hemoglobin in the bloodstream promotes the reduction of oxygen-active sites that provokes asphyxia. Nitrogen dioxide (NO_2) contributes to photochemical smog and acid rain, and is irritating to the eyes, skin and respiratory system. Nitrogen monoxide (NO) is toxic by inhalation and irritating to the eyes and skin. Polycyclic aromatic hydrocarbons (PAHS) have been identified as carcinogenic compounds. Other combustion products are: ash, formed mainly by particulate residues of components of the lubricating oil, and soot, combustible matter in the exhaust gas (smoke). Automotive catalytic converters have been developed precisely to make these products less toxic (Morterra et al, 1995; Ismagilov et al, 1998).
The development of catalysts only was possible with the improvement in automotive engines as the replacement of carburetion system for electronic injection and introduction of the catalyst in the exhaust systems (Kaspar et al, 2003).
The catalysts of three ways (TWC –Three-Way Catalyst) are advanced systems of emission treatment of gasoline vehicles that reduce significantly the emissions of carbon monoxide (CO), hydrocarbons (HC) and nitrogen oxides (NOx) in atmosphere (Collins & Twigg 2007). Nevertheless, there is a steady increase in world production of vehicles powered by gasoline and by other types of fuel such as alcohol, gas and mix of fuel, leading to constant research in order to improve the catalysts already known and to develop new models (Mizukami et al, 1991; Silva et al, 2009 & Sideris, 1997).
Automotive catalysts are generally available in the form of monolith ceramic as cordierite and zeolites or metal substrate. The catalyst substrates more used are composed of magnesium cordierite ($2MgO.2Al_2O_3.5SiO_2$) with a honeycomb structure, which provides a high geometric surface area, coated with γ-alumina (catalyst wash-coat). This wash-coat is designed to increase the specific surface area and is the support for precious metals, mainly platinum (Pt), palladium (Pd) and rhodium (Rh), which promotes the catalytic reduction and oxidation of

pollutants gases to more harmless ones, like carbon dioxide, water and nitrogen molecular. Cerium and zirconium oxides are used in the coating by their oxygen storage capability to improve catalytic efficiency. (Angelidis & Sklavounos, 1995). A variety of other additives also are used to stabilize the alumina wash-coat at high temperatures (operational temperature). Figure 1 illustrates the operation of a three-way catalyst. This catalyst transforms the toxic gas CO in CO_2, which although not so toxic contributes to the increase in greenhouse effect.

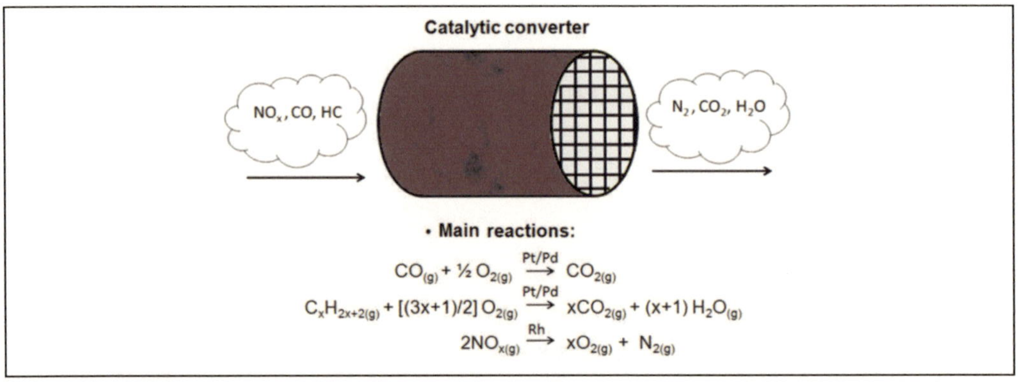

Fig. 1. Schematic illustration of the TWC operation.

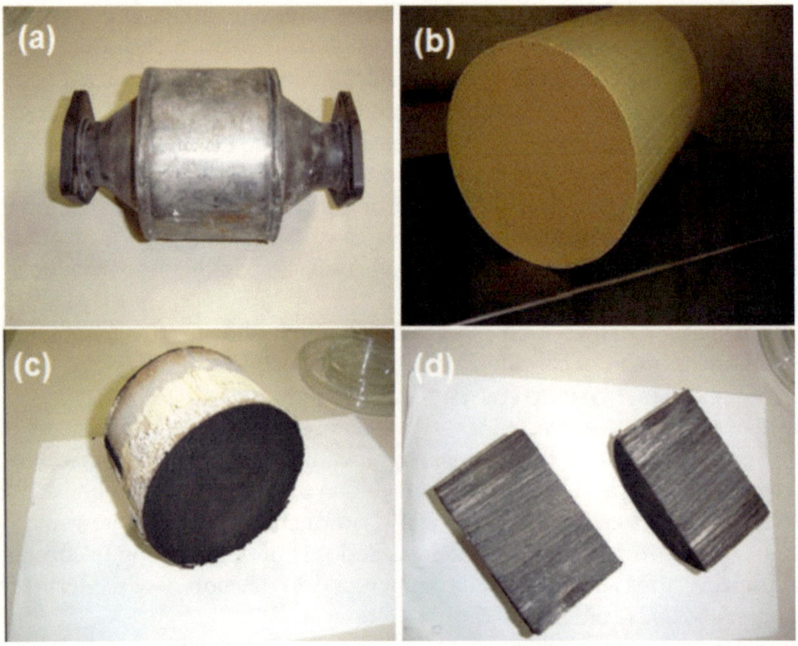

Fig. 2. (a) Catalyst module, (b) new catalyst (c) poisoned catalyst (d) samples of poisoned catalyst.

The catalyst can be deactivated by chemical, mechanical or thermal phenomena after some time of operation, depending on the composition of the used fuel and lubricants, and of the vehicle adjustment. The chemical deactivation can be promoted by poisoning due to the chemisorptions of compounds on the surface of the catalyst; and due to chemical reactions that produce volatile compounds or inactive phases. The thermal deactivation occurs due to the sintering process and active metal diffusion. The mechanical deactivation is due to the deposition of particles from the gas phase onto the pores and catalyst surface, and effects of abrasion caused by mechanical crushing of the catalyst.

Figure 2a shows a typical catalyst module formed by a metal housing containing the catalyst. Figure 2b shows a new catalyst and the visual effect of deactivation in a poisoned catalyst is shown in Figure 2c and 2d. The amount of soot deposited in an used catalyst depends of the fuel quality, since gasoline contains some amount of contaminants such as sulfur, and oxygen and nitrogen compounds. A new catalyst sample is shown in Figure 3a and 3b, and Figure 3c shows the clogging of the honeycomb structure caused by the poisoning of the catalyst.

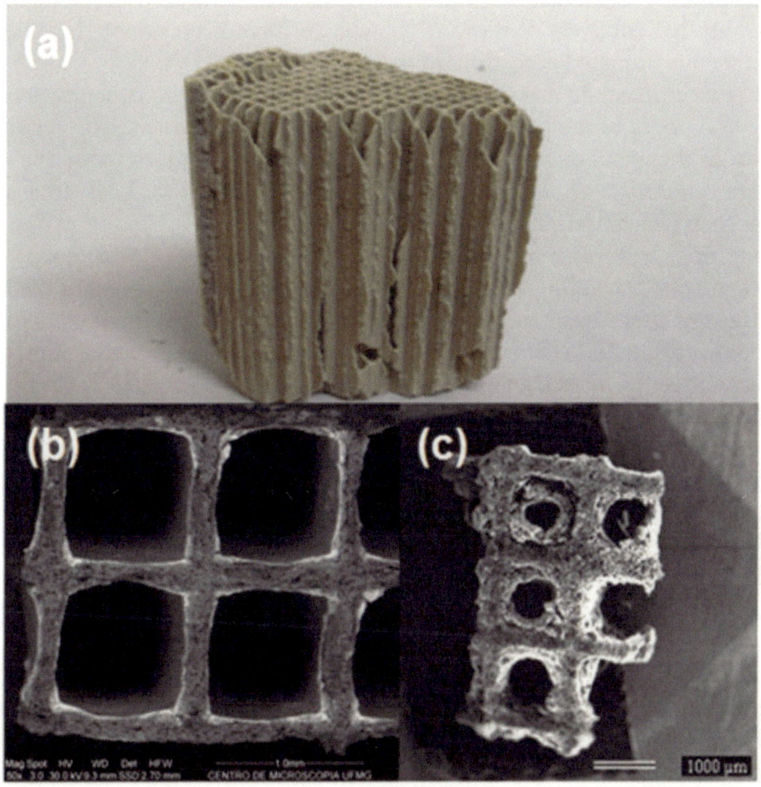

Fig. 3. (a) Photography of a catalyst sample, (b) SEM micrograph of a honeycomb structure, (c) clogging of the honeycomb structure caused by the catalyst poisoning.

Beyond the vehicles powered by gasoline, there has been a move for the utilization of other types of vehicles that have been developed to use different fuels that produce less CO_2, which cause the greenhouse effect. These fuels are the alcohol, renewable bio-fuel derived by sugar-cane or corn, gases such as liquefied petroleum gas (LPG) and compressed natural gas (CNG), or mix of fuels as the used in flex-fuel technology. These alternative fuels have a lower carbon to hydrogen ratio than gasoline, producing less CO_2 per travelled distance, and reduce the needed of fossil fuel consumption (Cohn, 2005).

Some technologies have been developed, adapting the engine for the mixtures of fuels like gasoline and ethanol with predetermined composition. Moreover, there are the new flex-fuel technology that is related to the flexibility of choice of the car fuel, where is possible to use only hydrated alcohol or gasoline, or a mixture of these fuels in any concentration (Delgado et al, 2007). The people can buy the cheapest fuel, whose prices depend on the economic moment. The flex-fuel technology is based on sensors that detect the concentration of the mixture of gasoline and hydrated alcohol, and in the subsequent automatic adjustment of the engine. The addition of ethanol in gasoline decreases the concentration of CO emissions, making this process a very interesting technology. Some countries are planning to employ this technology, since about 85% of the cars produced in Brazil are equipped with this technology.
All of these factors impact the design of TWC, since its geometric surface area until the alumina thin film formulations. It would be necessary a corresponding catalyst for each type of used fuel, leading in consideration the type of chemical reaction that occurs in the engine. But in the reality the catalysts of these new vehicles have been adapted without rigorous criteria, and they are adjusted according to need (Silva, 2008).
Other factors that influenced the development of TWCs were the economics ones, mainly the related to the prices of the platinum group metal and of the fuels. The constant increase and instability in the gasoline price led to the development of more economic engines that also need different design of catalyst. In this sense, various types of substrate as *zeolites* or metallic have been tested and/or used (Collins & Twigg 2007).
Actually, recycling and regeneration of catalysts are common practices. Regeneration consists in a controlled oxidation at high temperature to eliminate soot and convert sulfides to oxides. After this process, some catalysts also require additional treatments to recover the full activity. Non-regenerable catalysts have to be recycled for metals recuperation. This can be performed either by hydrometallurgy or pyrometallurgy (Angelidis et al, 1995; Silva, 2008; Dufresne, 2007 & Hirokazu, 1999).
In this chapter, textural, morphological and structural characteristics of selected new and used catalysts, analyzed by gas adsorption, pycnometry, X-ray diffractometry, thermal analyses and scanning electron microscopy, are shown. EDS and WDS electronic microprobe were used to detect the composition of the catalysts and their contaminants. Subsequently, we discuss the textural and morphological changes of automotive catalyst by effect of high temperatures, which lead to its deactivation. New commercial automotive catalysts were thermally treated at various temperatures. Micrographies and adsorption-desorption isotherms were used to verify the changes in the catalyst characteristics with thermal treatments. Finally, problems about gas emission and the soot present in exhaust gas are discussed, beyond some aspects about reuse and recycling are considered. Some solutions about this theme are shown.

2. Textural, morphological and structural characteristics of new and used catalysts

2.1 Experimental

Some new and used automotive catalysts of vehicles powered by gasoline, by alcohol, and by flex fuel, of diverse suppliers, have been analyzed.
The samples have been analyzed by X-ray diffractometry (Rigaku, Geigerflex 3034) with CuKα radiation, 40kV and 30mA, time constant of 0.5s and crystal graphite monochromator to identify the phases present (metals and transition metal oxides).

The composition, metal distribution on the alumina thin film and morphology of the catalysts have been evaluated by an electron microprobe (Jeol JXA, model 8900RL) with an energy dispersive and wavelength dispersive spectrometers (EDS/WDS), and by scanning electron microscopy (Quanta 200, FEG-FEI).
Density measurements of the catalysts have been obtained by helium picnometry (Quantachrome) and sample textural characteristics were determined by nitrogen gas adsorption (Autosorb - Quantachrome) at liquid nitrogen temperature. Nitrogen gas has been used with a 22-point adsorption-desorption cycle. The samples have been outgassed at 200 °C for 12 hours before each analysis. Experiments have been made in triplicate. Specific surface area and total pore volume have been obtained by the application of Brunauer-Emmett-Teller (BET) equation and the BJH method, respectively (Lowell & Shields, 2005).

2.2 Results and discussion

2.2.1 X-ray diffraction

The diffractogram of the new catalyst (Figure 4a) is characteristic of nano and/or porous materials and shows a good correspondence with the cordierite diffractogram standard,

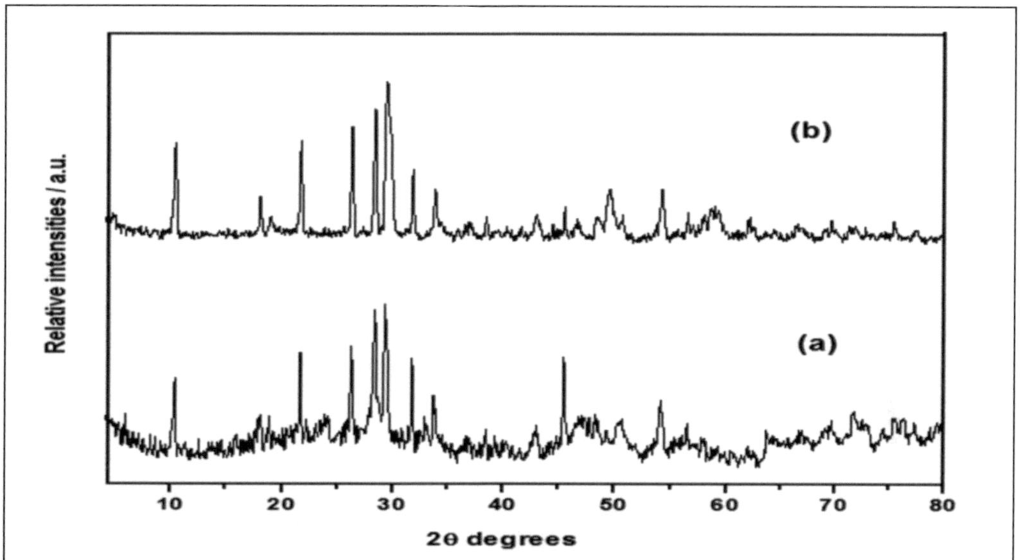

Fig. 4. X-ray diffraction patterns of (a) new and (b) used catalysts.

beyond characteristic peaks of the gamma-alumina film and of the metals dispersed in the wash-coat. A reasonable structural variation is evidenced in the diffractogram of the used catalyst (Figure 4b), that presents more crystalline behavior and characteristic peaks of precious metallic oxides.

2.2.2 Microanalysis and scanning electron microscopy

Fig. 5 shows an image of scanning electron microscopy of the catalyst obtained by back-scattering electrons. It is possible to observe the porous alumina thin film with precious metal heterogeneously dispersed (white dots) deposited on cordierite (macroporous material). The precious metal particle size varied from 1 to 15 μm. The chosen points of the

Fig. 5a have been analyzed with an EDS detector, confirming the expected basic cordierite compositions in region 1 (dark region), formed by Al, Mg and Si (Figure 6a). Region 2 also has the same composition of the cordierite, with some impurities such as TiO_2, Fe_2O_3, CaO and ZrO_2 (Figure 6b).

Fig. 5. (a) Backscattering SEM micrograph of a piece of a new automotive catalyst, and (b) detail of the alumina thin film on the cordierite.

The alumina wash-coat is pure (region 3 of Fig. 5a and Figure 6c) with metals and oxides dispersed such as cerium and zirconium oxide (Ce_2O and ZrO_2) in more quantity and traces of palladium (Pd) characterized by region 4 of Figure 5a and Figure 6d. Platinum and rhodium particles have been observed only by WDS detector because their minor quantity dispersed in the thin film.

After some time of utilization (months or years), the catalyst suffers poisoning due to the fuel and lubricant residues, chemical reactions and also effects of sintering due to the high operating temperatures, which generally can reach 900 °C. The images of Figure 7 show the morphological and textural comparison between the alumina films of a new and an used catalyst. The new catalyst surface (Fig. 7a) is porous with disperse precious metal particles, while the used (Fig. 7b) shows an eroded surface with agglomeration of the precious metal particles and the formation of microcraks. Texturally, the used catalyst shows a decrease in the porosity related to the new catalyst, due to the beginning of sintering caused by the operational temperature.

Figure 8 shows with more detail a morphological comparison of new and used catalysts of vehicles powered by gasoline. Column (a) shows a new cordierite substrate more macroporous and an alumina thin film more porous and preserved than those of the used catalyst (column b). It is possible to observe the precious metal diffusion inside the cordierite of the used catalyst, beyond an increase of the precious metal agglomerates also due to the diffusion process.

In general, used catalysts show a large quantity of ash and/or soot in the surface and inside of their pores. Figure 9a illustrates the obstruction of a catalyst by these contaminants. These

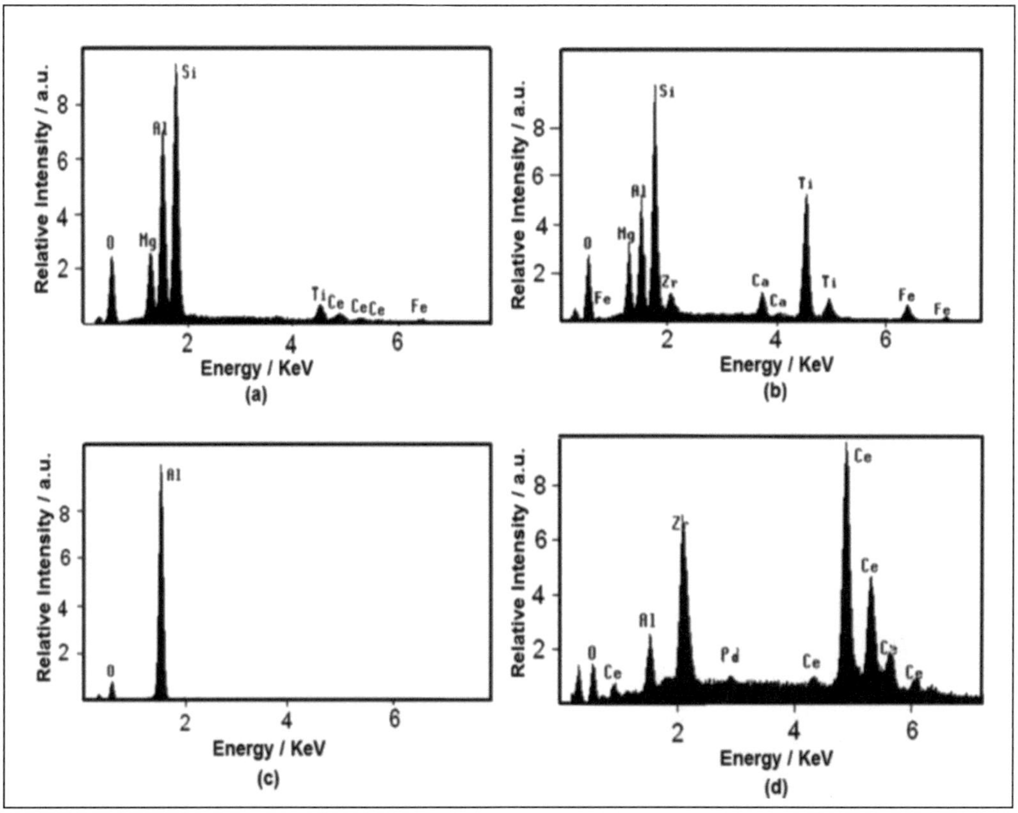

Fig. 6. EDS spectra of new automotive catalyst. a: cordierite (region 1), b: cordierite impurities (region 2), c: alumina film (region 3), d: active metals and oxides (region 4).

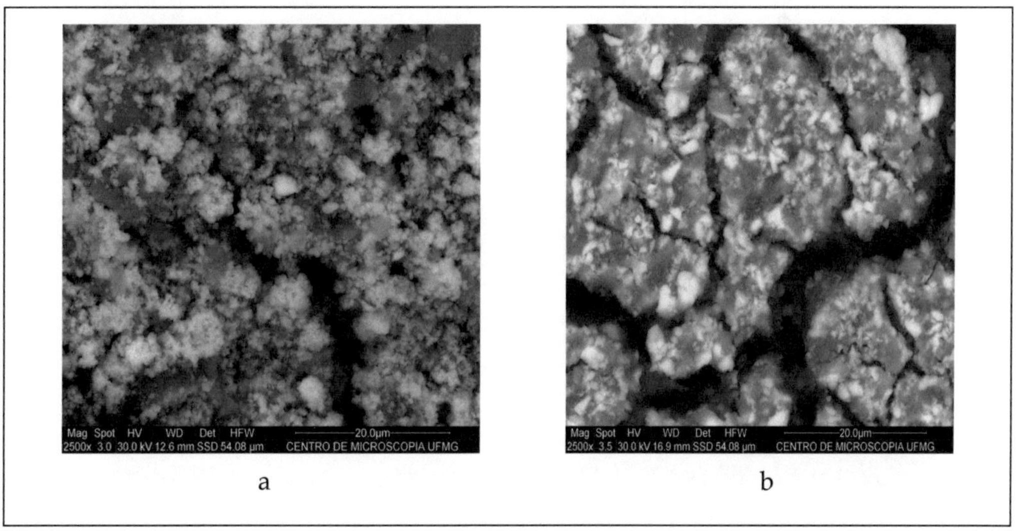

Fig. 7. Backscattering SEM micrograph of the alumina film of the (a) new and (b) used automotive catalyst.

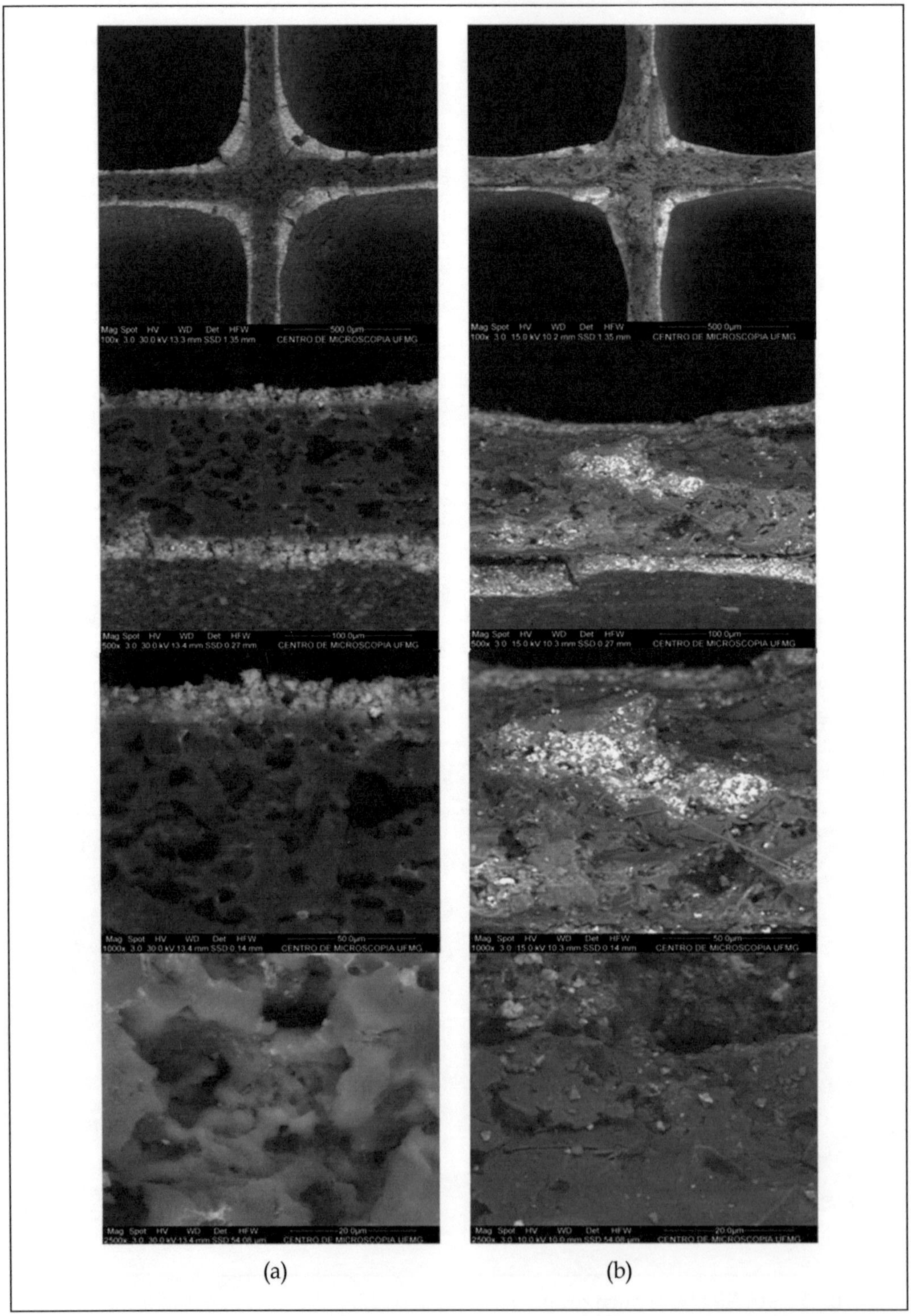

Fig. 8. Backscattering SEM micrographies of the (a) new and (b) used catalyst samples.

particles penetrate inside the pores of the catalyst, clogging the monolith cordierite channels (honeycomb structures). EDS analyses showed that the used catalysts has the same composition of the new catalysts, but also has a considerable amount of carbon, potassium, sulfur and chlorine that can come from of fuel and lubricating oil (Figure 9b). Particulate

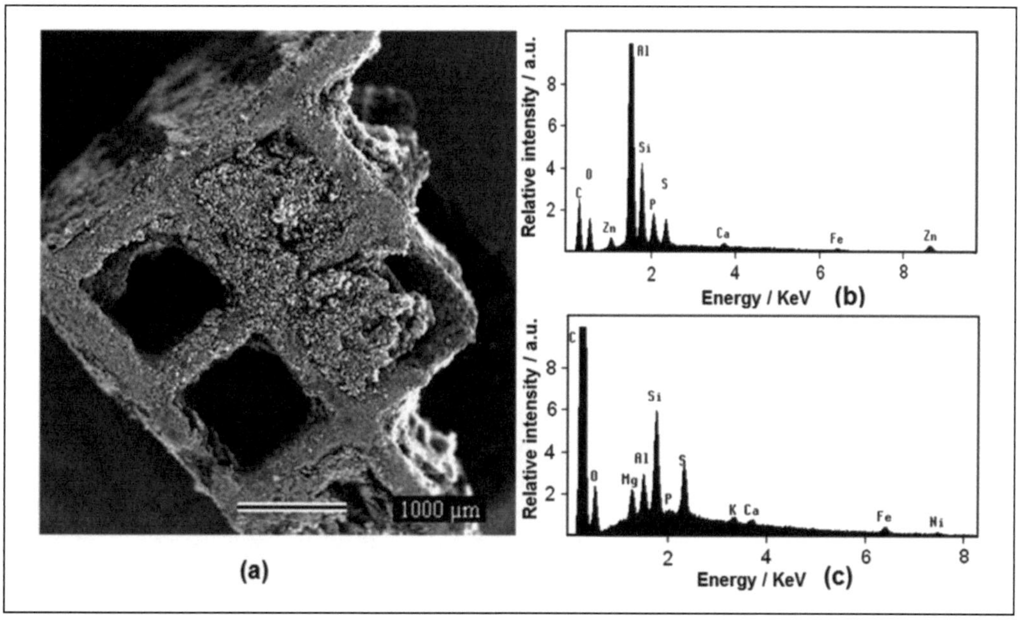

Fig. 9. (a) *SEM micrograph of the obstructed used catalyst; (b) EDS spectra of the used catalyst; and (c) EDS of the particulate material (ash and soot).*

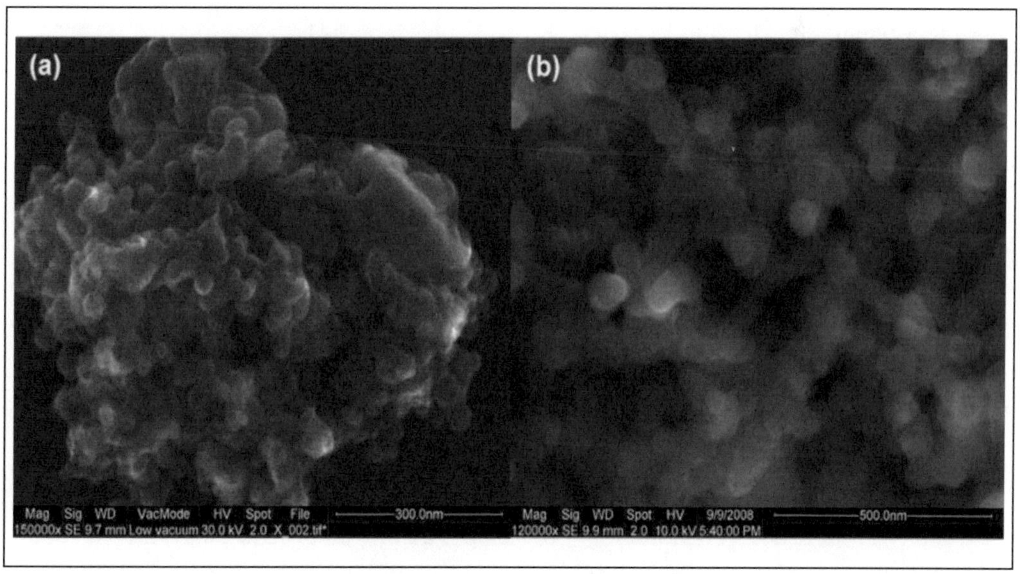

Fig. 10. SEM micrograph of the soot removed of a poisoned catalyst.

samples (ash and soot) collected of various *poisoned* catalysts were analyzed by EDS, and showed great amount of carbon, sulfur, silica, alumina and magnesia, as well as, smaller quantities of phosphorus, iron and nickel (Fig 9c).

Figure 10 shows micrographies of the ash and soot retired of a poisoned catalyst, which show characteristic of nanoparticulate material, with particle sizes about 45 ± 15 nm, forming agglomerates with size in the range from 500 to 2 μm. This type of material is very active due to its small particle size and, when inhaled, is harmful to the health, causing lung diseases. Actually, the filters used are not capable of retaining this kind of material that goes to atmosphere by the smoke.

2.2.3 Gas adsorption technique

The capacity of adsorption of new and used catalysts has been evaluated. Various types of catalysts have been analyzed and the results have been similar. The specific surface area varies with the type and model of the catalyst, but the decrease in the values is proportional. Table 1 shows the changes in density, specific surface area and total pore volume values of two catalysts that showed the lowest and largest specific surface area, one of a vehicle powered by gasoline and other of a flex-fuel vehicle. The used samples (poisoned catalysts) have been cleaned to eliminate the soot trapped in the beehive of the catalyst, remaining only the soot physically adsorbed in the pores of the catalyst. The density of the used catalyst is larger than the new catalyst, suggesting a densification process due to the high temperature of operation. Consequently, the used catalyst has textural values lower than those of the new catalyst, which causes its deactivation. It is observed increasing of about 85% in the specific surface area and 75% in the porosity.

Sample	Density $/g.cm^{-3}$	Specific Surface Area $/m^2.g^{-1}$	Total Pore Volume $/ 10^{-3} cm^3.g^{-1}$	Average Pore Size/nm
New (Gasoline)	3.1 ± 0.1	58	141	11
Used (Gasoline)	3.6 ± 0.1	9	36	4
New (Flex)	3.0 ± 0.1	282	100	12
Used (Flex)	3.4 ± 0.1	18	8	6

Table 1. Textural characteristics of new and used catalyst of a vehicles powered by gasoline and flex fuel (gasoline and alcohol).

The adsorption-desorption isotherms (Fig. 11 and Fig. 12) are characteristics of mesoporous materials (isotherm type IV, according to IUPAC classification) and show that the new catalyst sample adsorbs a higher volume of nitrogen when compared to the used catalyst. The catalyst of the vehicle powered by gasoline lost 75% of its adsorptive capacity and the catalyst of the flex-fuel vehicle lost 92%. The changes in the shape of the isotherms of the used catalysts show a large variation in pore shape, that together with the results in Table 1 prove the high variation on the textural characteristics of the used catalyst when compared with a new one.

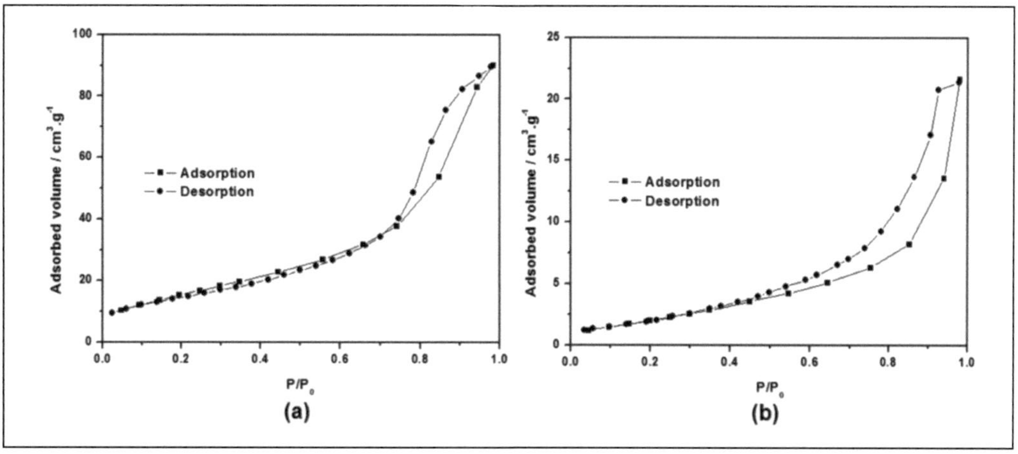

Fig. 11. Adsorption-desorption curves of (a) new and (b) used catalysts of a vehicle powered by gasoline.

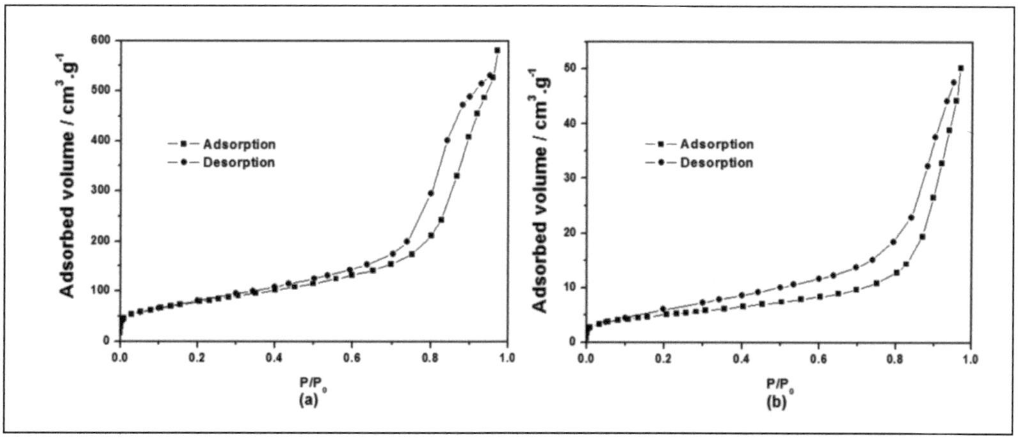

Fig. 12. Adsorption-desorption curves of (a) new and (b) used catalysts of a flex -fuel vehicle.

3. Study of the temperature effect on textural and morphological characteristics of automotive catalysts

3.1 Experimental

Selected new catalysts have been broken in pieces of 2 cm of side and have been thermally treated at 500, 700 and 900 °C during 5 hours to verify the changes in the textural, morphological and structural characteristics as a function of the temperature.

Simultaneous thermogravimetric and differential thermal analysis (TG-DTA) measurements have been performed in air and N_2 (TA Instrument SDT 2960). Samples have been heated from room temperature to 1400°C at 10 °C min^{-1}.

The variation on the sample morphologies have been observed by scanning electron microscopy in an equipment JEOL JSM, model 840 and in an equipment Quanta 200, FEG-FEI.

Variation in the true density has been evaluated by helium picnometry (Quantachrome) and the textural characteristics have been determined by nitrogen gas adsorption (Autosorb - Quantachrome) at liquid nitrogen temperature. The samples have been outgassed at 200 °C for 12 hours before each analysis.

3.2 Results and discussion

3.2.1 Thermal analysis

The analysis of the TG curves (Figure 13a) shows a significant loss of mass (about 20 %) for new automotive catalysts during the heating between 100 and 700 °C under air atmosphere. This loss corresponds probably to the oxidation, densification and crystallization processes, which is corroborated by exothermic events in the DTA curves in the same region, and by X-ray diffraction results obtained by new and used catalysts. TG and DTA curves (Figure 13b) of samples heated in N_2 atmosphere show a minor loss of mass (about 10 %) and events less exothermic. Considering that the automotive catalyst changes considerably with temperature up to 600 °C, we can conclude that the operational temperature of 900 °C is enough to deactivate partially the catalyst.

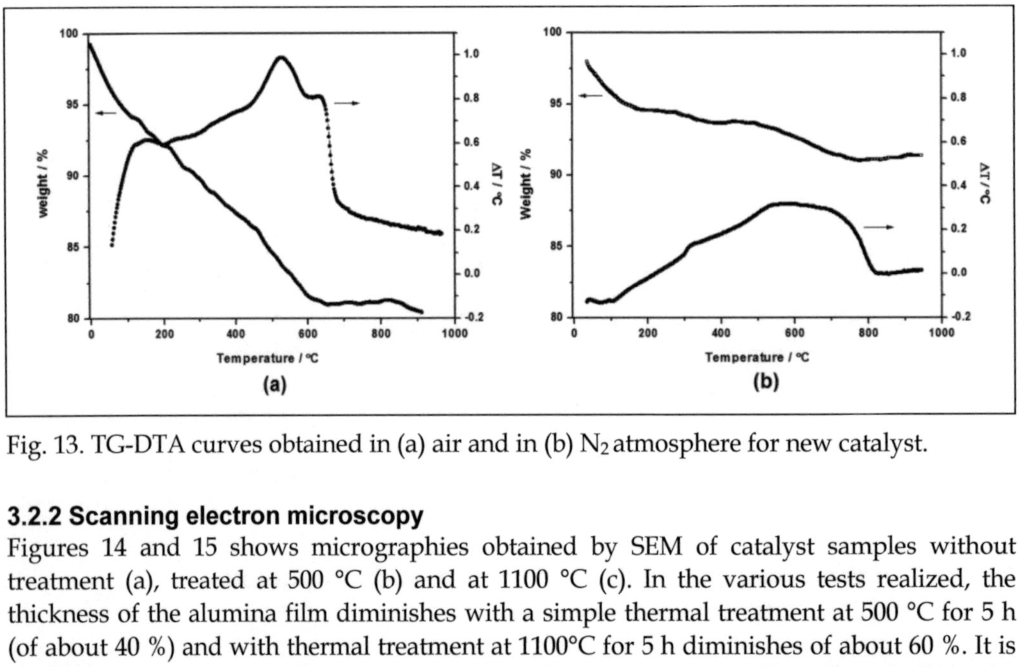

Fig. 13. TG-DTA curves obtained in (a) air and in (b) N_2 atmosphere for new catalyst.

3.2.2 Scanning electron microscopy

Figures 14 and 15 shows micrographies obtained by SEM of catalyst samples without treatment (a), treated at 500 °C (b) and at 1100 °C (c). In the various tests realized, the thickness of the alumina film diminishes with a simple thermal treatment at 500 °C for 5 h (of about 40 %) and with thermal treatment at 1100°C for 5 h diminishes of about 60 %. It is observed shrinkage and the appearance of cracks in the alumina films deposited on the cordierite due to the increasing in the heating temperature.

Figure 16 shows, with more detail, images of the heating effect in catalyst samples without treatment and treated at 500 and 900° C. Figure 16a shows the alumina film of a catalyst without treatment and Figure 16b shows the alumina film treated at 500 °C. It is possible to observe the beginning of densification of the film treated at 500 °C. Figures 16c and 16d show the cordierite without treatment, more porous, and treated at 500°C, respectively. Figures 16e and 16f show the diffusion of the precious metal and the sintering process of the catalyst treated at 900 °C, respectively.

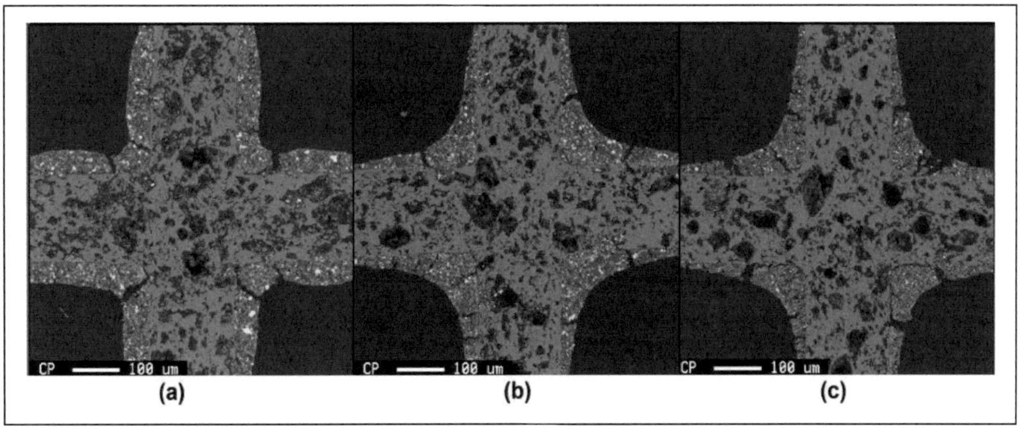

Fig. 14. Backscatter SEM images of (a) new catalyst without heating, (b) heated at 500 °C, and (c) heated at 1100 °C.

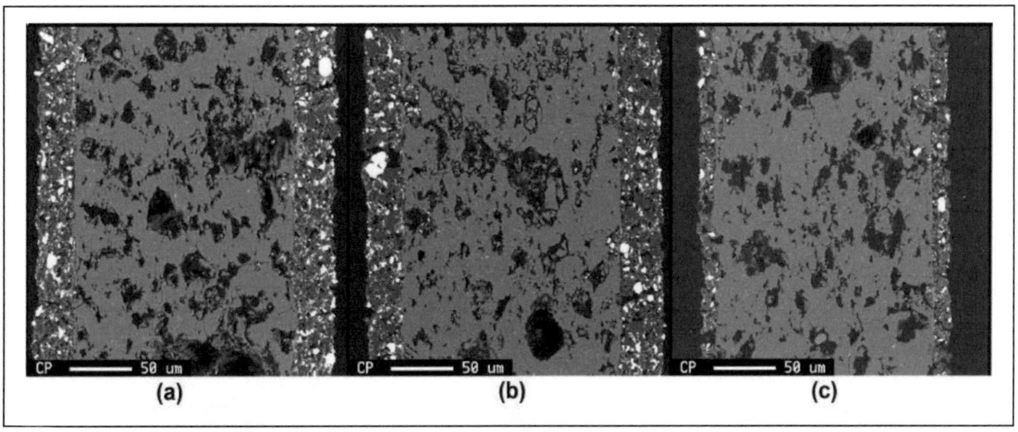

Fig. 15. Backscatter SEM images of (a) new catalyst without heating, (b) heated at 500 °C, and (c) heated at 1100 °C.

3.2.3 Gas adsorption technique

Table 2 shows the changes in density, specific surface area and total pore volume values of a catalyst heated at various temperatures. With the increasing in heating temperature occurred a drastic decreasing in the textural values up to 900 °C. These results suggest a densification process due to increase in the temperature next to the temperature of operation. The value obtained by the catalyst heated at 900 °C is similar to the obtained by the used catalysts.

The adsorption-desorption isotherms of samples treated between 500 and 900 °C (Fig. 17) are also characteristics of mesoporous materials and show a lower adsorptive capacity than the new catalyst without thermal treatment (Fig. 13a). The new catalysts lost 65% of their adsorptive capacity when heated at 500 °C for 5 hours and 72% when heated at 900 °C for 5 hours. The shape of hysteresis remained nearly constant, showing that there was little variation in pore size due to the densification process, while the amount of pores (porosity) decreased.

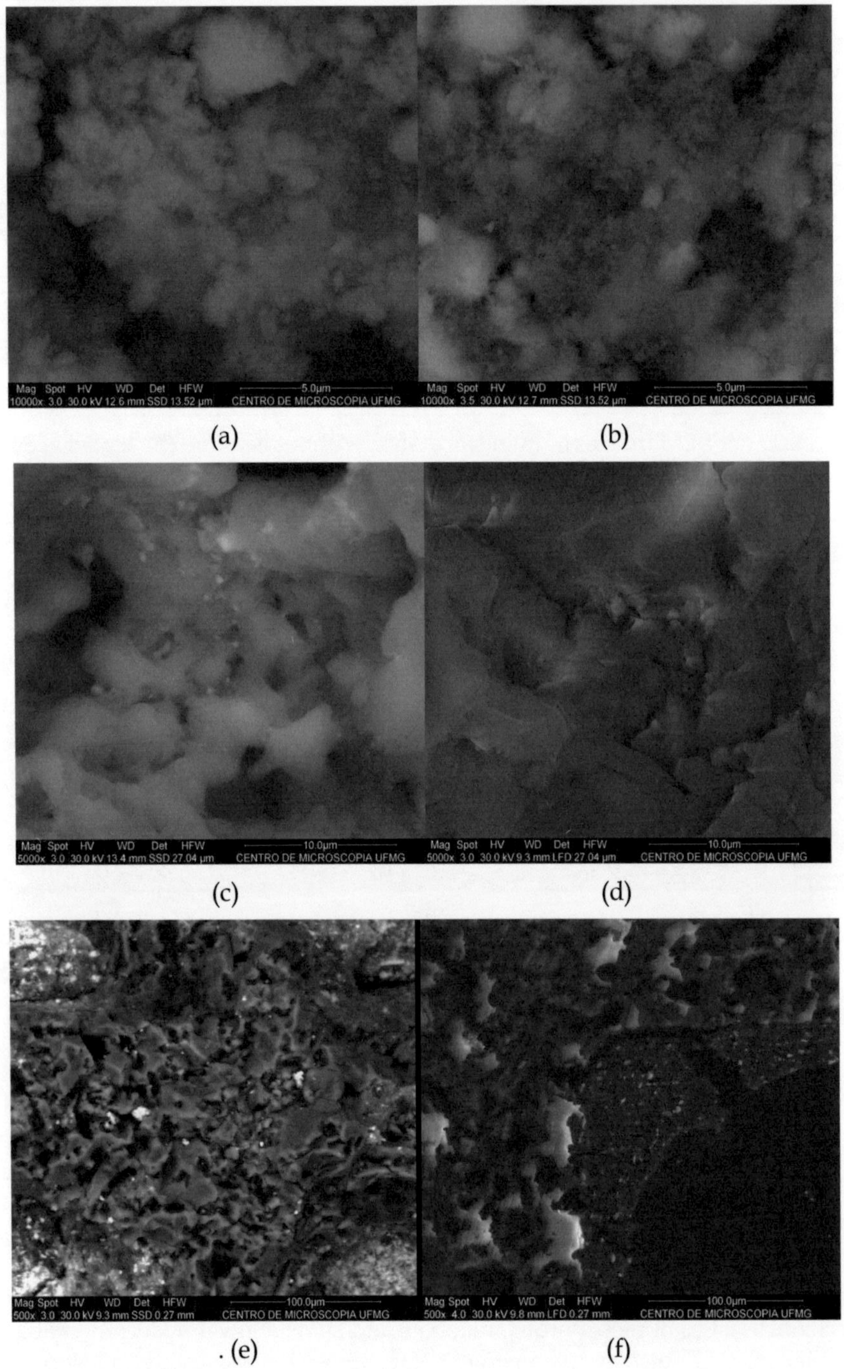

Fig. 16. SEM images of new catalyst (a and c) without heating, (b and d) heated at 500 °C, and (e and f) heated at 900 °C.

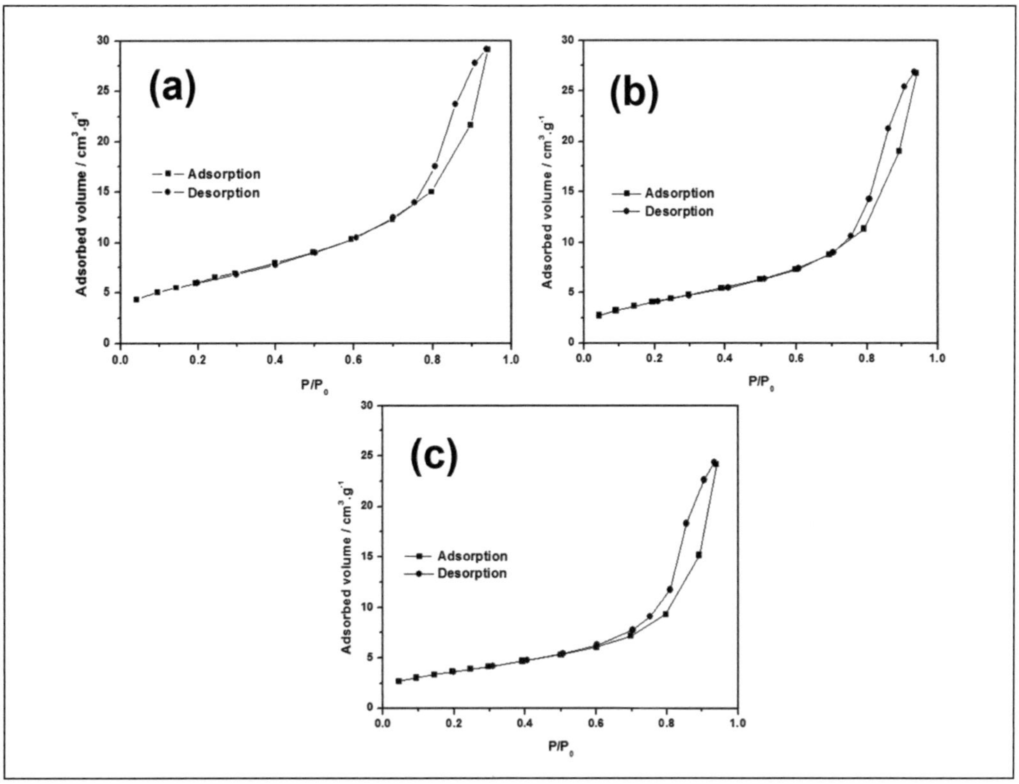

Fig. 17. Adsorption-desorption curves of new catalyst heated at: (a) 500 °C, (b) 700 °C and (c) 900 °C.

Heating Temperature New Catalyst/°C	Density /g.cm^{-3}	Specific Surface Area /m^2.g^{-1}	Total Pore Volume / 10^{-3} cm^3.g^{-1}	Average Pore Size /nm
New	3.1 ± 0.1	58	141	110
500	3.1 ± 0.1	30	48	37
700	3.2 ± 0.1	17	30	36
900	3.2 ± 0.1	12	27	30

Table 2. Textural characteristics of a new catalyst of vehicle powered by gasoline heated at various temperatures

The adsorption capacity of a catalyst decreases at high temperatures (operating temperature) due to the physical-chemistry alterations that occur mainly in ceramic catalyst surface, such as specific surface area decrease, changes in the phase of the alumina film and metal diffusion inside the ceramic matrix to generate large agglomerates. Fig. 18 shows a schematic illustration of the catalyst deactivation due to these thermal effects.

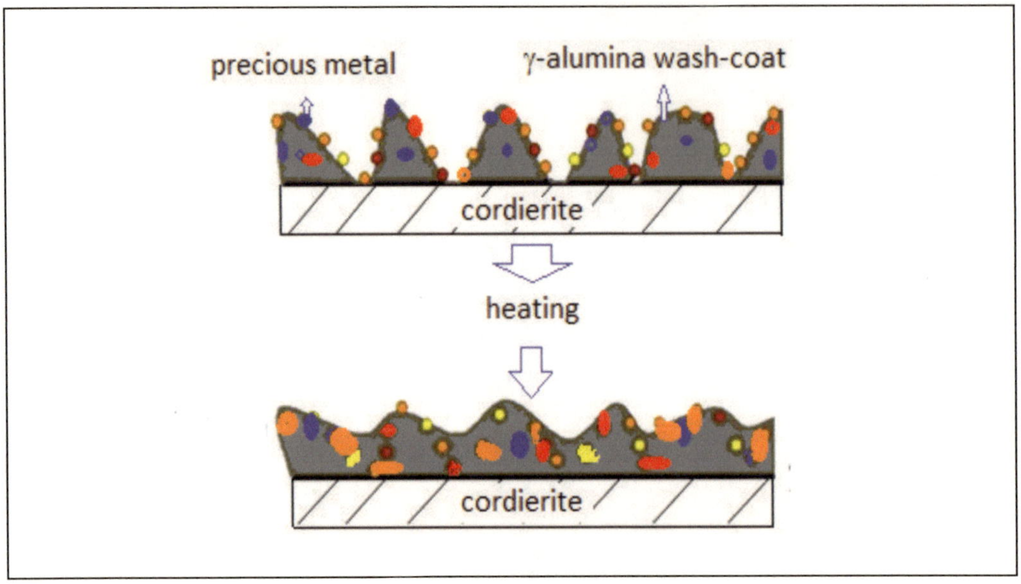

Fig. 18. Schematic illustration *of the catalyst deactivation due to the thermal effects.*

4. Conclusion

With the current increase in production of vehicles and consequent increase in gas emissions, especially CO_2, various solutions have been sought by scientists and governments around the world. The research on electric cars, already under implementation, the total or partial replacement of petrol by biofuels and gas, and the constant development of new types of automotive catalysts are examples of the attempt to contain the gaseous emissions.

Regarding the catalytic converter, new technologies have been developed but are not employed because the environment laws are not as stringent. With the increase in emissions and consequent global temperature increase, as has been happening during the year 2010, government decisions on the environment should be more stringent.

This work showed problems related to mechanical deactivation of catalysts, due to the soot deposited on their active surface and thermal deactivation involving the reduction of surface area, diffusion and agglomeration of the precious metals, with subsequent decrease of adsorption capacity of catalysts. Some results showed that a catalyst can be deactived in the first hours of operation, if the temperature reaches between 500 and 900 °C.

These problems can be solved by improving the precious metals distribution on the alumina films by nanotechnology practices, using alumina thin films more stable at high temperatures as sol-gel technology and various other new technologies available in the scientific literature.

5. Acknowledgements

This work was supported by CNPq and FAPEMIG (Brazilian funding agencies). The authors acknowledge the use of the infrastructure of the LMA and Center of Microscopy/UFMG and the help of their technicians.

6. References

Angelidis, T.N. & Sklavounos, S.A. (1995). A SEM-EDS study of new and used automotive catalysts. *Applied Catalysis A: General*, Vol. 133, pp 121-132

Angelidis, T.N. & Sklavounos, S.A. (1995). Partial regeneration of an aged commertial automotive catalyst. *Applied Catalysis B: Environmental*, Vol. 12, pp 193-206

Cohn, D. R.; Bromberg, L. & Heywood, J. B. (2005) Direct Injection ethanol boosted gasoline engines: Biofuel leveraguing for cost effective reduction of oil dependence and CO2 emission, Massachussets Int. Of Technology, Cambridge, MA 02139.

Collins N.R. & Twigg M. V. (2007). Three-way catalyssst emissions control technologies for spark ignition engines - Recent trends and future developments. *Topics in Catalysis*, Vol 42-43, pp 323-332.

Delgado, R. C. O. B.;Araújo, A. S. & Fernandes Jr., V. J. (2007). Properties of Brazilian gasoline mixed with hydrated ethanol for flex-fuel technology. Fuel Processing Technology vol. 88, pp 365-368.

Dufresne, P. (2007) Hydroprocessing catalysts regeneration and recycling. *Applied Catalysis A: General* Vol. 322, pp67-75.

Heck,R. M. & Farrauto,R. J. (1995*) Catalytic Air Pollution Control: Commercial Technology*, Van Nostrand Reinhold, New York.

HiroKazu, K. (1999) Off-site catalyst regeneration and catalyst recycling. *Petrotech* Vol. 22, pp 419-423

Ismagilov, Z.R.; Shkrabina,R.A.; Koryabkina N.A.; Arendarskii, D.A. & Shikina, N.V. (1998) Preparation of thermally stable wash-coat aluminas for automotive catalysts, in: N.

Kruse, A. Frennet, J.M. Bastin (Eds.*), Catalysis and Automotive Pollution Control,* vol. IV, Elsevier, Amsterdam, 1998, pp. 507–511.

Kaspar, J.; Fornasiero P. & Hickey, J. N. (2003). Automotive catalyst converters : States of art and perspectives. *Catalysis Today* Vol. 77 , pp 419-449.

Lowell, S. and Sields, S. (2005). *Powder Surface Area and Porosity*, Ed. Chapman and Hallm, NY.

Massad, E.; Saldiva, C. D.; Cardoso, L. M. N.; Silva, R.; Saldiva, P. H. N. & Böhm, G.M. (1985) Acute toxicity of gasoline and ethanol automobile engine exhaust gases. *Toxicol Lett* Vol. 26, pp 187-192.

Mizukami, F.; Maeda, K.; Watanabe, M.; Masuda, K.; Sano, T. & Kuno,K. (1991) Preparation Of thermostable high-surface-area aluminas and properties of the alumina-supported Pt catalysts, in: A. Crucg (Ed.*), Catalysis and Automotive Pollution Control,* vol. II, Elsevier, Amsterdam, pp. 557–568.

Morterra,C.; Magnacca G.; Bolis,V.; Cerrato, G. ; Barricco, M. A. ; & Fucale, M. (1995) Structural, morphological and surface chemical features of Al2O3 catalyst supports stabilized with CeO2, in: A. Frennet, J.M. Bastin (Eds.), *Catalysis and Automotive Pollution Control,* vol. III, Elsevier, Amsterdam, pp. 361–373.

Sideris, M. (1997). Methods for Monitoring and Diagnosing the Efficiency of Catalytic Converters: *A Patent Oriented Survey,* vol. 115, Elsevier, Amsterdam.

Silva, R. A. (2008) Estudo das características morfológicas, texturais e estruturais dos catalisadores automotivos. *Thesis*, Federal Universidty of Minas Gerais, Brazil

Silva, R. A.; Diniz C. F.; Viana, M.M.; Abreu, S. L.; Souza, T. G. F.; Mohallem, N. D. S. & Mohallem, T. D. S. (2009). Preparation of thin films using freeze drier alumina. *Brazilian J. Phys.* Vol 39, pp 236-238.

20

Materials in Automotive Application, State of the Art and Prospects

Elaheh Ghassemieh
University of Sheffield
UK

1. Introduction

This chapter gives a comprehensive account of the materials used in manufacturing vehicles. In the first section it explains the properties and characteristics that a suitable material should have to be accepted in automotive production. In later sections it reviews the history of development of the materials in automotive from the most traditional to the most recent ones. In the class of the metallic materials, steel, aluminium and magnesium and the most recent alloys of these used in the automotive are explained. Some of the properties, manufacturing and joining processes for these metals are described. The advantages and problems of using each of these materials are also reported. The potential application of these materials in different parts of a vehicle is identified. The other class of materials considered is composites and plastics with synthetic or natural fibre as reinforcement. Whilst the synthetic fibres are more traditional type of composites used, the natural fibre composites hold a relatively new place with substantial potential for growth due to the growing environmental concerns. With regard to the composite the cost is one of the most important barriers in use of these materials. Therefore a cost analysis is presented. Also the second barrier is a suitable manufacturing process for producing complex automotive parts. A review of the manufacturing process therefore is also offered for the composites with both synthetic and natural fibres. s

2. Requirements of the materials in automotive

The materials used in automotive industry need to fulfil several criteria before being approved. Some of the criteria are the results of regulation and legislation with the environmental and safety concerns and some are the requirements of the customers. In many occasions different factors are conflicting and therefore a successful design would only be possible through an optimised and balanced solution.

2.1 Lightweight

As there is a high emphasis on greenhouse gas reductions and improving fuel efficiency in the transportation sector, all car manufacturers, suppliers, assemblers, and component producers are investing significantly in lightweight materials Research and Development and commercialization. All are moving towards the objective of increasing the use of lightweight materials and to obtain more market penetration by manufacturing components and vehicle structures made from lightweight materials. Because the single main obstacle in application of lightweight materials is their high cost, priority is given to activities to reduce costs through development of new materials, forming technologies, and manufacturing processes.

Yet the weight reduction is still the most cost-effective means to reduce fuel consumption and greenhouse gases from the transportation sector. It has been estimated that for every 10% of weight eliminated from a vehicle's total weight, fuel economy improves by 7%. This also means that for every kilogram of weight reduced in a vehicle, there is about 20 kg of carbon dioxide reduction.

To achieve lightweight construction, without compensating on rigidity, automakers have been investigating the replacement of steel with aluminium, magnesium, composites, and foams. The recycling and recovery of end-of-life vehicles, which involves recovery targets of 85%, are driving the auto industry to adopt lightweight materials technology to meet these recovery targets.

Some of the facts about the lightweight materials are as follows: [McWilliams, 2007]

- The total global consumption of lightweight materials used in transportation equipment was 42.8 million tons/$80.5 billion in 2006 and will increase to 68.5 million tons/$106.4 billion by 2011, at a compound annual growth rate (CAGR) of 9.9% in tonnage terms and 5.7% in value terms between 2006 and 2011.
- High strength steel accounts for the largest percentage of total tons of lightweight materials consumed, followed by aluminium and plastics. In value terms, plastics with their relatively high unit prices are the largest market segment. Aluminium and high strength steel are the second and third largest product segments.
- Motor vehicles, particularly passenger cars and light trucks, are by far the largest end-user segment. Shipbuilding was the second largest consumer of lightweight materials, while the aircraft industry ranks second in the value of the lightweight materials consumed.

Car manufacturers are investigating the reduction of the weight in a viable economical way. For example study by Lotus Engineering concludes that a vehicle mass improvement of 38% versus a conventional mainstream vehicle can be achieved at only 3% cost. Comparison of weight reduction and associated costs for two future models are shown in Table 1. [Lotus Eng. Co., 2010]

Also a report from Corus indicates that the main structure – known as the Body In White (BIW) – is usually made of steel pressings welded together to form a strong and stiff frame.

This method of construction accounts for 99.9 per cent of all the cars produced in the world. The remaining 0.1 per cent is mostly constructed with aluminium BIW, while a very small number (less than 0.01 per cent) are constructed from carbon-fibre composite.

The material properties of steel (with its wide range of yield strength combined with high modulus) together with ease of manufacture and low cost, mean that steel intensive vehicles have by far the largest share of the market. The high cost of alternative materials such as aluminium or composites mean that steel's position as the first-choice material could be still secure. [Corus Automotive Eng., 2010] The BIW of a vehicle accounts for 20 per cent of the vehicle mass. The weight of the closures (doors, bonnet and boot/rear hatch), chassis (suspension parts) and driveline bring the total amount of steel and other ferrous metals to more than 60 per cent. In recent years, the amount of ferrous metal has declined, mostly driven by manufacturers replacing iron with aluminium for engine castings. The percentage of sheet steel per car has also dropped, mainly due to:

- Higher levels of equipment, trim and soundproofing.
- More aluminium used in wheels and suspension parts.
- More moulded plastics, especially under the bonnet.

Base Toyota Venza Excluding powertrain		Lotus Engineering Design			
		2020 Venza		2017 Venza	
System	Weight (kg)	% Mass Reduction	% Cost Factor	% Mass Reduction	% Cost Factor
Body	383	42%	135%	15%	98%
Closures/Fenders	143	41%	76%	25%	102%
Bumpers	18	11%	103%	11%	103%
Thermal	9.25	0%	100%	0%	100%
Electrical	23.6	36%	96%	29%	95%
Interior	252	39%	96%	27%	97%
Lighting	9.90	0%	100%	0%	100%
Suspension/Chassis	379	43%	95%	26%	100%
Glazing	43.7	0%	100%	0%	100%
Misc.	30.1	24%	99%	24%	99%
Totals	1290	38%	103%	21%s	98%

Table 1. An Assessment of Mass Reduction Opportunities for a 2017 - 2020 Model Year Vehicle Program by Lotus, [Source: Lotus Engineering, 2010]

The weight reduction versus the price increase by replacing steel by aluminium or magnesium for some of the parts is reported in Table 2.

	steel (kg)	Aluminium (kg)	Magnesium (kg)	% weight reduction (part)	(vehicle)	% cost increase (part)
Body In white (BIW)	285	218	N/A	23.5	3.90 Examples vehicle mass of 1700 kg	250
Bonnet (assembly)	14.8	8.3	N/A	44	0.48 Examples vehicle mass of 1350 kg	300
Door (assembly)	15.7	9.5	N/A	39	0.40 Examples vehicle mass of 1550kg	275
IP Beam (instrument panel support)	11.4	N/A	6.3	45	0.33 Examples vehicle mass of 1550kg	350

Table 2. Alternative materials, potential weight savings versus cost, [Source: Corus Automotive Eng., 2010 with permission]

2.2 Cost

One of the most important consumer driven factors in automotive industry is the cost. Since the cost of a new material is always compared to that presently employed in a product, it is one of the most important variables that determines whether any new material has an opportunity to be selected for a vehicle component. Cost includes three components: actual cost of raw materials, manufacturing value added, and the cost to design and test the product. This test cost can be large since it is only through successful vehicle testing that the product and manufacturing engineers can achieve a 'level of comfort" to choose newer materials for application in a high-volume production program.

Aluminium and magnesium alloys are certainly more costly than the currently used steel and cast irons that they might replace. The ability to approach the total cost of the competition, therefore, must be associated with lower component manufacturing costs. Compared to cast irons and steel, cast aluminium and magnesium components are potentially less costly. This is based on their reduced manufacturing cycle times, better machinability, , ability to have thinner and more variable wall dimensions, closer dimensional tolerances, reduced number of assemblies, more easily produced to near net shape(thus decreasing finishing costs, and less costly melting/metal-forming processes). However, wrought aluminium and magnesium components are almost always more costly to produce than their ferrous counterparts. Since cost may be higher, decisions to select light metals must be justified on the basis of improved functionality. Government regulations mandate reductions in exhaust emissions, improved occupant safety, enhanced fuel economy, reductions in workplace emissions, increased safety requirements, and requirements for toxic materials handling and disposal. Also the high cost is one of the major barriers in use of the composite materials.

2.3 Safety, crashworthiness

The ability to absorb impact energy and be survivable for the passengers is called the "crashworthiness" of the structure in vehicle. There are two important safety concepts in automotive industry to consider, crashworthiness and penetration resistance. Crashworthiness is defined as the potential of absorption of energy through controlled failure modes and mechanisms that provides a gradual decay in the load profile during absorption. However, penetration resistance is concerned with the total absorption without allowing projectile or fragment penetration. [Jacob et al, 2002]

The current legislation in design of the automobiles requires that, in the case of an impact at speeds up to 15.5 m/s (35 mph) with a solid, immovable object, the occupants of the passenger compartment should not experience a resulting force that produces a net deceleration greater than 20g.

The current trend of materials in car industry is towards replacing metal parts more and more by polymer composites in order to improve the fuel economy and reduce the weight of the vehicles. The behaviour of composite failure in compression is the opposite to metals. Most composites are generally characterized by a brittle rather than ductile response to load. While metal structures collapse under crush or impact by buckling and/or folding in accordion (concertina) type fashion involving extensive plastic deformation, composites fail through a sequence of fracture mechanisms involving fibre fracture, matrix crazing and cracking, fibre-matrix de-bonding, de-lamination and interply separation. The actual mechanisms and sequence of damage are highly dependent on the geometry of the structure, lamina orientation, and type of trigger and crush speed, all of which can be suitably designed to develop high energy absorbing mechanisms.

Several aspects are considered in design for improved crashworthiness including the geometrical and dimensional aspects which have key role in different stages of crash. However the materials deformation and progressive failure behaviour in terms of stiffness, yield, strain hardening, elongation and strain at break are also very important in the energy absorption capacity of the vehicle. [Witteman, 1999].

Thin-walled columns are basic components in the concept and design of automotive body structures. Their crashworthiness behaviour is of fundamental importance in the safety design of the whole vehicle because their plastic collapse is the mechanism that is used to dissipate the kinetic energy of the vehicle in an accident. The mechanism of plastic collapse should be reliable and its evolution during the crash regular so that the desired quantity of absorbed energy, a low load uniformity and the required level of deformation load can be achieved without increasing danger for the vehicle passengers. [Wallentowitz & Adam, 1996] To predict the characteristic values of automotive front structures energy absorption, e.g. weight specific energy absorption, load uniformity and structural effectiveness, the buckling of thin-walled columns, representing body front side members, normally are investigated. Geometries used for front side members like closed-hat, double-U and octagonal columns made by conventional steel, high-strength steel and light alloys could be joined with different joining methods e.g. spot-welding, press-joining and structural adhesive. The design parameters of the specimen (t/a-ratio, flange width, joining width, material thickness, etc.) are varied in a wide range. Axial and non-axial quasi-static tests and even dynamic tests with different collision speeds are normally carried out to evaluate the crash behaviour.

2.3.1 Crashworthiness tests

Apart from the test at the design stage explained above, there are more tests to be performed at the later stage of product evaluation. To determine crashworthiness — how well a vehicle protects its occupants in a crash — the Institute of highway safety rates vehicles good, acceptable, marginal, or poor based on performance in high-speed front and side crash tests, a rollover test, plus evaluations of seat/head restraints for protection against neck injuries in rear impacts. To earn Top Safety Pick for 2010 a vehicle must have good ratings in all four Institute tests. In addition, the winning vehicles must offer electronic stability control. [Insurance institute of highway safety, 2010]

Frontal offset crash test details

Today's passenger vehicles are designed to be more crashworthy than they used to be. Still, about 30,000 passenger vehicle occupants die in crashes on US roads each year. About half of the deaths occur in frontal crashes.

Since the late 1970s, the federal New Car Assessment Program has compared frontal crashworthiness among new passenger vehicles. This program, which involves 35 mph crash tests into a full-width rigid barrier, has been highly successful in providing consumers with comparative crashworthiness information. It also has been a major contributor to the crashworthiness improvements that characterize recent passenger vehicle models.

The very success of the New Car Assessment Program means remaining differences in performance among most new vehicles in full-width tests are small. This doesn't mean important crashworthiness differences no longer exist. They do exist, and additional crash test configurations can highlight these differences. One such test is the frontal offset crash.

Side impact crash testing/ratings criteria

Today's passenger vehicles are more crashworthy than they used to be, especially in frontal crashes. As occupant protection in frontal crashes improves, the relative importance of protection in side impacts increases. From the early 1980s until 2000, driver death rates per million cars registered decreased 47 percent. Most of this improvement was in frontal crashes, in which driver death rates decreased 52 percent. In contrast, the decrease in side impacts was only 24 percent.

These changes are attributable to two effects. There have been significant improvements in frontal crash protection – standard airbags, improved structural designs, and higher belt use rates, for example.

Since 1997 the federal New Car Assessment Program, which compares crashworthiness among new passenger vehicles, has included side impacts. In these tests, an impactor with a deformable front end representing the front of a car is used to strike the sides of the vehicles being assessed.

The Institute's side impact test is severe. Given the designs of today's vehicles, it's unlikely that people in real-world crashes as severe as this test would emerge uninjured. But with good side impact protection, people should be able to survive crashes of this severity without serious injuries.

Side impact crash test ratings can be compared across vehicle type and weight categories, while frontal crash test ratings will be compared only within weight categories.

Rollover evaluations

Rollover ratings assess vehicle roof strength for protection in rollover crashes. Roof strength is a new component of the *TOP SAFETY PICK* criteria for 2010 models.

To measure roof strength, a metal plate is pushed against one corner of a vehicle's roof at a constant speed. The maximum force sustained by the roof before 5 inches of crush is compared to the vehicle's weight to find the strength-to-weight ratio. This is a good assessment of vehicle structural protection in rollover crashes.

Other tests on the safety of the vehicle include the Rear crash protection/head restraint ratings which focus on how well seat/head restraint combinations protect against whiplash injury. The necessary first attribute of an effective head restraint is good geometry. If a head restraint isn't behind and close to the back of an occupant's head, it can't prevent a "whiplash" injury in a rear-end collision.

Also Electronic stability control (ESC) significantly reduces crash risk, especially the risk of fatal single-vehicle crashes, by helping drivers maintain control of their vehicles during emergency manoeuvres.

2.4 Recycling and life cycle considerations

One of the major growing concerns in all the industries including automotive, is an increased awareness for environment. Issues such as 'protection of resources', 'reduction of CO2 emissions', and 'recycling' are increasing the topics of consideration.

While the United States has not issued regulations concerning automotive end-of-life requirements, European Union (E.U.) and Asian countries have released stringent guidelines. European Union legislation implemented in 2006 dictates that a significant percentage of the vehicle should be re-used or recycled.

The End of Life Vehicles (ELV) Directive from environment agency aims to reduce the amount of waste produced from vehicles when they are scrapped. Around **two million vehicles** reach the end of their life in the UK each year. These vehicles are classed as hazardous waste until they have been fully treated. The directive requires ELV treatment sites to meet stricter environmental standards. This would mean that the last owner of a vehicle must be issued with a **Certificate of Destruction** for their vehicle and they must be able to dispose of their vehicle free of charge. Vehicle manufacturers and importers must cover all or most of the cost of the **free take-back system**. It also sets higher reuse, recycling and recovery targets and limits the use of hazardous substances in both new vehicles and replacement vehicle parts. [Environment agency, 2010]

As the results of the new legislations, no discussion of new materials in the automotive industry should conclude without a consideration of recycling. Considerable R&D efforts are now focused on developing materials with greater potential of recycling and re-use or developing ways of recycling and re-use of the current materials. This includes both metal and composite materials. The composition and forming processes of the metal materials are changing to accommodate this recycle and re-use demand. This also justifies the great attention towards natural fibre based composites and new high temperature resistant thermoplastic resins.

Figure 1 shows a simplified diagram of the vehicle life cycle. [Kumar & Sutherland, 2008] After the raw material is extracted, it is processed, manufactured into parts and products, and then sold to the consumer. It should be noted that there is a chain of part and component suppliers that contribute to the final manufacturing/assembly stage performed by the original equipment manufacturers (OEM). Traditionally the design of the part/components has mainly been the responsibility of the OEMs. Recently, however, manufacturers have begun to shift design responsibility to part/component suppliers. After the manufacturing stage, the vehicle spends a long period of time in the use phase of the life cycle. In the past, the median life of a vehicle has been reported to be 12-13 years [Libertiny, 1982, 1993]. Lately, the median life has increased to around 16 years with as many as 5% of the vehicles still remaining on the road after 30 years of operation [Davis, 2004].

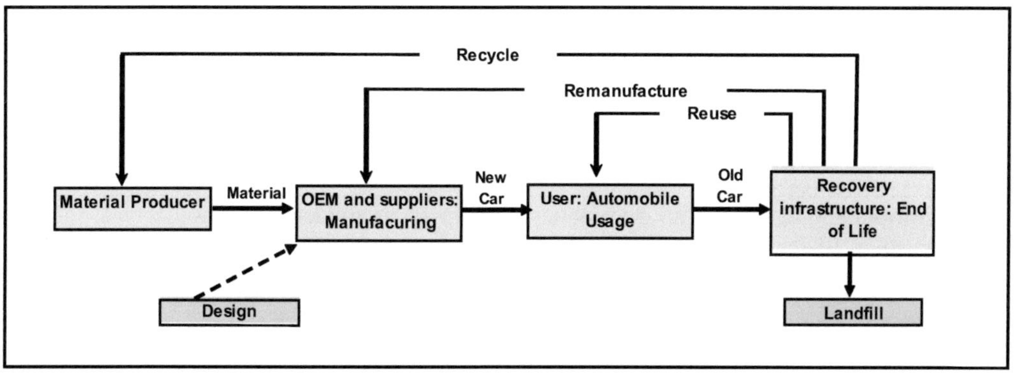

Fig. 1. Life cycle of automobile [Source: Kumar & Sutherland, 2008]

When a consumer decides that they no longer wish to use a given vehicle, there are a number of options available to them:

1. sell the whole vehicle to another user
2. disassemble the vehicle with some components/modules reused
3. remanufacture the vehicle
4. recycle the vehicle for material content, and/or
5. dispose the vehicle to a landfill (Zussmann et al., 1994; Kumar et al., 2005).

The option selected depends on the economic benefit to the parties/entities associated with each option. It should be noted that often times a vehicle undergoes multiple use cycles (i.e., the first option is selected repeatedly) before the user considers the other options. After multiple use cycles, the value of the vehicle to the user (and other potential users) approaches zero. This is because the vehicle no longer functions as desired - the vehicle is said to have reached the end of its life. However, from the perspective of a dismantler or shredder, the value of the vehicle is not zero because they are not directly interested in the vehicle function; rather, they are interested in the value of the vehicle subassemblies, parts, and materials (Kumar et al., 2007). It should be noted that the expression end of use product 'EOUP' is preferred over the commonly used end of life vehicle 'ELV' phrase, because as a whole the vehicle may no longer be functional, but certain subassemblies or parts may be functional and of course the material can always be recycled for the secondary materials market. Each life cycle stage shown in Figure 1 consumes certain amount of energy, and produces air/water emissions and municipal solid waste.

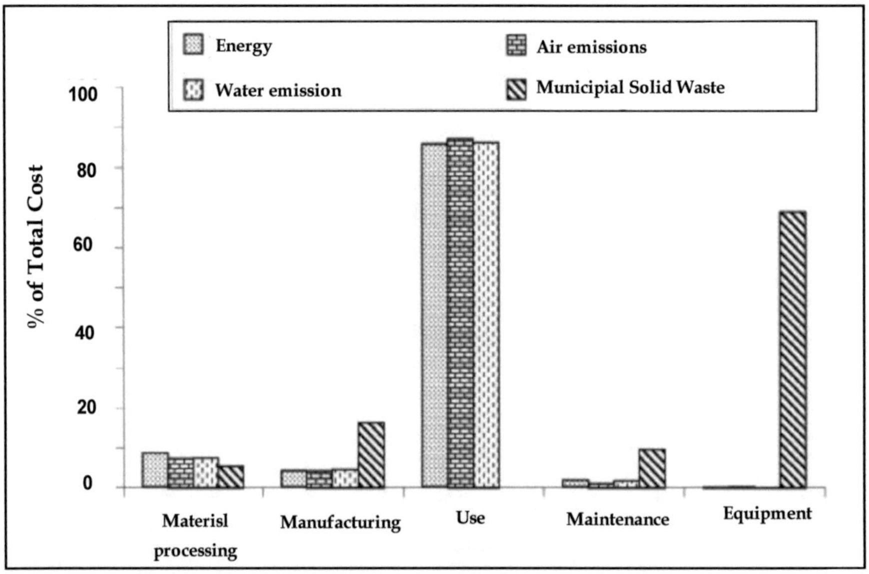

Fig. 2. Comparison of energy, air emission, water emission and municipal solid waste across the life cycle of 1995 generic vehicle [Source: Sullivan et al, 1998]

Figure 2 shows the contribution of each stage in terms of energy consumption, air and water emissions, and municipal solid waste. As can be seen from the figure 2, use stage is the most significant contributor in terms of energy consumption, and air and water emissions. This justifies the effort of the automotive manufacturers to use lighter-weight material and new powertrain technologies in its vehicles. However, while analysing the impact of use stage on the environmental performance of a vehicle, it is important to note that the associated energy consumption and emissions are distributed over a larger period of time as opposed to a shorter time periods during the other stages (Field et al., 2001). Thus, any conclusion with respect to the environmental performance should carefully include the temporally distributed nature of emissions. It can also be seen from the figure 2 that in terms of municipal solid waste, the EOUP stage was found to be the significant contributor.

It should be noted that the choice of material plays an important role on the fuel economy of the vehicle. For example, use of lighter-weight material such as aluminium as opposed to steel is usually considered a positive choice in terms of fuel economy (Ungureanu, 2007a). As an approximation, a 10% reduction in mass produces a 5% improvement in fuel economy. However, it should also be noted that lighter-weight materials might not always be the most sustainable choice from a total life cycle perspective. For example, in a study to determine the potential benefits of lighter materials such as aluminium, it was found that substantial change in the existing manufacturing and assembly technologies would be needed. It was found that the current equipment and processes are well suited for steel-based components and a complete redesign of this equipment and processes would be needed to manufacture aluminium components (Ungureanu, 2007b). Selection of material and manufacturing technology also impacts the automotive recovery infrastructure. It is suggested that replacing steel with aluminium or a composite material not only increases the fuel economy but also reduces the overall life cycle emissions and energy consumption. However, it is also likely that the total solid waste, which is usually landfilled will increase with such material substitutions. It is largely unknown what effect such substitutions will have on the generation of solid waste and thus, the environmental sustainability of the recovery infrastructure. Further detail analysis with consideration of interactions regionally is needed for better understanding of the life cycle issues.

3. Current materials in use and their future

The current materials used in automotive industry both categories of metals and composites are reviewed. The advantages and disadvantages of each material are discussed. Also the most recent topics of research on development of new compositions, manufacturing and characterisations of these materials for the specific purpose of use in automotive is presented.

3.1 Metals

3.1.1 Steel

Advanced iron and steel technologies have seen considerable development over the past decade and are frequently included into new designs and redesigns by all automakers. The steel industry and component suppliers are investing heavily in innovation. The result of the investment is numerous examples of successful, cost-effective use of stainless steel, new formulations of iron, high-strength steels, and an associated variety of new design, fabrication, and assembly techniques. Applications include not only vehicle bodies, but also engine, chassis, wheels and many other parts. The usages commonly demonstrate weight reduction plus simultaneous improvements in strength, stiffness, and other structural performance characteristics. Thus, a clear potential exists to affordably make vehicles lighter and safer at the same time. [DeCicco, 2005]

While body, chassis, engine and other powertrain components made of ferrous materials comprise the largest part of a vehicle by mass, lightweight steel and iron technologies compete with potential substitutes in all of these applications.

Mass reduction through advances in the use of iron and steel is significant (ferrous materials), because they are the dominant material (64% of a typical family vehicle).

Iron and steel form the critical elements of structure for the vast majority of vehicles, and are low-cost

materials with an extensive experience base and familiarity to the industry.

The past several years have seen steady increases in the use of high-strength steels (HSS), many versions of which are referred to as high-strength, low-alloy (HSLA) steels. These materials plus their associated advanced design and fabrication techniques (as well as improved design and fabrication using traditional steels) formed the basis of the American Iron and Steel Institute (AISI) Ultralight Steel Auto Body (ULSAB) series of studies and demonstration projects. The ULSAB car body demonstrated a 19% mass reduction in a body structure that had superior strength and structural performance (including crashworthiness) along with a reduced parts count and net manufacturing cost savings compared to a conventional steel body. [USAB, 1998] Comparable mass reductions and other benefits were achieved for doors, hoods, decklids, and hatchbacks. [Opbroek & Weissert, 1998] Improved steel materials and forming processes allow a significant optimization of vehicle body structures and components. [DeCicco, 2005]

The prime reason for using steel in the body structure of an automotive is its inherent capability to absorb impact energy in a crash situation [Marsh, 2000]. This, in combination with the good formability and joining capability, makes these materials often a first choice for the designer of the body-in-white (BIW) structure. [Magnusson et al, 2001]

New grades of steel and alloys

Materials are often described by properties such as yield- and tensile strength, elongation to fracture, anisotropy and Young's modulus but shape is not a material property. A sheet metal component is a material made into a certain shape through a forming process. Depending on loading condition, a material-and-shape combination resists the applied load best. Components in a BIW structure should also be able to absorb or transmit impact energy in a crash situation. Certain tests should be performed to decide about the suitability of the materials for automotive application.

In axial tensile loading of components, the shape is not as important as the cross-sectional area since all sections with the same area will carry the same stress. The strength of a component that should be under axial loading is related to the mechanical properties of the material [Meyers & Chawla, 1999].

In bending and torsion, both material and shape are important parameters for the efficiency of the component to carry the applied load [Ashby, 2000]. For bending, the elastic-plastic transition is a combination of shape and material properties. The strength of a beam under bending is related to the materials yield stress and Young's modulus. The stiffness is correlated to the materials Young's modulus and the shape of the component.

High-strength steel (HSS) is based on alloys that are categorized on the basis of yield strength. Standard HSS has a yield strength between 210 MPa and 550 MPa; ultra-high-strength steel (UHSS) has a yield strength higher than 550 MPa. High-strength steels can cost as much as 50% more than traditional mild steels, but they allow use of lower thicknesses than milder steels for achieving needed part performance specifications. Also, different grades of steel can be combined in tailored blanks (see below), so that the more costly or thicker materials can be placed only where needed. With HSS, there can be a trade-off between strength and formability; in other words, the stronger a steel is, e.g., in resisting stretching (tension), the more difficult it can be to forge into shapes, particularly the stylistically and aerodynamically optimized shapes needed for new vehicles. Steel suppliers are therefore developing steels with a range of properties that give engineers more flexibility in selecting an ideal grade of steel for any given application.[Heckelmann et al, 1999]

Stainless steel is a material of choice due to passivity and resistance to corrosion. Some of the stainless steel grades suggested for automotive are as follows: [Cunat, 2000]

a. Duplex austenitic-ferritic stainless steel

The most commonly used duplex grade is 0.02% C – 22% Cr – 5.5% Ni – 3% Mo – 0.15% N alloy, whose standard European designation is X2CrNiMoN22-5-3 / 1.4462.

b. Austenitic stainless steel

These steels have chromium (18 to 30 per cent) and nickel (6 to 20 per cent) as the major alloying elements. The austenitic phase is stabilised by the presence of a sufficient amount of

nickel. The principal characteristics are the ductile austenitic condition, rapid hardenability by cold working and excellent corrosion resistance.
One of the most commonly used grade for structural applications is the 0.02% C - 17.5% Cr - 7% Ni - 0.15% N alloy, whose standard European designation is X2CrNiN 18-7/1.4318.

Property	Duplex Stainless Steel (1)	Austenitic Stainless steel			6061 Aluminium Alloy		High Strength Steel
		Annealed	C850(2)	C1000(3)	T4(4)	T6(5)	HSLA
Density: $\rho(g/cm^3)$	7.8	7.9	7.9	7.9	2.7	2.7	7.83
Yield Stress: σ (N/mm^2)	640	370	600	880	130	275	410
Specific Strength $(N/mm^2/g/cm^3)$	82	46.8	76	111.4	48.1	100	52.4

(1) In the solution annealed condition, (2) In the cold worked condition C 850 (850<UTS (N/mm^2)<1000) (3) In the cold worked condition: C 1000 (1000<UTS (N/mm^2)<1150), (4) In the solution heat treated condition, (5) In the precipitation heat treated condition

Table 3: Specific Strength of Stainless Steels, 6061 Aluminium and High Strength Steel [Source: Cunat, 2000]

The specific stiffness of Stainless Steel is very similar to that of aluminium alloy and the HSLA steel, which means that the three materials can all be considered as "light materials".
The specific strength of the austenitic Stainless Steel in the cold worked condition, is much higher than the one for the other materials. The specific strength of different steel and aluminium are compared in the Table 3.
Crashworthiness energy absorption is a key property of the material used for structural components or complete structures so-called "space frames". Austenitic Stainless Steels i.e. Fe - Cr - Ni containing alloys have the advantage over aluminium alloys and carbon steels of being highly strain rate sensitive. This means that the faster the loading is applied the more the material will resist deformation. In addition to that, Stainless Steel has the capability to collapse progressively in a controlled and predetermined manner which is desirable in automotive application.

Advances in manufacturing and joining technique

Advances in fabrication and assembly technique are just as important as advances in materials. For lightweight steel technology, key process advances include *laser welding, hydroforming,* and *tailored blanks.* Both tailored blanks and hydroforming allow parts counts to be reduced, providing significant savings on tools and dies, simplifying later stages of assembly, and improving the integrity of components, subassemblies, and body structures. These processes can be combined in the production of any one component or subassembly.
Compared to conventional welding processes, ***laser welding*** creates a very clean and strong weld seam with minimum excess material. It is an important enabling technology used for multiple stages of steel materials fabrication and assembly. Laser welding permits production of new process input materials, such as tailored blanks, with smooth, high-integrity seams and minimal distortion or change in material properties surrounding the weld zone. It also improves strength, aesthetics, and overall quality of final assembled

structures. As automakers gain experience with laser welding and the structural design improvements it permits, they are reaping significant productivity savings as well A recently reported example is VW's use of the technology on the 2004 redesign of the Golf. [Kochan, 2003] Compared to the previous model, they reduced production time per car body by 25% while reducing weight.

A related innovation is greater use of steel tubes in place of shapes based on stampings of sheet steel. High-quality tubes are formed by bending sheets into a tubular shape with a laser welded seam. In addition to direct uses (e.g., in cross members and door beams), steel tubes also find broader use when further manipulated by hydroforming.

Hydroforming involves shaping a part in a die through the use of fluid pressure as opposed to stamping. Tube hydroforming permits the construction of relatively complex shapes with a single part that is stronger and lighter than the same part made as an assemblage of stampings. Although a number of challenges were identified, methods for overcoming them were found, with optimizations achieved using advanced CAD tools. Eight bodies were built for validation and testing purposes; the results demonstrated a small mass reduction with a 25% increase in torsional stiffness. Such results suggest that a lighter structure could have been built without as great an increase in stiffness. An economic analysis showed that the demonstrated hydroformed design could be implemented within pre-defined financial targets, in other words, would be cost-effective for the given application. Hydroforming is now coming into widespread use and is particularly valuable for optimizing the frames of light trucks. GM and Ford have both used hydroforming for frame components in their full-size pickups and vehicles sharing those platforms; again, however, the potential mass savings were sacrificed to provide further increases in stiffness and other structural performance attributes.

Tailored blanks combine different grades and thicknesses of steel into a single blank, referring to a piece of material that is inserted into a stamping press or other piece of forming equipment. They allow optimizations of strength, crash performance, and dent resistance with minimal material use and therefore lower weight than attempting to make a similarly performing part from a blank of uniform grade and thickness.[Kuroda, 2033] Tailored blanks also permit reduced major parts counts and simplified assembly. Instead of two or more different gauges being welded together to achieve the desired component, an integral component can be stamped or hydroformed from a tailored blank. This technique pushes some of the complexity upstream in the assembly process, but can do so with net cost savings and often substantial improvements in component performance (in terms of mass, stiffness, strength, etc.). Sandwich materials, involving a plastic core between thin sheets of a steel skin, are another innovation that can be used to save weight. Although sandwich steel cannot be welded, it can be formed and joined through many other common processes and is used in applications where bending stiffness is the principle performance need. One branded version of this material is "Quiet Steel®," which uses viscoelastic cores in a laminated steel composite to offer significant cost reduction opportunities and enhanced noise, vibration and hardness performance. [Materials science Co., 2004] A notable recent application is in the 2004 upgrade of Chrysler's Town & Country and Dodge Grand Caravan minivans. Driven by competition, Chrysler needed to add a stowable third-row seat and other refinements; steel sandwich material was used to make the tubs into which the foldable seats were stowed. [Kelly & Priddle, 2004] An overarching area of progress that enables further refinement in all aspects of iron and steel use is the major improvements in materials science, component characterization and modelling, and computerized simulation and design methods. Better techniques for measuring and modelling the properties of steels enable highly optimized designs. [Mahadevan et al, 2000] Such advances give engineers greater confidence in part performance, minimizing the "margin of error" that otherwise results in a larger or heavier part than needed. Extensive computer modelling development and validation work yields CAD/CAE/CAM34 techniques that enable many fewer adverse trade-offs in design, resulting in simultaneous progress in weight reduction, strength, stiffness, and energy absorption as needed, while cutting materials costs and waste and enhancing productivity in both design and manufacture. [Yoshimoto et al, 1999]

3.1.2 Aluminium

There are a broad range of opportunities for employing aluminium in automotive powertrain, chassis, and body structures. The use of aluminium offers considerable potential to reduce the weight of an automobile body. In current steel construction, the vehicle consists of stamped body panels spot welded together (body-in-white) to which stamped steel fenders, doors, hood, and deck lid are bolted. There are two methods of designing and manufacturing an aluminium body structure; one is similar to the current steel construction using stamped option and the other system which involves castings, extrusions, and stampings welded together, known as spaceframe. [Cole et al, 1995]

Adequate formability is one of the requirements for aluminium sheets to produce complex stampings at acceptable economical rates. This involves appreciation of the interaction of the crystallographic texture, sheet thickness and stamping die/lubricant parameters. In addition, the aluminium alloys chosen for exterior panels must have the ability of age hardening to provide suitable strength for dent resistance during the oven paint baking.

In order to use the Aluminium in automotive intake manifolds and transmission housings, it is essential that material shows the ability to be cast into leakproof components with well-defined inner passages for water and air flow, provides suitable thermal conductivity, and sufficient resistance to the mechanical forces at temperatures near 145'C. On the other hand, the components are exposed to high mechanical stresses from engine vibration and the thermal expansion loads. This can lead to thermal fatigue if the metallurgical structure is not sufficiently small and if the casting contains inclusions, oxide films, and porosity. This has initiated considerable research to aluminium castings with no defects to avoid fatigue and reduced impact resistance. Also control of solidification microstructure, dendrite arm spacing, grain size, and eutectic silicon morphology are the other areas that have become more and more under investigation. In addition to alloy chemistry and melt temperature, dissolved gas and nonmetallic inclusions must be controlled to limit porosity and stress-raising oxide films. Foundry practice to eliminate turbulence during pouring that can cause such films can be enhanced by computer-based heat-flow fluid flow and solidification modelling to design the location and geometry of sprue, ingates, and risers.

Aluminium usage in automotive applications has grown substantially within past years. A total of about 110 kg of aluminium: vehicle in 1996 is predicted to rise to 250 or 340 kg, with or without taking body panel or structure applications into account, by 2015 [Sears, 1997]. There are strong predictions for aluminium applications in hoods, trunk lids and doors hanging on a steel frame. Recent examples of aluminium applications in vehicles cover power trains, chassis, body structure and air conditioning. Aluminium castings have been applied to various automobile parts for a long period. As a key trend, the material for engine blocks, which is one of the heavier parts, is being switched from cast iron to aluminium resulting in significant weight reduction. Aluminium castings find the most widespread use in automobile. In automotive power train, aluminium castings have been used for almost 100% of pistons, about 75% of cylinder heads, 85% of intake manifolds and transmission (other parts-rear axle, differential housings and drive shafts etc.) For chassis applications, aluminium castings are used for about 40% of wheels, and for brackets, brake components, suspension (control arms, supports), steering components (air bag supports, steering shafts, knuckles, housings, wheels) and instrument panels. Recently, development effort to apply wrought aluminium is becoming more active than applying aluminium castings. Forged wheels have been used where the loading conditions are more extreme and where higher mechanical properties are required. Wrought aluminium is also finding applications in heat shields, bumper reinforcements, air bag housings, pneumatic systems, sumps, seat frames, side impact panels, to mention but a few. Aluminium alloys have also found extensive application in heat exchangers. Until 1970, automotive radiators and heaters were constructed from copper and brass using soldered joints. The oil crisis in 1974 triggered are-design to lighter-weight structures and heralded the use of aluminium. The market share of aluminium has grown steadily over the last 25 years and is now the material of choice for use in the automotive heat exchanger industry. Modern, high performance automobiles have many individual heat exchangers, e.g. engine and transmission cooling, charge air coolers (CACs), climate control. [Miller et al, 2000]

Aluminium alloys for body-in-white applications

Up to now the growth of aluminium in the automotive industry has been in the use of castings for engine, transmission and wheel applications, and in heat exchangers. The cost of aluminium and price stability remains its biggest impediment for its use in large-scale sheet applications. Aluminium industry has targeted the automotive industry for future growth and has devoted significant resources to support this effort. The body-in-white (BIW) offers the greatest scope for weight reduction with using large amount of aluminium.

Recent developments have shown that up to 50% weight saving for the BIW can be achieved by the substitution of steel by aluminium [Scott, 1995]. This can result in a 20–30% total vehicle weight reduction when added to other reduction opportunities. There are two types of design each of which has a different form philosophy in the use of aluminium. One is the extruded space frame exemplified by the Alcoa- Audi A8 , and the other is the conventional sheet monocoque architecture as used in most steel structures as by the Alcan-Ford aluminium intensive vehicle (AIV). Each type has its merits: the space frame offers lower tooling costs by eliminating some stampings, whereas the conventional sheet monocoque offers established processes and low piece costs. The updated examples of these two types are Ford P2000 and Audi AL2. Both of them could reduce weight about 40% on the BIW basis. The extruded space frame developed for Audi A8 is believed most appropriate for low volume production. The structure of Audi AL2 is a modified space frame with aluminium extrusions already developed for A8. Audi AL2 model is produced with an all aluminium body structure. In the AL2, there are fewer aluminium cast joints, which were extensively used in A8 since they are replaced with direct bonds. Aluminium extrusions in the AL2 are also made into as straight shape as possible. It is also clear that, as the automotive companies work more and more with aluminium, simplification of design results in lower overall cost.

Determining the right alloy for the body structure and hang-on panels has been the subject of considerable development effort [Bull, 1992] and most of the activity is now concentrated on a relatively small number of alloys.

For skin sheet material the emphasis is on achieving a good balance of formability, strength after the paint-bake, and a high surface quality after pressing and paint finish. Consequently, the bake hardening 6xxx alloys are the primary choice for these applications. For structural sheet materials, strength may be a limiting factor in certain areas, impact energy absorption and good deep drawing behaviour are often the most important.

To meet these requirements, 5xxx alloys are mostly used in North America. In Europe, 6xxx-T4 materials are still widely used. One obvious and significant difference between aluminium and steel is the outstanding bare metal corrosion of the 5xxx and 6xxx aluminium materials. Increasingly large amounts of steel are supplied zinccoated to achieve acceptable paint durability, this is not necessary for aluminium. However, the aluminium coil or sheet can be supplied with a range of pre-treatment and primer layers which can improve formability, surface quality and may eliminate the need for E-coating.

There is a wide range of aluminium materials and surface qualities, which can be chosen, and the growing design and process experience is enabling the aluminium industry to help the customer specifying the right material for the application.

There is a clear difference [Bottema et al, 1998] in the alloy choice and treatments for these applications between Europe and North America.

Aluminium alloys for brazing sheet applications

As mentioned earlier brazed aluminium components are used extensively in modern vehicles for engine and transmission cooling, charge air coolers and climate control. It consists of a core alloy which provides the strength and life cycle requirements of the heat exchanger and a clad layer which is of a low melting point aluminium silicon alloy. During the brazing process the Al-Si alloy melts and seals joints in the heat exchanger between the different sheet components. The brazing sheet can be clad on one or both sides with the Al-Si alloy and in some cases one side is clad with a different alloy to provide corrosion protection on the inner (water-side) of the a radiator.

During 1970 vacuum brazing [Miller, 1967] was developed to solve the problems associated with old techniques of dip brazing. It was an environmental friendly approach but requires significant capital investment. It became the main method for manufacturing heat exchangers in the 1980s and still remains the preferred brazing method for evaporators and charge air coolers. It is gradually being superseded by controlled atmosphere brazing (CAB). A main advantage of vacuum brazing over controlled atmosphere brazing is that high (0.3%) magnesium containing alloys can be used. Although, now in use for several decades the complete mechanisms behind the technique are still not fully understood. Since the introduction of Nocolok process by Alcan in 1978 [Cooke et. al, 1978], this process has become the workhorse in the brazing industry. It is a very attractive process since it can be operated continuously at low costs [Fortin, 1985]. Although the CAB process is very popular it has some constraints like, the flux can not tolerate high magnesium alloys [Bollingbroke, 1997] and the uniform application of the flux on the heat exchanger to be brazed can be very difficult to control.

3.1.3 Magnesium

Magnesium is 33% lighter than aluminium and 75% lighter than steel/cast-iron components. The corrosion resistance of modern, high-purity magnesium alloys is better than that of conventional aluminium die-cast alloys. As well, porosity-free die-cast AM501 AM60 can achieve 20% elongation, or over three times that of Al A380, leading to higher impact strength; but magnesium components have many mechanical/physical property disadvantages that require unique design for application to automotive products. Although its tensile yield strength is about the same, magnesium has lower ultimate tensile strength, fatigue strength, and creep strength compared to Aluminium. The modulus and hardness of magnesium alloys is lower than aluminium and the thermal expansion coefficient is greater. However, it should be noted that suitable ribbing and supports often can overcome the strength and modulus limitations.

Property	Magnesium	Aluminium	Iron
Crystal Structure	hcp	FCC	BCC
Density at 20^{0}C (g/cm^3)	1.74	2.70	7.86
Coefficient of thermal expansion 20-100^{0}C (*10^6/C)	25.2	23.6	11.7
Elastic modulus (10^6 MPa)	44.126	68.947	206.842
Tensile strength (MPa)	240	320	350
Melting point (^{0}C)	650	660	1.536

Table 4: Properties of Mg, Al, Fe [Source: Davies, 2003]

Despite the above issues Magnesium alloys have distinct advantages over aluminium that could not be dismissed. These include better manufacturability, longer die life and faster solidification due to lower latent heat. Therefore more castings can be produced per unit time compared to aluminium. Magnesium components have higher machinability. Magnesium components can be produced with improved dimensionality and surface quality, and smaller draft angles compared to aluminium. The capability of magnesium to be hot chamber die cast can reduce casting scrap by reducing dross and can limit gas and oxide inclusions while allowing more consistent melt temperature. A comparison of the properties of the Mg, Al and Fe is made in Table 4.

Mechanical properties of Mg alloys

Specific strength and specific stiffness of materials and structures are important for the design of weight saving components. Weight saving is particularly important for automotive bodies, components and other products where energy consumption and power limitations are a major concern [Tkachenko et. al, 2006]. The specific strength and specific stiffness of magnesium are compared with aluminium and iron in Figure 3. There is little difference between the specific stiffness between Mg, Al and Fe as seen in Figure 3. The specific stiffness of Al and Fe is higher than Mg only in the ratio of 0.69% and 3.7%, respectively. On the other hand, the specific strength of Mg is considerably higher than that of Al and Fe in the ratio of 14.1% and 67.7%, respectively. [Kulekci, 2008]

Because of its too low mechanical strength, pure magnesium must be alloyed with other elements, which confer improved properties. The Mg-Al-Zn group of alloys contains aluminium, manganese, and zinc. These are most common alloying elements for room temperature applications. Thorium, Cerium, and Zirconium (without aluminium) are used for elevated temperatures and form the Mg-Zn-Zr group. Thorium or cerium is added to improve strength at the temperatures of 260°C to 370°C. Mg-Al alloys are one major group among magnesium-based alloys. The strength of these alloys is improved [Aghion et al 2003, Pekguleryuz et al, 2003 a-b]. But they suffer from poor coherency, and high creep deformation at elevated temperature of >150 ^{0}C for long periods of time, the supersaturated Mg solid solution transforms to Mg matrix with coarsely dispersed Al (g) precipitates and contributes to grain boundary migration and creep deformation. Furthermore Al (g) is also prone to aging and has poor metallurgical stability, which limited its application in higher temperatures. Early developments in improving the creep properties of magnesium were made in the 1960s by Volkswagen [Medraj & Parvez, 2007]. It was based on Mg-Al-Si system. These alloys exhibit marginally improved creep resistance but are difficult to die-cast. Magnesium components are generally in the form of magnesium alloys. The addition of other alloying elements can strengthen and harden magnesium as well as alter its chemical reactivity.

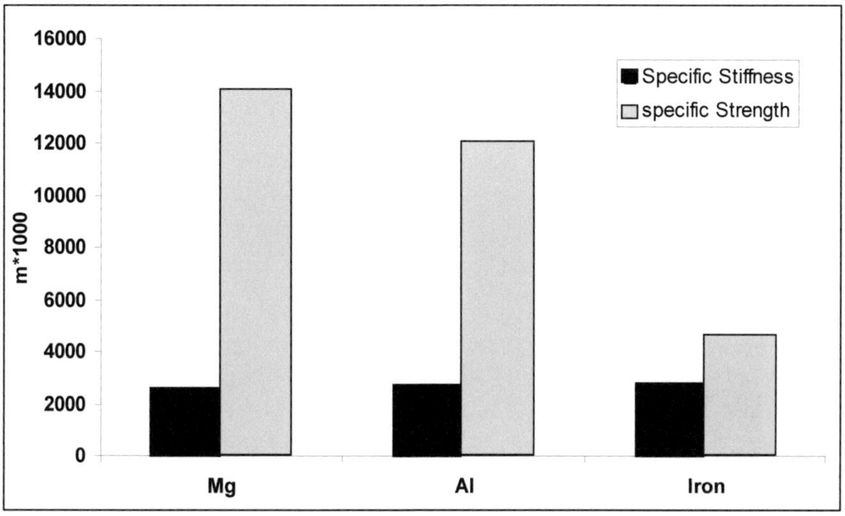

Fig. 3. comparison of specific stiffness and strength of the Mg, Al and Fe [Source: Kulekci, 2008]

AZ91D magnesium alloy has been shown to creep at ambient temperature under initial applied stress of only 39% of its yield stress [Grieve 2001]. The commonly used die-casting alloy AZ91, starts creep at temperatures above 100°C and has a maximum operating temperature at 125°C [Aghion et al, 2001].

Because of its creep behaviour, it is not convenient to use this alloy for power train and engine castings. Both of them operate at temperatures of 100°C or more and are fixed together with threaded fasteners so creep becomes a key issue for these applications [Pekguleryuz, 2003 a-b]]. The studies on AE42 alloy showed that AE42 has a greater percentage of initial compressive load than AZ91D as seen in Figure 4 [Aghion et al 2003, Pekguleryuz, 2003]. AE series alloys have better creep resistance with respect to AZ91D. Magnesium alloys for automotive applications must have good creep resistance property. These alloys should be thermally and metallurgically stable and resistance to flow during creep loading. Moreover, it should have adequate corrosion resistance, castability and strength. The AE42 (Mg-4 atomic percent Al-2 atomic percent rare earths) magnesium alloy has improved creep resistance over the other alloys as seen in Figure 4. Magnesium-thorium alloys display excellent creep properties at elevated temperature (350°C). However, these alloys have cast disadvantages due to expensive rare earth additions [Pekguleryuz, 2003, a-b]. The Mg-Al-Sr system is a recently developed alloy for the heat-resistant lightweight Mg alloys. The Mg- Al-Sr system is used by BMW for the manufacturing of die-cast engine blocks. This system has excellent mechanical properties, good corrosion resistance and excellent castability. Mg alloys with Sr addition have better creep resistance than other alloy systems as seen in Figure 4. Corrosion resistance of the Mg- Al-Sr alloys is similar to AZ91D and better than AE42, which indicates that strontium does not have adverse affect on corrosion properties [Medraj, 2007]. The addition of Al to Mg alloys provides good fluidity which adversely affects the creep resistance. Wrought alloys exhibit significantly better combination of strength and ductility compared with casting alloys. However wrought alloys are currently used to a very limited extent due to a lack of suitable alloys and some technological restrictions imposed by the hexagonal crystal structure of magnesium [Eliezer et al, 1998].

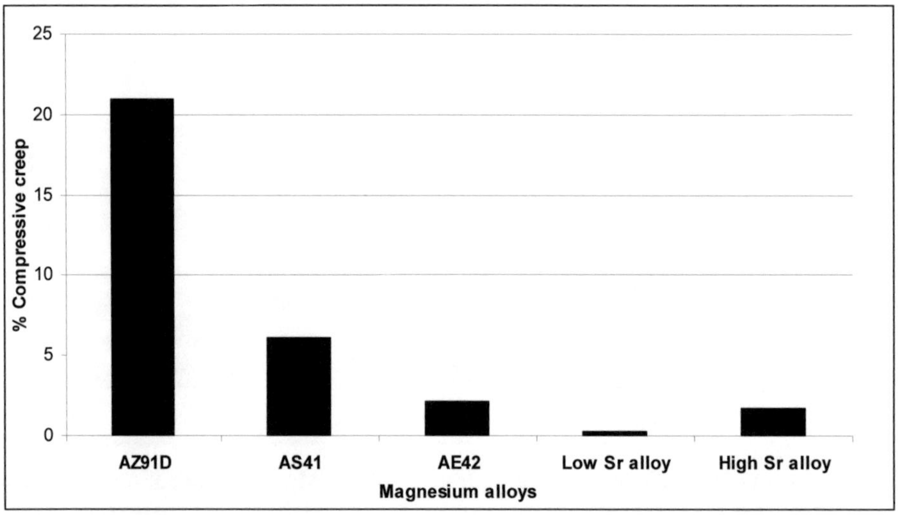

Fig. 4. Compressive creep of the magnesium alloys at 70 MPa, 150 ^{0}C after 200 hrs, [Source: Pekguleryuz et al, 2003]

Technical problems and solutions for use of magnesium alloys in automotive industry

The disadvantages of Mg alloys are high reactivity in the molten state, inferior fatigue and creep compared to aluminium and galvanic corrosion resistance. The problems in using magnesium alloys stem from their low melting points 650°C and their reactivity (inadequate corrosion resistance) [Haferkamp et al, 2001]. The main problem for Mg alloys encountered during fabrication and usage is the fire hazard/risk, especially in machining and grinding processes due to their relatively low melting point [Sreejith & Ngoi, 2000]. In roughing cuts the chips are generally thick and not likely to get hot enough to ignite. However, the thin chips produced in the finishing cuts are more likely to heat up and ignite. Similarly, the dust in grinding can ignite, even explode, if heated to melting temperatures. The fire hazard can be eliminated by avoiding fine cuts, dull tools, high speeds; using proper tool design to avoid heat build up; avoiding the accumulation of chips and dust on machines and cloths; and using coolants.

Magnesium is a reactive metal, so it is not found in the metallic state in nature. It is usually found in nature in the form of oxide, carbonate or silicate often in combination with calcium. Because of this reactivity the production of magnesium metal requires large amounts of energy. This situation makes magnesium an expensive metal. To prevent reactivity problems, protective finishes, such as anodic coating or paint are used [Shi et al, 2006]. Magnesium is attacked by inorganic acids. It is not attacked by alkalis and caustic soda.

Welding of Mg alloys can also present a fire risk if the hot/molten metal comes in contact with air. To overcome this problem, the welding region must be shielded by inert gas or flux. A larger amount of distortion relative to other metals may arise due to high thermal conductivity and coefficient of thermal expansion in welding of magnesium alloys if required precautions are not taken [Robots 4 welding, 2007]. Service temperatures must be well below the alloy melting points; otherwise the fire hazard might materialize. For example, it caused an engine fire in a DC-3 aircraft, resulting in a fatal crash. This particular aircraft was built during World War II, when aluminium shortages forced manufacturers to use of magnesium alloys as a replacement in some applications. The low creep properties of magnesium alloys limits the application of magnesium alloys to be used for critical parts, such as valve covers [Medraj, 2007]. The following are the main issues that need attention to increase creep properties of magnesium alloys: stress relaxation in bolted joints, the potential for creep at only moderately elevated temperatures, corrosion resistance, and the effects of recycled metal on properties.

Significant research is still needed on magnesium processing, alloy development, joining, surface treatment, corrosion resistance, and mechanical properties improvement.

Different coating methods are used to increase the corrosion resistance of magnesium alloys. Problems with contact corrosion can be minimized, on the one hand, by constructive measures and, on the other hand, by an appropriate choice of material couple or the use of protective coatings [Blawert et al, 2004]. Chromate coating of Mg alloys is hazardous and not environmentally friendly. A newly developed Teflon resin coating has been developed for Mg alloys [AIST, 2007]. The coating is obtained with an aluminium vapour deposition and finish treatment with a Teflon resin coating. The newly developed Teflon resin coating is a low cost, chromium-free corrosion resistant coating for magnesium alloys. The coating not only has corrosion resistant properties, but also good lubricity, high frictional-resistance and non-wetting properties. The main future of the coating is in the application of Teflon coating on Magnesium alloys.

3.2 Plastics and composites

Polymer composite materials have been a part of the automotive industry for several decades, with early application in the 1953 Corvette. These materials have been used for applications with low production volumes, because of their shortened lead times and lower investment costs relative to conventional steel fabrication. Important drivers of the growth of polymer composites have been the reduced weight and parts consolidation opportunities the material offers, as well as design flexibility, corrosion resistance, material anisotropy, and mechanical properties. Although these advantages are well known to the industry, polymer composite use has been impeded by high material costs, slow production rates, and to a lesser extent, concerns about recyclability. Several factors have hindered large scale automotive applications of polymer composites. Amongst these are concerns about crash energy absorption, recycling challenges, competitive and cost pressures, the industry's general lack of experience and comfort with the material.

The cost of composite materials is usually much higher (up to 10 times higher when using carbon fibres) than those of conventional metals. A comparison of the cost elements for the glass fibre composites and carbon fibre composites are made with the steel in figure 5. Therefore, the main targets for future development must be the use of hybrid composites (low-cost fibres to be used where possible and aramide and carbon fibres to be used only where they are required for damage tolerance or stiffness reasons), the evaluation of highly automated and rapid manufacturing processes including the application of intelligent preforms or half-finished goods, and the full use of the potential of composites for parts integration. Either glass or carbon, reinforced in the matrix of thermoset or thermoplastic polymer materials. The glass-reinforced thermoset composites are the most commonly used composite in automotive applications today, but with the development of very high

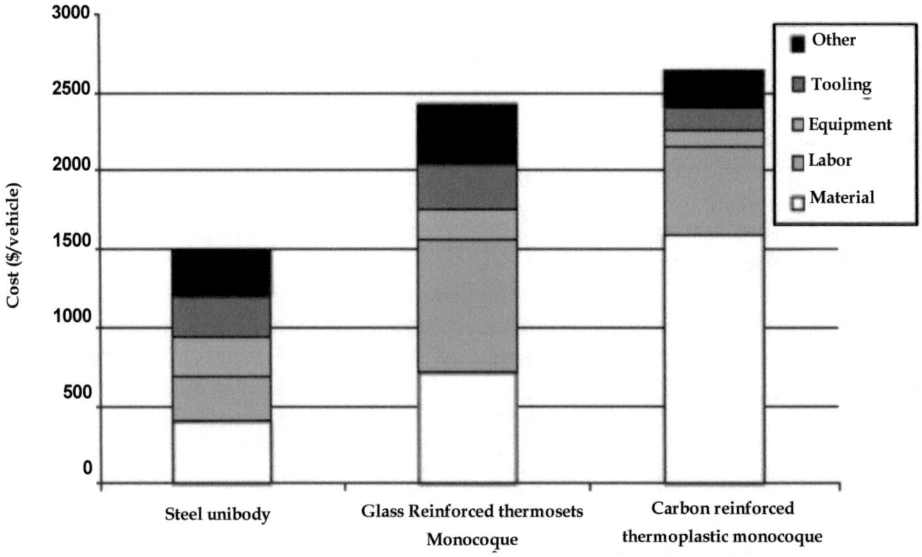

Fig. 5. Cost structure comparison of BIW designs [source: Dieffenbach et. al, 1996)]

3.2.1 Fabrication

The choice of a specific fabrication method depends on the costs and on the technical requirements of the component to be produced. In order to guarantee economic production, methods with a high throughput are absolutely necessary. High throughput can be achieved by means of low clock times or by means of high integrative parts. Table 5 compares the most commonly used composite fabrication processes available today, addressing their advantages, disadvantages, and cycle time.

The use of prepregs, which are reinforced with carbon or glass in fibre and fabric forms coated with epoxy resins, may be suitable for only limited automotive applications because of lower productivity. One of the chief obstacles in the way of achieving higher production volumes for structural composites is the time at the preforming stage required to place complex, properly oriented reinforcement in the moulding tools. This requirement results in long cycle times, high labour cost, and low productivity of the moulding tool investment. A recent study indicates that the cost of preforms contribute about 35% to the total composite BIW cost, compared to 50% for moulding and 15% for assembly (Mascarin 2000). Some of the approaches that are used for making preforms are specially knit fabric designed to drape properly for a given component; braided reinforcement over moulded foam cores; multiple ply vacuum preforming; and robotically applied chopped fibres known as P4 process The most broadly accepted reinforced thermoset composites used by automakers in today's market include sheet moulding composite (SMC), bulk moulding composite (BMC), reinforced reaction injection moulding (RRIM), and liquid composite moulding processes such as structural reaction injection moulding (SRIM) and resin transfer moulding (RTM).

SMC and RRIM are most widely used today, contributing to 48% and 40%, respectively to the total thermoset components used in the 2000 model year passenger cars (ACA 2000). RTM and SRIM composite moulding processes have been considered to provide the best economic balance for the automotive structural products. These processes have favourable cycle times with large parts and produce a surface quality corresponding to the automotive standard

Moulding process	Adavantages	Disadvantages	Cycle time
Prepreg	Better resin/fibre control	Labour intensive for large complex parts	5-10 hrs
Preforming	Good mouldability with complicated shapes and the elimination of trimming operation	Cost-effective only for large complicated shape parts and large scrap generated when fibre mats used	45-75 secs. (compform process) 4-5 mins (vacuum forming)
RTM	Inside and outside finish possible with thickness control, more complex parts possible with vacuum assisted	Low viscosity resin necessary and the possibility of voids formation without vacuum assisted	8-10 mins for large parts: 3-4 mins for vacuum assisted
Liquid compression moulding	Favoured method for mass production with high fibre volumes	Expensive set p cost for low production	1-2 mins
SMC	Cost effective for production volume 10K-80K/year	Minimum weight savings potential	50-100 secs
RIM	Low cost tolling where prototypes can be made with soft tools	Difficult to control the process	1-2 mins
BMC	Low cost base material	Low fibre content randomly oriented, low structural quality, poor surface finish	30-60 secs
Extrusion compression moulding	Fully automated variety of polymers and fibres can be used with fibre volumes up to 60% by weight	Not for surface finish parts without paint film or similar process	3-6 mins
Structural reaction injection moulding	Low tooling cost with the good finish capability	Difficult to control the process particularly with low viscosity resin and longer cure cycle time	4 mins

Table 5: A Comparison of the Most Commonly Used Composite Moulding Processes [Source: Das, 2000]

3.2.2 Cost

Reducing the cost of manufacturing automotive structural components from lighter weight composite materials so that they are competitive with the component (including life cycle) costs of other materials is a major focus. Although cost reduction is a pervasive factor in all composites R&D activities, most of the activities in this area are related to materials, the major factor affecting the viability of composites in automotive applications today.

3.2.3 Manufacturability

Methods for high-volume production of automotive components from lightweight materials have not been adequately developed. Composite processing technologies need to be developed that yield the required component shape and properties in a cost-effective, rapid, repeatable, and environmentally conscious manner. For instance, technologies for high-rate forming and moulding of composites for large structural components and high-volume production of continuous fibre preforms are needed.

It is essential that high-rate preforming techniques be developed to obtain chopped-fibre preforms with consistent fibre distribution and density at the volumes required by the automotive industry.

3.2.4 Design data/test methodologies

One of the major challenges for the commercialization of polymer composites is the lack of adequate design data (e.g., material property databases), test methods, analytical design tools (i.e., models), and durability data. DOE is focusing on the development of enabling technologies and property data to predict the response of materials in a given structural design after long-term loading, under exposure to different environments, and in crash events.

Theoretical and computational models are being developed for predicting energy absorption and dissipation in automotive composites. These models are tools designers need to minimize component weight while maximizing occupant safety.

3.2.5 Joining and inspection

High-volume, high-yielding technologies for joining composites to each other and to metal structures in an automotive assembly environment do not currently exist but are being developed. Current efforts concentrate on adhesive formulation, modelling, and processing. Significant work is being conducted to understand the synergistic effects of environmental stressors on adhesive joint integrity. The next five-year research focus is on the development of non-adhesive joining techniques such as chemical bonding of thermoset composites and the joining of carbon fibre based composites to a variety of materials. Fast, reliable, and affordable methods to test bond integrity and assembled structures are needed.

One of the major drawbacks in the use of composites for automotive applications is that technologies for cost-effective recycling and repair of advanced composite materials do not exist. Cost-effective methods for the separation and recycling of composite materials into high-value applications, as opposed to using them only as filler, need to be developed. Methods are being pursued for separating glass and carbon fibre from thermoset and thermoplastic resin systems. Efforts are also underway to identify alternate uses for post-consumer automotive grade composites.

3.3 Renewable materials, barriers and incentives in use of biocomposites

The lightweight, low cost natural fibres offer the possibility to replace a large portion of the glass and mineral fillers in several automotive interior and exterior parts. In the past decade, natural-fibre composites with thermoplastic and thermoset matrices have been embraced by European car manufacturers and suppliers for door panels, seat backs, headliners, package trays, dashboards, and interior parts. Natural fibres such as kenaf, hemp, flax, jute, and sisal are providing automobile part reinforcement due to such drivers as reductions in weight, cost, and CO2, less reliance on foreign oil sources, recyclability, and the added benefit that these fibre sources are "green" or ecofriendly. As a result, today most automakers are evaluating the environmental impact of a vehicle's entire lifecycle, from raw materials to manufacturing to disposal. At this time, glass-fibre-reinforced plastics have proven to meet the structural and durability demands of automobile interior and exterior parts. Good mechanical properties and a well-developed, installed manufacturing base have aided in the insertion of fibreglass-reinforced plastics within the automotive industry. However, glass-reinforced plastics show shortcomings such as relatively high fibre density (approximately 40% higher than natural fibres), difficulty to machine, and poor recycling properties, not to mention the potential health hazards posed by glass-fibre particulate.

Blast	Leaf	Seed	Fruit	Stalk	Wood Fibres
Flax Hemp Jute Kenaf Ramie Banana Rattan	Sisal Manila Curauna Banana Palm	Cotton Kapok	Coconut Coir	Bamboo Wheat Rice Grass Barley Corn	Hardwood Softwood

Table 6. A list of vegetable and cellulose fibre classifications [Source: Holbery & Houston, 2006]

An ecological evaluation, or eco-balance, of natural-fibre mat as compared to glass-fibre mat offers another perspective. The energy consumption to produce a flax-fibre mat (9.55 MJ/kg), including cultivation, harvesting, and fibre separation, amounts to approximately 17% of the energy to produce a glass-fibre mat 54.7 MJ/kg). [Patel et al, 2002] Though natural-fibre-reinforced plastic parts offer many benefits as compared to fibreglass, several major technical considerations must be addressed before the engineering, scientific, and commercial communities gain the confidence to enable wide-scale acceptance, particularly in exterior parts where a Class A surface finish is required. Challenges include the homogenization of the fibre's properties, and a full understanding of the degree of polymerization and crystallization, adhesion between the fibre and matrix, moisture repellence, and flame retardant properties, to name but a few. Technology for implementing natural fibre composites into interior trim continues to be developed by Tier I and Tier II automotive suppliers, typically in partnership with producers of natural fibre- based processing capabilities for mat or other material forms. Compression moulding, injection moulding, thermoforming, and structural reaction injection moulding are all processes utilized to process natural-fibre composites. [Holbery & Houston, 2006]

3.3.1 Thermoplastic/ thermoset polymers

The manufacture of natural-fibre composites includes the use of either a thermoset or thermoplastic polymer binder system combined with the natural fibre preform or mat. In automotive applications, the most common system used today is thermoplastic polypropylene, particularly for nonstructural components. Polypropylene is favoured due to its low density, excellent processability, mechanical properties, excellent electrical properties, and good dimensional stability and impact strength. [George et al, 2001], However, several synthetic thermoplastics are utilized including polyethylene, polystyrene, and polyamides (nylon 6 and 6, 6). The development of thermoplastic natural-fibre composites is constrained by two primary physical limits: the upper temperature at which the fibre can be processed and the significant difference between the surface energy of the wood and the polymer matrix. Process temperature is a limiting factor in natural fibre applications. The generally perceived upper limit before fibre degradation occurs is on the order of 150°C for long processing durations, although fibres may withstand short-term exposures to 220°C.

The result of prolonged high-temperature exposure may be discoloration, volatile release, poor interfacial adhesion, or embrittlement of the cellulose components. Therefore, it is important to obtain as rapid a reaction rate as possible during both surface treatment and polymer processing to limit exposure to cell wall components preventing degradation. The development of low-process-temperature surface

treatments with high service capabilities are viewed as enabling technology for the application of natural fibres in composite materials. Because the interfacial adhesion between the natural fibre and polymer matrix determines the composite physical properties, it is usually necessary to compatibilise or couple the blend. [Baille, 2004] Compatibilisation is any operation performed on the fibre and polymer that increases the wetting within the blend. Coupling is a process in which dissimilar polymers or fillers are made into an alloy by use of external agents called coupling agents. [Bledzki & Gassan, 1999] The result of properly applying a compatibiliser or coupling agent to the composite is an increase in physical properties and environmental durability. [Mohanty et al, 2005] The primary thermoset resins used today in natural-fibre composites for automotive applications are polyester, vinylester, and epoxy resins. [Mohanty et al, 2005] In natural fibres, polar groups are the main structural units and the primary contributor to mechanical properties; these also render cellulose more compatible with polar, acidic, or basic groups, as opposed to nonpolar polymers. Polyester resins are widely used, particularly the "unsaturated" type capable of cure from a liquid to a solid under a variety of conditions. Epoxy resins offer high performance and resistance to environmental degradation. Epoxies have wide appeal in industry, although in the automotive industry epoxies have not gained broad use due to longer cure schedules and high monomer cost. Vinylester resins is a relatively new addition in the family of thermosetting resins which combine excellent chemical resistance, good thermal and mechanical properties, and the relative ease of processing and rapid cure characteristics of polyester resins. These have better moisture resistance than epoxies when cured at room temperature. Vinylester resins are similar in their molecular structure to polyesters, but differ in that the reactive sites are positioned at the ends of the molecular chains, allowing for the chain to absorb energy. This results in a tougher material when compared to polyesters.

3.3.2 Composite processing

The primary drivers for the selection of the appropriate process technology for natural-fibre composite manufacture include the final desired product form, performance attributes, cost, and ease of manufacturing. Several factors must be considered in selecting a process. One must insure: that the fibre is distributed evenly within the matrix, that there is adequate compatibility between the hydrophobic matrix and hydrophilic fibres, that fibre attrition is minimized due to processing to insure reinforcement, that the desired fibre orientation effects will be imparted, that thermal stability of the fibre is maintained throughout the processing step, and that the moisture inherent within the fibre is at the desired level, minimizing problems with swelling or part distortion. The control of moisture in the fibre and the effect of moisture after moulding are primary considerations in natural-fibre composites in automobiles. Similarly, the ability to eliminate water absorption during service of natural-fibre-based composite components is paramount in industrial applications. For example, it has been shown in sisal fibre/unsaturated polyester composites that storage in water will result in a reduction of up to 50% in flexural modulus. [Zafeiropoulous et al, 2002]

Compounding processes that blend the natural fibres with a thermoplastic matrix are gaining wide acceptance due to the high degree of consistency feasible in the pellet form. The purpose of a compounding operation is to produce a pelletised feed stock that can be processed further, similar to any other thermoplastic processing technique, such as injection moulding, extrusion, or thermoforming. There are several types of compounding processes, including extrusion, kneading, and high-shear mixers.

Injection moulding is a versatile process and is the most widely used processing technique for making composite products, particularly where intricate shapes are needed in cyclic, high-volume production. The benefits include excellent dimensional tolerance and short cycle times coupled with few post-processing operations. According to BMW, it is possible to manufacture bio-based composites that are as much as 40 percent lighter than equivalent injection-moulded plastic parts.[Singh 1998] One of the challenges posed by injection moulding natural-fibre composites is to produce pellets of a consistent quality. This has been explored by both North American and European injection moulding equipment suppliers through a process called direct long-fibre thermoplastic (D-LFT) moulding. In this continuous process, first developed for glass fibres, the fibres are spooled and fed into a heating zone, where the thermoplastic is integrated with the fibre bundles. These bundles are then cut at a desired length and fed continuously into an injection moulding hopper, and parts are moulded continuously. It is reasonable to assume that the recent developments in producing continuous natural-fibre roving could be integrated on a large scale into the D-LFT process. Several companies are working in this development area.

Thermoforming is mainly used to produce natural-fibre-mat thermoplastic composites. The process takes pre-cut layers of fibre (or preformed mats that could comprise random fibres or roving) and polymer sheet that are inserted in a heated mould, and consolidates the material as heat is transferred through conduction to melt the thermoplastic. The thermoplastic flows to penetrate the fibre component, with pressure applied during the heating and cooling phases. After reaching the melt temperature in a hot press, the molten hybrid material is consolidated into a composite in a cold press, with very rapid processing times possible via combined heating-cooling presses in parallel. Compression moulding using thermoset polymer matrices is another major platform used to manufacture large parts for the automotive industry, producing light, strong, and thin panels and structures. The primary advantage of this process is low fibre attrition and process speed. A comparison of compression moulded unsaturated polyester composites reinforced with glass fibre and with natural fibres (flax) is provided in Table 7.

Property	Glass fibre (30% wt)	Natural Fibre (35% wt)
Flex Strength (MPa)	80	70
Flex Modulus (GPa)	6.0	6.0
Elongation at Break (%)	2.2	1.9
Impact Strength (KJ/mm^2)	38	20
Density (g/cm^3)	1.54	1.42

Table 7: Comparison of the properties of the flax fibre and glass fibre.

This indicates that the properties are comparable with properties with similar fibre loadings. Another method of compression moulding is the sheet moulding compound (SMC) process which has been used for glass composites for years. Many variations of compression moulding have been developed that are suitable for automotive application, and recent developments to combine extrusion and compression of thermoplastic composites, initially with glass fibres, are beginning to enter into the automotive industry. This process extrudes large thermoplastic fibre bundles, or pre-heated plugs, into a compression mould in-situ, and then the compression moulds the part. However, high capitalization costs will preclude this process from large-scale insertion into the Tier 1 supply chain in the near future. The foaming technique produces foamed products that can be used in upholsteries and in insulation applications. Finally, thermoset polymer composite manufacture via resin transfer and vacuum-assisted resin transfer moulding has gained interest from the automotive industry. The primary benefits of this processing platform include compounding at low shear and temperatures with minimal degradation of the cellulose fibre. Higher fibre loadings to 70% are possible, as well as good devolatilization. However, these processes are meeting resistance due to the high capital expenditure requirements.

4. Conclusions

The competition in the market of materials for automotive applications is substantial. This is due to the size and value of the market. In the more recent years the environmental concern has opened the need for lighter vehicle for lower fuel consumption and also for the need of recycling. These recent pressures have opened the door for introduction of new materials to the automotive market such as alternative metals and composites. However there are yet significant barriers in large scale use of these materials mainly due to the cost of the raw materials or the large capital investment need for transformation of the forming processes. Therefore the need for further research for suitable processes, properties and lower cost materials in this lucrative industry is at its peak. The more traditional materials such as steel producers are trying hard to keep their market by further innovations and improvements in their alloying and their processes in order to offer lighter material and structure option. But at the same time the newer materials such as alternative metals and composites are at the heart of the research and innovation for opening the possibility of the lighter and more environmentally friendly future vehicles.

5. References

Aghion E, Bronfin B, Eliezer D (2001) The role of the magnesium industry in protecting the environment. J Mater Process Technol 117(3):381–385

Aghion E, Bronfin B, Von Buch F, Schumann S, Friedrich H (2003) Newly developed magnesium alloys for powertrain applications. Jom-J Miner Metals Mater Society 55(11):A30–A33

AIST (2007) New technology for increasing the application for magnesium alloys. http://www.aist.go.jp/aist_e/latest_research/2001/20011203/20011203.html

American Honda Motor Co., (2005) "Body" and "Chassis" technical information for 2003 Honda Civic, Torrance, CA, February 6, 2002 (www.hondanews.com, accessed March 3, 2005).

Ashby, M. F. (2000) Materials Selection in Mechanical Design Butterworth-Heinemann ISBN 0-7506-4357-9

Automotive Composites Alliance (ACA) (2000) "2000 Model Year Passenger Car and Truck Thermoset Composite Components," Troy, MI.

Blawert C, Hort N, Kainer KV (2004) Automotive applications of magnesium and its alloys. Trans indian Inst Met 57(4):397–408

Heckelmann I.,Bode R., Engl B., Hartmann G., (1999) A new generation of surface-treated car body steel with improved tensile strength and high stretch forming ability, SAE Paper No. 1999-01-3199;

Bollingbroke R.K., Gray A., Lauzon D., (1997), Optimisation of nocolok brazing conditions for higher strength brazing sheet, VTM 3 Conf., Indianapolis, USA, SAE Paper 971861.

Bottema J., Lahaye C., Baartman R., Zhuang L., De Smet P., (1998) Recent developments in AA6016-T4 aluminium type body sheet product, SAE Technical Paper Series 981007.

Buchholz, K. (1998). "Lightweight Body Panel Materials," Automotive Engineering International, Vol. 106, No. 12, Dec., pp. 19-22.

Bull M., (1992), Aluminium sheet alloys for structural and skin applications, SAE Technical Paper Series 920669.

Chapman, G. B. (1999). "A Thermoplastic Approach to a Composite Automotive Body," SAE Paper No. 1999-01-3222, The Society of Automotive Engineers, Warrendale, PA.

Cole G.S., Sherman, A.M., 1995. Light weight materials for automotive applications. Mater. Charact. 35, 3–9.

Cooke W.E., Wright R.E., Hirschfiels J.A., (1978), Furnace brazing of Aluminium with non corrosive flux, SAE Paper 780300, March.

Corus Eng. Co., The multimaterial car, 2010, Corus Automotive - The multi-material car.htm

Cunat P.J., (2000) Stainless steel properties for structural automotive applications, Metal Bulletin International Automotive Materials Conference, Cologne, 21st to 23rd June.

Das S., (2001) The cost of automotive polymer composites: a review and assessment of DOE's lightweight materials composite research, January 2001, OAK RIDGE NATIONAL LABORATORY

Davies G (2003) Magnesium. Materials for automotive bodies, Elsevier, G. London, pp 91, 158, 159

Davis, S.C. (2004) Transportation Energy Data Book: Edition 24, ORNL-6973, Oak Ridge National Laboratory, Oak Ridge, TN.
DeCicco J.M., (2005), Steel and Iron Technologies for Automotive Lightweighting, Environmental Defense, March 3
Defosse, M. T. (1999) "Ford, GM move to composite truck beds", Modern Plastics, October 1.
Demmler, A. (1998). "Trends in Automotive Materials," Automotive Engineering International, Dec. 1, pp. 26-27.
Dieffenbach, J.R., P.D. Palmer, and A.E. Mascarin (1996a). "Making the PNGV Super Car a Reality with Carbon Fibre: Pragmatic Goal or Pipe Dream?" SAE Paper No. 960243, Society of Automotive Engineers, Warrendale, PA.
Eliezer D, Aghion E, Froes FH (1998) Magnesium science and technology. Adv Mat Performance 5:201–212
Environment-agency (2010), http://www.environment-agency.gov.uk.
Field, F.R., Kirchain, R. and Clark, J.P. (2001) 'Life-cycle assessment and temporal distributions of emissions', Journal of Industrial Ecology, Vol. 4, No. 2, pp.71–91.
Fortin P.E., (1985), Aluminium materials and processes for automotive heat exchangers applications, SAE Paper 85228.
George J., Sreekala M.S., and Thomas S., (2001), "A Review on Interface Modifi cation and Characterization of Natural Fibre Reinforced Plastic Composites," Poly. Eng. and Science, 41 (8), pp. 1471–1485.
Grieve DJ (2001) Magnesium Die Castings. http://www.tech. plym.ac.uk/sme/mech330/magcast.htm
Haferkamp H, Boehm R, Holzkamp U, Jaschik C, Kaese V, Niemeyer M (2001) Alloy development, processing and applications in magnesium lithium alloys. Mater Trans 42(7):1160–1166
Holbery J. and Houston D., (2006), Natural-Fibre-Reinforced Polymer Composites in Automotive Applications, 6 November • JOM
Insurance institute of highway safety, (2010), http://www.iihs.org/ratings/frontal_test_info.html
Jacob G.C., Fellers J.F. , Simunovic S. and Starbuck J.M.., (2002), Energy Absorption in Polymer Composites for Automotive Crashworthiness, Journal of Composite Materials; 36; 813
Kelly, K., and Priddle A., (2004), "Stow 'n Go" -- Seats fold flat for next-generation Chrysler minivans. Ward's AutoWorld, January, p. 9.
Kochan, A., (2003) ,"Laser technology is key to new VW Golf -- Streamlining in body shop cuts production time by 25 percent," Automotive News Europe, November 17.
Kulekci M. K., (2008), Magnesium and its alloys applications in automotive industry, Int J Adv Manuf Technol, 39:851–865
Kumar V. and Sutherland J.W., (2008), Sustainability of the automotive recycling infrastructure: review of current research and identification of future challenges, Int. J. Sustainable Manufacturing, Vol. 1, Nos. 1/2.
Kumar, V., Bee, D.J., Shirodkar, P.S., Bettig, B.P. and Sutherland, J.W. (2005) 'Towards sustainable 'product and material flow' cycles: identifying barriers to achieving product multi-use and zero waste', Proceedings of 2005 ASME International Mechanical Engineering Conference and Exposition, 5–11 November, Orlando, FL (IMECE2005-81347), Appeared on CD-ROM.

Kumar, V., Shirodkar, P.S., Camelio, J.A. and Sutherland, J.W. (2007) 'Value flow characterization during product life cycle to assist in recovery decisions', International Journal of Production Research, Vol. 45, Nos. 18–19, pp.4555–4572.

Kuroda, A., Ishizuka S., Tsuda H., (2003), Weight reduction and improvement of safety level in impacts with high-strength steels tailored blanks technology, SAE Paper No. 2003-01-2736.

Libertiny, G.Z. (1982) Estimating the Population of Passenger Cars in Service, SAE Technical Paper \#820164.

Libertiny, G.Z. (1993) A New Look at the Service Life Expectancy of Passenger Cars in the United States, SAE Technical Paper #930840, SAE International Congress and Exposition, 1–5 March, Detroit, Michigan.

Lotus engineering, co., (2010), An Assessment of Mass Reduction Opportunities for a 2017 – 2020 Model Year Vehicle Program, March 2010

Magnusson C. and Andersson R., (2001), Stainless Steel as a Lightweight Automotive Material R&D-Forming & Materials, Swedish Tool & Die Technology, Luleå, Sweden, 2001

Mahadevan, K., Fekete, J. R., Liang, P., (2000), Effect of Strain-rate in Full Vehicle Frontal Crash Analysis, SAE paper 2000-01-0625.

Marsh, M. (2000) Development of AutoBody Sheet Materials for Crash Performance. In: Proc. Of Conference of Materials and Structures for Energy Absorption, IMech, London May 9 2000.

Mascarin, A.E. (2000). Personal communication with Sujit Das, Oak Ridge National Laboratory, IBIS Associates, Wellesley, MA. Mar.

Materials Sciences Corp., (2004) ,"Quiet Steel®" (www.quietsteel.com). http://www.quietsteel.com/quietsteel.html.

McWilliams A., (2007), Advanced Materials, Lightweight Materials in Transportation, report, Report Code: AVM056A

Medraj M, Parvez A (2007) Analyse the importance of Magnesium- aluminium-strontium alloys for more fuel-efficient automobiles. Automotive 45–47

Meyers, M. A. and Chawla, K. K. (1999) Mechanical Behaviour of Materials Upper Saddle River, Prentice-Hall. ISBN 0-13-262817-1

Miel, Rhoda (2000). "BMW Puts New Design Idea on Carbon-Fibre Diet," Automotive News, vol. 75, no. 5900, pp. 36D, Oct. 23.

Miller C.J., US Patents 3,321,828, May 30, 1967; 3,322,-517, May 30, 1967; 3,373,482, March 19, 1968; 3,373,482, March 19, 1968, assigned to General Electric Company.

Miller W.S. , Zhuang L. , Bottema J., Wittebrood A. J., De Smet P. , Haszler A., Vieregge A., (2000), Recent development in aluminium alloys for the automotive industry Materials Science and Engineering A280, 37–49

Mohanty A.K, Misra M., and Drzal L.T., (2005), Natural Fibres, Biopolymers, and Biocomposites, (Boca Raton, FL: CRC Press).

Opbroek, E., and Weissert U., (1998), Ultralight steel auto closures project, SAE Paper No. 982308.

Patel M., Bastioli C., Marini L., Wurdinger E., (2002), Assessment of Bio-Based Polymers and Natural Fibres (Netherlands: Utrecht University).

Pekguleryuz MO, Baril E, Labelle P, Argo D (2003a) Creep resistant Mg-Al-Sr alloys. J Adv Mater 35(3):32–38

Pekguleryuz MO, Kaya AA (2003b) Creep resistant magnesium alloys for powertrain applications. Adv Eng Mater 5(12):866–878

Robots 4 welding (2007) Welding magnesium. http://www.robots 4welding.com/magnesium-welding-automation.htm

Scott C.E.,(1995), Materiaux and Techniques 10–11, 21–27.

Sears K., (1997), Automotive Engineering: Strategic Overview 2 (1) 55–68.

Shi ZM, Song GL, Atrens A (2006) Influence of anodising current on the corrosion resistance of anodised AZ91D magnesium alloy. Corros Sci 48(8):1939–1959

Singh B., Verma A., and Gupta A., (1998), "Studies on Adsorptive Interaction Between Natural Fibre and Coupling Agents," J. Appl. Polym. Sci., 70, pp. 1847–1858.

Sreejith PS, Ngoi BKA (2000) Dry machining: machining of the future. J Mater Process Technol 101:287–291

Sullivan, J.L., Williams, R.L., Yester, S., Cobas-Flores, E., Chubbs, S.T., Hentges, S.G. and Pomper, S.D. (1998) Life Cycle Inventory of a Generic US Family Sedan: Overview of Results USCAR AMP Project, SAE Technical paper #982160.

Tkachenko VG, Maksimchuk IN, Volosevich PY, Lashuk NK, Malka AN, Friezel VV (2006) Creep resistance and long-term strength of structural magnesium alloys. High Temp Mater Proc 25(1–2):97–107

Ultralight Steel Auto Body, (1998), Final Report. Southfield, MI: American Iron and Steel Institute, March.

Ungureanu, C.A., Das, S. and Jawahir, I.S. (2007a) 'Life-cycle cost analysis: aluminium versus steel in passenger cars', Proceedings of TMS Conference, ASM, February, Orlando, FL.

Ungureanu, C.A., Das, S. and Jawahir, I.S. (2007b) 'Development of a sustainability scoring method for manufactured automotive products: a case study of auto body panels', Proceedings of ASME international Mechanical Engineering Congress and Exposition,IMECE2007-43818.

Wallentowitz H.; Adam H., (1996), Predicting the crashworthiness of vehicle structures made by lightweight design materials and innovative joining methods, International Journal of Crashworthiness, 1754-2111, Volume 1, Issue 2, Pages 163 – 180

Witteman, W. J., (1999), Improved Vehicle Crashworthiness Design by Control of the Energy Absorption for Different Collision Situations, Thesis, 1999, Eindhoven University of Technology, ISBN 90-386-0880-2

Yoshimoto, T., Yu Q., Yajima H., Shiratori M., Motoyama K.., (1999), Multi-objective optimization for crash safety design of vehicles, SAE Paper No. 1999-01-3210.

Zafeiropoulous N.E., Williams D.R., Baillie C.A. ; Matthews F.L., (2002), "Engineering and Characterization of the Interface in Flax Fibre/ Propylene Composite Materials. Part I. Development and Investigation of Surface Treatments," Composites:Part A, 33, pp. 1083–1093.

Zussmann, E., Krivet, A. and Seliger, G. (1994) 'Disassembly oriented assessment methodology to support design for recycling', Annals of the CIRP, Vol. 34, pp.9–14.